Thirty Years of Photosynthesis

G.J. Kelly • E. Latzko

Grahame J. Kelly • Erwin Latzko

Thirty Years of Photosynthesis

1974–2004

With 61 Figures and 2 Tables

Springer

Dr. Grahame J. Kelly
Queensland University of Technology
Faculty of Science
School of Life Sciences
GPO Box 2434
Brisbane Qld 4001
Australia

Prof. Dr. Erwin Latzko
Kirchbergstr. 21
85402 Kranzberg
Germany

ISBN-10 3-540-28382-X Springer Berlin Heidelberg New York
ISBN-13 978-3-540-28382-9 Springer Berlin Heidelberg New York

Springer is a part of Springer Science+Business Media
springeronline.com

Printed in Germany

Editor: Dr. Dieter Czeschlik, Heidelberg, Germany
Desk editor: Ursula Gramm, Heidelberg, Germany
Cover design: *design & production* GmbH, Heidelberg, Germany
Production: Friedmut Kröner, Heidelberg, Germany

31/3152 YK – 5 4 3 2 1 0 – Printed on acid free paper

Foreword

"The path of carbon in photosynthesis" for PROGRESS IN BOTANY:
50 years of CALVIN-BENSON cycle – 30 years of KELLY-LATZKO reviews

While writing this Foreword and trying to focus my thoughts on the biochemistry of photosynthesis, a handsome slim hardcover booklet of 104 pages bound in dark blue linen is in front of me on my desk:

"The Path of Carbon in Photosynthesis"
J. A. BASSHAM and M. CALVIN, 1957

I acquired it in the month of my oral Ph.D.-exams, April 1960, to get prepared with the Nobel-laureate's text. In 2004 in his last swan-song review for PROGRESS IN BOTANY GRAHAME J. KELLY celebrated "The CALVIN cycle's golden jubilee" in an overview of 50 years of carbon flowing for the progress in botany. He had met ERWIN LATZKO in 1970 in another then foremost and now historic place of the biochemistry of photosynthesis, the laboratory of MARTIN GIBBS at Brandeis University, Massachusetts. Four years later LATZKO and KELLY (1974) published their first joint review on photosynthetic carbon metabolism, starting off a long flow of articles on the flow of carbon in the series PROGRESS IN BOTANY. Most faithfully they produced regular accounts of the progress in PROGRESS IN BOTANY every second year, and when ERWIN LATZKO decided to retire after the 1996 review GRAHAME KELLY carried on alone.

Most of the reviews are headed by quotations and mottos alluding to both the author's sympathetic sense of humour and their deep concern for the issue of photosynthesis harvesting the sun, and in this vein the quotation in the 1992 review (KELLY and LATZKO, 1992) best suits as a representative example:

"The first man I saw was of a meagre aspect, with sooty hands and face, his hair and beard long, ragged and singed in several places. His clothes, shirt and skin were all of the same colour. He had been eight years upon a project for extracting sunbeams out of cucumbers, which were to be put into vials hermetically sealed, and let out to warm the air in raw inclement summers. He told me, he did not doubt in eight years more, that he should be able to supply the Governor's garden with sunshine at a reasonable rate ..."
JONATHAN SWIFT in *Gulliver's Travels*, ca. 1727.

With the 2004 review GRAHAME KELLY now also decided to retire. He does it at the climax of the golden jubilee of the CALVIN cycle, and this now certainly ends an era of photosynthesis reviewing. Reviews in the future will have a different touch, in as much as progress advances into the molecular realm: *"The transformation of a pig with a spinach gene has now been described"* (Saeki et al. 2004; see notes added in proof to the vol. 64, 2002-review in PROGRESS IN BOTANY, in this volume).

Looking forward to this book presenting the 16 reviews of 30 years of photosynthesis in PROGRESS IN BOTANY has a historical twist as the book demonstrates the development of the field over the decades. Nevertheless, it is more than looking back. It presents the reader with an up-to-date account of the state of the field. Hence we owe thanks to GRAHAME KELLY and Springer-Verlag for presenting the entire collection of the 16 reviews in this volume. It will be stimulating reading for all interested in the biochemistry of photosynthesis. Great thanks are also due to URSULA GRAMM for her continuous great care over very many years in general for PROGRESS IN BOTANY at Springer Verlag in Heidelberg and particularly for her care for the present book.

GRAHAME KELLY has meticulously re-read all the reviews again to develop updating comments from the perspective of 2005. These are added at the ends of the individual reviews compiled in this volume. It is noteworthy that astonishingly little needed to be said which underlines the evergreen liveliness of each individual review.

I wish this book about the path and metabolism of carbon in photosynthesis smooth paths and a metabolic generation of new approaches, ideas and insights in the community interested in the biochemistry of photosynthesis.

Darmstadt, in August 2005 ULRICH LÜTTGE

References

Bassham JA, Calvin M (1957) The path of carbon in photosynthesis. Prentice Hall, Inc., Englewood Cliffs, N.J.

Kelly GJ, Latzko E (1992) Photosynthesis: Carbon metabolism. Twenty years of following carbon cycles in photosynthetic cells. Progress Bot 54: 174–200

Saeki et al. (2004) Functional expression of a Δ12 fatty acid desaturase gene from spinach in transgenic pigs. Proc Natl Acad Sci USA 101:6361–6366

Preface

My interest in photosynthesis began, I suspect, at an early age. As a child brought up in the small Australian town of Kempsey, one of my enjoyments was to create a vegetable garden behind the family house, and sell produce to my parents at 50 % of the prevailing market value. This involved weighing the produce, and I began to wonder how my garden plants could accumulate such weight. This curiosity, combined with an excessive fear of being asked to study the French language upon entering high school (perhaps I foresaw that it was to be Germany, not France, where I was to spend six years of my academic life), dictated that I enter the "B" class which was free of foreign languages but included "agriculture". I continued into "Agricultural Science" at the University of Sydney where, I can still recall, John F Turner attended to my childhood query by describing the famous willow-twig experiment of Van Helmont (who obviously asked similar questions as a child) and, more importantly, by outlining the Benson-Calvin cycle of photosynthetic CO_2 fixation. The biochemistry of plant growth became my passion, and subsequently coaxed me into Turner's laboratory in Sydney, followed by Martin Gibbs' laboratory in Waltham and Erwin Latzko's laboratories in Weihenstephan and Münster.

The above brief personal history is pertinent because this entire book is, in essence, an historical account. Erwin Latzko and I embarked on the production of this chronicle in 1974, although at the time we neither planned nor anticipated that it would continue uninterrupted for three decades. But it did, thus this resultant collection is unique: it offers the opportunity to read the 30-year history of a particular topic in photosynthetic carbon metabolism (e.g. the history of Rubisco, or the history of chloroplast starch synthesis and degradation) written "as it took place". Nothing was written in the context of later discoveries, because these were unknown at the time of writing! Readers can comfortably follow a topic by jumping from one review to the next, reading the appropriate section in each review as they proceed. An additional bonus is that, in our contortive efforts to make subsections of photosynthesis crystal-clear to all plant scientists, even anatomists and taxonomists, we redrew many metabolic pathways (some old, some new) such that all inputs and outputs were balanced. Twenty-three of these are distributed through the 16 reviews. For example, in Figure 2 of the 1980 review one can clearly see the successful photosynthetic generation of sucrose by a leaf simultaneously photorespiring 37 % of its fixed CO_2, a situation that is close to natural for most plants. Textbooks always show pathways, but seldom attend to such metabolic accountancy.

The collection of reviews acquired, along the way, a multifaceted character. Examples include: (1) the often overlooked menagerie of marine microalgae that perform close to half of Earth's CO_2 fixation was given prominence in 1984 after I was introduced to these colourful photosynthesizers by Shirley Jeffrey of the CSIRO Marine Laboratories in Hobart; (2) recognition of the emerging link between photosynthesis research and sociopolitical issues was made in 1990 after my brief visit to "Die Grünen" in Bonn; and (3) the inevitable claim that new molecular biotechnologies might be put to use to boost photosynthesis and increase humanity's food supply was critically questioned in 1994, 1998 (with inspiration from a little of the philosophy of David Walker) and 2002.

This history of photosynthetic carbon metabolism is not solely a production from Erwin Latzko and me. We recognise the thousands (literally) of curious researchers who have, over the years, slowly but surely untwined the intricacies of the biochemistry within the great variety of photosynthetic cells that exist in an equal variety of sunlit niches on Earth. This book is the story of their work. Having been drafted by us, we are then grateful to the editors (Hubert Ziegler; Ulrich Lüttge) who have polished what we have written, and to the capable staff at Springer-Verlag, most recently Dieter Czeschlik and Ursula Gramm (who, by the way, makes delicious coffee) for ensuring that it became available as printed pages for the readers. Finally, we hail the eternal bright support of our colleagues, friends and families over the years. As writer of this Preface, I particularly express my gratitude for the endurance of my long-time co-author Erwin Latzko and his delightful wife Sylvia, and for the patience of my beautiful wife Chusi and three charming daughters Madeleine (born in Freising during preparation of the 1978 review), Vivienne (born in Münster during preparation of the 1980 review), and Genevieve (born in Wollongong during submission of the 1984 review). Their patience has been rewarded by Springer-Verlag's decision to publish this unusual book which, we trust, will delight the minds of those 21st-century young scientists who become struck with a passion for contemplating the biochemistry of plant growth.

Grahame J Kelly

Contents

Photosynthesis
Carbon Metabolism

By ERWIN LATZKO and GRAHAM J. KELLY

1. Introduction

Photosynthetic organisms capture light energy which they use to form ATP and reducing power. Knowledge of the mechanisms involved in this process was surveyed last year by J. AMESZ. This year we shall examine carbon metabolism during photosynthesis with emphasis on the literature published since the reviews in Volume 34 of Fortschritte der Botanik.

It is now generally agreed that all photosynthetic organisms possess the RPP cycle (Calvin cycle) for the reductive assimilation of CO_2. Recent research has concentrated on regulatory aspects of this cycle, particularly at the level of the individual enzymes and also in relation to the movement of metabolites between the chloroplast and cytoplasm. In addition, the loss of considerable amounts of fixed carbon during photorespiration is a problem being extensively studied. Indeed, a group of plants which do not lose carbon from photorespiration is currently the subject of intensive research and the observations reported so far not only explain how photorespired CO_2 is retained, but also clearly demonstrate that the metabolism of carbon compounds during photosynthesis involves metabolic pathways which can differ in importance quite significantly from one type of photosynthetic tissue to another.

2. Enzymology of the Reductive Pentose Phosphate Cycle

The photosynthetic assimilation of CO_2 is catalysed by the twelve enzymes of the RPP cycle. Initially, RuDP is carboxylated to produce two molecules of PGA in an irreversible, rate-limiting reaction (BASSHAM and KRAUSE) catalysed by RuDP carboxylase. A presumed 6C intermediate, 2-carboxi-3-ketoribitol-1,5-diP, was recently snythesised by SIEGEL and LANE and shown to be most probably involved in the reaction.

In the past the observed affinity of isolated RuDP carboxylase for CO_2 was much lower than the affinity of chloroplasts for CO_2. Although the enzyme affinity was later shown to be increased by high concentrations of Mg^{2+} (BASSHAM et al., 3) or by the presence of fructose-6-P (BUCHANAN and SCHÜRMANN, 1), values reported were still only one-quarter that of the chloroplast. Studies by BAHR and JENSEN (1, 2)

Abbreviations: GAP, glyceraldehyde-3-P; PEP, P-enolpyruvate; PGA, 3-P-glycerate; RPP, reductive pentose phosphate; RuDP, ribulose-1,5-diP.

G.J. Kelly and E. Latzko:
Thirty Years of Photosynthesis

Originally published in Progress in Botany 36

have now confirmed that the enzymes from spinach and maize have affinities for CO_2 equal to that of the chloroplast, but are unstable and rapidly convert to forms with low affinity following isolation. A relatively high affinity was also recently observed for the spinach enzyme at pH 9 by LYTTLETON.

The instability of RuDP carboxylase and the effects of Mg^{2+} and fructose-6-P may be related to the complex subunit structure of this protein; the entire molecule possibly consists of eight large and eight small subunits (see review by SIEGEL et al.). Kinetic (NISHIMURA et al.) and immunological (GRAY and KEKWICK; NISHIMURA and AKAZAWA, 1) experiments indicate that the larger subunit contains the catalytic site for RuDP and CO_2, while the regulatory influence of Mg^{2+} is effected through the smaller subunit. NISHIMURA and AKAZAWA (2) are now able to reconstitute the spinach enzyme from the constituent subunits and show partial recovery of specific activity.

Other experiments have emphasised the somewhat unusual physical characteristics of RuDP carboxylase. Thus KWOK and WILDMAN found that the substrate RuDP greatly increased the solubility and altered the tertiary structure of the crystalline enzyme from tobacco. In addition, in at least one procaryotic organism RuDP carboxylase was found associated with polyhedral inclusions which are often observed in chemoautotrophic bacteria and blue-green algae (SHIVELY et al.).

The activity of RuDP carboxylase is stimulated by fructose-6-P (possibly derived in the chloroplast from a light-activated fructose diphosphatase, BUCHANAN and SCHÜRMANN, 1) and strongly inhibited by 6-P-gluconate (CHU and BASSHAM, 1; TABITA and McFADDEN); the latter effect may explain the inactivation of the carboxylation reaction in the dark when 6-P-gluconate levels in the leaf are higher (BASSHAM, 1). Reactivation in the light may involve a separate small protein (WILDNER and CRIDDLE), although an increased level of Mg^{2+} could also explain this phenomenon (WALKER, 1). Interestingly, although 6-P-gluconate inhibits the catalytic activity of the enzyme, this metabolite can activate the enzyme during preincubation (BUCHANAN and SCHÜRMANN, 1; CHU and BASSHAM, 2). Details of these regulatory effects have been recently reviewed (BASSHAM, 2; BUCHANAN and SCHÜRMANN, 2). Activation of RuDP carboxylase is also reported to be a component of the stimulation of CO_2 fixation in isolated chloroplasts by antimycin A (CHAMPIGNY and MIGINIAC-MASLOW; SCHACTER and BASSHAM).

Following carboxylation, the two subsequent steps of the RPP cycle effect the reduction of PGA to GAP utilising ATP and reduced NADP and catalysed by the enzymes PGA kinase and GAP dehydrogenase. A specific chloroplast PGA kinase has been demonstrated (ANDERSON and ADVANI) and shown to be regulated by the extent of phosphorylation of the adenylate pool (PACOLD and ANDERSON). Similar regulation of the cytoplasmic enzyme from other sources has been reported (KRIETSCH and BÜCHER).

The second enzyme, GAP dehydrogenase, presumably uses NADP during photosynthesis, although the isolated enzyme will use NAD as well as NADP. These two activities are probably associated with the same enzyme molecule (VACCHI et al.; McGOWAN and GIBBS). The chloroplast enzyme may be a modified form of the cytoplasmic NAD-linked GAP dehydrogenase (CERFF and QUAIL). However, McGOWAN and GIBBS have recently obtained evidence from immunological and isoelectric point data that the structure of the pea chloroplast enzyme is considerably different to that of the pea cytoplasmic enzyme and is therefore unlikely to be a modified form of the cytoplasmic enzyme.

GAP dehydrogenase in chloroplasts is activated upon illumination (ZIEGLER and ZIEGLER). Light activation does not require protein synthesis since only the latter is sensitive to X-rays (SEUBERLING). Rather, the effect of light may be related to the reported *in vitro* activation by reduced dithiols (MÜLLER et al.; ANDERSON and LIM). Indeed, there are indications that the enzymes from spinach (PAWLIZKI and LATZKO) and beans (BRADBEER et al.) gain NADP-linked activity and undergo dissociation in the presence of dithiothreitol. Natural dithiols may activate enzymes *in vivo*; AHMADI and TING have found that the chloroplast NADP-linked malate dehydrogenase is activated by reduced lipoate. Besides dithiols, NADP is also reported to cause subunit dissociation of GAP dehydrogenase, and simultaneously the affinity for NADP indreased (PUPILLO and PICCARI).

The remainder of the RPP cycle regenerates RuDP from GAP. This process could be regulated at the steps catalysed by fructose diphosphatase and ribulose-5-P kinase; both of these enzymes are activated during illumination. Light activation of ribulose-5-P kinase (LATZKO et al., 1) has now been examined in intact chloroplasts and suggested to be mediated by a photoproduced reductant (AVRON and GIBBS, 1); in the dark the residual activity remaining could be inhibited by 6-P-gluconate (ANDERSON).

Fructose diphosphatase activation may involve a reduction mediated by ferredoxin (BUCHANAN et al., 1). Alternatively, the greater activity could result from an elevated pH (HELDT et al., 2) and an increased concentration of Mg^{2+} (LIN and NOBEL) induced in the chloroplast stroma upon illumination. These factors may work together to activate the chloroplast fructose diphosphatase, which has ideal kinetic properties for such a response (GARNIER and LATZKO). The Mg^{2+}-induced decrease in the pH optimum recently reported for the crystalline enzyme (EL-BADRY) is also consistent with this proposal.

Reports have implicated a reductive carboxylic acid cycle as an alternative to the RPP cycle for the photoassimilation of CO_2 in some photosynthetic bacteria (BUCHANAN et al., 2). Although feasible, some enzymological criteria (BEUSCHER and GOTTSCHALK; BOTHE et al.) for this scheme may be questioned, and in particular one of these organisms does not lack RuDP carboxylase (TABITA et al.) as previously suspected (BUCHANAN et al., 2). A second proposal for the reductive assimilation of CO_2, this time involving a direct reduction of CO_2 to formate, was recently advanced to explain the anomalous labelling of some organic acids after $^{14}CO_2$ fixation by intact leaves (KENT; KENT et al.). While the reduction itself was not directly demonstrated, it has been shown possible in an aerobic species of *Pseudomonas* (HÖPNER et al.).

3. Studies with Isolated Chloroplasts

It is now possible to isolate chloroplasts capable of fixing CO_2 at rates comparable with *in vivo* rates of photosynthesis (see GIBBS). The rates of CO_2 fixation observed are very dependent on pH (AVRON and GIBBS, 2) and, although the chloroplast is impermeable to protons, the external pH influences the level of bicarbonate which accumulates in the stroma following the diffusion of CO_2 (WERDAN and HELDT), and probably influences the transport of other metabolites during photosynthesis. The pH of the interior of the chloroplast has also been recognised to influence CO_2 fixation: WERDAN and HELDT have proposed that the pH rise in the stroma which follows the light-induced transport of protons into the thylakoids (HELDT et al., 2) could act as a CO_2-pump by altering the equilibrium of the reaction catalysed by carbonic anhydrase in favour of bicarbonate accumulation. Carbonic anhydrase is a zinc-containing protein (see POCKER and NG) largely localised in chloroplasts (EVERSON and SLACK; POINCELOT).

Photosynthetic CO_2 fixation by isolated chloroplasts is inhibited by free Mg^{2+} in the reaction medium (AVRON and GIBBS, 2). Substances able to complex Mg^{2+}, such as pyrophosphate, will relieve this inhibition (FORTI and ROSA). The level of Mg^{2+} in the reaction medium can also affect product distribution, as shown by BALDRY and COOMBS with their technique of illuminating chloroplasts on filter paper discs.

Pyrophosphate can also influence CO_2 fixation through an effect other than Mg^{2+} chelation: the pyrophosphate may be hydrolysed by pyrophosphatase (released from ruptured chloroplasts) and so act as a source of inorganic phosphate for photosynthesis (SCHWENN et al.). The effects of a protein factor earlier reported to regulate photosynthesis (BASSHAM et al., 1) have now been attributed to pyrophosphatase activity present in this factor (LILLEY et al., 1; LEVINE and BASSHAM).

Illuminated chloroplasts show a delay of several minutes before a steady rate of CO_2 fixation is attained (TURNER et al.). This delay is believed to represent the time required to accumulate intermediates of the RPP cycle (LATZKO and GIBBS; WALKER et al.), and may ultimately relate to a short supply of photosynthetically generated ATP which is reported to be rate limiting for CO_2 fixation (HEBER, 1; HEBER and KIRK). Indeed, rapid attainment of steady state levels has been shown to depend in chloroplasts (SCHÜRMANN et al.) and in *Chlorella* cells (KLOB et al.) on ATP from cyclic photophosphorylation, possibly to boost the level of the CO_2 acceptor RuDP. This, in turn, may be dependent on light activation of ribulose-5-P kinase (LATZKO et al., 1). The actual carboxylation of RuDP is probably not a primary cause of the delay, although light activation of RuDP carboxylase does influence the process (WALKER, 1); results of more recent experiments (LILLEY et al., 2) suggest that the Mg^{2+} activation of RuDP carboxylase (BASSHAM et al., 3) expected following light-induced Mg^{2+} accumulation in the stroma (LIN and NOBEL) occurs more rapidly than the development of a steady rate of photosynthesis.

A new and interesting approach to the study of chloroplast metabolism is the development of the reconstituted chloroplast system (STOKES and WALKER). In this system osmotically shocked chloroplasts, supplemented with a soluble fraction derived from chloroplasts and catalytic quantities of cofactors, show light dependent O_2 evolution in the presence of PGA. More recently, BASSHAM et al. (2) have reported *in vivo* rates of CO_2 fixation in a reconstituted system. These authors provide data indicating that high rates of CO_2 fixation are more dependent on a high concentration of soluble chloroplast components in the vicinity of chlorophyll than on an intact chloroplast envelope.

4. Communication between Chloroplast and Cytoplasm

Of the intermediates of the RPP cycle only the pentose monophosphates, triose phosphates and PGA are now believed to pass freely between the chloroplast and cytoplasm (see WALKER, 2), although the absolute permeability of the chloroplast envelope can clearly be altered (NOBEL). An elegant series of experiments by HELDT and RAPLEY has demonstrated that the triose phosphates and PGA are actively transported by a translocator on the inner membrane (HELDT and SAUER) of the chloroplast envelope. Recently, DOUCE et al. found ATPase activity associated with isolated chloroplast envelopes and speculated that this ATPase may be involved in such active transport.

These observations and others showing accumulation of triose phosphates in the medium of photosynthesising chloroplasts indicate that assimilated carbon can be exported from chloroplasts as triose phosphates. These may then be metabolised to sucrose since sucrose-synthesising enzymes predominate in the cytoplasm (BIRD et al., 3); earlier conclusions to the contrary may have resulted from faults in (BIRD et al., 1) or improper use of (HEBER, 2) the technique of non-aqueous isolation of chloroplasts. The presence of a separate fructose diphosphatase in the cytoplasm of green leaves (LATZKO et al., 3) is consistent with the cytoplasmic location of sucrose biosynthesis. On the other hand, starch is synthesised inside the chloroplast in green leaves (see PREISS and KOSUGE) and is presumably metabolised to sugar phosphates when export from the chloroplast becomes necessary.

In the cytoplasm the complete glycolytic sequence may be restricted during photosynthesis because of inhibition of the phosphofructokinase reaction by ATP (HIRT and TANNER); plant phosphofructokinase is strongly inhibited by free ATP (KELLY and TURNER). However, the latter portion of glycolysis could continue to operate and convert triose phosphates exported from the chloroplast to pyruvate. This pyruvate may enter the tricarboxylic acid cycle which continues to operate in the light (RAVEN; CHAPMAN and GRAHAM), and probably supplies dicarboxylic acid carbon skeletons for the synthesis of amino acids reported to occur in chloroplasts (KIRK and LEECH; MAGALHAES et al.; see GIVAN and LEECH). In fact, these dicarboxylic acids could be actively transported into chloroplasts by a membrane-located dicarboxylate translocator (HELDT and RAPLEY).

In contrast to the free movement of some carbon compounds, photosynthetically generated ATP (HELDT et al., 1) and reduced NADP (ROBINSON and STOCKING) cannot cross the chloroplast envelope. Nevertheless, these substances can be indirectly transferred to the cytoplasm through a shuttle whereby PGA enters the chloroplast and is reduced to triose phosphate, which returns to the cytoplasm where it is converted back to PGA with the production of ATP (HEBER and SANTARIUS; KRAUSE) and reduced NAD (STOCKING and LARSON). Alternatively, reduced NADP may be produced in the cytoplasm when the triose phosphate is directly oxidised to PGA by the non reversible GAP dehydrogenase (KELLY and GIBBS, 2). The properties of this enzyme, particularly the high affinity for NADP (KELLY and GIBBS, 1), seem ideally suited to this role. Reducing equivalents may also be transferred via a malate-oxaloacetate shuttle which can move reducing power either into or out of the chloroplast (HEBER and KIRK).

Two detailed and thorough reviews on the flow of metabolites between the chloroplast and cytoplasm (HEBER, 2; WALKER, 2) are in press at the time of this writing.

5. Glycollate and Photorespiration

Large amounts of glycollate are often synthesised during photosynthesis and are believed to be the substrate for photorespiration (see ZELITCH, 1). The possibility of increasing net photosynthesis by preventing photorespiration was recently demonstrated by ZELITCH (4) who showed glycidate, an analogue of glycollate, could partially inhibit glycollate synthesis in tobacco leaves and simultaneously increase the net CO_2 fixation by up to 5o%. Quite recent reviews on the biosynthesis and metabolism of glycollate in higher plants and algae are available (BECK; TOLBERT; MERRETT and LORD).

Whether glycollate originates from oxidation of RuDP, from oxidation of a glycolaldehyde-transketolase complex, from a CO_2 plus CO_2 condensation or from some other source (see BECK; TOLBERT) remains an open question; more than one type of reaction may be involved (BASSHAM and KIRK; EICKENBUSCH and BECK; ZELITCH, 3). At present the first of these possibilities is receiving special attention, since the RuDP carboxylases from leaves (BOWES et al.) and *Chromatium* (TAKABE and AKAZAWA) catalyse the oxidative cleavage of RuDP to PGA and P-glycollate and the enzymes from maize (BAHR and JENSEN, 2), *Chlamydomonas* (BERRY and BOWES) and spinach (NISHIMURA and AKAZAWA, 1) show an RuDP-dependent uptake of O_2; in the latter case O_2 uptake was shown to depend on the larger subunit of the enzyme. Binding of O_2 to the active site is conceivable, since SO_3^{2-} apparently also binds to this site (ZIEGLER). If glycollate is formed from RuDP oxidation, the inhibition of glycollate synthesis in algae by blue light is not easily explained (MERRETT and LORD), while glycollate formation from a glycolaldehyde-transketolase complex (SHAIN and GIBBS) is consistent with this observation since blue light is reported to inactivate transketolase (CODD). On the other hand, the incorporation of $^{18}O_2$ into the carboxyl group of glycollate (LORIMER et al.) is at present more easily explained (TOLBERT) when glycollate originates from RuDP oxidation than from a glycolaldehyde-transketolase complex.

In most land plants and some algae oxidation of glycollate to glyoxylate is catalysed by glycollate oxidase in microbodies and is accompanied by H_2O_2 formation from O_2 (FREDERICK et al.), but in other algae a glycollate dehydrogenase which does not link directly to O_2 is used (MERRETT and LORD). A number of pathways have been proposed for the subsequent metabolism of glyoxylate. This substance may be returned to the chloroplast for decarboxylation (ZELITCH, 2) with H_2O_2 or $O_2^{\bar{}}$ (superoxide free radical ion) as the primary oxidant (ELSTNER and HEUPEL), or directly decarboxylated in microbodies (HALLIWELL and BUTT). Alternatively, TOLBERT has proposed that glyoxylate is converted to glycine which is partially decarboxylated in mitochondria. BIRD et al. (2) have evidence that this reaction is directly linked to oxidative phosphorylation and thus permits light-dependent synthesis of ATP in mitochondria. Serine formed in the reaction is converted in microbodies to glycerate which supposedly returns to the chloroplast for phosphorylation (TOLBERT), although HEBER et al. have questioned whether the activity of glycerate kinase is adequate for this pathway. In *Chlorella* glycerate is instead metabolised to pyruvate (LORD and MERRETT). An interesting variation for the conversion of glyoxylate to glycerate involving the intermediate tartronic semialdehyde (but still producing photorespiratory CO_2) was recently detected in the blue-green alga *Anabaena* (CODD and STEWART).

During early studies of photorespiration plants such as maize were found not to evolve photorespiratory CO_2 (FORRESTER et al.). Nevertheless, the metabolism of glycollate leading to the production of CO_2 does occur in these plants, but is located in a compact layer of cells around the vascular bundles (bundle-sheath cells, CHOLLET and OGREN, 1; MÜHLBACH and WEGMANN), which contain the major portion of microbodies and enzymes of glycollate metabolism (FREDERICK and NEWCOMB; LIU and BLACK; HUANG and BEEVERS). The photorespired CO_2 is not evolved, but rather captured in adjacent mesophyll cells by high levels of PEP carboxylase.

6. C_4 Metabolism

When $^{14}CO_2$ is fed to plant leaves in the light, radioactivity first appears in either the 3C-acid PGA (BASSHAM and CALVIN) or the 4C-dicarboxylic acids malate and aspartate (KORTSCHAK et al.; HATCH and SLACK, 1). These two groups are commonly distinguished as C_3 plants and C_4 plants. Certain C_4 plants such as tropical grasses and some saltbushes have leaves with two distinct cell types which cooperate in the assimilation of CO_2 by utilising the C_4 pathway during photosynthesis: the structural and metabolic characteristics of these plants and evidence for the C_4 pathway have been thoroughly reviewed by BLACK.

In the C_4 pathway CO_2 is captured by PEP carboxylase in the cytoplasm of mesophyll cells, and the oxaloacetate so formed is either reduced to malate in chloroplasts or converted to aspartate in the cytoplasm; both malate and aspartate may be formed in the one plant (KENNEDY and LAETSCH). These products are transported into bundle-sheath cells where decarboxylation occurs. Malate is decarboxylated by malic enzyme in chloroplasts; reducing power from this reaction could overcome any deficiency of photosystem II in bundle-sheath chloroplasts (OSMOND). Aspartate is first converted to oxaloacetate (HATCH and MAU) which is either decarboxylated by PEP carboxikinase (see BLACK) or reduced to malate which, in these plants, is decarboxylated by malic enzyme in mitochondria (HATCH and KAGAWA, 2). The CO_2 released from these decarboxylations is reassimilated by the RPP cycle in bundle-sheath chloroplasts. The 3C compound remaining is returned as pyruvate to the mesophyll and used for regeneration of CO_2-acceptor (PEP) by pyruvate P_i dikinase (HATCH and SLACK, 2; SUGIYAMA). Carbon assimilated in the bundle sheath may be further metabolised to starch in bundle-sheath chloroplasts (DE FEKETE and VIEWEG), or returned to the mesophyll where much of the sucrose synthesis is located (DOWNTON and HAWKER).

A number of enzymes involved in the C_4 pathway, including PEP carboxylase, pyruvate P_i dikinase, NADP-linked malate dehydrogenase and aspartate aminotransferase, are activated upon illumination (GRAHAM et al.; JOHNSON and HATCH; HATCH and MAU). The activity of PEP carboxylase is also regulated by changes in the levels of oxaloacetate (LOWE and SLACK), glucose-6-P (COOMBS et al., 1) and Mg^{2+} (COOMBS et al., 2). The enzyme from light-grown C_4 plants has a lower affinity for PEP compared to the enzyme from C_3 plants (TING and OSMOND) or from dark-grown C_4 plants (GOATLY and SMITH), reflecting the different role required of the enzyme in the C_4 pathway.

Good evidence for the C_4 pathway has been obtained from studies of isolated mesophyll and bundle-sheath cells, particularly with respect to the location of enzyme activities (CHEN et al.; KANAI and EDWARDS; see BLACK). In more recent experiments of this type both mesophyll cells (SALIN et al.) and mesophyll chloroplasts (KAGAWA and HATCH) were found to reduce oxaloacetate to malate in a light-dependent reaction linked to O_2 evolution. Bundle-sheath cells could decarboxylate malate (HUBER et al.) and transfer the β-carboxyl of this malate into products of the RPP cycle (DITTRICH et al.) which is clearly present in these cells (CHOLLET and OGREN, 2).

The considerable intercellular transport required by the C_4 pathway may be feasible, since every mesophyll cell is generally next to a bundle-sheath cell (CROOKSTON and MOSS) and cells are interconnected by plasmodesmata. However, an alternative scheme requiring far less metabolite movement has been proposed, whereby RuDP carboxylation can also occur in mesophyll cells and sucrose synthesised there is transported to bundle-sheath cells for starch synthesis (COOMBS). A critical question is whether mesophyll cells contain RuDP carboxylase; there is compelling evidence that they do not (HATCH and KAGAWA, 1; SALIN and BLACK; see BLACK), and indeed KAGAWA and HATCH have recently isolated mesophyll chloroplasts able to use light energy for the reduction of PGA to triose phosphate but unable to photosynthetically assimilate CO_2. Chloroplasts capable of light-dependent CO_2 fixation can be isolated from young maize leaves, and in this case CO_2 fixation has been shown to occur exclusively by carboxylation of RuDP (O'NEAL et al.). However, BLACK has argued that some of these chloroplasts may have originated from bundle-sheath cells.

A fundamental question is to what extent the C_4 pathway is dependent on the RPP cycle. As usually shown (see BLACK), the C_4 pathway requires the RPP cycle only as an acceptor of CO_2 from the decarboxylation step in the bundle-sheath cells. The experiments of NAGY et al. showing that PEP carboxylation was suppressed in a maize mutant with low RuDP carboxylase activity emphasise the importance of this connection. A second possible link, whereby the RPP cycle supplies the CO_2-acceptor (PEP) for the C_4 pathway, has recently been investigated: when maize leaves were fed $^{14}CO_2$ in preillumination experiments the distribution of label in malate (LABER et al., 1) and changes in the levels of RuDP and PEP (LABER et al., 2) demonstrated that the carboxylation of RuDP, at least under those conditions, could precede the carboxylation of PEP, indicating that PEP for the C_4 pathway can be obtained from PGA supplied by the RPP cycle. A similar conclusion has been reached by GALMICHE from short pulse chase experiments. High activities of PGA mutase and enolase required for PEP formation from PGA have been found in maize leaves (LATZKO et al., 2). The inhibition by O_2 of CO_2 fixation into 4C compounds in maize leaves (STAMIESZKIN et al.) is also consistent with PEP being supplied from the RPP cycle, since CO_2 fixation in bundle-sheath cells (where the RPP cycle is located) is more sensitive to O_2 than that in mesophyll cells (CHOLLET).

In the long term it is obvious that, to grow, a C_4 plant must derive some PEP from PGA produced from RuDP carboxylation. In a growing plant fixing CO_2 by carboxylation of an acceptor molecule, the reactions designed to regenerate the CO_2-acceptor must be arranged in a sequence which, in itself, permits growth in the level of this acceptor molecule (Fig. 1). In a recent publication, WALKER (3) has described how the reactions of the RPP cycle allow regeneration of more acceptor molecule (RuDP) than was available before carboxylation. This allows growth (Fig. 1). The C_4 pathway is not an alternative to the RPP cycle because the pathway in itself cannot regenerate more acceptor molecule (PEP) than was available before the carboxylation (WALKER, 3) and hence cannot grow (Fig. 1).

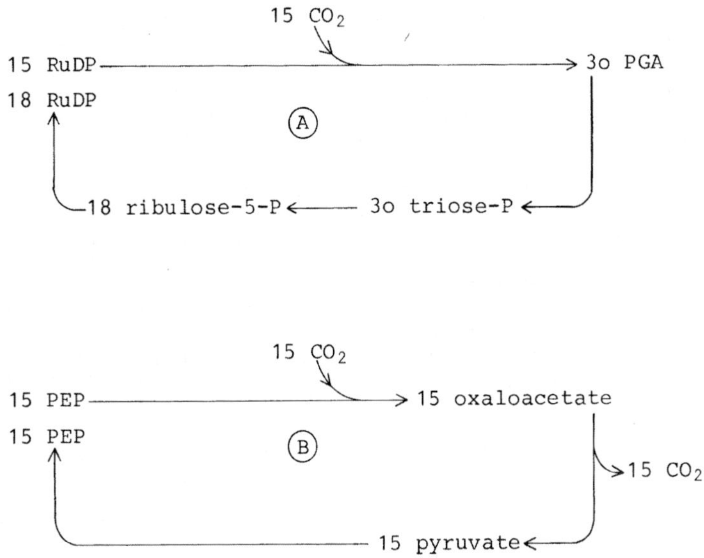

Fig. 1. Comparison of the abilities of (A) the RPP cycle and (B) the C_4 pathway to grow. Adapted from WALKER (3)

The value of C_4 metabolism probably lies in the varied and often quite remarkable ecological adaptions it permits. In some tropical grasses such as maize the pathway will recapture CO_2 evolved from photorespiration and hence allow more closure of stomates and reduced water loss from transpiration. An extreme case of this is *Triodia*, a grass which survives in the hot arid Australian inland by utilising C_4 metabolism in a leaf especially resistent to water loss (McWILLIAM and MISON). Other drought-adapted plants separate C_4 metabolism into distinct day and night phases (crassulacean acid metabolism: see review by M. KLUGE in this volume), so that water loss during the day is practically excluded. There is also evidence that C_4 metabolism is important for the adaption of *Chlorella* to certain growth conditions (GRAHAM and WHITTINGHAM; DÖHLER), for the metabolism of guard cells on plant leaves (WILLMER and DITTRICH) and for the adaption of salt tolerant plants to highly saline conditions (SHOMER-ILAN and WAISEL). Particularly interesting is evidence that barley, classically considered to be a C_3 plant, possesses C_4 metabolism in the photosynthetic pericarp (DUFFUS and ROSIE).

References

AHMADI, N., TING, I.P.: Plant Sci. Lett. 1, 11-14 (1973). - ANDERSON, L.E.: Biochim. Biophys. Acta 321, 484-488 (1973). - ANDERSON, L.E., ADVANI, V.R.: Plant Physiol. 45, 583-585 (197o). - ANDERSON, L.E., LIM, T.C.: FEBS Lett. 27, 189-191 (1972). - AVRON, M., GIBBS, M.: (1) Plant Physiol. 53, 136-139 (1974); - (2) Plant Physiol. 53, 14o-143 (1974).

BAHR, J.T., JENSEN, R.G.: (1) Plant Physiol. 53, 39-44 (1974); - (2) Biochem. Biophys. Res. Commun. 57, 118o-1185 (1974). - BALDRY, C.W., COOMBS, J.: Z. Pflanzenphysiol. 69, 213-216 (1973). - BASSHAM, J.A.: (1) Proc. Natl. Acad. Sci. US 68, 2877-2882 (1971); - (2) Symp. Soc. Exp. Biol. No. XXVII, 461-483 (1973). - BASSHAM, J.A., CALVIN, M.: The Path of Carbon in Photosynthesis, 1o4 pp. Englewood Cliffs, N.J.: Prentice Hall 1957. - BASSHAM, J.A., KIRK, M.: Plant Physiol. 52, 4o7-411 (1973). - BASSHAM, J.A., KRAUSE, G.H.: Biochim. Biophys. Acta 189, 2o7-221 (1969). - BASSHAM, J.A., EL-BADRY, A.M., KIRK, M.R., OTTENHEYM, H.C.J., SPRINGER-LEDERER, H.: (1) Biochim. Biophys. Acta 223, 261-274 (197o). - BASSHAM, J.A., LEVINE, G., FORGER, J.: (2) Plant Sci. Lett. 2, 15-21 (1974). - BASSHAM, J.A., SHARP, P., MORRIS, I.: (3) Biochim. Biophys. Acta 153, 898-9oo (1968). - BECK, E.: Fortschr. Botan. 34, 139-154 (1972). - BERRY, J.A., BOWES, G.: Carnegie Inst. Yearbook 4o5-4o7 (1973). - BEUSCHER, N., GOTTSCHALK, G.: Z. Naturforsch. 27b, 967-973 (1972). - BIRD, I.F., CORNELIUS, M.J., DYER, T.A., KEYS, A.J.: (1) J. Exp. Botany 24, 211-215 (1973). - BIRD, I.F., CORNELIUS, M.J., KEYS, A.J., WHITTINGHAM, C.P.: (2) Phytochemistry 11, 1587-1594 (1972); - (3) Phytochemistry 13, 59-64 (1974). - BLACK, C.C.: Ann. Rev. Plant Physiol. 24, 253-286 (1973). - BOTHE, H., FALKENBERG, B., NOLTEERNSTING, U.: Arch. Mikrobiol. 96, 291-3o4 (1974). - BOWES, G., OGREN, W.L., HAGEMAN, R.H.: Biochem. Biophys. Res. Commun. 45, 716-722 (1971). - BRADBEER, J.W., WARA-ASWAPATI, O., ARRON, G.: Proc. Intern. Biochem. Congr. 152 (1973). - BUCHANAN, B.B., SCHÜRMANN, P.: (1) J. Biol. Chem. 248, 4956-4964 (1973); - (2) Ribulose 1,5-Diphosphate Carboxylase: A Regulatory Enzyme in the Photosynthetic Assimilation of Carbon Dioxide, 1-2o. In: Current Topics in Cellular Regulation, eds. B.L. HORECKER, E.R. STADTMAN, Vol. 7. New York-London: Academic Press 1973. - BUCHANAN, B.B., SCHÜRMANN, P., KALBERER, P.P.: (1) J. Biol. Chem. 246, 5952-5959 (1971). - BUCHANAN, B.B., SCHÜRMANN, P., SHANMUGAM, K.T.: (2) Biochim. Biophys. Acta 283, 136-145 (1972).

CERFF, R., QUAIL, P.: Hoppe-Seylers Z. Physiol. Chem. 354, 1179 (1973). - CHAMPIGNY, M.L., MIGINIAC-MASLOW, M.: Biochim. Biophys. Acta 234, 335-343 (1971). - CHAPMAN, E.A., GRAHAM, D.: Plant Physiol., in press. - CHEN, T.M., CAMPBELL, W.H., DITTRICH, P., BLACK, C.C.: Biochem. Biophys. Res. Commun. 51, 461-467 (1973). - CHOLLET, R.: Biochem. Biophys. Res. Commun. 55, 85o-856 (1973). - CHOLLET, R., OGREN, W.L.: (1) Biochem. Biophys. Res. Commun. 48, 684-688 (1972); - (2) Plant Physiol. 51, 787-792 (1973). - CHU, D.K., BASSHAM, J.A.: (1) Plant Physiol. 5o, 224-227 (1972); - (2) Plant Physiol. 52, 373-379 (1973). - CODD, G.A.: Z. Naturforsch. 27b, 7o1-7o4 (1972). - CODD, G.A., STEWART, W.D.P.: Arch. Mikrobiol. 94, 11-28 (1973). - COOMBS, J.: Proc. Roy. Soc., (London) Ser. B 179, 221-235 (1971). - COOMBS, J., BALDRY, C.W., BUCKE, C.: (1) Planta 11o, 95-1o7 (1973). - COOMBS, J., MAW, S.L., BALDRY, C.W.: (2) Planta, in press. - CROOKSTON, R.K., MOSS, D.N.: Plant Physiol. 52, 397-4o2 (1973).

DE FEKETE, M.A.R., VIEWEG, G.H.: Planta 117, 83-91 (1974). - DITTRICH, P., SALIN, M.L., BLACK, C.C.: Biochem. Biophys. Res. Commun. 55, 1o4-11o (1973). - DÖHLER, G.: Z. Pflanzenphysiol. 71, 144-153 (1974). - DOUCE, R., HOLTZ, R.B., BENSON, A.A.: J. Biol. Chem. 248, 7215-7222 (1973). - DOWNTON, W.J.S., HAWKER, J.S.: Phytochemistry 12, 1551-1556 (1973). - DUFFUS, C.M., ROSIE, R.: Planta 114, 219-226 (1973).

EICKENBUSCH, J.D., BECK, E.: FEBS Lett. 31, 225-228 (1973). - EL-BADRY, A.M.: Biochim. Biophys. Acta 333, 366-377 (1974). - ELSTNER, E.F., HEUPEL, A.: Biochim. Biophys. Acta 325, 182-188 (1973). - EVERSON, R.G., SLACK, C.R.: Phytochemistry 7, 581-584 (1968).

FORRESTER, M.L., KROTKOV, G., NELSON, C.D.: Plant Physiol. 41, 428-431 (1966). - FORTI, G., ROSA, L.: Plant Sci. Lett. 2, 95-1oo (1974). - FREDERICK, S.E., NEWCOMB, E.H.: Planta 96, 152-174 (1971). - FREDERICK, S.E., GRUBER, P.J., TOLBERT, N.E.: Plant Physiol. 52, 318-323 (1973).

GALMICHE, J.M.: Plant Physiol. 51, 512-519 (1973). - GARNIER, R.V., LATZKO, E.: Proc. IInd. Intern. Congr. Photosynthesis Res., Vol. 3, 1839-1845. The Hague: Dr. W. Junk N.V. 1972. - GIBBS, M.: Carbohydrate Metabolism by Chloroplasts, 169-

214. In: Structure and Function of Chloroplasts, ed. M. GIBBS. Berlin-Heidelberg-New York: Springer 1971. - GIVAN, C.V., LEECH, R.M.: Biol. Rev. 46, 4o9-428 (1971). - GOATLY, M.B., SMITH, H.: Planta 117, 67-73 (1974). - GRAHAM, D., WHITTINGHAM, C.P.: Z. Pflanzenphysiol. 58, 418-427 (1968). - GRAHAM, D., HATCH, M.D., SLACK, C.R., SMILLIE, R.M.: Phytochemistry 9, 521-532 (197o). - GRAY, J.C., KEKWICK, R.G.O.: Europ. J. Biochem. 44, 481-489 (1974).

HALLIWELL, B., BUTT, V.S.: Biochem. J. 138, 217-224 (1974). - HATCH, M.D., KAGAWA, T.: (1) Arch. Biochem. Biophys. 159, 842-853 (1973); - (2) Arch. Biochem. Biophys. 16o, 346-349 (1974). - HATCH, M.D., MAU, S.L.: Arch. Biochem. Biophys. 156, 195-2o6 (1973). - HATCH, M.D., SLACK, C.R.: (1) Biochem. J. 1o1, 1o3-111 (1966); - (2) Biochem. J. 1o6, 141-146 (1968). - HEBER, U.: (1) Biochim. Biophys. Acta 3o5, 14o-152 (1973); - (2) Ann. Rev. Plant Physiol., in press. - HEBER, U., KIRK, M.R.: Biochim. Biophys. Acta, in press. - HEBER, U., SANTARIUS, K.A.: Z. Naturforsch. 25b, 718-728 (197o). - HEBER, U., KIRK, M.R., GIMMLER, H., SCHÄFER, G.: Planta, in press. - HELDT, H.W., RAPLEY, L.: FEBS Lett. 1o, 143-148 (197o). - HELDT, H.W., SAUER, F.: Biochim. Biophys. Acta 234, 83-91 (1971). - HELDT, H.W., SAUER, F., RAPLEY, L.: (1) Proc. IInd. Intern. Congr. Photosynthesis Res., Vol. 2, 1345-1355. The Hague: Dr. W. Junk N.V. 1972. - HELDT, H.W., WERDAN, K., MILOVANCEV, M., GELLER, G.: (2) Biochim. Biophys. Acta 314, 224-241 (1973). - HIRT, G., TANNER, W.: Proc. IInd. Intern. Congr. Photosynthesis Res., Vol. 3, 1989-1997. The Hague: Dr. W. Junk N.V. 1972. - HÖPNER, T., RUSCHIG, U., MÜLLER, U.: Hoppe-Seylers Z. Physiol. Chem. 354, 216 (1973). - HUANG, A.H.C., BEEVERS, H.: Plant Physiol. 5o, 242-248 (1972). - HUBER, S.C., KANAI, R., EDWARDS, G.E.: Planta 113, 53-66 (1973).

JOHNSON, H.S., HATCH, M.D.: Biochem. J. 119, 273-28o (197o).

KAGAWA, T., HATCH, M.D.: Australian J. Plant Physiol. 1, 51-64 (1974). - KANAI, R., EDWARDS, G.E.: Plant Physiol. 51, 1133-1137 (1973). - KELLY, G.J., GIBBS, M.: (1) Plant Physiol. 52, 111-118 (1973); - (2) Plant Physiol. 52, 674-676 (1973). - KELLY, G.J., TURNER, J.F.: Biochem. J. 115, 481-487 (1969). - KENNEDY, R.A., LAETSCH, W.M.: Planta 115, 113-124 (1973). - KENT, S.S.: J. Biol. Chem. 247, 7293-73o2 (1972). - KENT, S.S., PINKERTON, F.D., STROBEL, G.A.: Plant Physiol. 53, 491-495 (1974). - KIRK, P.R., LEECH, R.M.: Plant Physiol. 5o, 228-234 (1972). - KLOB, W., KANDLER, O., TANNER, W.: Plant Physiol. 51, 825-827 (1973). - KORTSCHAK, H.P., HARTT, C.E., BURR, G.O.: Plant Physiol. 4o, 2o9-213 (1965). - KRAUSE, G.H.: Z. Pflanzenphysiol. 65, 13-23 (1971). - KRIETSCH, W.K.G., BÜCHER, T.: Europ. J. Biochem. 17, 568-58o (197o). - KWOK, S.Y., WILDMAN, S.G.: Arch. Biochem. Biophys. 161, 354-359 (1974).

LABER, L.J., LATZKO, E., LEVI, C., GIBBS, M.: (1) Proc. IInd. Intern. Congr. Photosynthesis Res., Vol. 3, 1737-1744. The Hague: Dr. W. Junk N.V. 1972. - LABER, L.J., LATZKO, E., GIBBS, M.: (2) J. Biol. Chem., in press. - LATZKO, E., GIBBS, M.: Plant Physiol. 44, 396-4o2 (1969). - LATZKO, E., GARNIER, R.V., GIBBS, M.: (1) Biochem. Biophys. Res. Commun. 39, 114o-1144 (197o). - LATZKO, E., GIBBS, M., LABER, L.J.: (2) Proc. Ist Europ. Biophys. Congr., Vol. IV, 97-1o7, Wien 1971. - LATZKO, E., ZIMMERMANN, G., FELLER, U.: (3) Hoppe-Seylers Z. Physiol. Chem. 355, 321-326 (1974). - LEVINE, G., BASSHAM, J.A.: Biochim. Biophys. Acta 333, 136-14o (1974). - LILLEY, R.McC., SCHWENN, J.D., WALKER, D.A.: (1) Biochim. Biophys. Acta 325, 596-6o4 (1973). - LILLEY, R.McC., HOLBOROW, K., WALKER, D.A.: (2) New Phytologist, in press. - LIN, D.C., NOBEL, P.S.: Arch. Biochem. Biophys. 145, 622-632 (1971). - LIU, A.Y., BLACK, C.C.: Arch. Biochem. Biophys. 149, 269-28o (1972). - LORD, J.M., MERRETT, M.J.: New Phytologist 72, 249-252 (1973). - LORIMER, G.H., ANDREWS, T.J., TOLBERT, N.E.: Biochemistry 12, 18-23 (1973). - LOWE, J., SLACK, C.R.: Biochim. Biophys. Acta 235, 2o7-2o9 (1971). - LYTTLETON, J.W.: FEBS Lett. 38, 4-6 (1973).

MAGALHAES, A.C., NEYRA, C.A., HAGEMAN, R.H.: Plant Physiol. 53, 411-415 (1974). - McGOWAN, R.E., GIBBS, M.: Plant Physiol., in press. - McWILLIAM, J.R., MISON, K.:

Australian J. Plant Physiol. 1, 171-175 (1974). - MERRETT, M.J., LORD, J.M.: New Phytologist 72, 751-767 (1973). - MÜHLBACH, H.P., WEGMANN, K.: Hoppe Seylers Z. Physiol. Chem. 354, 1225-1226 (1973). - MÜLLER, B., ZIEGLER, I., ZIEGLER, H.: Europ. J. Biochem. 9, 1o1-1o6 (1969).

NAGY, A.H., GYURJÁN, I., SZÉKELY, S., DOMAN, N.G.: Photosynthetica 7, 87-92 (1973). - NISHIMURA, M., AKAZAWA, T.: (1) Biochemistry, in press; - (2) J. Biochem. (Tokyo), in press. NISHIMURA, M., TAKABE, T., SUGIYAMA, T., AKAZAWA, T.: J. Biochem. (Tokyo) 74, 945-954 (1973). - NOBEL, P.S.: Planta 115, 369-372 (1974).

O'NEAL, D., HEW, C.S., LATZKO, E., GIBBS, M.: Plant Physiol. 49, 6o7-614 (1972). - OSMOND, C.B.: Australian J. Plant Physiol. 1, 41-5o (1974).

PACOLD, I., ANDERSON, L.E.: Biochem. Biophys. Res. Commun. 51, 139-143 (1973). - PAWLIZKI, K., LATZKO, E.: FEBS Lett., in press. - POCKER, Y., NG, J.S.Y.: Biochemistry 12, 5127-5134 (1973). - POINCELOT, R.P.: Biochim. Biophys. Acta 258, 637-642 (1972). - PREISS, J., KOSUGE, T.: Ann. Rev. Plant Physiol. 21, 433-466 (197o). - PUPILLO, P., PICCARI, G.G.: Arch. Biochem. Biophys. 154, 324-331 (1973).

RAVEN, J.A.: New Phytologist 71, 227-247 (1972). - ROBINSON, J.M., STOCKING, C.R.: Plant Physiol. 43, 1597-16o4 (1968).

SALIN, M.L., BLACK, C.C.: Plant Sci. Lett. 2, 3o3-3o8 (1974). - SALIN, M.L., CAMPBELL, W.H., BLACK, C.C.: Proc. Natl. Acad. Sci. US 7o, 373o-3734 (1973). - SCHACTER, B., BASSHAM, J.A.: Plant Physiol. 49, 411-416 (1972). - SCHÜRMANN, P., BUCHANAN, B.B., ARNON, D.I.: Biochim. Biophys. Acta 267, 111-124 (1972). - SCHWENN, J.D., LILLEY, R.McC., WALKER, D.A.: Biochim. Biophys. Acta 325, 586-595 (1973). - SEUBERLING, H.B.: Planta 98, 27o-273 (1971). - SHAIN, Y., GIBBS, M.: Plant Physiol. 48, 325-33o (1971). - SHIVELY, J.M., BALL, F., BROWN, D.H., SAUNDERS, R.E.: Science 182, 584-586 (1973). - SHOMER-ILAN, A., WAISEL, Y.: Physiol. Plantarum 29, 19o-193 (1973). - SIEGEL, M.I., LANE, M.D.: J. Biol. Chem. 248, 5486-5498 (1973). - SIEGEL, M.I., WISHNICK, M., LANE, M.D.: Ribulose-1,5-diphosphate Carboxylase, 169-192. In: The Enzymes, ed. P.D. BOYER, Vol. 6. New York-London: Academic Press 1972. - STAMIESZKIN, I., MALESZEWSKI, S., POSKUTA, J.: Z. Pflanzenphysiol. 67, 18o-182 (1972). - STOCKING, C.R., LARSON, S.: Biochem. Biophys. Res. Commun. 37, 278-282 (1969). - STOKES, D.M., WALKER, D.A.: Plant Physiol. 48, 163-165 (1971). - SUGIYAMA, T.: Biochemistry 12, 2862-2868 (1973).

TABITA, F.R., McFADDEN, B.A.: Biochem. Biophys. Res. Commun. 48, 1153-1159 (1972). - TABITA, F.R., McFADDEN, B.A., PFENNIG, N.: Biochim. Biophys. Acta 341, 187-194 (1974). - TAKABE, T., AKAZAWA, T.: Biochem. Biophys. Res. Commun. 53, 1173-1179 (1973). - TING, I.P., OSMOND, C.B.: Plant Physiol. 51, 439-447 (1973). - TOLBERT, N.E.: Glycolate Biosynthesis, 21-5o. In: Current Topics in Cellular Regulation, eds. B.L. HORECKER, E.R. STADTMAN, Vol. 7. New York-London: Academic Press 1973. - TURNER, J.F., BLACK, C.C., GIBBS, M.: J. Biol. Chem. 237, 577-579 (1962).

VACCHI, C., PICCARI, G.G., PUPILLO, P.: Z. Pflanzenphysiol. 69, 351-358 (1973).

WALKER, D.A.: (1) New Phytologist 72, 2o9-235 (1973); - (2) Chloroplast and Cell-Concerning the Movement of Certain Key Metabolites etc. across the Chloroplast Envelope, 1-49. In: MTP International Review of Science, Biochemistry Section, ed. D.H. NORTHCOTE, Vol. 11, in press; - (3) Some Characteristics of a Primary Carboxylating Mechanism. In: Phytochemical Society Symposium on Plant Carbohydrate Biochemistry (Edinburgh, 1972). New York-London: Academic Press, in press. - WALKER, D.A., KOSCIUKIEWICZ, K., CASE, C.: New Phytologist 72, 237-247 (1973). - WERDAN, K., HELDT, H.W.: Biochim. Biophys. Acta 283, 43o-441 (1972). - WILDNER, G.F., CRIDDLE, R.S.: Biochem. Biophys. Res. Commun. 37, 952-96o (1969). - WILLMER, C.M., DITTRICH, P.: Planta 117, 123-132 (1974).

ZELITCH, I.: (1) Photosynthesis, Photorespiration and Plant Productivity, 347 pp. New York-London: Academic Press 1971; - (2) Arch. Biochem. Biophys. 15o, 698-7o7 (1972); - (3) Plant Physiol. 51, 299-3o5 (1973); - (4) Arch. Biochem. Biophys., in press. - ZIEGLER, I.: Planta 1o3, 155-163 (1972). - ZIEGLER, H., ZIEGLER, I.: Planta 65, 369-38o (1965).

Professor Dr. E. LATZKO
Abteilung Chemische Pflanzen-
physiologie
Technische Universität München
D-8o5o Weihenstephan

Dr. G.J. KELLY
Abteilung Chemische Pflanzen-
physiologie
Technische Universität München
D-8o5o Weihenstephan

Notes added in proof (2005):

Perhaps the separate small protein described by WILDNER and CRIDDLE (Sect. 2) was the Rubisco activase discovered in 1985 (see 1988 review)?
The chloroplast-envelope translocator that transports triose phosphates and PGA (Sect. 4) is not, strictly speaking, an "active transporter".

Photosynthesis. Biochemical and Physiological Aspects of Carbon Metabolism

By Erwin Latzko and Grahame J. Kelly

1. Introduction

In the two years since we last reviewed photosynthetic carbon metabolism in Volume 36 of Progress in Botany, significant advances have been made in explaining, at the biochemical and physiological levels, the direction and regulation of carbon flow in photosynthetic cells. Regulatory properties of enzymes of the RPP cycle (Calvin cycle), including mechanisms whereby some of these enzymes are activated upon illumination, have received almost as much attention as the controversial nature of C_4 metabolism and the persistent problem of the value of photorespiration. These and related subjects are reviewed below, covering predominantly that literature published since our last writing, which it is recommended the reader consult for a background to the research covered here.

2. Enzymology of the Reductive Pentose Phosphate Cycle

a) RuDP Carboxylase

The carboxylation step of photosynthesis is catalyzed by RuDP carboxylase, the most abundant protein in nature and the enzyme receiving more attention than any other of the RPP cycle at the present time. BAHR and JENSEN (1) stimulated research by demonstrating that the enzyme freshly released from isolated chloroplasts had a high affinity for CO_2, similar to that of the intact leaf. This high affinity form has now been stabilized by addition of bicarbonate and Mg^{2+} to enzyme preparations (LAING et al., 2; LYTTLETON; ANDREWS et al.) and the affinity even further increased by raising the pH of reaction mixtures (BOWES et al.). Preincubation of both the chloroplast (LYTTLETON) and *Rhodospirillum rubrum* (TABITA and McFADDEN, 1) enzymes with these two ions also increased the catalytic activity, and with the chloroplast enzyme the inhibition by physiological levels of RuDP was relieved (CHU and BASSHAM, 2). LORIMER et al. (2) have recently concluded that CO_2 (and not bicarbonate) is the activating species. These authors attribute the activation to the formation of an enzyme-CO_2-Mg ternary complex, and in a separate report (ANDREWS et al.) reached the conclusion that the enzyme with low affinity for CO_2 is not active, although it appears to have activity because bicarbonate and Mg^{2+} in assay mixtures converts some enzyme to the active high-affinity form. A refreshing insight concerning the kinetic constants of RuDP carboxylase was recently provided by LILLEY and WALKER (3) who pointed out that the calculated Km for CO_2 naturally depends on the observed maximum

Abbreviations: CAM, crassulacean acid metabolism; DTT, dithiothreitol; FDP, fructose-1,6-diphosphate; GAP, glyceraldehyde-3-P; PEP, phosphoenolpyruvate; 3PGA, 3-phosphoglycerate; RPP, reductive pentose phosphate; RuDP, ribulose-1,5-diphosphate.

G.J. Kelly and E. Latzko:
Thirty Years of Photosynthesis

Originally published in Progress in Botany 38

velocity which, however, was disturbingly low in some reports. Using an improved assay system (LILLEY and WALKER, 1), these authors found maximum activities close to 1000 μ moles per mg chlorophyll per h and, although this consequently increased the Km (CO_2) to about 50 μ M, activities with air levels of CO_2 were still sufficient to accommodate observed rates of leaf photosynthesis. Clearly, carboxylation in vivo is just as likely to be limited by the supply of CO_2 and RuDP (ROBINSON et al.) as by the amount of enzyme.

CHU and BASSHAM (1, 2) have continued to study the regulation of RuDP carboxylase activity by various metabolites and have found that FDP below 0.1 mM and NADPH increased activity, similar to the previously observed 6-P-gluconate regulation but perhaps of greater physiological significance. A model involving four types of regulatory sites was proposed (CHU and BASSHAM, 2). Nevertheless, a description of the observed enzyme activation and regulatory properties in terms of the comparatively complex arrangement of enzyme subunits and their interaction is far from complete. Studies utilizing the strong inhibition by 6-P-gluconate (CHU and BASSHAM, 2) and the very potent inhibition by xylitol-1,5-P_2 (RYAN et al., 1) could assist in relating the enzyme structure to the wealth of kinetic and regulatory properties.

RuDP carboxylase has been purified from photosynthetic microorganisms, including *Chlorella* (LORD and BROWN), *Euglena* (McFADDEN et al.; RABINOWITZ et al.) and *Thiobacillus* (McCARTHY and CHARLES), and the observed regulatory properties were found similar to those of the higher plant enzyme. An electron micrograph of the *Euglena* enzyme molecule supported the proposed structure of eight large and eight small subunits (McFADDEN et al.). The *Chromatium* enzyme, which TAKABE and AKAZAWA (2) have dissociated into subunits and then reassociated into active enzyme, was shown to be similar to the spinach enzyme by both immunological studies (TAKABE and AKAZAWA, 2) and the observed loss of enzyme activity following progressive masking of sulphydryl groups (TAKABE and AKAZAWA, 1). However, the *Rhodospirillum rubrum* enzyme was found to be immunologically distinct from the enzyme of algae and higher plants, and consists of only two large subunits (TABITA and McFADDEN, 2).

Blue-green algal carboxylases were immunologically similar to the *Euglena* (LORD et al., 2) and spinach (TAKABE et al.) enzymes. Nevertheless, the exact structure of the carboxylase from these algae is at present uncertain. TABITA et al. reported that the marine blue-green alga *Agmenellum quadruplicatum* contained eight large subunits but no small subunits, yet TAKABE et al. found both large and small subunits in two other blue-green algae. The indication that a marine blue-green algal carboxylase lacks small subunits (TABITA et al.) should be substantiated since it carries attractive evolutionary implications: in eucaryotes the large subunit of RuDP carboxylase is synthesized in the chloroplast (CRIDDLE et al.), the small subunit is synthesized in the cytoplasm (GRAY and KEKWICK; ROY et al.) and appears later to move across the chloroplast envelope (COBB and WELLBURN), and there is evidence that more of one subunit than the other can be synthesized (GIVAN; LORD et al., 1). With these observations in mind, the absence of the small subunit in a blue-green alga is consistent with the proposed evolution of chloroplasts from blue-green algae which had developed some form of symbiotic existence.

The separate synthesis of the small and large subunits of RuDP carboxylase in the cytoplasm and chloroplast respectively is likely to have practical value for plant breeding since the subunits provide excellent phenotypic markers for the nuclear and chloroplast genomes. CHEN et al. (1) have utilized the polypeptide composition of subunits to show that the direction of the cross which formed *Triticum aestivum* was *T. dicoccum* ♀ × *Aegilops squarrosa* ♂. Techniques to extend this approach have been recently developed (GRAY and WILDMAN).

b) Glyceraldehyde-3-P Dehydrogenase

The reductive step of photosynthesis is catalyzed by the NADP-linked activity of GAP dehydrogenase. Attempts have been made in the past to identify a regulatory role for this enzyme, but two considerations preclude a simple interpretation of results to date. Firstly, activities of GAP dehydrogenase reported for photosynthetic tissues are generally many orders of magnitude higher than the rate of leaf photosynthesis, thus reducing the likelihood of the enzyme exerting a rate-limiting effect. Secondly, there exist several potential impediments to correct estimation of the enzyme activity. These include the formation of complexes between the buffer base Tris and triose-P (SEGAL and BOYER), the powerful inhibition of the NADP-linked activity by a possible contaminant of cation-exchange resins (PAWLIZKI et al.), inhibition of the forward reaction by the non-physiological L-isomer of GAP (KELLY and GIBBS), inhibition of the reverse reaction by the product D-GAP (K.H. FISCHER, personal communication), and interference from glutathione reductase if reduced glutathione is used as sulphydryl protector.

Studies have also been complicated by the tendency of the enzyme to aggregate (McGOWAN and GIBBS; PUPILLO and PICCARI). Evidence consistent with the association of four enzyme molecules and concomitant loss of NADP-linked activity has been reported for spinach (PUPILLO and PICCARI) and *Scenedesmus* (O'BRIEN and POWLS) GAP dehydrogenases but it is doubtful whether these changes occur in vivo; the quaternary structure of non-plant GAP dehydrogenase is also unstable in vitro but this is not believed important as a regulatory mechanism in vivo owing to high local enzyme concentrations (MARKOVICH and KRAPIVINSKAYA). Present evidence suggests the chloroplast enzyme is a tetramer of 140,000 MW (McGOWAN and GIBBS; PAWLIZKI and LATZKO; GRISSOM and KAHN; PUPILLO and PICCARI; O'BRIEN and POWLS) with an NADP-linked activity several-fold (PUPILLO and PICCARI) to many times (IHLENFELDT and GIBSON) higher than the NAD-linked activity. *Euglena gracilis* contains an NADP-linked GAP dehydrogenase which is relatively hydrophobic (GRISSOM and KAHN). This property, if true for other chloroplast GAP dehydrogenases, could explain the observed enzyme aggregation in aqueous solutions. GRISSOM and KAHN pointed out that the environment around the chloroplast membrane is also hydrophobic, and it is interesting that SCHWARZ was able to extract twice as much enzyme activity from bean leaves by adding 1% Na cholate to extraction media. In animal systems, erythrocyte membranes are able to bind GAP dehydrogenase which can be specifically released by NADH (McDANIEL and KIRTLEY) or the tetrapyrrole derivative bilirubin (GIROTTI); the question may be asked whether the chloroplast enzyme binds to chloroplast membranes and can be released by NADPH.

Chloroplast GAP dehydrogenase is immunologically distinct from the cytoplasmic enzyme (McGOWAN and GIBBS), and although there is evidence that it is synthesized in the chloroplast (BATT and WOOLHOUSE) the possibility that the cytoplasmic enzyme may be a precursor of the chloroplast enzyme is supported by developmental studies with *Sinapis alba* (CERFF; CERFF and QUAIL).

c) Fructose-1,6-diphosphatase

A principal regulatory step of photosynthesis is catalyzed by FDPase. This conclusion is drawn from observations that the enzyme activity is similar to rates of chloroplast CO_2 fixation and responds to changes in the concentrations of the substrate FDP and cofactor Mg^{2+} in a sigmoid fashion (BAIER and LATZKO; LÁZARO et al., 2). It is also activated by DTT, and BAIER and LATZKO concluded that the combined effects of FDP, Mg^{2+} and DTT under estimated physiological conditions in the chloroplast stroma were consistent with almost full activity in the light but almost none in the dark.

Chloroplast FDPase, like the enzyme from other sources, is a tetramer (ZIMMERMANN et al.). It very easily dissociates into dimers (LÁZARO et al., 1; ZIMMERMANN et al.). The enzyme obtained after purification is inactive at pH 7.5 (although quite active at pH 8.8), but ZIMMERMANN et al. have now demonstrated that incubation of the enzyme with DTT induces almost full activity at pH 7.5 and, at the same time, increases the number of freely available sulphydryl groups three-fold. This observation has considerable implications for light activation (see below).

d) Other RPP Cycle Enzymes

The remaining nine enzymes of the RPP cycle have received little attention in the past two years. Chloroplast 3PGA kinase was found similar to the enzyme from most other sources; activity in the photosynthetic direction was strongly inhibited by ADP (a consequence of energy charge regulation) (PACOLD and ANDERSON). This inhibition was shown to be responsible for the decreased 3PGA-dependent O_2 evolution by a reconstituted chloroplast system upon addition of ADP (SLABAS and WALKER, 1). The other kinase of the RPP cycle, ribulose-5-P kinase, has been purified from *Hydrogenomonas eutropha* (ABDELAL and SCHLEGEL) and spinach leaves (LAVERGNE et al.). The bacterial enzyme possessed a sigmoid ATP saturation curve, while the spinach enzyme was regulated by energy charge, an observation in contrast to an earlier report (ANDERSON, 1). Chloroplast aldolase has been purified and found similar to other aldolases (ANDERSON et al., 1).

e) Light Mediated Activation of RPP Cycle Enzymes

There is evidence that five enzymes of the RPP cycle are activated upon illumination, viz RuDP carboxylase, GAP dehydrogenase, FDPase, ribulose-5-P kinase and, more recently, sedoheptulose-1,7-diphosphatase (ANDERSON, 2; SCHÜRMANN and BUCHANAN). Activation of RuDP carboxylase may result from the light-induced exchange of Mg^{2+} and protons between thylakoids and the stroma (LILLEY et al., 2; BOWES et al.) and CO_2 activation (LORIMER et al., 2). However, there is strong evidence, including activation by DTT in vitro (ANDERSON, 2, 3; DUGGAN and ANDERSON), that the other four enzymes are activated by reductive processes in vivo. CHAMPIGNY and BISMUTH recently demonstrated light-induced activation of all five enzymes in isolated intact chloroplasts and found that stimulators of photosynthetic electron transport, including antimycin A, increased the activations (except that of RuDP carboxylase) again implying that reduced agents were involved for the other four enzymes. Two proposed reductants are reduced ferredoxin and vicinal dithiols. SCHÜRMANN and BUCHANAN observed that sedoheptulose-1,7-diphosphatase was activated by addition of reduced ferredoxin to reaction mixtures in a similar manner to activation of FDPase in previous experiments. ANDERSON and AVRON, on the other hand, studied activation of GAP dehydrogenase, sedoheptulose-1,7-diphosphatase and ribulose-5-P kinase by illuminated chloroplast membranes and concluded that vicinal dithiols generated on the membranes were responsible for the activations. It is tempting to favor a role for dithiols since VALLEJOS and ANDREO have evidence for light-induced conformational changes of chloroplast membranes resulting in exposure of vicinal dithiols to the stroma. In addition, light activation of FDPase in isolated intact chloroplasts, first reported by CHAMPIGNY and BISMUTH, was found by KELLY et al. (2)

to parallel DTT activation in vitro in that increased activity could be detected only by assaying at pH values below the optimum for this enzyme's activity. In fact, the most marked activation was measured at precisely the same pH as that produced in the chloroplast stroma following light-induced movement of protons into the thylakoids (WERDAN et al.) hinting that regulatory effects produced by illumination are really quite closely interconnected.

f) Carbonic Anhydrase

The exact role of carbonic anhydrase in photosynthesis remains unclear. Evidence continues to accumulate that it is principally located in the chloroplast (TRIOLO et al.; CHANG; JACOBSON et al.) although a similar cytoplasmic enzyme has been detected (KACHRU and ANDERSON; WALK and METZNER). Studies with algae suggest the enzyme facilitates movement of bicarbonate and inorganic ions (FINDENEGG; RYBOVÁ and SLAVÍKOVÁ). It is interesting that the activity in blue-green algae (which lack chloroplasts) was significantly lower than in unicellular green algae, and was reduced even further by growth under high CO_2 levels (DÖHLER, 1; INGLE and COLMAN). JACOBSON et al. have discounted the concept that carbonic anhydrase is associated with the diffusion or concentration of CO_2 in leaves, but rather suggest that it has a protective function, such as against sudden and drastic localized pH changes which would otherwise cause protein denaturation and which might be expected in leaves intermittently shading each other.

3. Studies with Isolated Chloroplasts

It is becoming apparent that preparations of isolated chloroplasts do not necessarily contain a homogeneous population of organelles. Some chloroplasts are believed by LILLEY et al. (1) to rupture during isolation and then subsequently reseal; these chloroplasts do not reduce ferricyanide and thus are registered as intact but nevertheless have low rates of photosynthesis. LARSSON and ALBERTSSON separated chloroplasts by counter-current distribution and found that, besides intact and broken plastids, a third group was surrounded by a membrane-bound cytoplasmic layer containing mitochondria and microbodies. The location of mitochondria in deep invaginations of chloroplasts was recently seen by MONTES and BRADBEER.

Separation of intact chloroplasts from those stripped of their outer envelopes was achieved by MORGENTHALER et al. (2) with a density gradient centrifugation technique, which has since been further characterized (MORGENTHALER et al., 1). Short-time density gradient centrifugations were found effective by MIFLIN and BEEVERS to separate chloroplasts and other plastids from microbodies that have similar densities but, being smaller, take longer to move down the gradient. In other experiments, KAGAN-ZUR and LIPS found that CO_2 fixation by chloroplasts isolated on a density gradient was increased up to three-fold when a microbody-rich fraction from the same gradient was added. Greatest stimulation occurred when dark-pretreated leaves were used. One drawback to studies with isolated chloroplasts has always been that the metabolic interaction between plastids and the cytoplasm is cut off. Some interesting experiments may therefore follow the aquisition by NISHIMURA and AKAZAWA of spinach leaf protoplasts capable of high photosynthetic rates for up to nine hours.

Photosynthetic CO_2 fixation by chloroplasts isolated from C_3 plants has been known for over two decades, while the isolation of photosynthetically competent chloroplasts from a C_4 plant was reported several years ago (GIBBS et al.). Appreciable rates of CO_2 fixation by chloroplasts isolated from a CAM plant, *Kalanchoe diagremontiana*, has now been described by LEVI and GIBBS. The products of CO_2 fixation by these chloroplasts were, as for all others, indicative of RuDP carboxylation and metabolism through the RPP cycle.

Photosynthesis by isolated chloroplasts can be inhibited by 3PGA. CO_2 fixation was inhibited by addition of 3PGA at concentrations greater than 1 mM to reaction vessels, probably because it exchanged with chloroplast triose-P, which therefore became unavailable for regeneration of RuDP (ANDERSEN and GIBBS; BAMBERGER et al.); RuDP carboxylase inhibition also may have been involved (BAMBERGER and AVRON). 3PGA at 1 mM increased the enhancement of photosynthesis by 640 nm light, but at a lower concentration of 10 μM it abolished this effect, indicating that the Emerson enhancement effect reflects events of the RPP cycle as well as photochemistry (PEAVEY and GIBBS).

Regulatory aspects of CO_2 fixation have been examined with the reconstituted chloroplast system which has now been developed to the stage of fixing CO_2 autocatalytically at high rates (WALKER and LILLEY; WALKER and SLABAS). 3PGA reduction by this system was strongly inhibited by ADP (LILLEY and WALKER, 2) including the ADP formed when added ribose-5-P was converted to RuDP (SLABAS and WALKER, 2); the inhibition was localized at the 3PGA kinase reaction (SLABAS and WALKER, 1) and might be expected from energy charge regulation of this enzyme (PACOLD and ANDERSON). Clearly, CO_2 fixation could be quite sensitive to regulation by the levels of ADP and ATP in chloroplasts, and it is interesting that rapid transient changes in the chloroplast ATP level have been observed (COCKBURN). Rapid and sizeable increases in the energy charge of chloroplasts was recently reported to follow the illumination of wheat leaves under N_2, and similar but reciprocal changes occurred upon darkening in air (SELLAMI). However, somewhat smaller changes in the energy charge may not be detrimental to CO_2 fixation; antimycin A reduced photosynthetic phosphorylation in chloroplasts but nevertheless increased CO_2 fixation (MIGINIAC-MASLOW and CHAMPIGNY; MIGINIAC-MASLOW and HOARAU).

The influence of sulphite on photosynthesis is now known to occur at more than one site. At concentrations of less than 1 mM this anion enhanced chloroplast CO_2 fixation through an effect on the photochemical reactions (ZIEGLER and LIBERA). However, sulphite at higher concentrations reduced CO_2 fixation, at least partly due to RuDP carboxylase inhibition, and shifted carbon flow towards C_4 compounds (LIBERA et al.), which have been detected as a minor product of chloroplast CO_2 fixation (SCHEIBE and BECK).

The light-induced pH rise in the chloroplast stroma appears to be a primary requirement for CO_2 fixation. WERDAN et al. have now induced chloroplasts to fix CO_2 in darkness, although not yet at high rates, by artificially raising the stromal pH, adding a sulphydryl reducing agent and operating suitable shuttle systems to generate ATP and reoxidize NADPH. Under normal conditions the pH rise results from protons moving into the thylakoids (in which the pH correspondingly falls) in exchange for other cations, and although several ions including K^+ and Mg^{2+} were earlier considered to fill this role (BARBER et al., 2; HIND et al.; KRAUSE; PFLÜGER), more recent experiments using specific ionophores with intact chloroplasts indicate that only divalent cations, probably Mg^{2+}, are involved (BARBER et al., 1; TELFER et al.).

4. Communication between Chloroplast and Cytoplasm

Net photosynthesis by a growing plant requires not only that products of CO_2 fixation can easily leave the chloroplast but also that starting materials can freely enter. It is now widely believed that triose-P is the major product exported (HEBER and KIRK), hence CO_2 and Pi (and of course H_2O) must enter in corresponding amounts. Such transport is controlled by the selective permeability of the inner membrane of the chloroplast envelope; this changes noticeably during chloroplast development (COCKBURN and WELLBURN). The active export of triose-P, mediated by a chloroplast membrane-located phosphate translocator (HELDT and RAPLEY), requires the simultaneous entry of Pi. HEBER has pointed out that, since triose-P is further transformed in the cytoplasm to secondary products such as sucrose, the Pi is set free and recycled: in this way a balance is established, thus preventing a momentary export of large amounts of triose-P which would deplete the RPP cycle of substrate to regenerate RuDP. Such a depletion has been demonstrated with isolated chloroplasts (BAMBERGER et al.). It is interesting to contemplate that the Pi is released in the cytoplasm by the phosphatases hydrolyzing FDP and sucrose-P; the activity of these enzymes could therefore indirectly influence CO_2 fixation and reports of their regulatory properties will be eagerly awaited. In other experiments, chloroplasts responded to triose-P added externally by rapidly assimilating it into the RPP cycle and chloroplast starch (HEBER and KIRK), and apparently could utilize cyclic photophosphorylation to do so since reductive power was not required and indeed CO_2-dependent O_2 evolution was not observed (KAISER and URBACH).

Concepts concerning the entry of CO_2 into chloroplasts are currently less clear-cut than those of the triose-P/Pi exchange. POINCELOT (1) challenged the conclusion of WERDAN and HELDT that CO_2 is the entity transported into chloroplasts, and contends that bicarbonate is the transported species. Subsequent studies with isolated envelope membranes of spinach chloroplasts upheld this idea and supported earlier conclusions, from studies with whole chloroplasts, on the relative permeability of other metabolites (POINCELOT, 2). In recent experiments POINCELOT and DAY, studying chloroplast envelope membranes from three plants with different relative rates of photosynthesis and photorespiration, obtained some correlation between bicarbonate transport and the activity of a Mg^{2+}-dependent nonlatent ATPase. CO_2 flux to the chloroplast stroma may be limited at the site of transport across the chloroplast envelope: temperature-induced phase changes of the lipids of envelope membranes were suggested by CHAMPIGNY and MOYSE to influence the response of chloroplast CO_2 fixation to temperature.

5. Starch and Sucrose Metabolism

Two major products of photosynthesis in green leaves are starch and sucrose. Sucrose is synthesized in the cytoplasm from triose-P exported out of chloroplasts and is largely moved to the phloem for transport to other parts of the plant. Metabolic aspects of leaf sucrose synthesis have received little attention in recent literature. Flow of fixed carbon into sucrose was favored over that into glycine and serine (proposed intermediates of photorespiration) at higher levels of CO_2 (LEE and WHITTINGHAM; SNYDER and TOLBERT) and with higher temperatures (SAWADA and MIYACHI). BUCKE and OLIVER re-examined the location of sucrose synthesis in C_4 plants and concluded that both mesophyll and bundle-sheath cells were capable of forming the disaccharide. A curious observation was that of BEITLER and HENDRIX who found the tetrasaccharide stachyose was labeled faster than sucrose in squash leaves fed $^{14}CO_2$ in the light; these leaves transport much photosynthate as stachyose. The possibility that maltose is also a product of photosynthesis was recently strengthened by the detection of asymmetrically labeled maltose in leaves fed $^{14}CO_2$ during photosynthesis (LINDEN et al., 1). The distribution of label was consistent with the non-reducing glucose being formed from an activated hexose of the RPP cycle; maltose-P has now been reported in photosynthesizing chloroplasts (SCHILLING et al.).

The pathway of starch biosynthesis, which occurs inside chloroplasts and is principally regulated at the ADP-glucose pyrophosphorylase reaction (GHOSH and PREISS), is now well established. The regulatory reaction may be pulled toward starch synthesis by the chloroplast pyrophosphatase which has a higher substrate affinity than its cytoplasmic counterpart (KLEMME and JACOBI). Primer oligosaccharides necessary for starch synthesis might be formed by the glucosyl and glucosylglucan transferases demonstrated in leaves (LINDEN et al., 2); all leaf starch synthesis is presently believed to require a primer (HAWKER et al.). A significant study, complimenting the considerations of HEBER concerning chloroplast-cytoplasm communications, has been performed by SHEU-HWA et al. who found that photosynthetic starch formation was increased more than ten-fold by sequestration of cytoplasmic Pi. These authors suggested that the lack of external Pi may have prevented export of triose-P which was instead metabolized to starch within the chloroplast. The recent observation that Al^{3+} stimulated starch synthesis in epidermal strips of *Vicia faba* (SCHNABL) may not be inconsistent with sequestration of cytoplasmic Pi.

Chloroplast starch is of a transitory nature and so must eventually be metabolized to compounds able to leave the chloroplast; this is reported to require 12 days of darkness for a CAM plant (SARDA et al.). With the recent demonstration of phosphofructokinase activity in chloroplasts (KELLY and LATZKO) it is now realistic to consider the degradation of starch to easily exportable triose-P by reactions of the glycolytic sequence; the observed activity was just sufficient to accommodate the overnight dissimilation of reported levels of leaf starch (MILFORD and PEARMAN; FORDE et al.). Coupled with indications that the enzyme has allosteric regulatory properties, it seems likely that phosphofructokinase may compliment ADP-glucose pyrophosphorylase in regulating starch metabolism. However, triose-P may also be formed from starch by reactions of the oxidative pentose-P pathway in which case glucose-6-P dehydrogenase may well be the enzyme exerting a regulatory influence; the activity of this enzyme is controlled by the NADP/NADPH ratio (GROSSMAN and McGOWAN; LENDZIAN and BASSHAM; WILDNER) and partially inactivated following illumination (ANDERSON et al., 2; WILDNER). Starch degradation by the pentose-P pathway would have the advantage of generating NADPH useful for biosynthetic reactions, but in so doing half of the fixed carbon would be lost in the form of CO_2.

In the past it has sometimes been implied that photosynthesis, at a rate of 100 µmoles per mg chlorophyll per h, requires at least this activity of each of the RPP cycle enzymes. This is not strictly true. As shown in Figure 1, the direction in which the immediate products of photosynthesis flow will influence the minimum activities required of nine of the twelve enzymes of the RPP cycle (assuming saturating concentrations of substrates), a consideration which could have regulatory consequences. For example, a fall in the activity of chloroplast FDPase, which is never much greater than rates of CO_2 fixation, could favor sucrose synthesis over starch synthesis since the latter (Fig. 1C) requires 50% more chloroplast FDPase activity than the former

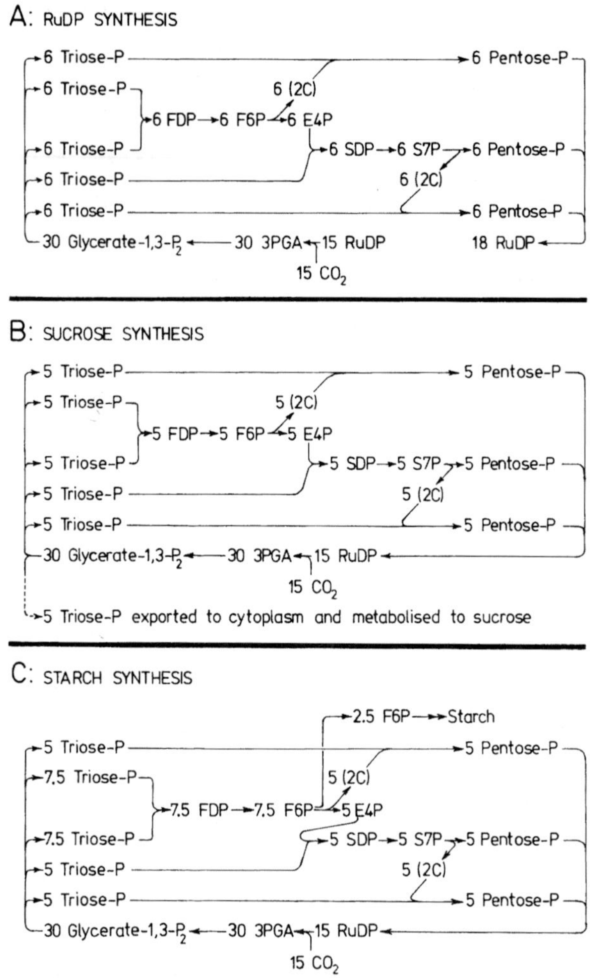

Fig. 1A-C. Minimum required activities of RPP cycle enzymes when either RuDP (A), sucrose (B) or starch (C) is synthesized from the incorporated CO_2. In A, three molecules of RuDP are synthesized from 15 molecules of CO_2, while in B and C the level of RuDP is maintained. Examples of minimum required activities (μ moles per mg chlorophyll per h when CO_2 fixation by RuDP carboxylase is 90 μ moles per mg chlorophyll per h): GAP dehydrogenase, 180 (A, B, C); FDPase, 36 (A), 30 (B), 45 (C); ribulose-5-P kinase, 108 (A), 90 (B and C). Additional abbreviations: *2C*: glycolaldehyde-transketolase intermediate; *E4P*: erythrose-4-P; *F6P*: fructose-6-P; *SDP*: sedoheptulose-1,7-P_2; *S7P*: sedoheptulose-7-P

(Fig. 1B). Since the light-induced activation of this enzyme is greater at higher light intensities (KELLY et al., 2) it is not inconceivable that the lack of chloroplast starch in plants grown under low light is related to this difference in activity requirements.

6. Photorespiration and Glycollate Metabolism

The light-dependent uptake of O_2 and release of CO_2 by photosynthetic tissues, termed photorespiration, continues to be a popular topic of research. Photorespiration is more obvious with C_3 plants than with C_4 plants because the latter efficiently recapture the respired CO_2 and consequently have very low CO_2 compensation points. However, there are reports of C_3 plants with unusually low CO_2 compensation points and reduced photorespiration (SALIN and HOMANN; STANLEY and NAYLOR; CHENG and COLMAN) and one of a high value in a C_4 plant (LESTER and GOLDSWORTHY). Despite an extensive survey by KRENZER et al. showing that all but two of several hundred plants had high or low (but not intermediate) compensation points, one of the two exceptions (*Panicum milioides*) is currently creating special interest since also in other respects it was found intermediate between C_3 and C_4 plants (KESTLER et al.).

a) Glycollate Synthesis

The mechanism of O_2 uptake is now widely believed to involve, at least in part, the oxidative cleavage of RuDP to one molecule each of 3PGA and P-glycollate, catalyzed by an oxygenase activity of RuDP carboxylase. The oxygenase activity demonstrated in extracts of freshly ruptured chloroplasts by BAHR and JENSEN (2) was sufficient to direct 40% of fixed carbon into glycollate. Oxygenase activity has now been demonstrated with the more simply structured carboxylase from *Rhodospirillum rubrum* (McFADDEN; RYAN et al., 2; CODD and TURNBULL), which can synthesize glycollate (CODD and SMITH). The oxygenase and carboxylase activities in leaves are mutually competitive (BADGER and ANDREWS; LAING et al., 1) and could therefore regulate net CO_2 fixation, especially with increased temperature which favors the oxygenase activity (LAING et al., 1) or higher pH which favors glycollate synthesis (KRAPF and JACOBI). A consequence of O_2 uptake during RuDP oxidation and release in photosynthesis is that chloroplasts may be considered to possess an O_2 compensation point. Recent experiments by HEBER et al. provided strong evidence for such a compensation point in chloroplasts performing high rates of photosynthesis. The theoretical considerations of PEISKER concerning the influence of O_2 on photosynthetic carboxylation may not be unrelated to this concept.

Genetic evidence has now been obtained with *Chlamydomonas reinhardi* that both the carboxylase and oxygenase reactions are catalyzed by the one protein molecule (NELSON and SURZYCKI). Impressive experiments by KUNG and MARSHO have confirmed, through genetic techniques, that the ratio of the activities is determined by the small subunits.

Hydrolysis of P-glycollate (formed from RuDP oxidation) is one means of glycollate production by plants and should be feasible at physiological pH (KERR and GEAR). A previously proposed alternative, the light-dependent oxidation of a glycolaldehyde transketolase complex, has been questioned since isolated maize bundle-sheath strands were found to synthesize glycollate in darkness (CHOLLET). Nevertheless, this alternative remains alive as the belief that glycollate normally originates from more than one reaction gains increasing support. D'AOUST and CANVIN concluded that the CO_2 produced by leaves in the light came from two sources, and FOCK et al. (1, 2) have proposed that *Chlorella* forms glycollate from both recent photosynthate and storage material. EICKENBUSCH et al. showed glycollate could be synthesized from either RuDP or fructose-6-P in a reconstituted chloroplast system but only the synthesis from RuDP was influenced by the O_2 level. Glycollate biosynthesis by isolated chloroplasts (ROBINSON and GIBBS) and *Chromatium* (ASAMI and AKAZAWA) could not be attributed to RuDP oxidation. In a recent review ZELITCH has pointed out that earlier experiments with ^{18}O (LORIMER et al., 1) did not prove that more than one-third of synthesized glycollate originated from an RuDP oxygenase reaction, even under 100% O_2.

At least three observations suggest that glycollate formation is distinct from the O_2 inhibition of photosynthesis (Warburg effect): (a) the two processes were found to have different temperature optima (CORNIC), (b) several metabolites had contrasting effects on the Warburg effect and production of glycollate during studies with isolated chloroplasts (ROBINSON and GIBBS), and (c) lowering O_2 did not modify the size of the post-illumination burst of CO_2 (an indicator of photorespiration) in a CAM plant (OSMOND and BJÖRKMAN).

b) Glycollate Utilization

Experiments with protoplasts indicate that glycollate is rapidly metabolized in plant cells (NISHIMURA et al.). Conversion of two molecules of glycollate to one each of 3PGA and CO_2 via glyoxylate, glycine, serine and hydroxypyruvate and involving reactions in peroxisomes, mitochondria and chloroplasts (TOLBERT) is thought to accommodate much of this metabolism and is supported by data denoting a requirement for transaminases (WAIDYANATHA et al.). Glycollate oxidase (KERR and GROVES) and serine-glyoxylate aminotransferase (CARPE and SMITH) have been purified from plant leaves. Curiously, the amount of carbon passing through this glycollate pathway seemed not to be affected by large changes in the CO_2 level (SNYDER and TOLBERT) or temperature (SAWADA and MIYACHI) which increased sucrose synthesis at least ten-fold. A further complication has arisen from the extensive studies of MAHON et al. (1, 2, 3) who concluded that serine in leaves from both C_3 and C_4 plants did not originate exclusively from glycollate. In fact, it was questioned whether photorespiration in C_4 plants was metabolically similar to that in C_3 plants (MAHON et al., 3), a viewpoint in agreement with the suggestion by WYNN et al., that the pools of substrate for the post-illumination burst of CO_2 (considered a remnant of photorespiration) differ between these two plant groups.

The pathway from glycollate to 3PGA is not the only one believed capable of releasing photorespiratory CO_2 in leaves. One alternative, the peroxisome-located oxidation of glycollate to glyoxylate which is further oxidized to formate and CO_2, has received substantial support from GRODZINSKI and BUTT who showed that, despite high quantities of catalase in peroxisomes, the often overlooked low affinity of this enzyme for H_2O_2 ensured that sufficient H_2O_2 remained for the proposed two oxidations. Further metabolism of formate in plant leaves appears possible (CROSTI).

Photosynthetic microorganisms provide interesting variations of glycollate metabolism. Following indications of an unusual glycollate oxidation system in the diatom *Thallassiosira pseudonana* (PAUL and VOLCANI, 1) and the possible mitochondrial location of glycollate oxidation in the algae *Chlamydomonas* and *Chlorogonium* (STABENAU, 1, 2), glycollate oxidation by mitochondria from diatoms (PAUL et al.) and green algae (COLLINS and MERRETT; PAUL and VOLCANI, 2) was confirmed and shown to be coupled to oxidative phosphorylation (COLLINS et al.); a P:O ratio of 1.7 suggested that glycollate was oxidized in the electron transport chain at the flavoprotein level. The blue-green alga *Anabaena cylindrica* contains a glycollate oxidizing system (CODD and STEWART) which SALLAL and CODD consider to be associated with the thylakoid membranes where oxidation is linked to O_2 via electron carriers, thus providing evidence for the potentially important prospect that these thylakoid membranes are a site of oxidative electron transport.

7. C_4 Metabolism

Substantial incorporation of CO_2 into 4C-acids through PEP carboxylation is closely associated with photosynthesis in many plants, including those with the C_4 pathway (C_4 plants) and those with CAM. Such plants are considered to possess C_4 metabolism, and the C_4 pathway has been distinguished as a special case of C_4 metabolism involving movement of carbon between mesophyll and bundle-sheath cells (KELLY et al., 1). Plants with C_4 metabolism are absolutely dependent on the RPP cycle for growth (see our preceding review).

The more classical examples of C_4 plants fix CO_2 into either malate or aspartate in mesophyll cells and are then believed to transport these acids to bundle-sheath cells, although the capacity of plasmodesmata to accommodate this transport is still under review (OLESEN). Decarboxylation in bundle-sheath cells is achieved through either NADP-malic enzyme in chloroplasts (see BLACK), PEP carboxykinase in chloroplasts (RATHNAM and EDWARDS) or NAD-malic enzyme in mitochondria (HATCH and KAGAWA). Correlations between the decarboxylating enzyme utilized by any one plant and that plant's arrangement of chloroplasts (GUTIERREZ et al.) and mitochondria (HATCH et al., 1) have been observed. An extensive enzyme-distribution pattern fully consistent with the C_4 pathway has been published by CHEN et al. (2), and portions of the pathway have been demonstrated with cells or organelles from mesophyll or bundle-sheaths (HUBER and EDWARDS, 1, 2; KAGAWA and HATCH, 1, 2; USUDA et al., 2). There is further evidence that plants with high NADP-malic enzyme activity have bundle-sheath chloroplasts deficient in photosystem II and must partially depend on the reductive decarboxylation of malate for NADPH (KU et al., 2; MAYNE et al.; FARINEAU; USUDA et al., 1). Despite these studies, the conclusion that the RPP cycle is located even predominantly in the bundle-sheath of C_4 plants has been further debated (KU et al., 1) and questioned (BUCKE and OLIVER). The ultimate source of the CO_2-acceptor PEP has also been disputed (USUDA et al., 3) although there is evidence that it can be derived from 3PGA (LABER et al.; HUBER and EDWARDS, 2), rather than from pyruvate, and it was emphasized in our last review that some PEP must be supplied from the RPP cycle to permit growth.

C_4 metabolism in CAM plants involves the nocturnal synthesis of the 4C-acid malate which is stored prior to decarboxylation the following day. This daytime decarboxylation can be catalyzed by any of the three decarboxylating enzymes found in C_4 plants; the latest identified was the mitochondrial NAD-malic enzyme which is widespread in the Crassulaceae family (DITTRICH). At night, malate formation is favored when water stress is not too severe (SZAREK and TING), photoperiods are short (ZABKA and CHATURVEDI) and leaves are older with larger cell vacuoles (JONES). Closer examinations of CAM have shown the presence of circadian rhythms which have been interpreted in terms of changing enzyme activities (QUEIROZ and MOREL) or osmotic gradients (LÜTTGE et al.). Malate formation during CAM results from carboxylation of PEP supplied from starch and other glucans via glycolysis (KLUGE et al.; SUTTON); such a carboxylation (in $^{14}CO_2$) would be expected to produce only C4-labeled malate, and an explanation for the earlier observed label in the C1 of malate is still being sought (KLUGE et al.; COCKBURN and McAULAY). One possibility, that PEP is derived from the products of a preceding RuDP carboxylation, has been demonstrated in a CAM plant near the end of a period of illumination (OSMOND and ALLAWAY).

The contribution of C_4 metabolism to salt tolerance in some halophytes appears related to a salt-induced increase in PEP carboxylase activity (TREICHEL et al.; BEER et al.) and malate accumulation (VON WILLERT, 2); increasing water stress by other means similarly increased the enzyme activity (WINTER). A salt-induced switch from C_3 to C_4 metabolism in the halophytic grass *Aeluropus litoralis*, considered an artifact by DOWNTON and TÖRÖKFALVY, has been rapidly defended by SHOMER-ILAN and WAISEL.

The participation of PEP carboxylase in C_4 metabolism results in less discrimination between the naturally occurring isotopes of carbon (^{13}C and ^{12}C) during CO_2 fixation than with RuDP carboxylase alone (KLUGE). This difference is reflected in the ratio of these isotopes in the plant material and has been used to categorize C_3, C_4 and CAM plants, a recent example being the recognition of both C_3 and CAM plants among tropical orchids (NEALES and HEW). However, TROUGHTON and CARD have recently cast some doubt on the conclusion that the carboxylation reactions are the exclusive site of carbon isotope fractionation, and a recent unique study by ZIEGLER et al. showing discrimination between hydrogen isotopes in plants, possibly at the level of ferredoxin, emphasizes the likely complexity of discrimination processes.

Of the enzymes involved in C_4 metabolism, only pyruvate, Pi dikinase, PEP carboxylase and the malic enzymes have received detailed investigations in the last two years. Earlier studies of pyruvate, Pi dikinase, which has now been detected in the photosynthetic bacteria *Chlorobium*, *Chromatium* and *Rhodospirillum* (BUCHANAN) and in the CAM plant *Kalanchoe* at high activity (SUGIYAMA and LAETSCH), have been supplemented by evidence for a proteinaceous factor involved in the reactivation of enzyme isolated from dark-treated maize leaves (SUGIYAMA). PEP carboxylase, located in the cytoplasm (COOMBS et al., 1) or possibly chloroplast envelope (RATHNAM and DAS), appears to be a modified form of the enzyme found in dark-grown C_4 plants (GOATLY et al.) and may have a higher affinity for the substrate MgPEP than previously believed (MIZIORKO et al.; MUKERJI). The other substrate is bicarbonate, as for the enzyme from C_3 plants (COOMBS et al., 2). Regulation of the PEP carboxylase activity has now been proposed to involve inhibition by high energy charge (COOMBS et al., 1), triose-P and 3PGA (COOMBS and BALDRY) and malate and aspartate (HUBER and EDWARDS, 3; VON WILLERT, 1). NADP-malic enzyme from C_4 plants has been characterized and found to have a greater affinity for malate than that from C_3 and CAM plants (NISHIKIDO and WADA; ZIEGLER); the CAM enzyme showed some degree of allosteric regulation (GARNIER-DARDART and QUEIROZ). The mitochondrial NAD-malic enzyme from C_4 and CAM plants was markedly stimulated by acetyl coenzyme A and coenzyme A, absolutely dependent on Mn^{2+} and commonly possessed sigmoid substrate saturation curves (HATCH et al., 2; DITTRICH).

One of the more fascinating developments from recent research on photosynthetic carbon metabolism has been the steady erosion of any clear-cut distinction between C_3 and C_4 plants. At one extreme, there is evidence that many classical C_3 plants exhibit some degree of C_4 metabolism at certain growth stages or in particular tissues: the list now includes blue-green and green algae (KAREKAR and JOSHI; DÖHLER, 2, 3; COLMAN et al.) and rust-infected wheat leaves (WAYGOOD et al.) as well as stomata (WILLMER and DITTRICH) and the barley pericarp (DUFFUS and ROSIE). More crucial, however, are challenges to the rigorous identification of C_4 plants. Wider surveys of the plant kingdom have not only extended the list of C_4 plants (OSMOND; DOWNTON) and their range to tree forms in the Hawaiian rainforest (PEARCY and TROUGHTON) and cool temperate species adapted to European coastal sand dunes (LONG et al.), but have also drawn attention to species difficult to classify. One of these is *Portulaca oleracea* which showed characteristics of C_3 plants (KENNEDY) and rapidly produced alanine during photosynthesis (KENNEDY and LAETSCH, 1) under some conditions, but clearly behaved as a C_4 plant under other conditions (HATCH). Others include C_4 plants deficient in Kranz anatomy (SHOMER-ILAN et al.; RATHNAM et al.) although the possibility of reversion from a Kranz arrangement, as suggested for certain Chenopodiaceae (CAROLIN et al.), might be considered here. Perhaps the most startling examples, however, are plants which appear truly intermediate between C_3 and C_4 on the basis of two or more of the four criteria of leaf anatomy, photorespiration, CO_2 compensation point, and products of CO_2 fixation. The first of these reported was *Mollugo verticillata* (KENNEDY and LAETSCH, 2), although more attention has been devoted to a second, *Panicum milioides* (BROWN and BROWN; KANAI and KASHIWAGI; KESTLER et al.; GOLDSTEIN et al.) for which the ratio of RuDP carboxylase to PEP carboxylase activities was also intermediate between C_3 and C_4 (KESTLER et al.). It is interesting that, in the C_3 plant *Antirrhinum majus*, this ratio increased from the youngest to the oldest tissue, indicating a tendency towards C_4 metabolism in the youngest tissue (HEDLEY and ROWLAND). Clearly, C_4 metabolism occurs not only as a major and highly elaborated system in C_4 plants, but also more subtly in many other tissues where it may be less easily detected. HEDLEY et al. have suggested that high levels of PEP carboxylase are induced when there is a net loss of CO_2, and that C_4 plants represent an evolutionary response to environments inducing high rates of mitochondrial respiration and photorespiration. Therefore, besides providing a means for storing CO_2 in the form of 4C-acids (CRÉACH et al.; LABER et al.), the increasing likelihood that a primary function of plant PEP carboxylase is to recapture respired CO_2, whether from photorespiration (BLACK) or dark respiration (HEDLEY et al.), could add a new touch of realism to current views of C_4 metabolism.

References

ABDELAL, A.T.H., SCHLEGEL, H.G.: Biochem. J. 139, 481-489 (1974). - ANDERSEN, W.R., GIBBS, M.: Biochem. Biophys. Res. Commun. 62, 953-956 (1975). - ANDERSON, L.E.: (1) Biochim. Biophys. Acta 321, 484-488 (1973); - (2) Biochem. Biophys. Res. Commun. 59, 907-913 (1974); - (3) Proc. 3rd Int. Congr. Photosyn., M. AVRON (ed.). Amsterdam: Elsevier 1975, 1393-1405. - ANDERSON, L.E., AVRON, M.: Plant Physiol. 57, 209-213 (1976). - ANDERSON, L.E., HEINRIKSON, R.L., NOYES, C.: (1) Arch. Biochem. Biophys. 169, 262-268 (1975). - ANDERSON, L.E., NG, T.-C.L., PARK, K.-E.Y.: (2) Plant Physiol. 53, 835-839 (1974). - ANDREWS, T.J., BADGER, M.R., LORIMER, G.H.: Arch. Biochem. Biophys. 171, 93-103 (1975). - ASAMI, S., AKAZAWA, T.: Plant Cell Physiol. 16, 631-642 (1975).

BADGER, M.R., ANDREWS, T.J.: Biochem. Biophys. Res. Commun. 60, 204-210 (1974). - BAHR, J.T., JENSEN, R.G.: (1) Plant Physiol. 53, 39-44 (1974); - (2) Arch. Biochem. Biophys. 164, 408-413 (1974). - BAIER, D., LATZKO, E.: Biochim. Biophys. Acta 396, 141-147 (1975). - BAMBERGER, E.S., AVRON, M.: Plant Physiol. 56, 481-485 (1975). - BAMBERGER, E.S., EHRLICH, B.A., GIBBS, M.: Plant Physiol. 55, 1023-1030 (1975). - BARBER, J., MILLS, J., NICOLSON, J.: (1) FEBS Lett. 49, 106-110 (1974). - BARBER, J., TELFER, A., NICOLSON, J.: (2) Biochim. Biophys. Acta 357, 161-165 (1974). - BATT, T., WOOLHOUSE, H.W.: J. Exp. Botany 26, 569-579 (1975). - BEER, S., SHOMER-ILAN, A., WAISEL, Y.: Physiol. Plant. 34, 293-295 (1975). - BEITLER, G.A., HENDRIX, J.E.: Plant Physiol. 53, 674-676 (1974). - BLACK, C.C.: Ann. Rev. Plant Physiol. 24, 253-286 (1973). -

BOWES, G., OGREN, W.L., HAGEMAN, R.H.: Plant Physiol. 56, 630-633 (1975). - BROWN, R.H., BROWN, W.V.: Crop Sci. 15, 681-685 (1975). - BUCHANAN, B.B.: J. Bacteriol. 119, 1066-1068 (1974). - BUCKE, C., OLIVER, I.R.: Planta 122, 45-52 (1975).

CAROLIN, R.C., JACOBS, S.W.L., VESK, M.: Bot. Jahrbuch 95, 226-255 (1975). - CARPE, A.I., SMITH, I.K.: Biochim. Biophys. Acta 370, 96-101 (1974). - CERFF, R.: Z. Pflanzenphysiol. 73, 109-118 (1974). - CERFF, R., QUAIL, P.H.: Plant Physiol. 54, 100-104 (1974). - CHAMPIGNY, M.-L., BISMUTH, E.: Physiol. Plant. 36, 95-100 (1976). - CHAMPIGNY, M.-L., MOYSE, A.: Biochem. Physiol. Pflanzen 168, 575-583 (1975). - CHANG, C.W.: Phytochemistry 14, 119-121 (1975). - CHEN, K., GRAY, J.C., WILDMAN, S.G.: (1) Science 190, 1304-1306 (1975). - CHEN, T.M., DITTRICH, P., CAMPBELL, W.H., BLACK, C.C.: (2) Arch. Biochem. Biophys. 163, 246-262 (1974). - CHENG, K.H., COLMAN, B.: Planta 115, 207-212 (1974). - CHOLLET, R.: Arch. Biochem. Biophys. 163, 521-529 (1974). - CHU, D.K., BASSHAM, J.A.: (1) Plant Physiol. 54, 556-559 (1974); - (2) Plant Physiol. 55, 720-726 (1975). - COBB, A.H., WELLBURN, A.R.: Planta 129, 127-131 (1976). - COCKBURN, B.J., WELLBURN, A.R.: J. Exp. Botany 25, 36-49 (1974). - COCKBURN, W.: J. Exp. Botany 25, 111-120 (1974). - COCKBURN, W., McAULAY, A.: Plant Physiol. 55, 87-89 (1975). - CODD, G. A., SMITH, B.M.: FEBS Lett. 48, 105-108 (1974). - CODD, G.A., STEWART, W.D.P.: Plant Sci. Lett. 3, 199-205 (1974). - CODD, G.A., TURNBULL, F.: Arch. Microbiol. 104, 155-158 (1975). - COLLINS, N., BROWN, R.H., MERRETT, M.J.: Biochem. J. 150, 373-377 (1975). - COLLINS, N., MERRETT, M.J.: Biochem. J. 148, 321-328 (1975). - COLMAN, B., CHENG, K.H., INGLE, R.K.: Plant Sci. Lett. 6, 123-127 (1976). - COOMBS, J., BALDRY, C.W.: Planta 124, 153-158 (1975). - COOMBS, J., MAW, S.L., BALDRY, C.W.: (1) Planta 117, 279-292 (1974); - (2) Plant Sci. Lett. 4, 97-102 (1975). - CORNIC, G.: Physiol. Vége. 12, 83-94 (1974). - CRÉACH, E., MICHEL, J.P., THIBAULT, P.: Planta 118, 91-100 (1974). - CRIDDLE, R.S., DAU, B., KLEINKOPF, G.E., HUFFAKER, R.C.: Biochem. Biophys. Res. Commun. 41, 621-627 (1970). - CROSTI, P.: Ital. J. Biochem. 23, 72-86 (1974).

D'AOUST, A.L., CANVIN, D.T.: Physiol. Vége. 12, 545-560 (1974). - DITTRICH, P.: Plant Physiol. 57, 310-314 (1976). - DÖHLER, G.: (1) Planta 117, 97-99 (1974); - (2) Z. Pflanzenphysiol. 73, 283-295 (1974); - (3) Planta 118, 259-269 (1974). - DOWNTON, W.J.S.: Photosynthetica 9, 96-105 (1975). - DOWNTON, W.J.S., TÖRÖKFALVY, E.: Z. Pflanzenphysiol. 75, 143-150 (1975). - DUFFUS, C.M., ROSIE, R.: Planta 114, 219-226 (1973). - DUGGAN, J.X., ANDERSON, L.E.: Planta 122, 293-297 (1975).

EICKENBUSCH, J.D., SCHEIBE, R., BECK, E.: Z. Pflanzenphysiol. 75, 375-380 (1975).

FARINEAU, J.: Physiol. Plant. 33, 300-309 (1975). - FINDENEGG, G.R.: Planta 116, 123-131 (1974). - FOCK, H., BATE, G.C., EGLE, K.: (1) Planta 121, 9-16 (1974); - (2) Ber. Deut. Botan. Ges. 87, 239-247 (1974). - FORDE, B.J., WHITEHEAD, H.C.M., ROWLEY, J.A.: Australian J. Plant Physiol. 2, 185-195 (1975).

GARNIER-DARDART, J., QUEIROZ, O.: Phytochemistry 13, 1695-1702 (1974). - GHOSH, H. P., PREISS, J.: J. Biol. Chem. 241, 4491-4504 (1966). - GIBBS, M., LATZKO, E., O'NEAL, D., HEW, C.-S.: Biochem. Biophys. Res. Commun. 40, 1356-1361 (1970). - GIROTTI, A.W.: Arch. Biochem. Biophys. 173, 210-218 (1976). - GIVAN, A.L.: Planta 120, 181-188 (1974). - GOATLY, M.B., COOMBS, J., SMITH, H.: Planta 125, 15-24 (1975). - GOLDSTEIN, L.D., RAY, T.B., KESTLER, D.P., MAYNE, B.C., BROWN, R.H., BLACK, C.C.: Plant Sci. Lett. 6, 85-90 (1976). - GRAY, J.C., KEKWICK, R.G.O.: Europ. J. Biochem. 44, 491-500 (1974). - GRAY, J.C., WILDMAN, S.G.: Plant Sci. Lett. 6, 91-96 (1976). - GRISSOM, F.E., KAHN, J.S.: Arch. Biochem. Biophys. 171, 444-458 (1975). - GRODZINSKI, B., BUTT, V.S.: Planta 128,

225-231 (1976). - GROSSMAN, A., McGOWAN, R.E.: Plant Physiol. 55, 658-662 (1975). - GUTIERREZ, M., GRACEN, V.E., EDWARDS, G.E.: Planta 119, 279-300 (1974).

HATCH, M.D.: Planta 125, 273-279 (1975). - HATCH, M.D., KAGAWA, T.: Australian J. Plant Physiol. 1, 357-369 (1974). - HATCH, M.D., KAGAWA, T., CRAIG, S.: (1) Australian J. Plant Physiol. 2, 111-128 (1975). - HATCH, M.D., MAU, S.-L., KAGAWA, T.: (2) Arch. Biochem. Biophys. 165, 188-200 (1974). - HAWKER, J.S., OZBUN, J.L., OZAKI, H., GREENBERG, E., PREISS, J.: Arch. Biochem. Biophys. 160, 530-551 (1974). - HEBER, U.: Physiol. Vége. 13, 875-877 (1975). - HEBER, U., ANDREWS, T.J., BOARDMAN, N.K.: Plant Physiol. 57, 277-283 (1976). - HEBER, U., KIRK, M.R.: Biochem. Physiol. Pflanz. 168, 211-223 (1975). - HEDLEY, C.L., HARVEY, D.M., KEELY, R.J.: Nature 258, 352-354 (1975). - HEDLEY, C.L., ROWLAND, A.O.: Plant Sci. Lett. 5, 119-126 (1975). - HELDT, H.W., RAPLEY, L.: FEBS Lett. 10, 143-148 (1970). - HIND, G., NAKATANI, H.Y., IZAWA, S.: Proc. Nat. Acad. Sci. US 71, 1484-1488 (1974). - HUBER, S.C., EDWARDS, G.E.: (1) Plant Physiol. 55, 835-844 (1975); - (2) Plant Physiol. 56, 324-331 (1975); - (3) Canadian J. Bot. 53, 1925-1933 (1975).

IHLENFELDT, M.J.A., GIBSON, J.: Arch. Microbiol. 102, 13-21 (1975). - INGLE, R.K., COLMAN, B.: Canadian J. Bot. 53, 2385-2387 (1975).

JACOBSON, B.S., FONG, F., HEATH, R.L.: Plant Physiol. 55, 468-474 (1975). - JONES, M.B.: Planta 123, 91-96 (1975).

KACHRU, R.B., ANDERSON, L.E.: Planta 118, 235-240 (1974). - KAGAN-ZUR, V., LIPS, S.H.: Europ. J. Biochem. 59, 17-23 (1975). - KAGAWA, T., HATCH, M.D.: (1) Biochem. Biophys. Res. Commun. 59, 1326-1332 (1974); - (2) Arch. Biochem. Biophys. 167, 687-696 (1975). - KAISER, W., URBACH, W.: Biochim. Biophys. Acta 423, 91-102 (1976). - KANAI, R., KASHIWAGI, M.: Plant Cell Physiol. 16, 669-679 (1975). - KAREKAR, M.D., JOSHI, G.V.: Bot. Marina 16, 216-220 (1973). - KELLY, G.J., GIBBS, M.: Plant Physiol. 52, 111-118 (1973). - KELLY, G.J., LATZKO, E.: Nature 256, 429-430 (1975). - KELLY, G.J., LATZKO, E., GIBBS, M.: (1) Ann. Rev. Plant Physiol. 27, in press (1976). - KELLY, G.J., ZIMMERMANN, G., LATZKO, E.: (2) Biochem. Biophys. Res. Commun., in press (1976). - KENNEDY, R.A.: Planta 128, 149-154 (1976). - KENNEDY, R.A., LAETSCH, W.M.: (1) Plant Physiol. 54, 608-611 (1974); - (2) Science 184, 1087-1089 (1974). - KERR, M.W., GEAR, C.F.: Biochem. Soc. Trans. 2, 338-340 (1974). - KERR, M.W., GROVES, D.: Phytochemistry 14, 359-362 (1975). - KESTLER, D.P., MAYNE, B.C., RAY, T.B., GOLDSTEIN, L.D., BROWN, R.H., BLACK, C.C.: Biochem. Biophys. Res. Commun. 66, 1439-1446 (1975). - KLEMME, B., JACOBI, G.: Planta 120, 147-153 (1974). - KLUGE, M.: Fortschr. Botan. 36, 90-98 (1974). - KLUGE, M., KRIEBITZSCH, C., VON WILLERT, D.J.: Z. Pflanzenphysiol. 72, 460-465 (1974). - KRAPF, G., JACOBI, G.: Plant Sci. Lett. 5, 67-71 (1975). - KRAUSE, G.H.: Biochim. Biophys. Acta 333, 301-313 (1974). - KRENZER, E.G., MOSS, D.N., CROOKSTON, R.K.: Plant Physiol. 56, 194-206 (1975). - KU, S.B., GUTIERREZ, M., EDWARDS, G.E.: (1) Planta 119, 267-278 (1974). - KU, S.B., GUTIERREZ, M., KANAI, R., EDWARDS, G.E.: (2) Z. Pflanzenphysiol. 72, 320-337 (1974). - KUNG, S.D., MARSHO, T.V.: Nature 259, 325-326 (1976).

LABER, L.J., LATZKO, E., GIBBS, M.: J. Biol. Chem. 249, 3436-3441 (1974). - LAING, W.A., OGREN, W.L., HAGEMAN, R.H.: (1) Plant Physiol. 54, 678-685 (1974); - (2) Biochemistry 14, 2269-2275 (1975). - LARSSON, C., ALBERTSSON, P.-Å.: Biochim. Biophys. Acta 357, 412-419 (1974). - LAVERGNE, D., BISMUTH, E., CHAMPIGNY, M.L.: Plant Sci. Lett. 3, 391-397 (1974). - LÁZARO, J.J., CHUECA, A., LÓPEZ GORGÉ, J., MAYOR, F.: (1) Phytochemistry 13, 2455-2461 (1974); - (2) Phytochemistry 14, 2579-2583 (1975). - LEE, R.B., WHITTINGHAM, C.P.: J. Exp. Botany 25, 277-287 (1974). - LENDZIAN, K., BASSHAM, J.A.: Biochim. Biophys. Acta 396, 260-275 (1975).

- LESTER, J.N., GOLDSWORTHY, A.: J. Exp. Botany 24, 1031-1034 (1973). - LEVI, C., GIBBS, M.: Plant Physiol. 56, 164-166 (1975). - LIBERA, W., ZIEGLER, I., ZIEGLER, H.: Z. Pflanzenphysiol. 74, 420-433 (1975). - LILLEY, R.McC., FITZGERALD, M.P., RIENITS, K.G., WALKER, D.A.: (1) New Phytologist 75, 1-10 (1975). - LILLEY, R.McC., HOLBOROW, K., WALKER, D.A.: (2) New Phytologist 73, 657-662 (1974). - LILLEY, R.McC., WALKER, D.A.: (1) Biochim. Biophys. Acta 358, 226-229 (1974); - (2) Biochim. Biophys. Acta 368, 269-278 (1974); - (3) Plant Physiol. 55, 1087-1092 (1975). - LINDEN, J.C., SCHILLING, N., BRACKENHOFER, H., KANDLER, O.: (1) Z. Pflanzenphysiol. 76, 176-181 (1975). - LINDEN, J.C., TANNER, W., KANDLER, O.: (2) Plant Physiol. 54, 752-757 (1974). - LONG, S.P., INCOLL, L.D., WOOLHOUSE, H.W.: Nature 257, 622-624 (1975). - LORD, J.M., ARMITAGE, T.L., MERRETT, M.J.: (1) Plant Physiol. 56, 600-604 (1975). - LORD, J.M., BROWN, R.H.: Plant Physiol. 55, 360-364 (1975). - LORD, J.M., CODD, G.A., STEWART, W.D.P.: (2) Plant Sci. Lett. 4, 377-383 (1975). - LORIMER, G.H., ANDREWS, T.J., TOLBERT, N.E.: (1) Biochemistry 12, 18-23 (1973). - LORIMER, G.H., BADGER, M.R., ANDREWS, T.J.: (2) Biochemistry 15, 529-536 (1976). - LÜTTGE, U., KLUGE, M., BALL, E.: Plant Physiol. 56, 613-616 (1975). - LYTTLETON, J.W.: Plant Sci. Lett. 4, 385-389 (1975).

MAHON, J.D., FOCK, H., CANVIN, D.T.: (1) Planta 120, 125-134 (1974); - (2) Planta 120, 245-254 (1974). - MAHON, J.D., FOCK, H., HÖHLER, T., CANVIN, D.T.: (3) Planta 120, 113-123 (1974). - MARKOVICH, D.S., KRAPIVINSKAYA, L.D.: FEBS Lett. 58, 138-140 (1975). - MAYNE, B.C., DEE, A.M., EDWARDS, G.E.: Z. Pflanzenphysiol. 74, 275-291 (1974). - McCARTHY, J.T., CHARLES, A.M.: Arch. Microbiol. 105, 51-59 (1975). - McDANIEL, C.F., KIRTLEY, M.E.: Biochem. Biophys. Res. Commun. 65, 1196-1200 (1975). - McFADDEN, B.A.: Biochem. Biophys. Res. Commun. 60, 312-317 (1974). - McFADDEN, B.A., LORD, J.M., ROWE, A., DILKS, S.: Europ. J. Biochem. 54, 195-206 (1975). - McGOWAN, R.E., GIBBS, M.: Plant Physiol. 54, 312-319 (1974). - MIFLIN, B.J., BEEVERS, H.: Plant Physiol. 53, 870-874 (1974). - MIGINIAC-MASLOW, M., CHAMPIGNY, M.-L.: Plant Physiol. 53, 856-862 (1974). - MIGINIAC-MASLOW, M., HOARAU, A.: Physiol. Plant. 35, 186-190 (1975). - MILFORD, G.F.J., PEARMAN, I.: Photosynthetica 9, 78-83 (1975). - MIZIORKO, H.M., NOWAK, T., MILDVAN, A.S.: Arch. Biochem. Biophys. 163, 378-389 (1974). - MONTES, G., BRADBEER, J.W.: Plant Sci. Lett. 6, 35-41 (1976). - MORGENTHALER, J.-J., MARSDEN, M.P.F., PRICE, C.A.: (1) Arch. Biochem. Biophys. 168, 289-301 (1975). - MORGENTHALER, J.-J., PRICE, C.A., ROBINSON, J.M., GIBBS, M.: (2) Plant Physiol. 54, 532-534 (1974). - MUKERJI, S.K.: Plant Sci. Lett. 2, 243-248 (1974).

NEALES, T.F., HEW, C.S.: Planta 123, 303-306 (1975). - NELSON, P.E., SURZYCKI, S.J.: Europ. J. Biochem. 61, 475-480 (1976). - NISHIKIDO, T., WADA, T.: Biochem. Biophys. Res. Commun. 61, 243-249 (1974). - NISHIMURA, M., AKAZAWA, T.: Plant Physiol. 55, 712-716 (1975). - NISHIMURA, M., GRAHAM, D., AKAZAWA, T.: Plant Physiol. 56, 718-722 (1975).

O'BRIEN, M.J., POWLS, R.: Europ. J. Biochem. 63, 155-161 (1976). - OLESEN, P.: Planta 123, 199-202 (1975). - OSMOND, C.B.: Australian J. Botany 22, 39-44 (1974). - OSMOND, C.B., ALLAWAY, W.G.: Australian J. Plant Physiol. 1, 503-511 (1974). - OSMOND, C.B., BJÖRKMAN, O.: Australian J. Plant Physiol. 2, 155-162 (1975).

PACOLD, I., ANDERSON, L.E.: Plant Physiol. 55, 168-171 (1975). - PAUL, J.S., VOLCANI, B.E.: (1) Arch. Microbiol. 101, 115-120 (1974); - (2) Planta 129, 59-61 (1976). - PAUL, J.S., SULLIVAN, C.W., VOLCANI, B.E.: Arch. Biochem. Biophys. 169, 152-159 (1975). - PAWLIZKI, K.-H., KELLY, G.J., LATZKO, E.: Anal. Biochem., in press. - PAWLIZKI, K., LATZKO, E.: FEBS Lett. 42, 285-288 (1974). - PEARCY, R.W., TROUGHTON, J.: Plant Physiol. 55, 1054-1056 (1975). - PEAVEY, D.G., GIBBS, M.: Plant Physiol. 55, 799-802 (1975). - PEISKER, M.: Photosynthetica 8, 47-50 (1974). -

PFLÜGER, R.: Ber. Deut. Botan. Ges. 87, 383-388 (1974). - POINCELOT, R.P.: (1) Plant Physiol. 54, 520-526 (1974); - (2) Plant Physiol. 55, 849-852 (1975). - POINCELOT, R.P., DAY, P.R.: Plant Physiol. 57, 334-338 (1976). - PUPILLO, P., PICCARI, G.G.: Europ. J. Biochem. 51, 475-482 (1975).

QUEIROZ, O., MOREL, C.: Plant Physiol. 53, 596-602 (1974).

RABINOWITZ, H., REISFELD, A., SAGHER, D., EDELMAN, M.: Plant Physiol. 56, 345-350 (1975). - RATHNAM, C.K.M., DAS, V.S.R.: Z. Pflanzenphysiol. 75, 360-364 (1975). - RATHNAM, C.K.M., EDWARDS, G.E.: Arch. Biochem. Biophys. 171, 214-225 (1975). - RATHNAM, C.K.M., RAGHAVENDRA, A.S., DAS, V.S.R.: Z. Pflanzenphysiol. 77, 283-291 (1976). - ROBINSON, J.M., GIBBS, M.: Plant Physiol. 53, 790-797 (1974). - ROBINSON, J.M., LATZKO, E., GIBBS, M.: Plant Physiol. 55, 12-14 (1975). - ROY, H., PATTERSON, R., JAGENDORF, A.T.: Arch. Biochem. Biophys. 172, 64-73 (1976). - RYAN, F.J., BARKER, R., TOLBERT, N.E.: (1) Biochem. Biophys. Res. Commun. 65, 39-46 (1975). - RYAN, F.J., JOLLY, S.O., TOLBERT, N.E.: (2) Biochem. Biophys. Res. Commun. 59, 1233-1241 (1974). - RYBOVÁ, R., SLAVÍKOVÁ, M.: Z. Pflanzenphysiol. 72, 287-296 (1974).

SALIN, M.L., HOMANN, P.H.: Canadian J. Bot. 51, 1857-1865 (1973). - SALLAL, A.-K.J., CODD, G.A.: FEBS Lett. 56, 230-234 (1975). - SARDA, C., PRIOUL, J.-L., MOYSE, A.: Physiol. Végé. 13, 563-577 (1975). - SAWADA, S., MIYACHI, S.: Plant Cell Physiol. 15, 225-238 (1974). - SCHEIBE, R., BECK, E.: Planta 125, 63-67 (1975). - SCHILLING, N., SCHEIBE, R., BECK, E., KANDLER, O.: FEBS Lett. 61, 192-193 (1976). - SCHNABL, H.: Z. Pflanzenphysiol. 77, 167-173 (1976). - SCHÜRMANN, P., BUCHANAN, B.B.: Biochim. Biophys. Acta 376, 189-192 (1975). - SCHWARZ, Z.: Biochem. Physiol. Pflanzen 166, 525-536 (1974). - SEGAL, H.L., BOYER, P.D.: J. Biol. Chem. 204, 265-281 (1953). - SELLAMI, A.: Biochim. Biophys. Acta 423, 524-539 (1976). - SHEU-HWA, C.-S., LEWIS, D.H., WALKER, D.A.: New Phytologist 74, 383-392 (1975). - SHOMER-ILAN, A., WAISEL, Y.: Z. Pflanzenphysiol. 77, 272-273 (1976). - SHOMER-ILAN, A., BEER, S., WAISEL, Y.: Plant Physiol. 56, 676-679 (1975). - SLABAS, A. R., WALKER, D.A.: (1) Biochem. J. 154, 185-192 (1976); - (2) Biochim. Biophys. Acta 430, 154-164 (1976). - SNYDER, F.W., TOLBERT, N.E.: Plant Physiol. 53, 514-515 (1974). - STABENAU, H.: (1) Planta 118, 35-42 (1974); - (2) Plant Physiol. 54, 921-924 (1974). - STANLEY, R.A., NAYLOR, A.W.: Physiol. Plant. 29, 60-63 (1973). - SUGIYAMA, T.: Plant Cell Physiol. 15, 723-726 (1974). - SUGIYAMA, T., LAETSCH, W.M.: Plant Physiol. 56, 605-607 (1975). - SUTTON, B.G.: Australian J. Plant Physiol. 2, 377-387 (1975). - SZAREK, S.R., TING, I.P.: Plant Physiol. 54, 76-81 (1974).

TABITA, F.R., McFADDEN, B.A.: (1) J. Biol. Chem. 249, 3453-3458 (1974); - (2) J. Biol. Chem. 249, 3459-3464 (1974). - TABITA, F.R., STEVENS, S.E., QUIJANO, R.: Biochem. Biophys. Res. Commun. 61, 45-52 (1974). - TAKABE, T., AKAZAWA, T.: (1) Arch. Biochem. Biophys. 169, 686-694 (1975); - (2) Biochemistry 14, 46-50 (1975). - TAKABE, T., NISHIMURA, M., AKAZAWA, T.: Biochem. Biophys. Res. Commun. 68, 537-544 (1976). - TELFER, A., BARBER, J., NICOLSON, J.: Biochim. Biophys. Acta 396, 301-309 (1975). - TOLBERT, N.E.: Ann. Rev. Plant Physiol. 22, 45-74 (1971). - TREICHEL, S.P., KIRST, G.O., VON WILLERT, D.J.: Z. Pflanzenphysiol. 71, 437-449 (1974). - TRIOLO, L., BAGNARA, D., ANSELMI, L., BASSANELLI, C.: Physiol. Plant. 31, 86-89 (1974). - TROUGHTON, J.H., CARD, K.A.: Planta 123, 185-190 (1975).

USUDA, H., KANAI, R., MIYACHI, S.: (1) Plant Cell Physiol. 16, 485-494 (1975). - USUDA, H., MATSUSHIMA, H., MIYACHI, S.: (2) Plant Cell Physiol. 15, 517-526 (1974). - USUDA, H., SAMEJIMA, M., MIYACHI, S.: (3) Plant Cell Physiol. 14, 423-426 (1973).

VALLEJOS, R.H., ANDREO, C.S.: FEBS Lett. 61, 95-99 (1976).

WAIDYANATHA, U.P.deS., KEYS, A.J., WHITTINGHAM, C.P.: J. Exp. Botany 26, 15-26 (1975). - WALK, R.-A., METZNER, H.: Hoppe-Seylers Z. Physiol. Chem. 356, 1733-1741 (1975). - WALKER, D.A., LILLEY, R.McC.: Plant Physiol. 54, 950-952 (1974). - WALKER, D.A., SLABAS, A.R.: Plant Physiol. 57, 203-208 (1976). - WAYGOOD, E.R., PAO, L.Y., GODAVARI, H.R.: Experientia 30, 986-988 (1974). - WERDAN, K., HELDT, H.W.: Biochim. Biophys. Acta 283, 430-441 (1972). - WERDAN, K., HELDT, H.W., MILOVANCEV, M.: Biochim. Biophys. Acta 396, 276-292 (1975). - WILDNER, G.F.: Z. Naturforsch. 30c, 756-760 (1975). - VON WILLERT, D.J.: (1) Planta 122, 273-280 (1975); - (2) Z. Pflanzenphysiol. 76, 44-50 (1975). - WILLMER, C.M., DITTRICH, P.: Planta 117, 123-132 (1974). - WINTER, K.: Planta 121, 147-153 (1974). - WYNN, T., BROWN, H., CAMPBELL, W.H., BLACK, C.C.: Plant Physiol. 52, 288-291 (1973).

ZABKA, G.G., CHATURVEDI, S.N.: Plant Physiol. 55, 532-535 (1975). - ZELITCH, I.: Ann. Rev. Biochem. 44, 123-145 (1975). - ZIEGLER, H., OSMOND, C.B., STICHLER, W., TRIMBORN, P.: Planta 128, 85-92 (1976). - ZIEGLER, I.: Biochim. Biophys. Acta 364, 28-37 (1974). - ZIEGLER, I., LIBERA, W.: Z. Naturforsch. 30c, 634-637 (1975). - ZIMMERMANN, G., KELLY, G.J., LATZKO, E.: Europ. J. Biochem., in press.

Professor Dr. E. LATZKO
Abteilung Chemische Pflanzenphysiologie, Technische Universität München
D 8050 Weihenstephan

Dr. G.J. KELLY
Abteilung Chemische Pflanzenphysiologie, Technische Universität München
D 8050 Weihenstephan

Photosynthesis
Control of Carbon Metabolism Through Enzyme Regulation and Membrane-Mediated Metabolite Transport*

By ERWIN LATZKO and GRAHAME J. KELLY

1. Introduction

Research on photosynthetic carbon metabolism is passing through a period of change. The emphasis given in previous years to the comparative physiology and biochemistry of C_3 and C_4 plants has declined, while the mechanisms and regulation of intracellular carbon flow have emerged as one of the subjects most intensively investigated at present. As it so happens, the latter is receiving much benefit from the former; methods for the separation of leaf protoplasts and chloroplasts, originally developed for the study of C_4 plants, are now being adopted and refined so that acquisition of cellular organelles from a variety of plants is at last permitting a comparative investigation of the metabolism and transport of carbon compounds during photosynthesis (EDWARDS and HUBER). This review covers the literature of the last two years (since our last review in Volume 38 of Progress in Botany) with this trend in mind.

2. The Reductive Pentose Phosphate Cycle (Calvin Cycle)

The path followed by carbon in the Calvin cycle of photosynthetic CO_2 fixation is generally accepted as that outlined in the 1950s (see BASSHAM). In recent years more attention has been given to the form in which the net production of reduced carbon is removed from the cycle. To the possibilities of starch synthesis in the chloroplast, and glycolate and triose-P export to the cytoplasm, P-hydroxypyruvate and/or P-serine (derived from glycerate-3-P) might be additionally implicated, based on the observations of DALEY and BIDWELL, and CHAPMAN and LEECH. There is increasing evidence that leaf serine is formed from glycerate-3-P by a more direct route than that involving the Calvin cycle plus glycolate pathway (PLATT et al., 1; CHAPMAN and LEECH).

It is worth repeating the basic precept that, in higher plants and algae, the Calvin cycle is the only known photosynthetic carbon fixation pathway that permits plant growth (see LATZKO and KELLY). However, evidence continues to accumulate against operation of the cycle in certain photosynthetic bacteria (BUCHANAN and SIREVÅG; SIREVÅG et al.; TAKABE and AKAZAWA) in which the alternative of a reductive carboxylic acid cycle has been proposed (BUCHANAN et al., 1).

* Abbreviations: CAM, crassulacean acid metabolism; FBP, fructose-1,6-P_2; GAP, glyceraldehyde-3-P; PEP, P-enolpyruvate; RuBP, ribulose-1,5-P_2; SBP, sedoheptulose-1,7-P_2.

G.J. Kelly and E. Latzko:
Thirty Years of Photosynthesis

Originally published in Progress in Botany 40

a) Enzymes of the Cycle

When PLATT et al. (1) increased the amount of CO_2 available to alfalfa leaflets performing steady-state photosynthesis there followed a doubling in the rate of CO_2 fixation but only slight increases in the pool sizes of Calvin cycle intermediates. This strong evidence that the cycle is particularly well regulated warrents increased attention toward the participating enzymes, but apart from RuBP carboxylase, studies on these enzymes over the past two years have been rather meager in number.

It is now well established that RuBP carboxylase molecules, from higher plants and green algae, are composed of eight large and eight small subunits [the large subunits contain three almost identical polypeptides, while the small subunits contain one or more types of polypeptide chain; when more than one, clear differences are seen (CHEN et al.; GRAY et al.)]. The structure of the enzyme from autotrophic bacteria is less clear-cut: RuBP carboxylase from *Alcaligenes eutrophus* (BOWIEN et al.), *Hydrogenomonas eutropha* (PUROHIT and McFADDEN), and two species of *Rhodopseudomonas* (GIBSON and TABITA, 1, 2) were composed of eight large and eight small subunits, but in *H. eutropha* two slightly different large subunits were detected, while in the *Rhodopseudomonas spp.* an additional enzyme completely lacking small subunits was found. The latter observation is reminiscent of the report that four blue-green algal RuBP carboxylases lacked small subunits (TABITA et al.), a conclusion challenged on the grounds that the small subunit is found when an acid precipitation step is omitted from the enzyme purification procedure (CODD and STEWART, 3; STEWART et al.).

A new aspect related to structure is the state of RuBP carboxylase in vivo. Not all of it may be in solution. Crystals in the marine green alga *Dasycladus clavaeformis* (PARADIES et al.) and polyhedral bodies in the blue-green alga *Anabaena cylindrica* (CODD and STEWART, 2) are claimed to contain RuBP carboxylase activity, while in higher plants enzyme activity was postulated for crystalline inclusions found in tobacco chloroplast stroma (BAKER et al.) and even inside thylakoids of chloroplasts from "stressed" spinach leaves (SPREY and LAMBERT). These observations might reflect the fact that 30%-50% of leaf protein is RuBP carboxylase, and a third role for this enzyme, i.e., as a nitrogen reserve available for degradation and redistribution from older to younger parts of the plant, has received experimental support from PEOPLES and DALLING.

Previous confusion concerning the kinetic and regulatory properties of RuBP carboxylase has been replaced by the picture of an enzyme which has adequate affinity for substrates and ample catalytic activity to support photosynthesis when activated by CO_2 and Mg^{2+} (LORIMER et al., 1, 2; LAING and CHRISTELLER). The enzyme, which slowly reverts to the inactive form in the absence of CO_2 and Mg^{2+}, is still able to bind RuBP, which thus stabilizes this inactive form in that it must be released again before activation by CO_2 and Mg^{2+} can begin (LAING and CHRISTELLER). BAHR and JENSEN resently returned to the intact chloroplast and demonstrated a similar activation in vivo by first supplying the chloroplasts with CO_2 directly and Mg^{2+} indirectly (by illumination, which induces Mg^{2+} to move from thylakoids to the stroma), and then measuring activity immediately after chloroplast breakage. Interestingly, these workers detected enzyme inactivation in the absence of CO_2 with chloroplasts held at 25°, but not at 0°. All these observations might eventually be related to the considerable enzyme conformational changes observed fluorimetrically upon addition of RuBP or Mg^{2+}, or temperature change (WILDNER; WILDNER and HENKEL, 2; VATER et al.; CHOLLET and ANDERSON, 2), and, on the practical side, to the greater stability of RuBP carboxylase from leaves of cold-hardened plants (e. g., *Secale cereale*) (HUNER and MacDOWALL).

Catalysis occurs on the large subunit of RuBP carboxylase, but a role for the small subunit is only gradually becoming apparent. LORIMER et al. (1) originally pointed out that the activating CO_2, which did not seem to act as substrate, possibly bound to an uncharged amino group on the enzyme, and PAECH et al. have recently shown that eight ε-amino groups (of lysyl residues) on the enzyme are required to bind the activating CO_2, or the inactivating RuBP in the absence of CO_2. It now seems likely that these eight amino groups are on the eight small subunits, since AKAZAWA has found that antisera directed against the small subunit markedly reduces the enzyme activity measured after preincubation with CO_2 plus Mg^{2+}. Each small subunit of the tobacco enzyme contains about seven lysyl residues (GRAY et al.).

RuBP carboxylase also catalyzes an oxygenase reaction, producing the photorespiratory substrate P-glycolate (see Sect. 5a). The enzyme must be similarly activated with CO_2 and Mg^{2+} before this oxygenase activity can be realistically evaluated (BADGER and LORIMER). It is undoubtedly in the minds of researchers to specifically inhibit the oxygenase activity (since photorespiration is an apparently wasteful process), and indeed WILDNER and HENKEL (1) claim to have achieved this goal. It is surprising that their results have not been extended. Certainly the carboxylase and oxygenase activities respond very similarly to temperature and effector molecules (CHOLLET and ANDERSON, 1).

The biosynthesis of RuBP carboxylase continues to be an exciting field. Synthesis of the small subunit involves formation of a precursor polypeptide about 20% larger than the final subunit polypeptide. The extra length is required for transport across the chloroplast envelope (DOBBERSTEIN et al.) and is in fact hydrolyzed off by a protease in the envelope (HIGHFIELD and ELLIS). The large subunit is synthesized in the chloroplast from messenger RNA supplied by chloroplast genes; restriction fragments containing the gene have been cloned in *Escherichia coli* and used for in vitro synthesis of the large subunit (GELVIN et al.; COEN et al.).

The nature of chloroplast GAP dehydrogenase is still much of a mystery. Six recent reports all speak of a range of molecular weights from about 140,000 to 1.5 million (WOLOSIUK and BUCHANAN, 1; SCHWARZ et al.; O'BRIEN et al., 1, 2; CERFF; FERRI et al.). It is nevertheless clear that the enzyme subunits have a molecular weight of about 35,000, and rather certain that at least two different subunits are present in the spinach (FERRI et al.) and *Sinapis alba* (CERFF) enzymes. Possibly, the molecular weight problem will be clarified by CERFF, who concluded that the enzyme becomes "sticky" in the presence of NAD^+ or NADH. Comparative studies with algal enzymes, which in some respects are distinctly different from the higher plant enzyme (O'BRIEN et al., 1), should also be of benefit.

Little attention has been given to other enzymes of the Calvin cycle. Both glycerate-3-P kinase (CAVELL and SCOPES) and ribose-5-P isomerase (KAWASHIMA and TANABE) have been purified from leaves to homogeneity. Chloroplast FBPase has been further studied, and the reductive activation of the enzyme shown to occur relatively slowly (SCHÜRMANN and WOLOSIUK). Interestingly, this enzyme acquires SBPase activity upon dissociation into two halves (BUCHANAN et al., 2), although the level of SBPase activity was well below that required to support observed rates of photosynthesis.

b) Light-Mediated Effects on Enzyme Activity

Possibly the most significant event during the last two years of research on light-mediated changes in the activities of several chloroplast enzymes was the discovery by WOLOSIUK and BUCHANAN (2) that the protein factor required for activation of FBPase and SBPase by reduced ferredoxin consists of two components (SCHÜRMANN et al.) identified as thioredoxin and the enzyme ferredoxin-thioredoxin reductase. Chloroplast NADP-linked malate dehydrogenase was also activated in this system (WOLOSIUK et al.). Unfortunately, the activities of the activated enzymes in these studies were rather low, tempting one to search for alternative activating systems. Although similar protein factors were found necessary for light-mediated activation of the NADP-linked malate dehydrogenase from bean and spinach leaves (JACQUOT et al., 1, 2) and maize mesophyll (HATCH), KAGAWA and HATCH have brought attention to the fact that these contrast with the protein factor required for light-mediated regulation of pyruvate Pi dikinase (SUGIYAMA).

The concept of dithiol-containing "light effect mediators" associated with the photosynthetic electron transport system has been expanded in a report by ANDERSON and DUGGAN describing how pea-leaf chloroplast glucose-6-P dehydrogenase is inactivated by these mediators following illumination. A subsequent report claiming that this system can also decrease the activity of the cytoplasmic glucose-6-P dehydrogenase (ANDERSON and NEHRLICH) would infer some mechanism currently difficult to comprehend. Nevertheless, the involvement of membrane-located dithiols in the light-mediated control of enzyme activity remains attractive since VALLEJOS and ANDREO have more evidence for the exposure of vicinal dithiols by light-induced conformational changes of membranes, including the coupling factor, which is known to undergo a light-induced conformational change (RYRIE and JAGENDORF) and is conveniently located on stroma lamellae and on the outer, non-appressed membranes of grana stacks (see JAGENDORF).

Enzymes activated in vivo by light-mediated processes are deactivated in darkness. The mechanism by which this deactivation occurs is not clear. WOLOSIUK and BUCHANAN (2) suggested that the level of oxidized glutathione increases in darkness and is involved in the oxidative deactivation of enzymes such as FBPase, but HALLIWELL and FOYER pointed out that, even in darkness, over 90% of the chloroplast glutathione is in the reduced form. In addition, these workers found that oxidized glutathione at the maximum possible physiologic concentration had no effect on FBPase activity.

3. The Chloroplast

The chloroplast is a persnickety organelle. After being isolated from the surrounding cytoplasm, chloroplasts present a formidable challenge to anyone attempting to select the right medium in which they will carry out sustained photosynthetic CO_2 fixation at rates comparable to their in vivo capacity. Moreover, a medium suitable for chloroplasts from one plant may be unsuitable or even ditrimental to chloroplasts from another plant, and it is becoming clear that this is partly the reason why spinach has enjoyed a virtual monopoly in supplying active chloroplasts over the past decade. However, this decade has not been wasted; the accumulated understanding of the spinach chloroplast, particularly in relation to metabolite transport across the envelope (see HELDT), and the struggle of the chloroplast to keep within its envelope at least five-sixths of the carbon per Calvin cycle revolution (see WALKER and HEROLD), is the foundation upon which investigations with functional chloroplasts from other plants are being based.

a) The Chloroplast Envelope

Central to this controlled metabolite transport are the specific translocators on the inner membrane of the chloroplast envelope (HELDT). Spinach chloroplasts have active "phosphate translocators" (which catalyze counterexchange between any two of dihydroxyacetone-P, glycerate-3-P, and Pi) and "dicarboxylate translocators" (for counterexchange of certain dicarboxylic acids, including some amino acids), but only poorly active "adenylate translocators" (for counterexchange of ATP and ADP) (HELDT; LEHNER and HELDT). An elegant microscale ion-exchange chromatographic method developed by LILLEY et al. has demonstrated that photosynthetically fixed carbon appears to be more easily exported through the phosphate translocator as triose-P than as glycerate-3-P, possibly because of the extra negative charge carried by the latter. First attempts to characterize this translocator have been reported by FLÜGGE and HELDT. Operation of the translocator can have wide-reaching effects, e.g., when the external triose-P concentration is several-fold greater than the level of external Pi, the latter will be poorly taken up and a decreased internal ATP/ADP ratio will ensue (KAISER and URBACH).

A clear-cut demonstration that all chloroplasts are not necessarily like spinach chloroplasts was outlined during 1977 by ROBINSON and WISKICH (1-4). These authors, on finding that pea chloroplasts photosynthesized better when supplied with small quantities of ATP (ROBINSON and WISKICH, 1), subsequently observed pyrophosphate inhibition of CO_2 fixation by pea chloroplasts and attributed this to depletion of the chloroplast's ATP by counterexchange with external pyrophosphate in a process mediated by the adenylate translocator, which, in contrast to spinach chloroplasts, is considerably more active in pea chloroplasts (ROBINSON and WISKICH, 2, 3). STANKOVIC and WALKER substantially agreed with these conclusions. It is ironic that pyrophosphate, which was chosen at a moment of serendipity during efforts to make spinach chloroplasts fix CO_2, turned out to be surreptitiously inhibiting photosynthesis by pea chloroplasts, and possibly other chloroplasts as well; HUBER and EDWARDS (1) found an adenylate translocator, at least six times (and possibly 20 times) more active than that in spinach chloroplasts, in the mesophyll chloroplasts of the C_4 plant *Digitaria sanguinalis*.

The mechanism of the adenylate translocator is by no means clear, although MURAKAMI and STROTMANN have suggested involvment of an envelope-located Mg^{2+}-dependent ATPase, which they have partly characterized. Such an ATPase has already been implicated in the transfer of CO_2/HCO_3^- into chloroplasts (POINCELOT and DAY; CHAMPIGNY and BISMUTH).

The special problems faced by the mesophyll chloroplasts of C_4 plants have been approached by HUBER and EDWARDS (2, 3), who found that these chloroplasts from *Digitaria sanguinalis* were capable of taking up both pyruvate (in a carrier-mediated process) and Pi, the latter in counterexchange for P-enolpyruvate (see EDWARDS and HUBER). These observations are quite significant since they not only confirm that the pyruvate Pi dikinase in these chloroplasts can obtain pyruvate and Pi from the cytoplasm and return PEP to the cytoplasm, but also provide another instance of variety in chloroplast character: spinach chloroplasts do not show high rates of pyruvate uptake (HUBER and EDWARDS, 2) and have a Pi translocator that only poorly recognizes P-enolpyruvate (HELDT).

However, chloroplast isolation techniques are no less important than chloroplast variability; most chloroplasts are battered when mechanical isolation from tough and fibrous leaves, such as grass leaves, is attempted. To circumvent this problem, gentle isolation of chloroplasts from entire protoplasts, previously isolated by enzymic digestion of the cell wall and intervening material, has been attempted (NISHIMURA et al.; RATHNAM and EDWARDS, 1; EDWARDS and HUBER). Active chloroplasts from the grasses *Triticum aestivum*, *Hordeum vulgare*, *Agropyron repens*, and *Panicum bisulcatum* were obtained by RATHNAM and EDWARDS (1) using this procedure. A fuller description of the properties of chloroplasts from the cereal grasses will be eagerly awaited.

These impressive advances in the study of chloroplasts by no means indicate that all problems have been solved. Improved procedures for the isolation of chloroplasts from spinach (NAKATANI and BARBER) and CAM plants (NISHIDA and SANADA) have been reported, and the sensitivity of the chloroplast envelope was further emphasized by DELANEY and WALKER, who found that isolated sunflower chloroplasts, despite being intact and in possession of the necessary enzyme activities, were still incapable of appreciable CO_2 fixation. Of course, one way to get around the problem of the chloroplast envelope is to remove it, and this is part of the rationale for the development of reconstituted chloroplast systems from sunflower (DELANEY and WALKER) and spinach (LENDZIAN and BASSHAM; SLABAS and WALKER; WALKER et al.).

b) Stroma and Thylakoids

Ion transport across the thylakoid membrane is as important as metabolite transport across the chloroplast envelope for the regulation of photosynthetic CO_2 fixation. The principal event in ion transport is the light-induced counterexchange between protons, which move from the stroma into thylakoids (thereby increasing the stromal pH), and Mg^{2+}, which moves out of thylakoids into the stroma. These ion movements are reversed as soon as illumination ceases. [An analogous light-induced alkalization of the cytoplasm in blue-green algae has recently been reported (FALKNER et al.; MASAMOTO and NISHIMURA).] Compared to the total amount of Mg^{2+} in the chloroplast, the light-induced concentration increase is relatively small, estimated with an ionophore to be 1-3 mM (PORTIS and HELDT) and 7 mM (MIGINIAC-MASLOW and HOARAU) and spectroscopically to be 2-5 mM (KRAUSE). Much of the total chloroplast Mg^{2+} is thought to be bound, and it is interesting that BEN-HAYYIM, using isolated thylakoids, recently concluded that all of the Mg^{2+} released upon illumination remained attached to the thylakoid membrane. Chloroplasts also contain considerable amounts of Ca^{2+}, but MIGINIAC-MASLOW and HOARAU did not detect any free Ca^{2+} in the stroma, which is perhaps all to the good since this cation is a potent inhibitor of chloroplast FBPase (LATZKO, unpublished).

Both the increased stromal pH and the increased stromal Mg^{2+} concentration are important for the maintenance of chloroplast CO_2 fixation. Artificial reduction of the stromal Mg^{2+} concentration (PORTIS and HELDT; PORTIS et al.) or of the stromal pH (PURCZELD et al.) inhibited photosynthesis, and analysis of the levels of Calvin cycle intermediates indicated in both instances that the reactions catalized by FBPase and SBPase were hindered. In the case of FBPase, this agrees remarkably well with the observed relationships between the activity of this enzyme and the Mg^{2+} concentration and pH (BAIER and LATZKO). Similar evidence implicating FBPase and SBPase, and also ribulose-5-P kinase, in the light-dark regulation of CO_2 fixation by the blue-green alga *Aphanocapsa* has been reported by PELROY et al.

The number of known ways in which thylakoids influence chloroplast carbon metabolism, in addition to being the source of ATP and NADPH, is clearly increasing. Besides mediating the proton-Mg^{2+} counterexchange, and the indirect effects of light on enzyme activity (see Sect.

2b), the effects of other electron transport products such as H_2O_2 are becoming more apparent. KAISER found that CO_2 fixation by isolated chloroplasts was reduced 50% by 10 µM H_2O_2, presumably due to inhibition of transketolase. Since chloroplasts always contain O_2 (STEIGER et al.), and so can form H_2O_2, a method for the removal of this substance must exist. Catalase cannot be considered since it is absent from chloroplasts (ALLEN). One likely contender is a membrane-bound ascorbate peroxidase, which has a suitably high affinity for H_2O_2 (GRODEN and BECK).

4. Products of Photosynthesis

Three major end products of CO_2 fixation may be distinguished in photosynthetic cells: starch, sucrose, and amino acids. Starch is synthesized in the chloroplast, sucrose is formed in the cytoplasm, while the synthesis of animo acids involves most cell compartments.

a) Starch

The mechanism of starch biosynthesis, and its control through allosteric regulation of ADP-glucose pyrophosphorylase, have been understood for over a decade; PREISS and LEVI recently emphasized that it is the ratio of the concentrations of the activator glycerate-3-P and the inhibitor Pi that is the critical factor in the regulation of ADP-glucose pyrophosphorylase. The regulatory properties of this enzyme from the blue-green alga *Synechococcus* were found to be similar to those of the enzyme from higher plants and green algae (LEVI and PREIS), although appropriately different regulatory properties exist for the enzyme from certain groups of photosynthetic bacteria (PREISS and LEVI).

Once deposited in chloroplasts, starch remains rigidly fixed (see POPEYE in WALKER) until enzymes metabolize it to soluble sugar compounds that can leave the chloroplast. Controversy surrounds the nature of this process. There is evidence that both phosphorylase and amylase are involved in the initial degradation steps since all of maltose, glucose, and sugar phosphates have been found as products of chloroplast starch degradation (LEVI and GIBBS; PEAVEY et al.; HELDT et al.), although maltose and glucose were often in lesser amounts (HELDT et al.), especially when Pi levels were higher (PEAVEY et al.). Phosphorylase activity has been confirmed in leaves (RICHARDSON and MATHESON) and chloroplasts (PREISS and LEVI), but the origin of maltose and glucose is unclear since PREISS and LEVI did not obtain convincing evidence that pea chloroplasts contain β-amylase or maltase, although there was evidently much β-amylase outside the chloroplast.

Whether maltose, glucose, or glucose-1-P is formed from starch, it is somewhat better established that further metabolism to triose-P and/or glycerate-3-P in the chloroplast is necessary before export to the cytoplasm (LEVI and GIBBS; PEAVEY et al.; HELDT et al.). The glycolytic pathway could accommodate this metabolism since all enzymes are present in chloroplasts. One of these, phosphofructokinase, is present with an activity just sufficient to support observed rates of starch degradation (KELLY and LATZKO, 1), and this property, coupled to the wealth of allosteric regulatory properties (KELLY and LATZKO, 2), signifies a potential role in the regulation of chloroplast starch degradation. Interestingly, the enzyme appeared to be absent from the chloroplasts of *Allium schoenoprasum*, which do not synthesize starch (KELLY and LATZKO, 1).

If glucose is formed from chloroplast starch, its fate is also in question. It might be directed into glycolysis since chloroplasts contain hexokinase (PEAVEY et al.; PREISS and LEVI), but the not inconsiderable ability of free glucose to diffuse in and out of the chloroplast, as evidenced by recent experiments with spinach (SCHÄFER et al.) and *Hydrodictyon africanum* (RAVEN), points to this as an alternative.

The influence of Pi on starch metabolism is particularly relevant. With this anion at low concentrations, or effectively absent, starch synthesis in the light is promoted, due to the inability of chloroplasts to export triose-P in the absence of Pi (see Sect. 3a), and/or to the absence of any Pi inhibition of ADP-glucose pyrophosphorylase (HEROLD et al.; STEUP et al.; HELDT et al.). In contrast, starch degradation in darkness requires ample Pi that is required by phosphorylase (STEUP et al.) and will act as the counterexchanged anion for exported triose-P; in addition, it may inhibit ADP-glucose pyrophosphorylase, and relieve any inhibition of phosphofructokinase by strong inhibitors (KELLY and LATZKO, 2).

b) Sucrose

Not much attention has been given to the biochemistry of sucrose synthesis in photosynthetic cells during the past two years. Certain aspects have been reviewed by WALKER and HEROLD, who point out that the photosynthesizing chloroplast is dependent on cytoplasmic sucrose formation in that the Pi released during this process must be returned to the chloroplast in order that export of triose-P into the cytoplasm can continue. Of the enzymes involved in cytoplasmic sucrose biosynthesis, sucrose-P synthetase and FBPase are two which are likely to have a regulatory role in vivo since their activities are not much greater than the rate of sucrose synthesis. The cytoplasmic FBPase has been purified from spinach leaves and shown to be quite different from the chloroplast enzyme; activity was allosterically inhibited by AMP (ZIMMERMANN et al.).

c) Amino Acids

Confirming previous observations made with *Chlorella*, PLATT et al. (2) found that alfalfa (*Medicago sativa*) leaf discs supplied with NH_4^+ during photosynthesis immediately produced amino acids at the expense of sucrose formation. A similar result was obtained with maize by BLACKWOOD and MIFLIN, who showed that both starch and sucrose formation were reduced in the presence of NH_4^+ or NO_3^-. These indications that photosynthetic tissues quickly incorporate inorganic nitrogen into amino acids may be related to the possible toxic effects of higher concentrations of NH_4^+ [which can uncouple photosynthetic phosphorylation (see JAGENDORF)] and NO_3^- [which can reduce the chloroplast stromal pH (PURCZELD et al.)].

Amino acid biosynthesis from NH_4^+ by chloroplasts was once a puzzle because the only known mechanism was that utilizing glutamate dehydrogenase, which, however, had a notoriously poor affinity for NH_4^+. The problem was solved when, several years ago, the enzymes glutamine synthetase (HAYSTEAD; O'NEAL and JOY) and a ferredoxin-linked glutamine-α-oxoglutarate aminotransferase (popularly known as GOGAT) (LEA and MIFLIN) were demonstrated to be present in chloroplasts. Together, these seemed capable of effectively assimilating NH_4^+ into glutamate (Fig. 1) since GOGAT has a high affinity for NH_4^+.

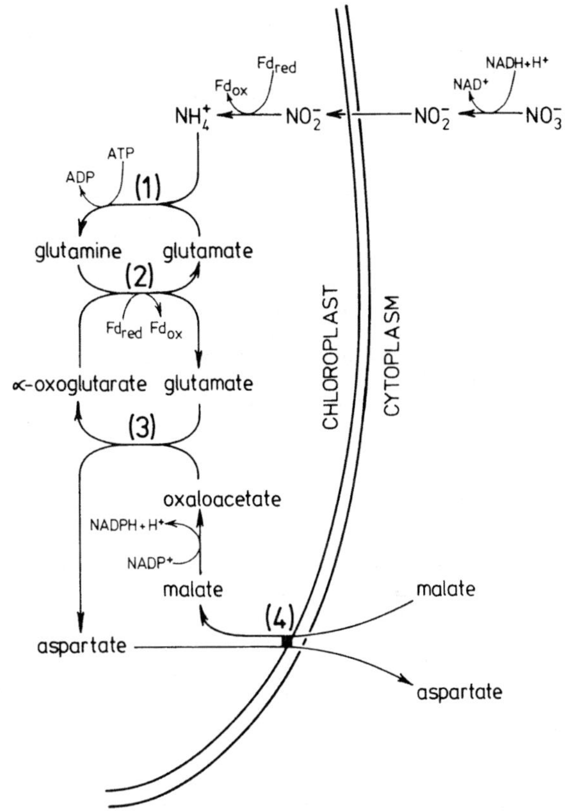

Fig. 1. Amino acids as end products of CO_2 fixation. Nitrogen in NH_4^+ is incorporated into glutamate by glutamine synthase (1) and glutamate synthase (GOGAT) (2) before transamination (3). The product amino acid is transported by the dicarboxylate translocator (4) in exchange for a carboxylic acid. Adapted from LEA and MIFLIN. $\underline{Fd_{ox}}$ and $\underline{Fd_{red}}$, oxidized and reduced ferredoxin, respectively

Light-dependent synthesis of glutamine [reaction (1), Fig. 1] by pea chloroplasts was subsequently demonstrated (MITCHELL and STOCKING; GIVAN), and GOGAT activity [reaction (2), Fig. 1] linked to photosynthetic O_2 evolution via ferredoxin (ANDERSON and DONE, 1) was more recently observed. Finally, these two reactions were combined to give light-dependent O_2 evolution with NH_4^+ plus α-oxoglutarate as electron acceptor (ANDERSON and DONE, 2). Chloroplasts are able to synthesize α-oxoglutarate from isocitrate, although the source of the latter acid is unclear (ELIAS and GIVAN). Of course, during NH_4^+ assimilation, α-oxoglutarate is recycled after transamination [reaction (3), Fig. 1] between glutamate and an acceptor acid that moves into the chloroplast in exchange for a product amino acid (Fig. 1); based on reports showing that aspartate and malate are rapidly transported by the dicarboxylate translocator (LEHNER and HELDT) and that aspartate is a major product of photosynthesis by alfalfa leaves supplied with NH_4^+ (PLATT et al., 2), it is tempting to suggest that malate is contransported with aspartate in this final phase of NH_4^+ assimilation (Fig. 1).

Leaf GOGAT has been purified and partially characterized (WALLSGROVE et al.). The level of GOGAT activity, in contrast to that of glutamate dehydrogenase, closely followed photosynthetic activity in wheat leaves (NICKLISCH et al.); however, the latter enzyme is not yet completely excluded from consideration, especially since a form with higher affinity for NH_4^+ was reported in the marine green alga *Caulerpa simpliciuscula* (GAYLER and MORGAN).

Less is known about a second possible mechanism of nitrogen assimilation: amination of P-hydroxypyruvate (formed by oxidation of glycerate-3-P supplied by the Calvin cycle) to form P-serine (CHAPMAN and LEECH; DALEY and BIDWELL; PLATT et al., 1). Although these two reactions are generally viewed as being in the cytoplasm (e.g., see HALLIWELL), they might also occur in the chloroplast since CHAPMAN and LEECH found that P-serine was the major amino acid formed by photosynthesizing *Zea mays* chloroplasts.

5. Photorespiration

a) The Pathway

Oxidation of RuBP to 3PGA and P-glycolate, and subsequent metabolism of the latter compound through the glycolate pathway (Fig. 2), is now more popular than ever as an explanation for the overall process of photorespiration, i.e., the light-dependent release of CO_2 by photosynthetic cells. This process may involve, but is not necessarily identical with, the inhibition of photosynthesis by O_2; with isolated chloroplasts, this O_2 inhibition was greater at higher pH (ROBINSON et al.; KOW et al.), whereas with isolated leaf cells photorespiration decreased with increasing pH (SERVAITES and OGREN, 2).

LORIMER et al. (3) and KRAUSE et al. (2) argue, from experiments with isolated chloroplasts, that essentially the sole source of glycolate is RuBP oxidation and that the alternative of transketolase-bound glycolaldehyde oxidation plays no substantial role. This alternative has also been dropped from consideration as a source of glucolate in *Chlorella* and *Chromatium* (LORIMER et al., 4), although it was seriously contemplated shortly before (ASAMI and AKAZAWA). However, from experiments with barley, PASSERA and ALBUZIO suggest that a source of glycolate in addition to RuBP oxidation might exist in older leaves.

Glycolate is oxidized to glyoxylate by glycolate oxidase in leaf peroxisomes; in green algae a mitochondrial glycolate dehydrogenase, located in the outer mitochondrial compartment (BEEZLEY et al.), is usually involved. However, peroxisome-located glycolate oxidase, as in higher plants, has now been found in the filamentous green algae *Spirogyra* (STABENAU) and *Klebsormidium* (GRUBER and FREDERICK).

Further metabolism of glyoxylate can involve non-enzymic oxidation by H_2O_2 at relatively high temperatures (e.g., 35°) in leaf peroxisomes to form formate and CO_2 (GRODZINSKI and BUTT), or conversion via tartronic semialdehyde to glycerate in algae such as *Anabaena cylindrica* (CODD and STEWART, 1) and *Cylindrotheca fusiformis* (PAUL and VOLCANI). However, with most plants under normal conditions, the glycolate pathway (Fig. 2) is considered to further metabolize most of the glyoxylate. In recent years much attention has been devoted to glycine, serine, and their interconversion. Appropriate rates of glycine decarboxylation by mitochondria from the leaves of several plants, and localization of the responsible enzyme near the inner mitochondrial membrane, have been confirmed by WOO and OSMOND, and by MOORE et al. Experiments with inhibitors indicated that this reaction is the source of photorespiratory CO_2 (KUMARASINGHE et al.). There remains, however, the

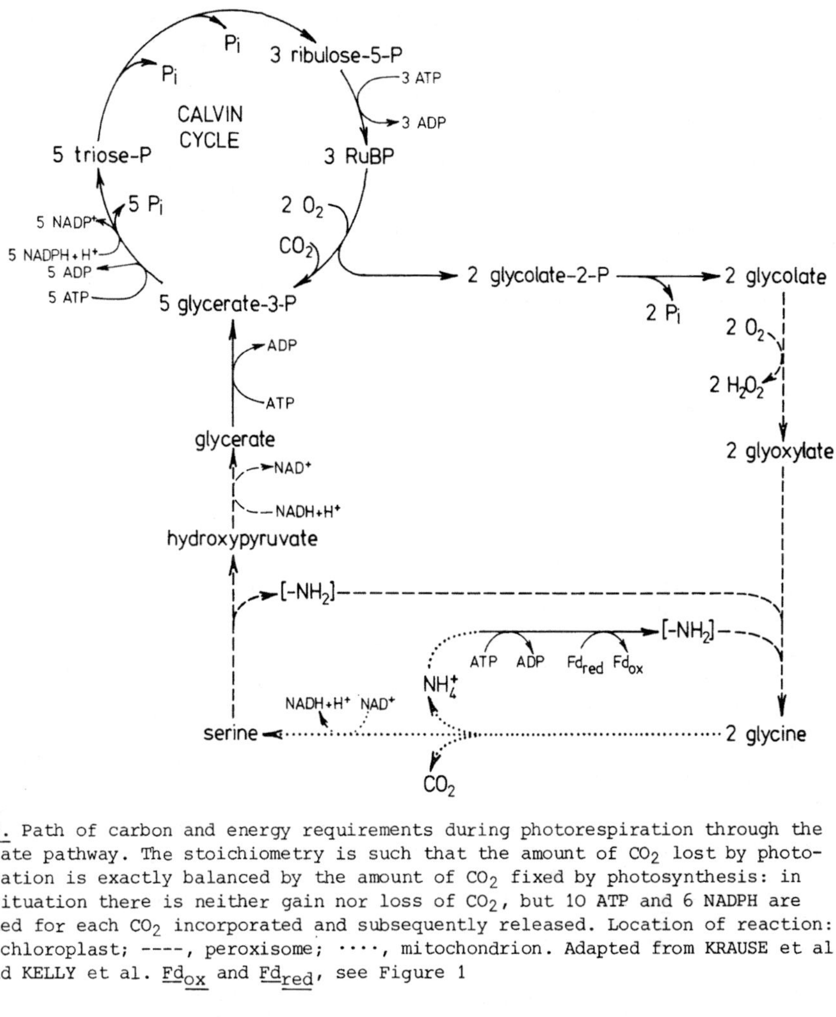

Fig. 2. Path of carbon and energy requirements during photorespiration through the glycolate pathway. The stoichiometry is such that the amount of CO_2 lost by photorespiration is exactly balanced by the amount of CO_2 fixed by photosynthesis: in this situation there is neither gain nor loss of CO_2, but 10 ATP and 6 NADPH are utilized for each CO_2 incorporated and subsequently released. Location of reaction: ——, chloroplast; ----, peroxisome; ····, mitochondrion. Adapted from KRAUSE et al. (1) and KELLY et al. Fd_{ox} and Fd_{red}, see Figure 1

problem that both glycine (LAWLOR and FOCK; CANVIN et al.) and serine (PLATT et al., 1; SERVAITES and OGREN, 1) in leaves are not exclusively involved in the glycolate pathway, and it will be a challenge to elucidate the nature and regulation of the metabolism of these two amino acids, especially since the content of glycine in leaves seems to be strictly controlled (CANVIN et al.). The ability of metabolites such as glyoxylate, P-enolpyruvate, and glutamate to influence glycolate accumulation in leaf discs (OLIVER and ZELITCH, 1, 2) may in some way be connected with these regulatory aspects.

The glycolate pathway requires that glycolate moves from chloroplasts to peroxisomes (Fig. 2); hence it is interesting that close association of these organelles has been observed in vivo (SCHOPFER et al.), and even in vitro in the presence of Pi (SCHNARRENBERGER and BURKHARD). In the greening cytoledons of fat-storing seeds, peroxisomes are thought to be derived from similar microbodies, the glyoxysomes (BURKE and TRELEASE; SCHOPFER et al.; THEIMER; GERHARDT and BETSCHE), although separate development of these organelles is also considered (BROWN and MERRETT).

The amount of attention given to photorespiration is not matched by that given to its counterpart, dark respiration, in photosynthetic tissues. Information so far indicates that dark respiration is to a greater or lesser extent inhibited in the light (IMAFUKU and KATOH; PRZYBYLLA and FOCK; RAVEN). The relationship between the Calvin cycle and the tricarboxylic acid cycle does not appear to be straightforward (KENT). Perhaps future research on this subject will increase now that a method for the isolation of good quality mitochondria from green leaves has been developed (DOUCE et al.).

b) The Purpose

Photorespiration is often referred to as an "apparently wasteful process" because it uses much photosynthetically derived energy to drive a series of reactions, which, in the extreme case, can release CO_2 in an amount equal to that fixed during photosynthesis. Although this is clearly unfavorable to the carbon economy of the plant, the suggestion has been made that photorespiration is nevertheless valuble because of the energy it utilizes, and it is consequently looked upon as an "energy dissipation" process, the argument being that this energy represents excess photosynthetic energy that might otherwise damage the cell by photooxidative reactions (KRAUSE et al., 1). The magnitude of this energy dissipation can be calculated from Figure 2, in which the Calvin cycle and glycolate pathway are linked up so that, in effect, one molecule of CO_2 is "cycled" between incorporation by RuBP carboxylase and release by glycine decarboxylase: Each "CO_2 cycle" uses up 6 NADPH and 10 ATP. In view of reports of decreased leaf CO_2 fixation (POSKUTA et al.) and damage to membranes (probably by reduced species of oxygen, see HALLIWELL) by high amounts of light and/or oxygen, this proposed role of photorespiration deserves attention. However, a number of observations are difficult to reconcile with this interpretation; these relate to the survival of plants in which photorespiration is low or absent. Photorespiration in bryophytes is rather low (DILKS) and has been reported to be absent in the chlorophyll-containing spores of the moss *Funaria hygrometrica* (CHEVALLIER and DOUCE). More startling are recent reports by LLOYD et al., and by BIDWELL, claiming that freshwater and marine algae do not exhibit photorespiration under normal conditions of growth. Finally, much evidence, including recent studies by KUBOWICZ et al., implies that the mesophyll tissue of C_4 plants lacks photorespiration; the insignificant levels of glycine decarboxylase in the mitochondria from this tissue (WOO and OSMOND) tend to confirm this. Yet the mesophyll tissue of C_4 plants might be thought to be in most need of a mechanism such as photorespiration to dispose of excess photosynthetic energy since these plants are so often found in habitats with high illumination.

It is worth pointing out that energy dissipation by photorespiration as depicted in Figure 2 is at its maximum; if the ratio of RuBP oxidized to that carboxylated exceeds the value of two as shown, the RuBP level in the Calvin cycle would become depleted. Evidence for such a situation has been experimentally obtained with leaf cells (SERVAITES and OGREN, 1), and it has been demonstrated with isolated chloroplasts (KIRK and HEBER). It is interesting to contemplate that almost the same amount of "energy dissipation" as that shown in Figure 2 could be achieved by incorporating and reducing three molecules of CO_2 to the level of triose-P in the Calvin cycle (and regenerating the acceptor RuBP) rather than incorporating one molecule of CO_2 and two of O_2, and then releasing the CO_2 again. Obviously, for a plant to grow, it must strive toward the former alternative; utilization of all energy by the latter, as in Figure 2, would not permit growth. One wonders why plants are not more successful in making use of CO_2 reduction, rather than photorespiration, as a means of using up available energy. In fact, the more efficient carboxylation proposed for *Panicum milioides* (KECK and OGREN), which has been described as a C_3 plant with reduced photorespiration (QUEBEDEAUX and CHOLLET), may be an example of just such a successful plant.

6. C_4 Metabolism

a) C_4 Plants

C_4 plants are those possessing the C_4 pathway, which involves a traffic of carbon compounds between mesophyll and bundle sheath cells. CO_2 is fixed by a cytoplasmic PEP carboxylase in mesophyll cells, and the resultant 4C-skeleton transported as either malate or asparate to bundle sheath cells where it is decarboxylated by one of three enzymes (chloroplast NADP-linked malic enzyme, chloroplast PEP carboxykinase, or mitochondrial NAD-linked malic enzyme), depending on the plant species. The CO_2 released is incorporated by the Calvin cycle in bundle sheath chloroplasts, while the 3C-skeleton is returned to the mesophyll chloroplast for PEP synthesis by pyruvate Pi dikinase. It now seems most unlikely that mesophyll chloroplasts possess any Calvin cycle since no trace of RuBP carboxylase protein could be found in mesophyll cells (HATTERSLEY et al.; MATSUMOTO et al.) and chloroplasts (HUBER et al.; WALBOT). A major protein component of mesophyll cells is PEP carboxylase (HUBER et al.), which represents 10% of the soluble protein (UEDAN and SUGIYAMA). This enzyme has been purified and characterized from *Zea mays* leaves, but reported properties differed somewhat (UEDAN and SUGIYAMA; MUKERJI); the presence of several forms of this enzyme in plants (VIDAL et al.) may help explain these discrepancies.

The C_4 pathway requires that mesophyll and bundle sheath cells be in close contact; in C_4 plants such as *Triodia irritans* (CRAIG and GOODCHILD) and *Calligonum persicum* (WINTER et al.), this arrangement is not easily observed. Metabolite transport between the cells via plasmodesmata has been judged possible by EVERT et al., and RAGHAVENDRA and DAS (1, 2) have evidence for active transport of 4C acids involving an ATPase.

Irrespective of whether malate or aspartate is transported into bundle sheath cells (KENNEDY and WILLIAMS), the capacity of these cells to release CO_2 (for the Calvin cycle) from these 4C acids has been well proved; in cases where malate is decarboxylated in the chloroplast by NADP-linked malic enzyme, a close connection between NADPH production by this enzyme, and utilization by chloroplast GAP dehydrogenase, was observed (HATCH and KAGAWA; RAY and BLACK, 3; RATHNAM and EDWARDS, 2; USUDA and MIYACHI). Specific inhibition of decarboxylating enzymes [PEP carboxykinase by 3-mercaptopicolinic acid (RAY and BLACK, 1), and NAD-linked malic enzyme by oxalic acid (RATHNAM and EDWARDS, 3)] has been used by RATHNAM and EDWARDS (3) to identify which of the three decarboxylating enzymes operates in any particular C_4 plant. There seems to be a correlation between use of NAD-linked malic enzyme and cold tolerance (CALDWELL et al.), although some activity of this enzyme is present (but probably has no function in photosynthesis) in *Zea mays* (HATCH and MAU, 1).

PEP carboxykinase, which is the predominant decarboxylating enzyme in the C_4 grasses *Panicum maximum* (RAY and BLACK, 1) and *Chloris gayana* (GIROLAMI and CAVALIÉ), has received considerable attention over the past two years. The enzyme has been purified and studied from these two grasses (RAY and BLACK, 2; HATCH and MAU, 2), as well as from the CAM plant *Ananas comosus* (DALEY et al.), the diatom *Phaeodactylum tricornutum* (HOLDSWORTH and BRUCK), and the bacterium *Rhodospirillum rubrum* (KLEMME). Properties of the enzyme from the C_4 and CAM plants were consistent with catalysis of oxaloacetate decarboxylation during C_4 metabolism, but the diatom enzyme seemed designed for the reverse reaction.

C_4 plants are often thought of as more productive than C_3 plants, although GIFFORD pointed out some time ago that no difference exists between the best examples of each type growing in their own preferred environments. This viewpoint is strengthened by the recent observation that an ephemeral desert plant, *Camissonia claviformis*, is a C_3 plant that can fix CO_2 at a rate of 95 $mg/dm^2/h$ (MOONEY et al.), well above the range of 40-80 $mg/dm^2/h$ quoted for C_4 plants (BLACK).

b) Crassulacean Acid Metabolism

In contrast to C_4 plants (in which the formation and decarboxylation of 4C acids occur more or less simultaneously but in different groups of cells), plants with CAM synthesize the 4C acid malate at night and accumulate this in the vacuole until the following day, when decarboxylation in the same cell is carried out. An ATPase pump has been suggested to assist the transport of 4C acids between cytoplasm and vacuole (DALEY and VINES). BUSER and MATILE have isolated vacuoles from *Bryophyllum* leaves and confirmed that they contain the bulk of the cellular malic acid.

On feeding $^{14}CO_2$ to CAM plants, it would be expected that ^{14}C would appear only in the C-4 of malate if PEP carboxylase and malate dehydrogenase were the only two enzymes involved. Using *Kalanchoë fedtschenkoi*, DITTRICH has shown that this is true only during the first few seconds of dark CO_2 fixation; thereafter the labeled malate exchanges rapidly enough between the vacuolor reserve and a smaller cytoplasmic pool that label is equilibrated by the action of mitochondrial fumarase so that, after 20 min, about one-third of the ^{14}C is in C-1 and two-thirds in C-4. A previous explanation for this labeling pattern, involving RuBP carboxylation in the dark, is no longer widely considered, and in addition, techniques used to obtain some earlier labeling data have been brought into question (WINTER and KANDLER). The substrate PEP used in CAM is supplied from the glycolytic breakdown of insoluble sugars, and during this process glucose-6-P activation of PEP carboxylase and PEP inhibition of P-fructokinase are believed to have regulatory roles (COCKBURN and McAULAY). This glycolytic pathway seems to be distinct from a second pathway that breaks down soluble sugars for respiration (DELEENS and GARNIER-DARDART).

CAM is popularly associated with plants growing in hot arid regions, but it also occurs in species found in the cold arid environment of the Andes (MEDINA and DELGADO). HARTSOCK and NOBEL have induced *Agave deserti* to adjust from CAM to daytime CO_2 uptake by prolonged watering, but it is not thought that the short-term seasonal changes that prevail under natural conditions ever induce significant daytime CO_2 uptake (SUTTON et al.; SZAREK and THROUGHTON). A comprehensive list of plant species exhibiting CAM has been compiled by SZAREK and TING; to this list *Seyrigia humbertii* (Cucurbitaceae) might be added (DE LUCA et al.).

A special group of CAM plants are certain halophytes that display CAM only when under water stress or with relatively high levels of NaCl in the soil water. Investigations by VON WILLERT et al. (1, 2) with *Mesembryanthemum crystallinum* have shown that CAM development occurs principally in the mature leaves and is associated with the appearance of a third form (as distinguished by electrophoresis) of PEP carboxylase.

Another species recently found to develop CAM under saline growth conditions is *Portulacaria afra* (TING and HANSCOM).

c) C_4 Metabolism

Conspicuous participation of PEP carboxylase in a number of photosynthetic tissues that lack both the C_4 pathway and CAM was mentioned previously (KELLY et al.). A number of new developments have occurred in this field. A Massachusetts population of the C_3-C_4 intermediate plant *Mollugo verticillata* was found to be more C_3 than a Kansas population (SAYRE and KENNEDY). A tissue culture from the C_4 plant *Portulaca oleracea* continued to display C_4 metabolism although it lacked differentation of cells into mesophyll and bundle sheath (and therefore C_4 pathway) (KENNEDY et al.). Even in spinach leaf slices suspended under appropriate conditions PEP carboxylase will incorporate some CO_2 into 4C compounds, although this is not evidence for a C_4 pathway (BÖCHER and KLUGE). C_4 metabolism is a central component of stomatal operation (RASCHKE and DITTRICH). Earlier evidence for C_4 metabolism in marine algae has been challenged (KREMER and KÜPPERS), but DÖHLER has shown a clear swing toward C_4 metabolism by *Anacystis* grown under red light, and WEATHERS and ALLEN found that C_4 metabolism in blue-green algae increased when the stationary growth phase was attained; these authors suggested that this response could compensate for the incomplete tricarboxylic acid cycle in these organisms. Aquatic angiosperms show a clear tendency for C_4 metabolism (BENEDICT and SCOTT; DE GROOTE and KENNEDY; BOWES et al.); in the case of *Hydrilla verticillata* the ratio of the activities of RuBP carboxylase to PEP carboxylase increased from 0.25 in summer to 2.74 in winter (BOWES et al.). The ratio of these enzyme activities, together with products of short-term CO_2 fixation, has been used to demonstrate that C_4 metabolism exists in the tissues surrounding the grains of wheat and oat (WIRTH et al.).

Observations such as these emphasize that a biochemical distinction between C_3 plants, C_4 plants, and CAM plants is not easily realized. Until recently, one likely candidate was the enzyme pyruvate Pi dikinase, which had not been confirmed in C_3 plants, but an early indication that the enzyme was present in the green pericarp of barley (DUFFUS and ROSIE) recently has been confirmed by MEYER et al., who partially purified the enzyme from wheat grains and concluded that the activity in vivo exceeds 20 μmoles PEP formed/mg chlorophyll/h.

References

AKAZAWA, T.: Photosynthesis 1977. Proc. 4th Int. Congr. Photosyn. HALL, D.O., COOMBS, J., GOODWIN, T.W. (eds.), pp. 447-456. London: Biochem. Soc. 1978. - ALLEN, J.F.: FEBS Lett. 84, 221-224 (1977). - ANDERSON, J.W., DONE, J.: (1) Plant Physiol. 60, 354-359 (1977); - (2) Plant Physiol. 60, 504-508 (1977). - ANDERSON, L.E., DUGGAN, J.X.: Plant Physiol. 58, 135-139 (1976). - ANDERSON, L.E., NEHRLICH, S.C.: FEBS Lett. 76, 64-66 (1977). - ASAMI, S., AKAZAWA, T.: Biochemistry 16, 2202-2207 (1977).

BADGER, M.R., LORIMER, G.H.: Arch. Biochem. Biophys. 175, 723-729 (1976). - BAHR, J.T., JENSEN, R.G.: Arch. Biochem. Biophys. 185, 39-48 (1978). - BAIER, D., LATZKO, E.: Biochim. Biophys. Acta 396, 141-147 (1975). - BAKER, T.S., SUH, S.W., EISENBERG, D.: Proc. Natl. Acad. Sci. US 74, 1037-1041 (1977). - BASSHAM, J.A.: Science 197, 630-638 (1977). BEEZLEY, B.B., GRUBER, P.J., FREDERICK, S.E.: Plant Physiol. 58, 315-319

(1976). - BENEDICT, C.R., SCOTT, J.R.: Plant Physiol. 57, 876-880 (1976). - BEN-HAYYIM, G.: Europ. J. Biochem. 83, 99-104 (1978). - BIDWELL, R.G.S.: Can. J. Botany 55, 809-818 (1977). - BLACK, C.C.: Ann. Rev. Plant Physiol. 24, 253-286 (1973). - BLACKWOOD, G.C., MIFLIN, B.J.: J. Expt. Botany 27, 735-747 (1976). - BÖCHER, M., KLUGE, M.: Z. Pflanzenphysiol. 83, 347-361 (1977). - BOWES, G., HOLADAY, A.S., VAN, T.K., HALLER, W.T.: Photosynthesis 1977. Proc. 4th Int. Congr. Photosyn. HALL, D.O., COOMBS, J., GOODWIN, T.W. (eds.), pp. 289-298. London: Biochem. Soc. 1978. - BOWIEN, B., MAYER, F., CODD, G.A., SCHLEGEL, H.G.: Arch. Microbiol. 110, 157-166 (1976). - BROWN, R.H., MERRETT, M.J.: New Phytologist 79, 73-81 (1977). - BUCHANAN, B.B., SCHÜRMANN, P., SHANMUGAM, K.T.: (1) Biochim. Biophys. Acta 283, 136-145 (1972). - BUCHANAN, B.B., SCHÜRMANN, P., WOLOSIUK, R.A.: (2) Biochem. Biophys. Res. Commun. 69, 970-978 (1976). - BUCHANAN, B.B., SIREVÅG, R.: Arch. Microbiol. 109, 15-19 (1976). - BURKE, J.J., TRELEASE, R.N.: Plant Physiol. 56, 710-717 (1975). - BUSER, C., MATILE, P.: Z. Pflanzenphysiol. 82, 462-466 (1977).

CALDWELL, M.M., OSMOND, C.B., NOTT, D.L.: Plant Physiol. 60, 157-164 (1977). - CANVIN, D.T., FOCK, H., LLOYD, N.D.H.: Phytosynthesis 1977. Proc. 4th Int. Congr. Photosyn. HALL, D.O., COOMBS, J., GOODWIN, T.W. (eds.), pp. 323-334. London: Biochem. Soc. 1978. - CAVELL, S., SCOPES, R.K.: Europ. J. Biochem. 63, 483-490 (1976). - CERFF, R.: Europ. J. Biochem. 82, 45-53 (1978). - CHAMPIGNY, M.-L., BISMUTH, A.: Photosynthetic Organelles. Special Issue of Plant Cell Physiol. MIYACHI, S., KATOH, S., FUJITA, Y., SHIBATA, K. (eds.), pp. 365-375. Japan, Jap. Soc. Plant Physiol. 1977. - CHAPMAN, D.J., LEECH, R.M.: FEBS Lett. 68, 160-164 (1976). - CHEN, K., KUNG, S.D., GRAY, J.C., WILDMAN, S.G.: Plant Sci. Lett. 7, 429-434 (1976). - CHEVALLIER, D., DOUCE, R.: Plant Physiol. 57, 400-402 (1976). - CHOLLET, R., ANDERSON, L.L.: (1) Arch. Biochem. Biophys. 176, 344-351 (1976); - (2) Biochim. Biophys. Acta 482, 228-240 (1977). - COCKBURN, W., McAULAY, A.: Plant Physiol. 59, 455-458 (1977). - CODD, G.A., STEWART, W.D.P.: (1) Arch. Mikrobiol. 94, 11-28 (1973); - (2) Planta 130, 323-326 (1976); - (3) Arch Microbiol. 113, 105-110 (1977). - COEN, D.M., BEDBROOK, J.R., BOGORAD, L., RICH, A.: Proc. Natl. Acad. Sci. US 74, 5487-5491 (1977). - CRAIG, S., GOODCHILD, D.J.: Aust. J. Botany 25, 277-290 (1977).

DALEY, L.S., BIDWELL, R.G.S.: Plant Physiol. 60, 109-114 (1977). - DALEY, L.S., RAY, T.B., VINES, H.M., BLACK, C.C.: Plant Physiol. 59, 618-622 (1977). - DALEY, L.S., VINES, H.M.: Plant Sci. Lett. 10, 289-298 (1977). - DeGROOTE, D., KENNEDY, R.A.: Plant Physiol. 59, 1133-1135 (1977). - DELANEY, M.E., WALKER, D.A.: Plant Sci. Lett. 7, 285-294 (1976). - DELEENS, E., GARNIER-DARDART, J.: Planta 135, 241-248 (1977). - DeLUCA, P., ALFANI, A., VIRZO DE SANTO, A.: Botan. Gaz. 138, 474-478 (1977). - DILKS, T.J.K.: J. Expt. Botany 27, 98-104 (1976). - DITTRICH, P.: Plant Physiol. 58, 288-291 (1976). - DOBBERSTEIN, B., BLOBEL, G., CHUA, N.-H.: Proc. Natl. Acad. Sci. US 74, 1082-1085 (1977). - DÖHLER, G.: Planta 131, 129-133 (1976). - DOUCE, R., MOORE, A.L., NEUBURGER, M.: Plant Physiol. 60, 625-628 (1977). - DUFFUS, C.M., ROSIE, R.: Planta 114, 219-226 (1973).

EDWARDS, G.E., HUBER, S.C.: Photosynthesis 1977. Proc. 4th Int. Congr. Photosyn. HALL, D.O., COOMBS, J., GOODWIN, T.W. (eds.), pp. 95-106. London: Biochem. Soc. 1978. - ELIAS, B.A., GIVAN, C.V.: Plant Physiol. 59, 738-740 (1977). - EVERT, R.F., ESCHRICH, W., HEYSER, W.: Planta 136, 77-89 (1977).

FALKNER, G., HORNER, F., WERDAN, K., HELDT, H.W.: Plant Physiol. 58, 717-718 (1976). - FERRI, G., COMERIO, G., IADAROLA, P., ZAPPONI, M.C., SPERANZA, M.L.: Biochim. Biophys. Acta 522, 19-31 (1978). - FLÜGGE, U.I., HELDT, H.W.: FEBS Lett. 68, 259-262 (1976).

GAYLER, K.R., MORGAN, W.R.: Plant Physiol. 58, 283-287 (1976). - GELVIN, S., HEIZMANN, P., HOWELL, S.H.: Proc. Natl. Acad. Sci. US 74, 3193-3197 (1977). - GERHARDT, B., BETSCHE, T.: Ber. Deut. Botan. Ges. 89, 321-334 (1976). - GIBSON, J.L., TABITA, F.R.: (1) J. Biol. Chem. 252, 943-949 (1977); - (2) J. Bacteriol. 132, 818-823 (1977). - GIFFORD, R.M.: Australian J. Plant Physiol. 1, 107-117 (1974). - GIROLAMI, J.-P., CAVALIÉ, G.: Physiol. Vége. 15, 453-467 (1977). - GIVAN, C.V.: Planta 122, 281-291 (1975). - GRAY, J.C., KUNG, S.D., WILDMAN, S.G.: Arch. Biochem. Biophys. 185, 272-281 (1978). - GRODEN, D., BECK, E.: Abstr. 4th Int. Congr. Photosyn. Reading, England, 139 (1977). - GRODZINSKI, B., BUTT, V.S.: Planta 133, 261-266 (1977). - GRUBER, P.J., FREDERICK, S.E.: Planta 135, 45-49 (1977).

HALLIWELL, B.: Progr. Biophys. Mol. Biol. 33, 1-54 (1978). - HALLIWELL, B., FOYER, C.H.: Planta 139, 9-17 (1978). - HARTSOCK, T.L., NOBEL, P.S.: Nature 262, 574-576 (1976). - HATCH, M.D.: Photosynthetic Organelles. Special Issue of Plant Cell Physiol. MIYACHI, S., KATOH, S., FUJITA, Y., SHIBATA, K. (eds.), pp. 311-314. Japan, Jap. Soc. Plant Physiol. 1977. - HATCH, M.D., KAGAWA, T.: Arch. Biochem. Biophys. 175, 39-53 (1976). - HATCH, M.D., MAU, S.-L.: (1) Arch. Biochem. Biophys. 179, 361-369 (1977); - (2) Australian J. Plant Physiol. 4, 207-216 (1977). - HATTERSLEY, P.W., WATSON, L., OSMOND, C.B.: Australian J. Plant Physiol. 4, 523-539 (1977). - HAYSTEAD, A.: Planta 111, 271-274 (1973). - HELDT, H.W.: Encyclopedia of Plant Physiology, N.S., Vol. 3. HEBER, U., STOCKING, C.R. (eds.), pp. 137-143. Berlin-Heidelberg-New York: Springer 1976.-HELDT, H.W., CHON, C.J., MARONDE, D., HEROLD, A., STANKOVIC, Z.S., WALKER, D.A., KRAMINER, A., KIRK, M.R., HEBER, U.: Plant Physiol. 59, 1146-1155 (1977). - HEROLD, A., LEWIS, D.H., WALKER, D.A.: New Phytologist 76, 397-407 (1976). - HIGHFIELD, P.E., ELLIS, R.J.: Nature 271, 420-424 (1978). - HOLDSWORTH, E.S., BRUCK, K.: Arch. Biochem. Biophys. 182, 87-94 (1977). - HUBER, S.C., EDWARDS, G.E.: (1) Biochim. Biophys. Acta 440, 675-687 (1976); - (2) Biochim. Biophys. Acta 462, 583-602 (1977); - (3) Biochim. Biophys. Acta 462, 603-612 (1977). - HUBER, S.C., HALL, T.C., EDWARDS, G.E.: Plant Physiol. 57, 730-733 (1976). - HUNER, N.P.A., MacDOWALL, F.D.H.: Biochim. Biophys. Res. Commun 73, 411-420 (1976).

IMAFUKU, H., KATOH, T.: Plant Cell Physiol. 17, 515-524 (1976).

JACQUOT, J.P., VIDAL, J., GADAL, P.: (1) FEBS Lett. 71, 223-227 (1976); - (2) Planta 137, 89-90 (1977). - JAGENDORF, A.T.: Encyclopedia of Plant Physiology, N.S., Vol. 5. TREBST, A., AVRON, M. (eds.), pp. 307-337. Berlin-Heidelberg-New York: Springer 1977.

KAGAWA, T., HATCH, M.D.: Arch. Biochem. Biophys. 184, 290-297 (1977). - KAISER, W.: Biochim. Biophys. Acta 440, 476-482 (1976). - KAISER, W., URBACH, W.: Biochim. Biophys. Acta 459, 337-346 (1977). - KAWASHIMA, N., TANABE, Y.: Plant Cell Physiol. 17, 757-764 (1976). - KECK, R.W., OGREN, W.L.: Plant Physiol. 58, 552-555 (1976). - KELLY, G.J., LATZKO, E.: (1) Plant Physiol. 60, 290-294 (1977); - (2) Plant Physiol. 60, 295-299 (1977). - KELLY, G.J., LATZKO, E., GIBBS, M.: Ann. Rev. Plant Physiol. 27, 181-205 (1976). - KENNEDY, R.A., BARNES, J.E., LAETSCH, W.M.: Plant Physiol. 59, 600-603 (1977). - KENNEDY, R.A., WILLIAMS, L.E.: Plant Physiol. 59, 207-210 (1977). - KENT, S.S.: Plant Physiol. 60, 274-276 (1977). - KIRK, M.R., HEBER, U.: Planta 132, 131-141 (1976). - KLEMME, J.-H.: Arch. Microbiol. 107, 189-192 (1976). - KOW, Y.W., ROBINSON, J.M., GIBBS, M.: Plant Physiol. 60, 492-495 (1977). - KRAUSE, G.H.: Biochim. Biophys. Acta 460, 500-510 (1977). - KRAUSE, G.H., LORIMER, G.H., HEBER, U., KIRK, M.R.: (1) Photosynthesis 1977. Proc. 4th Int. Congr. Photosyn. HALL, D.O., COOMBS, J., GOODWIN, T.W. (eds.), pp. 299-310. London: Biochem. Soc. 1978. - KRAUSE, G.H., THORNE, S.W., LORIMER, G.H.: (2) Arch. Biochem. Biophys. 183, 471-479 (1977). -

KREMER, B.P., KUPPERS, U.: Planta 133, 191-196 (1977). - KUBOWICZ, D., MALESZEWSKI, S., POSKUTA, J.: Z. Pflanzenphysiol. 81, 141-146 (1977). - KUMARASINGHE, K.S., KEYS, A.J., WHITTINGHAM, C.P.: J. Exp. Botany 28, 1163-1168 (1977).

LAING, W.A., CHRISTELLER, J.T.: Biochem. J. 159, 563-570 (1976). - LATZKO, E., KELLY, G.J.: Progr. Botany 36, 77-89 (1974). - LAWLOR, D. W., FOCK, H.: J. Exp. Botany 28, 329-337 (1977). - LEA, P.J., MIFLIN, B.J.: Nature 251, 614-616 (1974). - LEHNER, K., HELDT, H.W.: Biochim. Biophys. Acta 501, 531-544 (1978). - LENDZIAN, K., BASSHAM, J.A.: Biochim. Biophys. Acta 430, 478-489 (1976). - LEVI, C., GIBBS, M.: Plant Physiol. 57, 933-935 (1976). - LEVI, C., PREISS, J.: Plant Physiol. 58, 753-756 (1976). - LILLEY, R.McC., CHON, C.J., MOSBACH, A., HELDT, H.W.: Biochim. Biophys. Acta 460, 259-272 (1977). - LLOYD, N.D.H., CANVIN, D.T., CULVER, D.A.: Plant Physiol. 59, 936-940 (1977). - LORIMER, G.H., BADGER, M.R., ANDREWS, T.J.: (1) Biochemistry 15, 529-536 (1976); - (2) Anal. Biochem. 78, 66-75 (1977). - LORIMER, G.H., KRAUSE, G.H., BERRY, J.A.: (3) FEBS Lett. 78, 199-202 (1977). - LORIMER, G.H., OSMOND, C.B., AKAZAWA, T., ASAMI, S.: (4) Arch. Biochem. Biophys. 185, 49-56 (1978).

MASAMOTO, K., NISHIMURA, M.: J. Biochem. 82, 483-487 (1977). - MATSUMOTO, K., NISHIMURA, M., AKAZAWA, T.: Plant Cell Physiol. 18, 1281-1290 (1977). - MEDINA, E., DELGADO, M.: Photosynthetica 10, 155-163 (1976). - MEYER, A.O., KELLY, G.J., LATZKO, E.: Plant Sci. Lett. 12, 35-40 (1978). - MIGINIAC-MASLOW, M., HOARAU, A.: Plant Sci. Lett. 9, 7-15 (1977). - MITCHELL, C.A., STOCKING, C.R.: Plant Physiol. 55, 59-63 (1975). - MOONEY, H.A., EHLERINGER, J., BERRY, J.A.: Science 194, 322-324 (1976). - MOORE, A.L., JACKSON, C., HALLIWELL, B., DENCH, J.E., HALL, D.O.: Biochem. Biophys. Res. Commun. 78, 483-491 (1977). - MUKERJI, S.K.: Arch. Biochem. Biophys. 182, 352-359 (1977). - MURAKAMI, S., STROTMANN, H.: Arch. Biochem. Biophys. 185, 30-38 (1978).

NAKATANI, H.Y., BARBER, J.: Biochem. Biophys. Acta 461, 510-512 (1977). - NICKLISCH, A., GESKE, W., KOHL, J.-G.: Biochem. Physiol. Pflanzen 170, 85-90 (1976). - NISHIDA, K., SANADA, Y.: Photosynthetic Organelles. Special Issue of Plant Cell Physiol. MIYACHI, S., KATOH, S., FUJITA, Y., SHIBATA, K. (eds.), pp. 341-346. Japan, Jap. Soc. Plant Physiol. 1977. - NISHIMURA, M., GRAHAM, D., AKAZAWA, T.: Plant Physiol. 58, 309-314 (1976).

O'BRIEN, M.J., EASTERBY, J.S., POWLS, R.: (1) Biochim. Biophys. Acta 449, 209-223 (1976); - (2) Biochim. Biophys. Acta 481, 348-358 (1977). - OLIVER, D.J., ZELITCH, I.: (1) Science 196, 1450-1451 (1977); - (2) Plant Physiol. 59, 688-694 (1977). - O'NEAL, D., JOY, K.W.: Nature New Biol. 246, 61-62 (1973).

PAECH, C., RYAN, F.J., TOLBERT, N.E.: Arch. Biochem. Biophys. 179, 279-288 (1977). - PARADIES, H.H., ZIMMER, B., WERZ, G.: Biochem. Biophys. Res. Commun. 74, 397-404 (1977). - PASSERA, C., ALBUZIO, A.: Biol. Plant. 19, 448-452 (1977). - PAUL, J.S., VOLCANI, B.E.: Arch. Microbiol. 110, 247-252 (1976). - PEAVEY, D.G., STEUP, M., GIBBS, M.: Plant Physiol. 60, 305-308 (1977). - PELROY, R.A., LEVINE, G.A., BASSHAM, J.A.: J. Bacteriol. 128, 633-643 (1976). - PEOPLES, M.B., DALLING, M.J.: Planta 138, 153-160 (1978). - PLATT, S.G., PLAUT, Z., BASSHAM, J.A.: (1) Plant Physiol. 60, 230-234 (1977); - (2) Plant Physiol. 60, 739-742 (1977). - POINCELOT, R.P., DAY, P.R.: Plant Physiol. 57, 334-338 (1976). - PORTIS, A.R., CHON, C.J., MOSBACH, A., HELDT, H.W.: Biochim. Biophys. Acta 461, 313-325 (1977). - PORTIS, A.R., HELDT, H.W.: Biochim. Biophys. Acta 449, 434-446 (1976). - POSKUTA, J., WRÓBLEWSKA, B., MIKULSKA, M.: Z. Pflanzenphysiol. 78, 396-402 (1976). - PREISS, J., LEVI, C.: Photosynthesis 1977. Proc. 4th Int. Congr. Photosyn. HALL, D.O., COOMBS, J., GOODWIN, T.W. (eds.), pp. 457-468. London: Biochem. Soc. 1978. - PRZYBYLLA, K.-R., FOCK, H.: Ber. Deut. Botan. Ges. 89, 651-661 (1976). - PURCZELD, P., CHON, C.J., PORTIS, A.R., HELDT, H.W., HEBER, U.: Biochim. Biophys. Acta 501, 488-498 (1978). - PUROHIT, K., McFADDEN, B.A.: Biochem. Biophys. Res. Commun. 71, 1220-1227 (1976).

QUEBEDEAUX, B., CHOLLET, R.: Plant Physiol. 59, 42-44 (1977).

RAGHAVENDRA, A.S., DAS, V.S.R.: (1) New Phytologist 79, 481-487 (1977); - (2) J. Exp. Botany 29, 39-47 (1978). - RASCHKE, K., DITTRICH, P.: Planta 134, 69-75 (1977). - RATHNAM, C.K.M., EDWARDS, G.E.: (1) Plant Cell Physiol. 17, 177-186 (1976); - (2) Arch. Biochem. Biophys. 182, 1-13 (1977); - (3) Plant Cell Physiol. 18, 963-968 (1977). - RAVEN, J.A.: New Phytologist 76, 195-204 (1976). - RAY, T.B., BLACK, C.C.: (1) J. Biol. Chem. 251, 5824-5826 (1976); - (2) Plant Physiol. 58, 603-607 (1976); - (3) Plant Physiol. 60, 193-196 (1977). - RICHARDSON, R.H., MATHESON, N.K.: Phytochemistry 16, 1875-1879 (1977). - ROBINSON, J.M., GIBBS, M., COTLER, D.N.: Plant Physiol. 59, 530-534 (1977). - ROBINSON, S.P., WISKICH, J.T.: (1) Plant Physiol. 58, 156-162 (1976); - (2) Plant Physiol. 59, 422-427 (1977); - (3) Biochim. Biophys. Acta 461, 131-140 (1977); - (4) Arch. Biochem. Biophys. 184, 546-554 (1977). - RYRIE, I.J., JAGENDORF, A.T.: J. Biol. Chem. 246, 3771-3774 (1971).

SAYRE, R.T., KENNEDY, R.A.: Planta 134, 257-262 (1977). - SCHÄFER, G., HEBER, U., HELDT, H.W.: Plant Physiol. 60, 286-289 (1977). - SCHNARRENBERGER, C., BURKHARD, C.: Planta 134, 109-114 (1977). - SCHOPFER, P., BAJRACHARYA, D., BERGFELD, R., FALK, H.: Planta 133, 73-80 (1976). - SCHÜRMANN, P., WOLOSIUK, R.A.: Biochim. Biophys. Acta 522, 130-138 (1978). - SCHÜRMANN, P., WOLOSIUK, R.A., BREAZEALE, V.D., BUCHANAN, B.B.: Nature 263, 257-258 (1976). - SCHWARZ, Z., MARETZKI, D., SCHÖNHERR, J.: Biochem. Physiol. Pflanzen 170, 37-50 (1976). - SERVAITES, J.C., OGREN, W.L.: (1) Plant Physiol. 60, 461-466 (1977); - (2) Plant Physiol. 60, 693-696 (1977). - SIREVÅG, R., BUCHANAN, B.B., BERRY, J. A., TROUGHTON, J.H.: Arch. Microbiol. 112, 35-38 (1977). - SLABAS, A. R., WALKER, D.A.: Arch. Biochem. Biophys. 175, 590-597 (1976). - SPREY, B., LAMBERT, C.: Z. Pflanzenphysiol. 83, 227-247 (1977). - STABENAU, H.: Plant Physiol. 58, 693-695 (1976). - STANKOVIC, Z.S., WALKER, D.A.: Plant Physiol. 59, 428-432 (1977). - STEIGER, H.M., BECK, E., BECK, R.: Plant Physiol. 60, 903-906 (1977). - STEUP, M., PEAVEY, D.G., GIBBS, M.: Biochem. Biophys. Res. Commun. 72, 1554-1561 (1976). - STEWART, R., AUCHTERLONIE, C.C., CODD, G.A.: Planta 136, 61-64 (1977). - SUGIYAMA, T.: Plant Cell Physiol. 15, 723-726 (1974). - SUTTON, B.G., TING, I.P., TROUGHTON, J.H.: Nature 261, 42-43 (1976). - SZAREK, S.R., TING, I.P.: Photosynthetica 11, 330-342 (1977). - SZAREK, S.R., TROUGHTON, J.H.: Plant Physiol. 58, 367-370 (1976).

TABITA, F.R., STEVENS, S.E., GIBSON, J.L.: J. Bacteriol. 125, 531-539 (1976). - TAKABE, T., AKAZAWA, T.: Plant Cell Physiol. 18, 753-765 (1977). - THEIMER, R.R.: FEBS Lett. 62, 297-300 (1976). - TING, I.P., HANSCOM, Z.: Plant Physiol. 59, 511-514 (1977).

UEDAN, K., SUGIYAMA, T.: Plant Physiol. 57, 906-910 (1976). - USUDA, H., MIYACHI, S.: Plant Cell Physiol. 18, 1109-1120 (1977).

VALLEJOS, R.H., ANDREO, C.S.: FEBS Lett. 61, 95-99 (1976). - VATER, J., SALNIKOW, J., KLEINKAUF, H.: Biochem. Biophys. Res. Commun. 74, 1618-1625 (1977). - VIDAL, J., CAVALIE, G., GADAL, P.: Plant Sci. Lett. 7, 265-270 (1976). - VON WILLERT, D.J., KIRST, G.O., TREICHEL, S., VON WILLERT, K.: (1) Plant Sci. Lett. 7, 341-346 (1976). - VON WILLERT, D.J., TREICHEL, S., KIRST, G.O., CURDTS, E.: (2) Phytochemistry 15, 1435-1436 (1976).

WALBOT, V.: Plant Physiol. 60, 102-108 (1977). - WALKER, D.A.: Encyclopedia of Plant Physiology, N.S., Vol. 3. HEBER, U., STOCKING, C.R. (eds.)., pp. 85-136. Berlin-Heidelberg-New York: Springer 1976. - WALKER, D.A., HEROLD, A.: Photosynthetic Organelles. Special Issue of Plant Cell Physiol. MIYACHI, S., KATOH, S., FUJITA, Y., SHIBATA, K. (eds.), pp. 295-310. Japan, Jap. Soc. Plant Physiol. 1977. - WALKER, D.A., SLABAS, A.R., FITZGERALD, M.P.: Biochem. Biophys. Acta 440, 147-162 (1976). - WALLSGROVE, R.M., HAREL, E., LEA, P.J., MIFLIN, B.J.: J. Exp. Botany 28, 588-596 (1977). - WEATHERS, P.J., ALLEN, M.M.: Arch. Microbiol. 116, 231-234 (1978). - WILDNER, G.F.: Z. Naturforsch. 31c, 267-271 (1976). - WILDNER, G.F., HENKEL, J.: (1) Biochem. Biophys. Res. Commun. 69, 268-275 (1976); - (2) Z. Naturforsch. 32c, 226-228 (1977). - WINTER, J., KANDLER, O.: Z. Pflanzenphysiol. 78, 103-112 (1976). - WINTER, K., KRAMER, D., TROUGHTON, J.H., CARD, K.A., FISCHER, K.: Z. Pflanzenphysiol. 81, 341-346 (1977). - WIRTH, E., KELLY, G.J., FISCHBECK, G., LATZKO, E.: Z. Pflanzenphysiol. 82, 78-87 (1977). - WOLOSIUK, R.A., BUCHANAN, B.B.: (1) J. Biol. Chem. 251, 6456-6461 (1976); - (2) Nature 266, 565-567 (1977). - WOLOSIUK, R.A., BUCHANAN, B.B., CRAWFORD, N.A.: FEBS Lett. 81, 253-258 (1977). - WOO, K.C., OSMOND, C.B.: Photosynthetic Organelles. Special Issue of Plant Cell Physiol. MIYACHI, S., KATOH, S., FUJITA, Y., SHIBATA, K. (eds.), pp. 315-323. Japan, Jap. Soc. Plant Physiol. 1977.

ZIMMERMANN, G., KELLY, G.J., LATZKO, E.: J. Biol. Chem. (in press, 1978).

Professor Dr. ERWIN LATZKO
Dr. GRAHAME J. KELLY
Botanisches Institut der
Universität Münster
Schloßgarten 3
D 4400 Münster

Note added in proof (2005):
Lines 11–13 of Sect. 2a should read "subunits [the large subunits are one or other of three almost identical polypeptides, while the small subunits are one or other of several more variable polypeptides (Chen et al."

Photosynthesis

Carbon Metabolism: Chloroplast Capability and the Uncertain Fate of CO_2

By ERWIN LATZKO and GRAHAME J. KELLY

"One fusses about style. One tries to write better. One takes pains to be simple, clear and succinct. One aims at rhythm and balance. One reads a sentence aloud to see that it sounds well. One sweats one's guts out."

W. SOMERSET MAUGHAM

1. Introduction

Following elucidation of the Calvin cycle of photosynthetic CO_2 fixation in the early 1950s, the carbon economy of plants was viewed for many years as being a simple balance between fixation by chloroplast RuBP carboxylase and release by mitochondrial respiration. However, during the late 1960s and early 1970s, details became clear of the paths followed by carbon in the newly-discovered processes of CAM[1], the C_4 pathway of photosynthesis, and photorespiration. These added new dimensions to the biochemical description of carbon balance in that other carboxylation and decarboxylation reactions were recognized as being quantitatively significant in determining this balance.

Related, but more recent, investigations have revealed that carboxylation by the chloroplast is just one of the essential tasks performed by this organelle. It is now clear that the capability of the chloroplast extends into the domains of nitrogen assimilation and fatty acid synthesis. This review, which continues where our last (Volume 40 of Progress in Botany) left off, is devoted to the twin themes of how plants strive to capture and keep the CO_2 which enters their cells, and to the variety of functions (including the fixation and reduction of CO_2) that occupy the average chloroplast.

2. Fixation of CO_2 by RuBP Carboxylase

a) Movement of CO_2 to the Enzyme

RuBP carboxylase utilizes CO_2, and not HCO_3^-, as substrate. However, the pH of the chloroplast stroma (where the carboxylase is located) is about

[1] Abbreviations: CAM, crassulacean acid metabolism; CoA, coenzyme A; FBP, fructose-1,6-P_2; GAP, glyceraldehyde-3-P; OAA, oxaloacetate; PEP, P-enolpyruvate; RuBP, ribulose-1,5-P_2; SBP, sedoheptulose-1,7-P_2.

G.J. Kelly and E. Latzko:
Thirty Years of Photosynthesis

Originally published in Progress in Botany 42

8.0 during photosynthesis; at this pH the equilibrium for the hydration of CO_2 to HCO_3^- determines that only 5% exists as free CO_2 (POINCELOT, 1979). It is clear, however, that this equilibrium will be rapidly regained after removal of CO_2 by RuBP carboxylase, since remarkably high activities of carbonic anhydrase are present both in the chloroplast and in the cytosol of leaf cells (POINCELOT, 1979; REED, 1979). Whether this is an essential role for the chloroplast carbonic anhydrase still remains obscure, since the rate of spontaneous dehydration of HCO_3^- is also appreciable. The role of the cytosolic enzyme is also unclear, although it is worth noting that the chloroplast envelope, like RuBP carboxylase, appears to be specific for CO_2 in that this is the form of carbon that crosses from the cytosol to the chloroplast stroma (WERDAN and HELDT, 1972; SHIRAIWA and MIYACHI, 1978). There is, however, evidence that carbonic anhydrase is essential for photosynthetic carbon assimilation by microorganisms. Observations with *Scenedesmus*, *Chlorella*, and *Coccochloris* indicate that inorganic carbon is taken up as HCO_3^- (via a HCO_3^-/OH^- exchange across the plasmalemma), and when cells grown on high levels of CO_2 are transferred to air levels of CO_2, photosynthesis temporarily ceases until the carbonic anhydrase activity (which is quite low in high-CO_2 grown cells) increases to a normal level (FINDENEGG, 1976, 1979; REED and GRAHAM, 1977; HOGETSU and MIYACHI, 1979; MILLER and COLMAN, 1980).

b) The Enzyme: RuBP Carboxylase

Two penetrating reviews (AKAZAWA, 1979; WILDMAN, 1979) and a symposium devoted almost exclusively to the enzyme (SIEGELMAN and HIND, 1978) leave no doubt that RuBP carboxylase has maintained its position of being the most widely investigated of the enzymes of photosynthetic carbon metabolism. Particular attention has been given to the slow activation by CO_2 and Mg^{2+} (BADGER, 1980; LAWLIS et al., 1979; MÄCHLER et al., 1980). Most RuBP carboxylases consist of large and small subunits and carry the catalytic site on the large subunit. It was therefore natural to consider a major role for the small subunit in this activation process, but it now seems clear that the site which binds the activating CO_2, although distinct from the catalytic site (LORIMER, 1979; MIZIORKO, 1979), nevertheless is on the large subunit (PAECH and TOLBERT, 1978; GIBSON and TABITA, 1979; KOBAYASHI et al., 1979; WHITMAN et al., 1979). Formation of a carbamate between an enzyme amino group and the activating CO_2 has been confirmed (O'LEARY et al., 1979). However, the small subunits are not without a role: GIBSON and TABITA (1979) believe that they influence the rate of activation of the *Rhodopseudomonas sphaeroides* enzyme, while KOBAYASHI et al. (1979) have evidence that they help sustain a conformation of the large subunit of the spinach enzyme consistent with high affinity for the substrate CO_2. Further observations of conformational changes of RuBP carboxylase have been made (GREBANIER et al., 1978).

Binding of CO_2 to the activator site must be followed by binding of Mg^{2+} before activation is achieved. This is now favoured as the principal, and perhaps only role of Mg^{2+} in the RuBP carboxylase reaction (PUROHIT and McFADDEN, 1979; MIZIORKO and SEALY, 1980), although a function in stabilizing a C_6 intermediate has also been proposed (PIERCE et al., 1980a).

RuBP carboxylase can use O_2 instead of CO_2 as substrate (see Sect. 4a), but there has been no suggestion that O_2 can also act as an activator until WILDNER and HENKEL (1979b) recently proposed that O_2 activation does indeed exist. In this case the role of O_2 appeared to be additional to that of CO_2, rather than a substitute for it.

A number of metabolites such as NADPH, SBP and 6-P-gluconate stimulate and/or inhibit RuBP carboxylase activity (LENDZIAN, 1978a; SALUJA and McFADDEN, 1978; TABITA and COLLETTI, 1979; VATER and SALNIKOW, 1979; WHITMAN et al., 1979), but apart from the possible advantage of NADPH activation, the metabolic significance of these effects is questionable (AKAZAWA, 1979). A regulatory role for RuBP also exists, but is no doubt complicated since this substrate can bind to an allosteric site as well as to the catalytic site (VATER and SALNIKOW, 1979), and in addition, in the absence of CO_2 and Mg^{2+}, it causes slow inactivation of the enzyme (LAING and CHRISTELLER, 1976). The concentration of RuBP binding sites in vivo probably exceeds the concentration of RuBP, at least in darkness (FARQUHAR, 1979; SICHER and JENSEN, 1979).

Despite instances in the past where claims were made that the subunit structures of certain microbial RuBP carboxylases deviated from the "classical" structure of the higher plant enzyme (eight large and eight small subunits), and where such claims were subsequently challenged [the most recent case relates to *Thiobacillus intermedius*; cf. PUROHIT et al. (1976) and BOWMAN and CHOLLET, 1980], evidence for unconventional structures of algal and bacterial enzymes continues to appear. Alongside the now well established simple two-subunit structure of *Rhodospirillum rubrum* RuBP carboxylase (AKAZAWA, 1979), LAWLIS et al. (1979) and ANDREWS and ABEL (1980) have independently proposed a structure of six large plus six small subunits for the carboxylases from *Pseudomonas oxalaticus* and *Synechococcus* (marine cyanobacterium) respectively. Other recently reported variants were an enzyme of eight large subunits but no small subunits in *Rhodopseudomonas sphaeroides* (GIBSON and TABITA, 1979), and another with two types of small subunits in *Thiocapsa roseopersicina* (PUROHIT et al., 1979). In addition to these structural disparities, accumulating differences in kinetic properties strengthen the view that there are fundamental differences between the RuBP carboxylases of prokaryotes and eukaryotes (ANDREWS and ABEL, 1980; BADGER, 1980; GORDON et al., 1980).

Investigations of the higher plant RuBP carboxylase have progressed to the point where the amino acids lysine, arginine and tyrosine have been identified at the catalytic site (PAECH and TOLBERT, 1978; CHOLLET, 1978a; CHOLLET and ANDERSON, 1978; ROBISON and TABITA, 1979). In quite another field, HUNER and MacDOWALL (1978, 1979a,b) have continued to examine the changes which occur in the enzmye in the leaves of Puma rye during cold hardening, and have shown that the enzyme's conformation is altered to a form with doubled specific activity and resistance to cold inactivation.

c) Other Enzymes of the Calvin Cycle

Of the remaining enzymes of the Calvin cycle, only GAP dehydrogenase and the phosphatases (FBPase and SBPase) have been given significant attention recently. Three laboratories report that the chloroplast GAP dehydrogenase is quite distinct from the two cytosolic GAP dehydrogenases (CERFF and CHAMBERS, 1979; MANN and MECKE, 1979; PUPILLO and FAGGIANI, 1979). The chloroplast enzyme has been purified from a range of plant leaves and commonly found to contain two slightly different subunits (A and B) and to exist in two forms: A_4 and A_2B_2. In the absence of $NADP^+$ these tetramers have a tendency to aggregate into a high molecular weight oligomer; a separate "binding protein" may be involved in this process (CERFF, 1978, 1979; CERFF and CHAMBERS, 1978, 1979; GOTO, 1979; PUPILLO and FAGGIANI, 1979). Reversible aggregation of *Scenedesmus obliquus* GAP dehydrogenase has also been observed; formation of tetramer was favoured by Pi and dithiothreitol, but this tetramer was unstable in the absence of NADPH (O'BRIEN et al., 1979; WOODROW et al., 1979).

Accumulating evidence for a regulatory role of chloroplast FBPase in CO_2 fixation (LATZKO and KELLY, 1979) has been supplemented by reports that the enzyme undergoes pH-dependent conformational changes (TAKABE et al., 1979) and that this FBPase, as well as SBPase, is susceptible to inhibition by Pi (CHARLES and HALLIWELL, 1980; FURBANK and LILLEY, 1980). Curiously, the prokaryote *Anacystis nidulans* contains an FBPase with properties intermediate between those of the higher plant chloroplast and cytoplasmic FBPases (BISHOP, 1979).

SBPase remains the enigma of the Calvin cycle. BREAZEALE et al. (1978) now believe that a separate, specific SBPase hydrolyzes SBP during CO_2 fixation, but neither these authors nor any others have yet reported an SBPase activity sufficient to accommodate observed rates of CO_2 fixation.

d) Light-Mediated Regulation

There is currently a proliferation of both mechanisms proposed to explain light-mediated regulation of enzyme activity, and enzymes proposed to be regulated. The most recent additions to the list of enzymes are cyanobacterial glycogen phosphorylase (LEHMANN and WÖBER, 1978) and the PEP carboxylases of C_4 and CAM plants (SAMEJIMA and MIYACHI, 1978; VON WILLERT and VON WILLERT, 1979). New proposals for light-mediated activation of RuBP carboxylase (DALEY et al., 1978; HELDT et al., 1978) are of doubtful significance to the in vivo situation (ROBINSON et al., 1979b). Light-mediated regulation of glucose-6-P dehydrogenase and of glycerate-3-P kinase are now considered to be consequences of the influence of altered pool sizes of intermediates on these enzymes (LENDZIAN, 1978b, 1980; ROBINSON and WALKER, 1979a).

Photosynthesis by intact chloroplasts is inhibited by external Mg^{2+}, and HUBER (1978, 1979a,b) has used barley chloroplasts to show that this Mg^{2+} prevents the light-mediated activation of GAP dehydrogenase, FBPase and ribulose-5-P kinase. At first it was thought that an increase in the concentration of Pi in the stroma mediated this effect, but more recent experiments (DEMMIG and GIMMLER, 1979; HUBER and MAURY, 1980) indicate that the culprit may be a decreased stromal pH which results from Mg^{2+} activation of a $(Na^+)K^+/H^+$ exchange across the chloroplast envelope.

There is abundant evidence that light-mediated activation is a reductive process. MISZALSKI and ZIEGLER (1979) found that both GAP dehydrogenase activity and the number of exposed thiol groups increased following application of SO_2. The inhibition of CO_2 fixation by H_2O_2 has now been attributed by KAISER (1979) to H_2O_2 interference with sulfhydryl groups involved in light-mediated enzyme activation.

Some workers argue that reductive activation involves transfer of photosynthetically generated electrons from ferredoxin to enzymes via thioredoxin. A ferredoxin-thioredoxin reductase has been characterized (DE LA TORRE et al., 1979), and a number of forms of thioredoxin, each relatively specific for the enzyme being regulated, have been reported (JACQUOT et al., 1978, 1979; WOLOSIUK et al., 1979; SUSKE et al., 1979). It is worth noting that not all forms were located in the chloroplast. Activations of GAP dehydrogenase and of ribulose-5-P kinase by the ferredoxin/thioredoxin system have been recorded, but rates were quite low, even after activation (BUCHANAN et al., 1978; WOLOSIUK and BUCHANAN, 1978a), and the authors themselves pointed out the difficulty in relating their observations to others in which much higher enzyme activities were detected without any apparent need for activation (WOLO-

SIUK and BUCHANAN, 1978b). Moderate improvements in rates obtained in these activation experiments have resulted from inclusion of β-mercaptoethanol and, in the case of FBPase, FBP (SCHÜRMANN and JACQUOT, 1979; WOLOSIUK et al., 1980). More recently it has been shown that the δ subunit of the chloroplast coupling factor can partially replace thioredoxin (McKINNEY et al., 1979), and in addition, a new activation system independent of both ferredoxin and thioredoxin has been described by LARA et al. (1980). It is another protein factor, and may be added to a list of protein factors [e.g. those associated with activation of bean and maize leaf NADP-linked malate dehydrogenase (VIDAL et al., 1978a,b)] involved in light-mediated regulation.

In contrast to the soluble systems described above, activation of enzymes by thylakoid membrane-bound components has been proposed (ANDERSON, 1979) and evidence for its presence in the CAM plant *Kalanchoë* provided (GUPTA and ANDERSON, 1978). At least two types of component exist, only one of which activates FBPase (ANDERSON et al., 1979). It is argued that activation involves thiol-disulfide exchange rather than any net reduction of disulfide (ANDERSON et al., 1978). Despite the rather meagre activations of Calvin cycle enzymes reported [average of 2.1-fold, highest value 7.7-fold (ANDERSON, 1979)], the system is interesting since the association of enzymes with membrane structures in vivo is more common than is perhaps generally recognized (THOMAS, 1976), and SCHEIBE and BECK (1979) have recently reported an activation of spinach chloroplast malate dehydrogenase that was quantitatively consistent with activation of only that portion (28%) of the enzyme found to be bound to the thylakoid membranes. Another enzyme found associated with membranes is ribulose-5-P kinase (FISCHER and LATZKO, 1979); this kinase joins chloroplast FBPase in having the distinction of undergoing "infinite" light-mediated activation in the sense that the activity recorded before illumination was zero.

The regulatory significance of light-mediated activation is self-evident for light/dark transitions, but remains to be evaluated for conditions of steady-state CO_2 fixation. It may simply reflect a device for ensuring that enzyme sulfhydryl groups are maintained in a reduced state (LEEGOOD and WALKER, 1980). If so, a correlation between enzyme oxidation (and consequent inactivation) and water-stress may exist, since more severe water stress results in a decrease of sulfhydryl-dependent activities in the plant (TOMATI and GALLI, 1979).

3. Products of CO_2 Fixation: A Consideration of Chloroplast Capability

Viewed through a standard light microscope, the average plant photosynthetic cell's most conspicuous attribute is its content of numerous green chloroplasts. However, despite their predominance, chloroplasts have long been considered to be almost exclusively occupied with the fixation and photosynthetic reduction of CO_2. Events of recent years are now demanding that this outlook be modified, since chloroplasts have been shown to be capable not only of achieving net CO_2 fixation, but also of reducing nitrogen from the level of nitrite to ammonium and of incorporating this ammonium into glutamate, of reducing sulfate and incorporating it into cysteine (SCHMIDT, 1979), of synthesizing a number of amino acids, and of synthesizing fatty acids (Fig. 1). Moreover, chloroplasts incapable of net CO_2 fixation exist: the mesophyll chloroplasts of C_4 plants lack RuBP carboxylase and ribulose-5-P kinase (RAY and BLACK, 1979), and OUTLAW et al. (1979) recently confirmed that the chloroplasts of stomatal guard cells lacked not only these two enzymes,

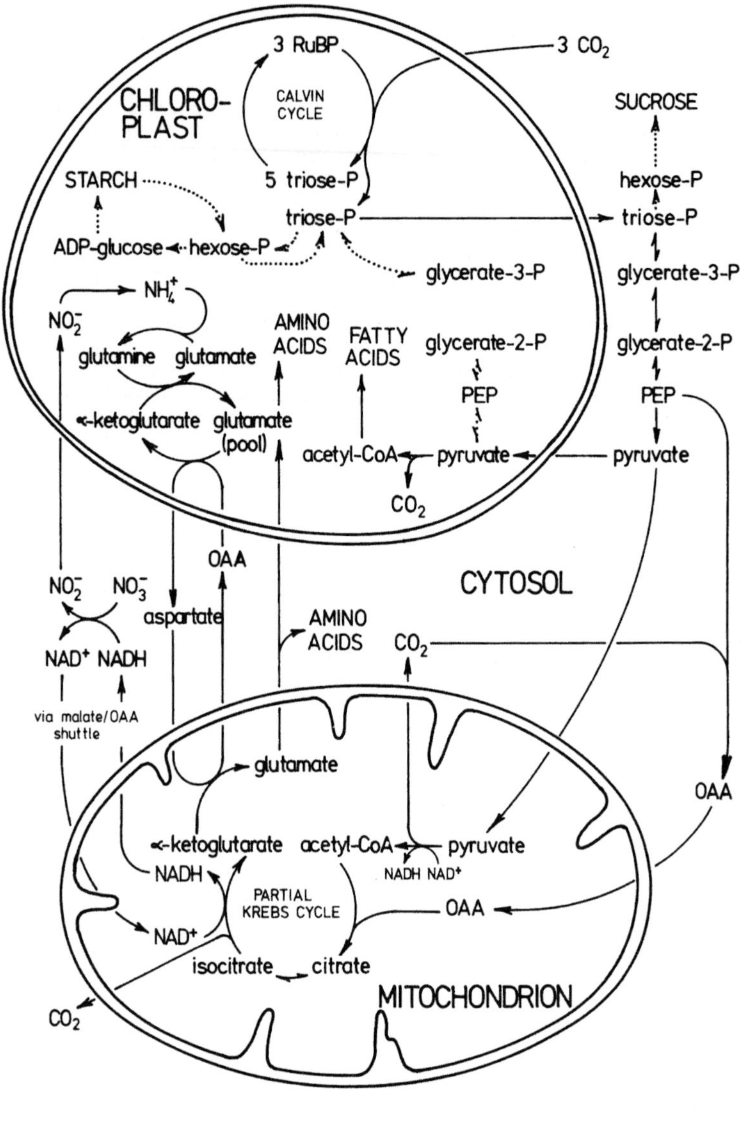

Fig. 1. Capability of the chloroplast to synthesize sugars, fatty acids, and amino acids. Dependence on the cytosol for pyruvate, and on the mitochondrion for the carbon skeleton of glutamate (α-ketoglutarate) is shown. Dotted lines show (chloroplast) starch metabolism, and what exists of the glycolytic sequence between triose-P and pyruvate, and (cytosol) sucrose synthesis. According to the scheme shown, the carbon economy of glutamate biosynthesis is: 6 $CO_2 + NH_4^+ \rightarrow$ glutamate + CO_2. Adapted from LEA and MIFLIN, 1974; STITT and apREES (1979), WOO and CANVIN (1980b), and WILLIAMS and RANDALL (1979)

but also (rather surprisingly) GAP dehydrogenase; this is the first report of a chloroplast lacking the enzyme which, in effect, reduces CO_2 to carbohydrate.

One of the greater methodological advances during the last decade has been the development of techniques for the acquisition of high-quality preparations of isolated chloroplasts and other organelles from enzymatically released leaf protoplasts (RATHNAM and EDWARDS, 1976). These methods have enabled EDWARDS et al. (1978) and LEEGOOD and WALKER (1979) to study wheat and sunflower chloroplasts; the sensitivity of these chloroplasts to factors in their environment, such as Pi concentration, serves to illustrate the sophistication of the control mechanisms utilized during CO_2 fixation. Much of the control is mediated by the phosphate translocator which catalyses the exchange across the chloroplast envelope between any two of Pi, triose-P, and glycerate-3-P (FLIEGE et al., 1978); a small membrane protein has been identified as a constituent of this translocator (FLÜGGE and HELDT, 1978). Inhibition of photosynthesis by linolenic acid has been attributed to interference by this fatty acid with the phosphate translocator (MVÉ AKAMBA and SIEGENTHALER, 1979a,b).

Two quantitatively important carbohydrate products of CO_2 fixation are sucrose and starch. ROBINSON and WALKER (1979b,c) have developed an elegant procedure for rapidly isolating chloroplast and cytoplasmic fractions from protoplasts and used it to confirm that sucrose is synthesized outside the chloroplast, probably in the cytosol. Enzymic cycling methods now permit estimation of the amount of sucrose in individual cells (JONES et al., 1977). Sucrose synthesis seems to be limited by the activity of sucrose-P synthetase which is relatively inactive during leaf expansion (GIAQUINTA, 1978; SILVIUS et al., 1978). Poor illumination also seems to hinder the development of activity (SILVIUS et al., 1979).

Further indications that the rate of starch deposition in chloroplasts during photosynthesis is regulated by the level of glycerate-3-P (which activates ADP-glucose formation) have been obtained (KAISER and BASSHAM, 1979a). Most recent research on chloroplast starch metabolism, however, has concentrated on the pathway of breakdown. Most workers favour glucose-1-P formation by starch phosphorylase as the initial step (LEVI and PREISS, 1978; STITT et al., 1978; STITT and apREES, 1980), although these and other reports (PONGRATZ and BECK, 1978; OKITA et al., 1979) indicate that spinach chloroplasts, but not pea chlorolasts, contain amylase activity. An unexpected observation was that much of the total cellular phosphorylase is located outside the chloroplasts of spinach and pea leaves (LEVI and PREISS, 1978; OKITA et al., 1979) in a form distinct from that of the chloroplast enzyme (STEUP and LATZKO, 1979; STEUP et al., 1979). The non-chloroplast phosphorylase has been purified to homogeneity (STEUP et al., 1980). The function of this non-chloroplast enzyme is unknown.

Subsequent metabolism of the hexose-P produced from starch could occur by glycolysis or the oxidative pentose-P pathway, but at present the relative contribution of each remains a subject of controversy (STITT and apREES, 1978, 1980; KAISER and BASSHAM, 1979b). From the point of view of the plant's carbon economy, the oxidative pentose-P pathway should be second choice since it loses, as CO_2, half the carbon previously fixed and incorporated into starch. Earlier and recent measurements of enzyme activities confirm that the chloroplast could use either pathway to produce triose-P and glycerate-3-P from hexose-P (HERBERT et al., 1979; STITT and apREES, 1979). However, it still seems unlikely that chloroplasts can form pyruvate glycolytically since they appear to lack P-glyceromutase activity (STITT and apREES, 1979).

Besides being used to form starch and sucrose, photosynthetically generated triose-P may also be used to synthesize amino acids (Fig. 1). It had been thought (PAUL et al., 1978) that NH_4^+ stimulated amino acid biosynthesis at the expense of sucrose synthesis, but it now seems more likely that NH_4^+ somehow stimulates PEP carboxylase which leads to increased levels of organic and amino acids without necessarily influencing the rate of sucrose formation (HAMMEL et al., 1979; WOO and CANVIN, 1980a). Further work on nitrogen assimilation has somewhat clarified other problems. It has been known for some time that NO_3^- reduction does not occur in darkness. SAWHNEY et al. (1978a,b) discovered that it does occur in darkness if the atmosphere is oxygen-free, and proposed that the inhibition of mitochondrial electron transport under this anaerobic atmosphere resulted in diversion of NADH from oxidation in electron transport to oxidation in other reactions, including NO_3^- reduction. Presumably, during illumination some product of photosynthesis (high ATP/ADP ratio?) has the same effect of inhibiting mitochondrial electron transport and making NADH free for NO_3^- reduction in the cytosol. This concept clashes somewhat with earlier ideas that NADH for NO_3^- reduction originates from chloroplasts via shuttle systems (MANN et al., 1978; RATHNAM, 1978), but has nevertheless received strong support by a Canadian team (CANVIN and WOO, 1979; WOO and CANVIN, 1980b; WOO et al., 1980) who have demonstrated NO_3^- reduction in a reconstituted system of mitochondria and supernatant from spinach leaves. At its simplest (Fig. 1) it can convert two molecules of PEP and one of NO_3^- into one molecule each of glutamate and CO_2; one of the two NADH formed during conversion of pyruvate plus OAA to α-ketoglutarate is used by the cytosolic nitrate reductase. It is noteworthy that only one-sixth of the carbon (originally fixed by the Calvin cycle) is lost as CO_2 during this formation of α-ketoglutarate.

A second problem now partly resolved relates to the question of whether or not the Krebs cycle operates during photosynthesis (GRAHAM and CHAPMAN, 1979). Some earlier conflicting results may now be explained by the conclusion that a portion of the cycle operates primarily to generate α-ketoglutarate for amino acid biosynthesis (KENT, 1979; WOO and CANVIN, 1980b), but that the NADH generated is oxidized externally. Mitochondrial production of α-ketoglutarate seems essential since chloroplasts lack citrate synthase, aconitase (ELIAS and GIVAN, 1977, 1978) and the ability to make some α-keto acids (LARSSON, 1979); α-ketoglutarate, together with pyruvate and OAA, is the predominant source of the carbon skeletons for almost all the non-aromatic amino acids (KELLY and LATZKO, 1980). The presence of aspartate:α-ketoglutarate transaminase in most cell compartments (HUANG et al., 1976) would allow NH_4^+, incorporated in the chloroplast, to be transported in the form of aspartate to the mitochondrion.

Reduction of NO_2^- and assimilation of NH_4^+ (Fig. 1) by chloroplasts was outlined in our most recent review. In C_4 plants this process, in contrast to net CO_2 fixation, is largely restricted to the mesophyll (NEYRA and HAGEMANN, 1978; MOORE and BLACK, 1979; VENKATARAMANA and DAS, 1979). In *Lemna minor* asparagine appeared to be closely involved in NH_4^+ assimilation (RHODES et al., 1980). The three enzymes involved in the conversion of NO_2^- to glutamate have been further investigated. Nitrite reductase, which has now been purified (IDA, 1977), appears to be induced by NO_3^- (ANDERSON and DONE, 1978). Two procaryotic glutamine synthetases have been examined but in only one case was control via adenylylation detected (JOHANSSON and GEST, 1977; STACEY et al., 1979). In leaves, the enzyme occurs not only in the chloroplast, but also as a separate form in the cytosol (MANN et al., 1979; WALLSGROVE et al., 1979). This second form would seem worthwile since glutamine does not easily cross the chloroplast envelope (BARBER and THURMAN, 1978). Leaf glutamate synthase has been purified and shown to possess both great specificity and high affinity for its substrates (WALLSGROVE et al., 1977; MATOH et al., 1979).

Besides glutamate and glutamine, there are some first indications that other amino acids, including aspartate, threonine and cysteine, can be synthesized by chloroplasts (BRYAN et al., 1977; NG and ANDERSON, 1978; LEA et al., 1979). However, it seems unlikey that chloroplasts synthesize much serine (LARSSON and ALBERTSSON, 1979). The extent to which chloroplasts are capable of synthesizing the other amino acids has been little investigated. It is worth recalling that the chloroplasts are capable of incorporating amino acids into protein, and despite their low allocation of genetic information, actually synthesize proteins in an amount quantitatively comparable to that of the cytosol (KUNG, 1977).

A major achievement during the past two years has been the clarification that all the fatty acids of leaves are synthesized in the chloroplast (Fig. 1). This conclusion, supported by earlier work (ROUGHAN et al., 1976; BOLTON and HARWOOD, 1977; MURPHY and LEECH, 1978), rests on the demonstration that all the acyl carrier protein of the spinach leaf cell is in the chloroplast (OHLROGGE et al., 1979). Chloroplasts contain other components of the fatty acid synthetase complex; recently the *Euglena* complex has been partially characterized (HENDREN and BLOCH, 1980). The acetyl-CoA required to synthesize fatty acids might be generated by acetyl-CoA synthetase (ROUGHAN et al., 1979) from acetate originating external to the chloroplast (presumably from the mitochondrion). However, ELIAS and GIVAN (1979) and WILLIAMS and RANDALL (1979) have both recently reported the presence of the pyruvate dehydrogenase complex in pea chloroplasts, hence it is likely that the acetyl-CoA is obtained from pyruvate in the chloroplast. This pyruvate must be supplied by cytosolic glycolysis (Fig. 1), if chloroplasts lack P-glyceromutase (STITT and apREES, 1979); pyruvate can be taken up by some chloroplasts (HUBER and EDWARDS, 1977). A point to note is that one third of the carbon fixed by the Calvin cycle is lost as CO_2 in the process of synthesizing fatty acids from triose-P (Fig. 1). Perhaps this is one reason for the predominance of PEP carboxylase (able to recapture this CO_2) in young expanding leaves (HEDLEY and ROWLAND, 1975; LAVERGNE et al., 1979; WILLIAMS and KENNEDY, 1978) which no doubt have higher rates of fatty acid synthesis.

4. Release of CO_2 by Photorespiration

a) The O_2-Uptake Reaction

Photorespiration, like dark respiration, involves uptake of O_2 and release of CO_2. However, the metabolic pathway involved is quite different to that of dark respiration (see Fig. 2 in our last review). The most widely studied O_2-uptake reaction is that in which O_2 replaces CO_2 as the substrate for RuBP carboxylase; the enzyme is then referred to as RuBP oxygenase, and the products are one molecule each of glycerate-3-P and glycolate-2-P (BERRY et al., 1978; HARRIS and STERN, 1978; PIERCE et al., 1980b). The affinity of the enzyme for CO_2 is far greater than that for O_2, but this difference is roughly balanced by the much higher concentration of O_2 compared to that of CO_2 in air. Consequently, under normal conditions, the chances of the enzyme utilizing O_2 or CO_2 are about equal. This is reflected by the fact that, in C_3 plants, the extent of photorespiration approaches that of photosynthesis (GERBAUD and ANDRÉ, 1979). It used to be thought that increasing the level of CO_2 would decrease photorespiration, but it is now believed that resultant alterations in pool sizes can counterbalance the effect of competition between CO_2 and O_2 for the enzyme, and in fact the actual rate of photorespiration is unaffected, although photosynthesis itself increases (OLIVER, 1979a; BRAVDO and CANVIN, 1979; FOCK et al., 1979). In addition, photorespiratory O_2-uptake mechanisms alternative to RuBP oxidation, and therefore not necessarily affected by the CO_2 concentration, are confidently believed to exist (BECK, 1979).

Despite an indication by McCASHIN and CANVIN (1979) that the possibility of reducing photorespiration without influencing photosynthesis is very low [indeed, earlier claims that compounds such as glycidate could reduce photorespiration have been brought into question (CHOLLET, 1978)], a number of workers have investigated the dual capability of RuBP carboxylase/oxygenase with this view in mind. Unequal variation of the two activities with leaf age was observed by diMARCO et al. (1979), but not by THOMAS et al. (1978). A distinct proclivity towards the carboxylase activity by the enzyme in tetraploid ryegrass (*Lolium perenne*), as compared to the enzyme in the diploid cultivar was detected by GARRETT (1978) and upheld by RATHNAM and CHOLLET (1980). Agreement was also reached by BHAGWAT et al. (1978) and OKABE et al. (1979) who found that preincubation of enzyme with hydroxylamine greatly increased the ratio of carboxylase to oxygenase activities. Several studies indicate that the oxygenase activity is favoured when the Mg^{2+} cofactor is replaced by Mn^{2+} or Co^{2+} (WILDNER and HENKEL, 1978, 1979a; CHRISTELLER and LAING, 1979; ROBISON et al., 1979).

A molecular spanner was thrown into the workings of RuBP carboxylase/oxygenase by BRÄNDÉN (1978) who reported that the carboxylase and oxygenase activities were in fact catalysed by two separate enzymes which the author claimed to have partially separated from parsley leaves. McCURRY et al. (1978) hastily purchased some parsley from the local grocery store and attempted to confirm the separation, but could not. Undaunted, BRÄNDÉN et al. (1980) have quite recently proposed an alternative: that some of the glycerate-3-P produced by RuBP carboxylase is in the L-enantiomeric form, and that a second enzyme (which normally co-purifies with the carboxylase) catalyses the oxidative decarboxylation of this L-glycerate-3-P to (probably) glycolate-P. Further consideration of this new concept will require additional evidence.

In closing this section, it should be noted that there is increasing evidence that O_2 inhibits photosynthesis by mechanisms other than RuBP oxidation (GLACOLEVA et al., 1978; DOEHLERT et al., 1979). Two possible mechanisms are inhibition by H_2O_2 (a product of photosynthetic O_2 reduction) and possibly direct O_2 inhibition of other Calvin cycle enzymes (KAISER, 1979; ROBINSON et al., 1980).

b) CO_2-Release Reactions

Glycolate-P produced by the RuBP oxygenase reaction is hydrolysed by a phosphatase (CHRISTELLER and TOLBERT, 1978; SOMERVILLE and OGREN, 1979) to produce glycolate which moves to peroxisomes where it is oxidized to glyoxylate in both higher plants and algae; in algae glycolate can also be oxidized by mitochondria, but this process does not seem to be related to photorespiration (YOKOTA et al., 1978). Cyanobacteria utilize a thylakoid-membrane located enzyme to oxidize glycolate (CODD and SALLAL, 1978).

Glyoxylate is generally considered to be converted to glycine which is the main substrate for the decarboxylation reaction of photorespiration, but several observations infer that this is not the only destiny of glyoxylate (HALLDAL and HOLMEN, 1979; NILSEN et al., 1979). One long-recognized alternative is oxidation to oxalate (ZINDLER-FRANK, 1976; KPODAR et al., 1978) which may represent a carbon reserve during winter dormancy for some plants (CALMÉS and PIQUEMAL, 1977). A second alternative is decarboxylation to formate; under some conditions this reaction may contribute somewhat to the total photorespiratory CO_2 production (GRODZINSKI, 1978; OLIVER, 1979b). The formate is utilized by formyl-tetrahydrofolate synthetase (GRODZINSKI, 1979). Yet another possible fate for glyoxylate is inherent in the suggestion by NAIK and SINGH (1980) that it combines with succinate to form isocitrate which might then be converted to α-ketoglutarate in a reaction producing CO_2. However, the reported activity of isocitrate lyase in leaves (HUNT and FLETCHER, 1977) would not permit this pathway to be quantitatively significant.

Amination of glyoxylate produces glycine which can move to the mitochondrion, and the current consensus is that most photorespiratory CO_2 originates from glycine decarboxylation in this organelle (TOLBERT, 1979; ARRON et al., 1979). The other carbon of the glycine is transferred by serine hydroxymethyltransferase, an enzyme located behind the inner membrane of the mitochondrion (WOO, 1979), to a second molecule of glycine to form serine. A molecule of NH_3 is released and one NAD^+ reduced: the former is believed to be assimilated by the cytosolic glutamine synthetase (KEYS et al., 1978) while the NADH could be shuttled into the peroxisome (WOO et al., 1980) where it is required to reduce the hydroxypyruvate formed when the serine is further metabolized (see Fig. 2 in our previous review). The first report of a compound (glycine hydroxamate) able to reversibly inhibit the conversion of glycine to serine plus CO_2 has been supplied by LAWYER and ZELITCH (1979). An impressive observation is that the ability to oxidize glycine is restricted to the mitochondria of chloroplast-containing cells; those in the leaf stalks and veins lack the ability (GARDESTRÖM et al., 1980).

c) Why Do Plants Photorespire?

The purpose of photorespiration remains imperspicuous. Perhaps the most realistic role suggested so far is that of dissipating excess photosynthetically generated energy, and experiments with attached leaves, isolated cells and isolated chloroplasts have demonstrated that the photosynthetic apparatus is damaged following illumination in the absence of CO_2 and at low partial pressures of oxygen (KRAUSE et al., 1978; POWLES and OSMOND, 1978; POWLES et al., 1979). Damage was largely prevented by increasing the oxygen level (in the absence of CO_2), presumably because the oxygen permitted internal CO_2 production by photorespiration (POWLES et al., 1979). Mosses, which normally select shady situations, show evidence of damage by excessive illumination even in normal air (ARO and VALANNE, 1979). The mechanism of damage in these cases is not known. Involvement of the superoxide ion (formed when O_2 acts as an electron acceptor) is not ruled out, although chloroplasts contain superoxide dismutase and hence can quickly convert this ion to H_2O_2 (ELSTNER, 1979), and cells contain ascorbate peroxidases capable of disposing of the H_2O_2 (GRODEN and BECK, 1979; KELLY and LATZKO, 1979). Under normal conditions chloroplasts do not produce toxic levels of H_2O_2 (ALLEN, 1978). Nevertheless, under conditions of stress involving a transitory interruption in the supply of CO_2, photorespiration might be safer than oxygen reduction as a means of dealing with the energy supplied by photosynthetic electron transport.

One problem in accepting this role of protection against photooxidative damage is that the mesophyll cells of C_4 plants lack photorespiration, and yet still seem to avoid being photooxidatively damaged (ZELITCH, 1979) [a similar situation for algal cells now seems unlikely since these cells do photorespire (FINDENEGG and FISCHER, 1978; BIRMINGHAM and COLMAN, 1979)]. A second more philosophical question relates to the continual operation of photorespiration under normal conditions (air levels of CO_2 and O_2). It is difficult to accept that here, too, it is dissipating energy: DELANEY and WALKER (1978) have shown that the sunflower has a capacity for photosynthetic electron transport more than 50% greater than five other C_3 plants (including Good King Henry), yet it is obviously able to deal with much of this "extra" energy by utilizing it usefully, as evidenced by its superior photosynthetic performance.

Another role suggested for photorespiration is that of draining fixed carbon toward sucrose synthesis in the cytosol (CHAMPIGNY and MOYSE). Certainly carbon flowing through the glycolate pool can enter sucrose via serine (KUMARASINGHE et al., 1977), but two difficulties arise with any proposal that this pathway could especially benefit sucrose synthesis. Firstly, HEBER et al. (1974) have reported that glycerate kinase, which is on the pathway between glycolate and sucrose, is mostly in the chloroplast, hence carbon drained into photorespiration must presumably re-enter the chloroplast before being used for sucrose synthesis (Fig. 2). FRY and BIDWELL (1977) have obtained data consistent with this scheme. Secondly, even assuming that adequate glycerate kinase activity exists in the cytosol, any suggestion that the pathway from glycolate to sucrose may bypass the chloroplast (KUMARASINGHE et al., 1977; CHAMPIGNY and MOYSE, 1979) is quite disturbing, because, as far as is known, there would be nothing to prevent all the glycolate being converted to CO_2 and sucrose. This would be disastrous. In Figure 2 (for example), conversion of all glycolate to sucrose and CO_2 would not permit regeneration of the 33 RuBP; even in the absence of sucrose synthesis from Calvin cycle products, there would be a shortfall of 1.8 molecules of the CO_2-acceptor. The situation would be worse with greater photorespiration [which is usual (ZELITCH, 1979)] and with additional sucrose being synthesized directly from Calvin cycle products (CHAMPIGNY and MOYSE, 1979).

In fact, one is tempted to believe that a plant with glycerate kinase in the leaf cytosol would fail to survive. Clearly, a prime objective of the chloroplast is to maintain its RuBP pool, and to do this it must be forbidden to indiscriminately use glycolate, which freely diffuses out of the chloroplast (HELDT, 1976). This is most probably achieved by returning a product of glycolate metabolism [probably glycerate (TOLBERT, 1979)] to the chloroplast to be further metabolized into the triose-P pool. As has been so often emphasized, the export of triose-P for cytosolic sucrose synthesis is elegantly and exactly regulated by specifically transporting it, via the phosphate translocator, in exchange for a Pi (WALKER and ROBINSON, 1978; see Sect. 3).

A typical C_3 plant cell loses 30-50% of its fixed carbon as CO_2 evolved from photorespiration, while most of the remainder is diverted into starch or sucrose. The type of carbon flow envisaged for this "standard" situation could be that of Figure 2, where sucrose is produced concomitant with a 37% loss of fixed carbon as photorespiratory CO_2. The approximately equal utilization of O_2 and CO_2 by RuBP carboxylase/oxygenase, as evidenced by the measurements of GERBAUD and ANDRÉ (1979), is predicted by this scheme. Why the cell chooses this approach to synthesize one molecule of sucrose, rather than carboxylating twelve mole-

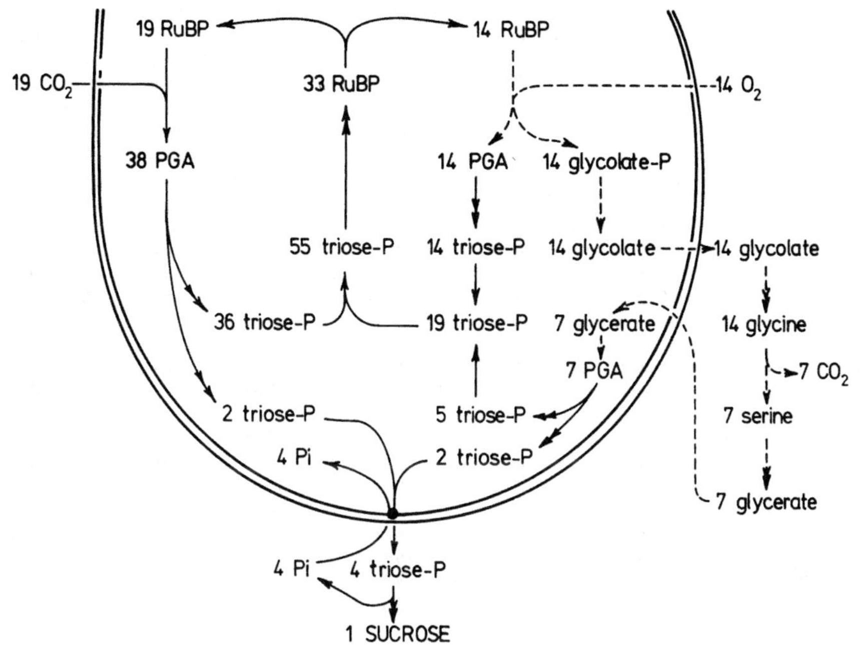

Fig. 2. Simultaneous operation of photosynthesis and photorespiration. The pathways are balanced so that, of the carbon fixed, 63% is used for the net synthesis of one molecule of sucrose, while 37% is lost again in photorespiration. Double-headed arrows indicate the involvement of two or more reactions. ●, phosphate translocator (see Sect. 1e). Abbreviation: PGA, glycerate-3-P. Adapted from KUMARASINGHE et al. (1977), and CHAMPIGNY and MOYSE (1979). ————, Calvin cycle and sucrose synthesis; - - - - -, glycolate pathway

cules of RuBP and dispensing altogether with photorespiration, remains an open question.

5. Capture and Recapture of CO_2 by PEP Carboxylase

a) C_4 Metabolism

Plant PEP carboxylase performs a dual function of capturing CO_2 entering from the air, and of recapturing respired CO_2. The importance of both these functions to C_4 and CAM plants is well established (see below). However, it is becoming clear that other plants rely considerably on this enzyme. A comparison of the $\delta^{13}C$ values of the products of RuBP carboxylase and of the constituents of the classical C_3 plants wheat and oat confirms that, even in these plants, carboxylation of PEP is more than a relatively minor process (ESTEP et al., 1978; WONG et al., 1979; WINKLER et al., 1978). It is a major component of the mechanism that controls stomatal aperture (RASCHKE and DITTRICH, 1979; THORPE et al., 1978; OUTLAW et al., 1979), and it probably recaptures some of the CO_2 respired by developing fruits (DAVIES, 1979). It is also claimed to be an important factor in capturing CO_2 in large submersed aquatic plants such as *Hydrilla verticillata* (HOLADAY and BOWES, 1980); these plants, although being C_3 plants possessing photorespiration, exhibit low CO_2 compensation points and high resistances to CO_2 diffusion (LLOYD et al., 1977; WINTER, 1978; ANDREWS and ABEL, 1979; BENEDICT et al., 1980).

b) C_4 Pathway

Our previous reviews contain descriptions of the pathway by which a group of plants, called C_4 plants, capture all CO_2 intended for photosynthesis by means of PEP carboxylase in mesophyll cells and then move the C_4 product into bundle-sheath cells where it is decarboxylated to produce CO_2 for the Calvin cycle. The mesophyll cells surround the bundle-sheath cells, but lack RuBP carboxylase/oxygenase and hence neither photosynthesize nor photorespire. Recently variants of this cell arrangement have been described for *Suaeda monoica* and *Arundinella hirta* (SHOMER-ILAN et al., 1979; REGER and YATES, 1979).

The level of PEP carboxylase in C_4 plant leaves is even higher than that previously reported (HATCH and OLIVER, 1978) and hence the enzyme might confidently be expected to capture all CO_2 initially entering the leaf. RAGHAVENDRA and DAS (1977) have data to support this concept, but CREACH (1979a,b) has reinvestigated light-enhanced dark CO_2 fixation by tomato and maize leaves and concluded that RuBP may, after all, be a primary acceptor of CO_2 in C_4 plant leaves under normal conditions. Earlier evidence to the contrary was attributed to an O_2-dependent loss of RuBP upon darkening; this loss was prevented by carrying out experiments at a low level of O_2.

A number of recent studies (MBAKU et al., 1978; HATCH, 1979a; KU et al., 1979) have served to establish the validity of the C_4 pathway as it is currently formulated (RAY and BLACK, 1979). Intact and functional chloroplasts can now be obtained and separated from both mesophyll and bundle-sheath cells (HORVÁTH et al., 1978; EDWARDS et al., 1979). Additional information on participating enzymes has been obtained. PEP carboxylase is susceptible to pyrophosphate inhibition (RAGHAVENDRA and VALLEJOS, 1980) and appears to be rapidly inactivated in darkness (SAMEJIMA and MIYACHI, 1978). Maize chloroplast NADP-linked malic enzyme has been purified (ASAMI et al., 1979) and a lower activity of a cytosolic form of the enzyme identified (PUPILLO and BOSSI, 1979). The cold-induced inactivation of pyruvate Pi dikinase has been shown to involve enzyme dissociation in the absence of substrates (SHIRAHASHI et al., 1978; HATCH, 1979b). The susceptibility to cold varies from species to species (SUGIYAMA et al., 1979); in *Spartina townsendii* the enzyme is apparently quite cold-resistant (LONG and WOOLHOUSE, 1978).

The benefit conveyed to C_4 plants by their possession of the C_4 pathway is primarily derived from an increase in the concentration of CO_2 in the vicinity of the bundle-sheath RuBP carboxylase. This increase could theoretically accelerate RuBP carboxylation, not only directly (i.e. it represents an increase in the enzyme's substrate concentration), but also indirectly in that CO_2 at this higher concentration would compete more effectively with O_2 for the active site and photorespiration would be reduced (RAY and BLACK, 1979; CANVIN, 1979; KU and EDWARDS, 1980). Bundle-sheath cells clearly have the capacity for photorespiration, but evolution of photorespiratory CO_2 is never detected (CANVIN, 1979; ZELITCH, 1979; ZSCHOCHE and TING, 1977); it is not clear whether this reflects complete suppression of photorespiration by the increased CO_2 level, or whether some photorespiratory CO_2 is evolved, only to be recaptured by PEP carboxylase in the mesophyll cells (CANVIN, 1979; RATHNAM, 1979b). The latter alternative is conceivable in view of the reports that a number of so-called C_3-C_4 intermediate plants belonging to the genera *Panicum* and *Mollugo* (RAGHAVENDRA et al., 1978; MORGAN and BROWN, 1979; SAYRE et al., 1979) photorespire much less than normal C_3 plants although, like C_3 plants, they possess RuBP carboxylase in both mesophyll and bundle-sheath cells (RATHNAM and CHOLLET, 1979a; SAYRE et al., 1979). This reduced photorespiration has been attributed to the efficient recapture of respired CO_2 (RATHNAM and CHOLLET, 1979b; BROWN, 1980).

c) Crassulacean Acid Metabolism (CAM)

Succulents such as the cactus are curious plants in that, unlike others, they capture the bulk of their CO_2 during the night (cf. OSMOND, 1978). This nocturnal CO_2 fixation is catalysed by PEP carboxylase. The product is reserved as malate in the vacuole until the next day when it is decarboxylated behind closed stomates; the resultant CO_2 is then incorporated by the Calvin cycle. RuBP carboxylase is thus relegated to the role of recapturing the CO_2 (not unlike its role in C_4 plants).

The development of CAM is associated with the appearance of a second form of PEP carboxylase in the leaves (GREENWAY et al., 1978; BRULFERT et al., 1979; KNOPF and KLUGE, 1979). Measurements made with *Aridaria* sp. and *Prenia sladeniana* in their natural habitat showed the total enzyme activity (measured at pH 7) was greater at night, and less susceptible to feedback inhibition by malate (VON WILLERT et al., 1979). QUEIROZ (1979; see also MOREL and QUEIROZ, 1978) has given considerable attention to such oscillations of several enzymes and concluded that an endogenous circadian rhythmicity is involved. In the case of one enzyme, the presence of such a rhythm has been disputed (USUDA, 1980).

New methods for isolating CAM plant cells, protoplasts and organelles (GERWICK et al., 1978; HERBERT et al., 1978; SPALDING et al., 1980; BUSER and MATILE, 1977) have enabled investigations that have provided additional support to the present concept of carbon flow during CAM: nocturnal accumulation of malate by vacuoles has been demonstrated (KENYON et al., 1978), the importance of the mitochondrial NAD-linked malic enzyme (one of the three possible decarboxylating enzymes) has been evaluated (DAY, 1980; SPALDING et al., 1980), and the intracellular location of several enzymes examined. In this last case, the location of PEP carboxylase remains unresolved: SPALDING et al. (1979) reported this to be a cytosolic enzyme (as in C_4 plants) in *Sedum praealtum*, but SCHNARRENBERGER et al. (1980) found the enzyme associated with chloroplasts of *Bryophyllum calycinum* and *Crassula lycopodioides*.

Nocturnal CO_2 capture by CAM plants must require a steady supply of PEP. This PEP originates via glycolysis from chloroplast starch; amylases, phosphorylase (SCHILLING and DITTRICH, 1979) and other necessary enzymes of glycolysis (DITTRICH, 1979; PIERRE and QUEIROZ, 1979) have been detected in CAM plants. An impressive observation is that pineapple leaves contain a pyrophosphate-linked phosphofructokinase with activity more than 10 times that of the conventional ATP-linked enzyme (CARNAL and BLACK, 1979). Daytime release of CO_2 from malate also releases (in the form of pyruvate or PEP) the three carbons of the originally-carboxylated PEP. This three-carbon compound is understood to re-enter the starch pool via gluconeogenesis (OSMOND, 1978; KLUGE, 1979), although ANDRÉ et al. (1979) believe that much of it (e.g. more than half in *Sempervivum grandifolium*) may be oxidized in the mitochondria.

Continuous operation of CAM is not obligatory for growth of CAM plants, hence it is not surprising that the contribution of CAM to the total CO_2 fixation varies. CAM is less obvious in the late afternoon

(OSMOND, 1978). It predominates in older leaves (NISHIDA, 1978). In fact, CAM plants form a spectrum, ranging from those that possess so little CAM (e.g. *Portulaca oleracea*) that no net nocturnal CO_2 uptake is achieved, through those (e.g. *Portulacaria afra*) that display CAM only when water-stressed, to the cacti (e.g. *Zygocactus truncatus*) which more-or-less continuously operate with CAM (HANSCOM and TING, 1978; KOCH and KENNEDY, 1980). One advantage of CAM to the cactus is well known: it permits stomates to be closed during the day, thus greatly reducing transpiration which, in turn, confers on the plant the ability to survive in areas of low moisture availability. A remarkable observation by COCKBURN et al. (1979) provides evidence of a second advantage. These authors found that decarboxylation behind closed stomates leads to high internal CO_2 levels, sometimes exceeding 2% (i.e. approaching 100 times that of air). SPALDING et al. (1979) responded to this observation and, after establishing that stomate closure only slightly affected the internal O_2 level, concluded that CAM, by developing a high CO_2/O_2 ratio, substantially reduces photorespiration [although the capacity for photorespiration exists (HERBERT et al., 1978; KLUGE, 1979)].

It is ironic that this most effective means of suppressing photorespiration is apparently an inadvertant consequence of the structural and metabolic alterations evolved by these succulent plants for the purpose of growing in arid environments. Or is it? Could it be that the ancestors of succulents evolved such alterations primarily to conquer photorespiration, and that the ability to survive in dry regions was an unintentional bonus?

References

AKAZAWA, T.: Ribulose-1,5-bisphosphate carboxylase, 208-229. In: Encyclopedia of Plant Physiology, N.S., Vol. 6, eds. M. GIBBS, E. LATZKO. Berlin-Heidelberg-New York: Springer 1979. - ALLEN, J.F.: Plant Sci. Lett. 12, 151-159 (1978). - ANDERSON, J.W., DONE, J.: Plant Physiol. 61, 692-697 (1978). - ANDERSON, L.E.: Interaction between photochemistry and activity of enzymes, 271-281. In: Encyclopedia of Plant Physiology, N.S., Vol. 6, eds. M. GIBBS, E. LATZKO. Berlin-Heidelberg-New York: Springer 1979. - ANDERSON, L.E., NEHRLICH, S.C., CHAMPIGNY, M.-L.: Plant Physiol. 61, 601-605 (1978). - ANDERSON, L.E., CHIN, H.-M., GUPTA, V.K.: Plant Physiol. 64, 491-494 (1979). - ANDRÉ, M., THOMAS, D.A., VON WILLERT, D.J., GERBAUD, A.: Planta 147, 141-144 (1979). - ANDREWS, T.J., ABEL, K.M.: Plant Physiol. 63, 650-656 (1979); - Proc. Australian Biochem. Soc. 13, 133 (1980). - ARO, E.-M., VALANNE, N.: Physiol. Plant. 45, 460-466 (1979). - ARRON, G.P., SPALDING, M.H., EDWARDS, G.E.: Biochem. J. 184, 457-460 (1979). - ASAMI, S., INOUE, K., MATSUMOTO, K., MURACHI, A., AKAZAWA, T.: Arch. Biochem. Biophys. 194, 503-510 (1979).

BADGER, M.R.: Arch. Biochem. Biophys. 201, 247-254 (1980). - BARBER, D.J., THURMAN, D.A.: Plant Cell Environ. 1, 297-303 (1978). - BECK, E.: Glycolic acid and photorespiration, 327-337. In: Encyclopedia of Plant Physiology, N.S., Vol. 6, eds. M. GIBBS, E. LATZKO. Berlin-Heidelberg-New York: Springer 1979. - BENEDICT, C.R., WONG, W.W.L., WONG, J.H.H.: Plant Physiol. 65, 512-517 (1980). - BERRY, J.A., OSMOND, C.B., LORIMER, G.H.: Plant Physiol. 62, 954-967 (1978). - BHAGWAT, A.S., RAMAKRISHNA, J., SANE, P.V.: Biochem. Biophys. Res. Commun. 83, 954-962 (1978). - BIRMINGHAM, B.C., COLMAN, B.: Plant Physiol. 64, 892-895 (1979). - BISHOP, R.H.: Arch. Biochem. Biophys. 196, 295-300 (1979). - BOLTON, P., HARWOOD, J.L.: Biochem. J. 168, 261-269 (1977). - BOWMAN, L.H., CHOLLET, R.: J. Bacteriol. 141, 652-657 (1980). - BRÄNDÉN, R.: Biochem.

Biophys. Res. Commun. 81, 539-546 (1978). - BRÄNDÉN, R., NILSSON, T., STYRING, S., ÅNGSTRÖM, J.: Biochem. Biophys. Res. Commun. 92, 1306-1312 (1980). - BRAVDO, B.-A., CANVIN, D.: Plant Physiol. 63, 399-401 (1979). - BREAZEALE, V.D., BUCHANAN, B.B., WOLOSIUK, R.A.: Z. Naturforsch. 33c, 521-528 (1978). - BROWN, R.H.: Plant Physiol. 65, 346-349 (1980). - BRULFERT, J., ARRABAÇA, M.C., GUERRIER, D., QUEIROZ, O.: Planta 146, 129-133 (1979). - BRYAN, J.K., LISSIK, E.A., MATTHEWS, B.F.: Plant Physiol. 59, 673-679 (1977). - BUCHANAN, B.B., CRAWFORD, N.A., WOLOSIUK, R.A.: Plant Sci. Lett. 12, 257-264 (1978). - BUSER, CH., MATILE, PH.: Z. Pflanzenphysiol. 82, 462-466 (1977).

CALMÉS, J., PIQUEMAL, M.: Can. J. Bot. 55, 2075-2078 (1977). - CANVIN, D.T.: Photorespiration: Comparison between C_3 and C_4 Plants, 368-396. In: Encyclopedia of Plant Physiology, N.S., Vol. 6, eds. M. GIBBS, E. LATZKO. Berlin-Heidelberg-New York: Springer 1979. - CANVIN, D.T., WOO, K.C.: Can. J. Bot. 57, 1155-1160 (1979). - CARNAL, N.W., BLACK, C.C.: Biochem. Biophys. Res. Commun. 86, 20-26 (1979). - CERFF, R.: Plant Physiol. 61, 369-372 (1978); - Europ. J. Biochem. 94, 243-247 (1979). - CERFF, R., CHAMBERS, S.E.: Hoppe-Seyler's Z. Physiol. Chem. 359, 769-772 (1978); - J. Biol. Chem. 254, 6094-6098 (1979). - CHAMPIGNY, M.-L., MOYSE, A.: Plant Cell Physiol. 20, 1167-1178 (1979). - CHARLES, S.A., HALLIWELL, B.: Biochem. J. 185, 689-693 (1980). - CHOLLET, R.: Biochem. Biophys. Res. Commun. 83, 1267-1274 (1978a); - Plant Physiol. 61, 929-932 (1978b). - CHOLLET, R., ANDERSON, L.L.: Biochim. Biophys. Acta 525, 455-467 (1978). - CHRISTELLER, J.T., LAING, W.A.: Biochem. J. 183, 747-750 (1979). - CHRISTELLER, J.T., TOLBERT, N.E.: J. Biol. Chem. 253, 1780-1785 (1978). - COCKBURN, W., TING, I.P., STERNBERG, L.O.: Plant Physiol. 63, 1029-1032 (1979). - CODD, G.A., SALLAL, A.-K.J.: Planta 139, 177-181 (1978). - CREACH, E.: Plant Physiol. 63, 788-791 (1979a); ibid. 64, 435-438 (1979b).

DALEY, L.S., DAILEY, F., CRIDDLE, R.S.: Plant Physiol. 62, 718-722 (1978). - DAVIES, D.D.: Annu. Rev. Plant Physiol. 30, 131-158 (1979). - DAY, D.A.: Plant Physiol. 65, 675-679 (1980). - DELANEY, M.E., WALKER, D.A.: Biochem. J. 171, 477-482 (1978). - DEMMIG, B., GIMMLER, H.: Z. Naturforsch. 34c, 233-241 (1979). - diMARCO, G., GREGO, S., TRICOLI, D.: J. Exp. Bot. 30, 851-861 (1979). - DITTRICH, P.: Enzymes of crassulacean acid metabolism, 263-270. In: Encyclopedia of Plant Physiology, N.S., Vol. 6, eds. M. GIBBS, E. LATZKO. Berlin-Heidelberg-New York: Springer 1979. - DOEHLERT, D.C., KU, M.S.B., EDWARDS, G.E.: Physiol. Plant. 46, 299-306 (1979).

EDWARDS, G.E., ROBINSON, S.P., TYLER, N.J.C., WALKER, D.A.: Arch. Biochem. Biophys. 190, 421-433 (1978). - EDWARDS, G.E., LILLEY, R.McC., CRAIG, S., HATCH, M.D.: Plant Physiol. 63, 821-827 (1979). - ELIAS, B.A., GIVAN, C.V.: Plant Physiol. 59, 738-740 (1977); - Planta 142, 317-320 (1978); - Plant Sci. Lett. 17, 115-122 (1979). - ELSTNER, E.F.: Oxygen activation and superoxide dismutase in chloroplasts, 410-415. In: Encyclopedia of Plant Physiology, N.S., Vol. 6, eds. M. GIBBS, E. LATZKO. Berlin-Heidelberg-New York: Springer 1979. - ESTEP, M.F., TABITA, F.R., PARKER, P.L., BAALEN, C.V.: Plant Physiol. 61, 680-687 (1978).

FARQUHAR, G.D.: Arch. Biochem. Biophys. 193, 456-468 (1979). - FINDENEGG, G.R.: Z. Pflanzenphysiol. 79, 428-437 (1976); - Plant Sci. Lett. 17, 101-108 (1979). - FINDENEGG, G.R., FISCHER, K.: Z. Pflanzenphysiol. 89, 363-371 (1978). - FISCHER, K.H., LATZKO, E.: Biochem. Biophys. Res. Commun. 89, 300-306 (1979). - FLIEGE, R., FLÜGGE, U.-I., WERDAN, K., HELDT, H.W.: Biochim. Biophys. Acta 502, 232-247 (1978). - FLÜGGE, U.-I., HELDT, H.W.: Biochem. Biophys. Res. Commun. 84, 37-44 (1978). - FOCK, H., KLUG, K., CANVIN, D.T.: Planta 145, 219-223 (1979). - FRY,

S.C., BIDWELL, R.G.S.: Can. J. Bot. 55, 1457-1464 (1977). - FURBANK, R.T., LILLEY, R.McC.: Proc. Australian Biochem. Soc. 13, 68 (1980).

GARDESTRÖM, P., BERGMAN, A., ERICSON, I.: Plant Physiol. 65, 389-391 (1980). - GARRETT, M.K.: Nature 274, 913-915 (1978). - GERBAUD, A., ANDRÉ, M.: Plant Physiol. 64, 735-738 (1979). - GERWICK, B.C., WILLIAMS, G.J., SPALDING, M.H., EDWARDS, G.E.: Plant Sci. Lett. 13, 389-396 (1978). - GIAQUINTA, R.: Plant Physiol. 61, 380-385 (1978). - GIBSON, J.L., TABITA, F.R.: J. Bacteriol. 140, 1023-1027 (1979). - GLACOLEVA, T.A., ZALENSKY, O.V., MOKRONOSOV, A.T.: Plant Physiol. 62, 204-209 (1978). - GORDON, G.L.R., LAWLIS, V.B., McFADDEN, B.A.: Arch. Biochem. Biophys. 199, 400-412 (1980). - GOTO, K.: Plant Cell Physiol. 20, 523-532 (1979). - GRAHAM, D., CHAPMAN, E.A.: Interactions between photosynthesis and respiration in higher plants, 150-162. In: Encyclopedia of Plant Physiology, N.S., Vol. 6, eds. M. GIBBS, E. LATZKO. Berlin-Heidelberg-New York: Springer 1979. - GREBANIER, A.E., CHAMPAGNE, D., ROY, H.: Biochemistry 17, 5150-5155 (1978). - GREENWAY, H., WINTER, K., LÜTTGE, U.: J. Exp. Bot. 29, 547-559 (1978). - GRODEN, D., BECK, E.: Biochim. Biophys. Acta 546, 426-435 (1979). - GRODZINSKI, B.: Planta 144, 31-37 (1978); - Plant Physiol. 63, 289-293 (1979). - GUPTA, V.K., ANDERSON, L.E.: Plant Physiol. 61, 469-471 (1978).

HALLDAL, P., HOLMEN, A.T.: Plant Cell Physiol. 20, 757-763 (1979). - HAMMEL, K.E., CORNWELL, K.L., BASSHAM, J.A.: Plant Cell Physiol. 20, 1523-1529 (1979). - HANSCOM, Z., TING, I.P.: Plant Physiol. 61, 327-330 (1978). - HARRIS, G.C., STERN, A.I.: J. Exp. Bot. 29, 561-566 (1978). - HATCH, M.D.: Arch. Biochem. Biophys. 194, 117-127 (1979a); - Australian J. Plant Physiol. 6, 607-619 (1979b). - HATCH, M.D., OLIVER, I.R.: Australian J. Plant Physiol. 5, 571-580 (1978). - HEBER, U., KIRK, M.R., GIMMLER, H., SCHÄFER, G.: Planta 120, 31-46 (1974). - HEDLEY, C.L., ROWLAND, A.O.: Plant Sci. Lett. 5, 119-126 (1975). - HELDT, H.W.: 215-234. In: The Intact Chloroplast, ed. J. BARBER. Amsterdam: Elsevier/North-Holland 1976. - HELDT, H.W., CHON, C.J., LORIMER, G.H.: FEBS Lett. 92, 234-240 (1978). - HENDREN, R.W., BLOCH, K.: J. Biol. Chem. 255, 1504-1508 (1980). - HERBERT, M., BURKHARD, C., SCHNARRENBERGER, C.: Planta 143, 279-284 (1978); - ibid. 145, 95-104 (1979). - HOGETSU, D., MIYACHI, S.: Plant Cell Physiol. 20, 747-756 (1979). - HOLADAY, A.S., BOWES, G.: Plant Physiol. 65, 331-335 (1980). - HORVÁTH, G., DROPPA, M., MUSTÁRDY, L.A., FALUDI-DÁNIEL, Á.: Planta 141, 239-244, (1978). - HUANG, A.H.C., LIU, K.D.F., YOULE, R.J.: Plant Physiol. 58, 110-113 (1976). - HUBER, S.C.: Plant Physiol. 62, 321-325 (1978); - ibid. 63, 754-757 (1979a); - Biochim. Biophys. Acta 545, 131-140 (1979b). - HUBER, S.C., EDWARDS, G.E.: Biochim. Biophys. Acta 462, 583-602 (1977). - HUBER, S.C., MAURY, W.: Plant Physiol. 65, 350-354 (1980). - HUNER, N.P.A., MacDOWALL, F.D.H.: Can. J. Biochem. 56, 1154-1161 (1978); - ibid. 57, 155-164 (1979a); - ibid. 57, 1036-1041 (1979b). - HUNT, L., FLETCHER, J.: Plant Sci. Lett. 10, 243-247 (1977).

IDA, S.: J. Biochem. 82, 915-918 (1977).

JACQUOT, J.-P., VIDAL, J., GADAL, P., SCHÜRMANN, P.: FEBS Lett. 96, 243-246 (1978). - JACQUOT, J.-P., MAUDINAS, B., GADAL, P.: Biochem. Biophys. Res. Commun. 91, 1371-1376 (1979). - JOHANSSON, B.C., GEST, H.: Eur. J. Biochem. 81, 365-371 (1977). - JONES, M.G.K., OUTLAW, W.H., LOWRY, O.H.: Plant Physiol. 60, 379-383 (1977).

KAISER, W.M.: Planta 145, 377-382 (1979). - KAISER, W.M., BASSHAM, J.A.: Plant Physiol. 63, 109-113 (1979a); - Planta 144, 193-200 (1979b). - KELLY, G.J., LATZKO, E.: Naturwissenschaften 66, 617-618 (1979); - 183-208 in: The Biochemistry of Plants, Vol. 1, ed. N.E. TOLBERT. New York: Academic Press 1980. - KENT, S.S.: Plant Physiol. 64, 159-161

(1979). - KENYON, W.H., KRINGSTAD, R., BLACK, C.C.: FEBS Lett. 94, 281-283 (1978). - KEYS, A.J., BIRD, I.F., CORNELIUS, M.J., LEA, P.J., WALLSGROVE, R.M., MIFLIN, B.J.: Nature 275, 741-743 (1978). - KLUGE, M.: The flow of carbon in crassulacean acid metabolism (CAM), 113-125. In: Encyclopedia of Plant Physiology, N.S., Vol. 6, eds. M. GIBBS, E. LATZKO. Berlin-Heidelberg-New York: Springer 1979. - KNOPF, O., KLUGE, M.: Plant Cell Environ. 2, 73-78 (1979). - KOBAYASHI, H., TAKABE, T., NISHIMURA, M., AKAZAWA, T.: J. Biochem. 85, 923-930 (1979). - KOCH, K., KENNEDY, R.A.: Plant Physiol. 65, 193-197 (1980). - KPODAR, M.P., PIQUEMAL, M., CALMÈS, J., LATCHÉ, J.-C.: Physiol. Vég. 16, 117-130 (1978). - KRAUSE, G.H., KIRK, M., HEBER, U., OSMOND, C.B.: Planta 142, 229-233 (1978). - KU, M.S.B., EDWARDS, G.E.: Planta 147, 277-282 (1980). - KU, M.S.B., SCHMITT, M.R., EDWARDS, G.E.: J. Exp. Bot. 30, 89-98 (1979). - KUMARASINGHE, K.S., KEYS, A.J., WHITTINGHAM, C.P.: J. Exp. Bot. 28, 1247-1257 (1977). - KUNG, S.-D.: Ann. Rev. Plant Physiol. 28, 401-437 (1977).

LAING, W.A., CHRISTELLER, J.T.: Biochem. J. 159, 563-570 (1976). - LARA, C., DE LA TORRE, A., BUCHANAN, B.B.: Biochem. Biophys. Res. Commun. 93, 544-551 (1980). - LARSSON, C.: Physiol. Plant. 46, 221-226 (1979). - LARSSON, C., ALBERTSSON, E.: Physiol. Plant. 45, 7-10 (1979). - LATZKO, E., KELLY, G.J.: Enzymes of the reductive pentose phosphate cycle, 239-250. In: Encyclopedia of Plant Physiology, N.S., Vol. 6, eds. M. GIBBS, E. LATZKO. Berlin-Heidelberg-New York: Springer 1979. - LAVERGNE, D., BISMUTH, E., SARDA, C., CHAMPIGNY, M.L.: Z. Pflanzenphysiol. 93, 159-170 (1979). - LAWLIS, V.B., GORDON, G.L.R., McFADDEN, B.A.: J. Bacteriol. 139, 287-298 (1979). - LAWYER, A.L., ZELITCH, I.: Plant Physiol. 64, 706-711 (1979). - LEA, P.J., MIFLIN, B.J.: Nature 251, 614-616 (1974). - LEA, P.J., MILLS, W.R., MIFLIN, B.J.: FEBS Lett. 98, 165-168 (1979). - LEEGOOD, R.C., WALKER, D.A.: Plant Physiol. 63, 1212-1214 (1979); - Arch. Biochem. Biophys. 200, 575-582 (1980). - LEHMANN, M., WÖBER, G.: Ber. Dtsch. Bot. Ges. 91, 527-538 (1978). - LENDZIAN, K.J.: Planta 143, 291-296 (1978a); - ibid. 141, 105-110 (1978b); - ibid. 148, 1-6 (1980). - LEVI, C., PREISS, J.: Plant Physiol. 61, 218-220 (1978). - LLOYD, N.D.H., CANVIN, D.T., BRISTOW, J.M.: Can. J. Bot. 55, 3001-3005 (1977). - LONG, S.P., WOOLHOUSE, H.W.: J. Exp. Bot. 29, 803-814 (1978). - LORIMER, G.H.: J. Biol. Chem. 254, 5599-5601 (1979).

MÄCHLER, F., KEYS, A.J., CORNELIUS, M.J.: J. Exp. Bot. 31, 7-14 (1980). - MANN, A.F., HUCKLESBY, D.P., HEWITT, E.J.: Planta 140, 261-263 (1978). - MANN, A.F., FENTEM, P.A., STEWART, G.R.: Biochem. Biophys. Res. Commun. 88, 515-521 (1979). - MANN, K., MECKE, D.: Nature 282, 535-536 (1979). - MATOH, T., SUZUKI, F., IDA, S.: Plant Cell Physiol. 20, 1329-1340 (1979). - MBAKU, S.B., FRITZ, G.J., BOWES, G.: Plant Physiol. 62, 510-515 (1978). - McCASHIN, B.G., CANVIN, D.T.: Plant Physiol. 64, 354-360 (1979). - McCURRY, S.D., HALL, N.P., PIERCE, J., PAECH, C., TOLBERT, N.E.: Biochem. Biophys. Res. Commun. 84, 895-900 (1978). - McKINNEY, D.W., BUCHANAN, B.B., WOLOSIUK, R.A.: Biochem. Biophys. Res. Commun. 86, 1178-1184 (1979). - MILLER, A.G., COLMAN, B.: Plant Physiol. 65, 397-402 (1980). - MISZALSKI, Z., ZIEGLER, I.: Planta 145, 383-387 (1979). - MIZIORKO, H.M.: J. Biol. Chem. 254, 270-272 (1979). - MIZIORKO, H.M., SEALY, R.C.: Biochemistry 19, 1167-1171 (1980). - MOORE, R., BLACK, C.C.: Plant Physiol. 64, 309-313 (1979). - MOREL, C., QUEIROZ, O.: Plant Cell Environ. 1, 141-149 (1978). - MORGAN, J.A., BROWN, R.H.: Plant Physiol. 64, 257-262 (1979). - MURPHY, D.J., LEECH, R.M.: FEBS Lett. 88, 192-196 (1978). - MVÉ AKAMBA, L., SIEGENTHALER, P.-A.: Plant Cell Physiol. 20, 405-411 (1979a); - Physiol. Vég. 17, 587-595 (1979b).

NAIK, M.S., SINGH, P.: FEBS Lett. 111, 277-280 (1980). - NEYRA, C.A., HAGEMAN, R.H.: Plant Physiol. 62, 618-621 (1978). - NG, B.H., ANDERSON,

J.W.: Phytochemistry 17, 879-885 (1978). - NILSEN, S., HAUGSTAD, M., HOLMEN, A.T.: Physiol. Plant. 47, 19-24 (1979). - NISHIDA, K.: Plant Cell Physiol. 19, 935-941 (1978).

O'BRIEN, M.J., WOODROW, S., EASTERBY, J.S., POWLS, R.: Arch. Microbiol. 122, 313-319 (1979). - OHLROGGE, J.B., KUHN, D.N., STUMPF, P.K.: Proc. Natl. Acad. Sci. USA 76, 1194-1198 (1979). - OKABE, K.-I., CODD, G.A., STEWART, W.D.P.: Nature 279, 525-527 (1979). - OKITA, T.W., GREENBERG, E., KUHN, D.N., PREISS, J.: Plant Physiol. 64, 187-192 (1979). - O'LEARY, M.H., JAWORSKI, R.J., HARTMAN, F.C. Proc. Natl. Acad. Sci. USA 76, 673-675 (1979). - OLIVER, D.J.: Plant Sci. Lett. 15, 35-40 (1979a); - Plant Physiol. 64, 1048-1052 (1979b). - OSMOND, C.B.: Annu. Rev. Plant Physiol. 29, 379-414 (1978). - OUTLAW, W.H., MANCHESTER, J., DiCAMELLI, C.A., RANDALL, D.D., RAPP, B., VEITH, G.M.: Proc. Natl. Acad. Sci. USA 76, 6371-6375 (1979).

PAECH, C., TOLBERT, N.E.: J. Biol. Chem. 253, 7864-7873 (1978). - PAUL, J.S., CORNWELL, K.L., BASSHAM, J.A.: Planta 142, 49-54 (1978). - PIERCE, J., TOLBERT, N.E., BARKER, R.: Biochemistry 19, 934-942 (1980a); - J. Biol. Chem. 255, 509-511 (1980b). - PIERRE, J.N., QUEIROZ, O.: Planta 144, 143-151 (1979). - POINCELOT, R.P.: Carbonic anhydrase, 230-238. In: Encyclopedia of Plant Physiology, N.S., Vol. 6, eds. M. GIBBS, E. LATZKO. Berlin-Heidelberg-New York: Springer 1979. - PONGRATZ, P., BECK, E.: Plant Physiol. 62, 687-689 (1978). - POWLES, S.B., OSMOND, C.B.: Australian J. Plant Physiol. 5, 619-629 (1978). - POWLES, S.B., OSMOND, C.B., THORNE, S.W.: Plant Physiol. 64, 982-988 (1979). - PUPILLO, P., BOSSI, P.: Planta 144, 283-289 (1979). - PUPILLO, P., FAGGIANI, R.: Arch. Biochem. Biophys. 194, 581-592 (1979). - PUROHIT, K., McFADDEN, B.A.: Arch. Biochem. Biophys. 194, 101-106 (1979). - PUROHIT, K., McFADDEN, B.A., COHEN, A.L.: J. Bacteriol. 127, 505-515 (1976). - PUROHIT, K., McFADDEN, B.A., LAWLIS, V.B.: Arch. Microbiol. 121, 75-82 (1979).

QUEIROZ, O.: CAM: Rhythms of enzyme capacity and activity as adaptive mechanisms, 126-139. In: Encyclopedia of Plant Physiology, N.S., Vol. 6, eds. M. GIBBS, E. LATZKO. Berlin-Heidelberg-New York: Springer 1979.

RAGHAVENDRA, A.S., DAS, V.S.R.: Australian J. Plant Physiol. 4, 833-841 (1977). - RAGHAVENDRA, A.S., VALLEJOS, R.H.: Arch. Biochem. Biophys. 201, 356-358 (1980). - RAGHAVENDRA, A.S., RAJENDRUDU, G., DAS, V.S.R.: Nature 273, 143-144 (1978). - RASCHKE, K., DITTRICH, P.: Planta 134, 69-75 (1977). - RATHNAM, C.K.M.: Plant Physiol. 62, 220-223 (1978); - Planta 145, 13-23 (1979). - RATHNAM, C.K.M., CHOLLET, R.: Biochim. Biophys. Acta 548, 500-519 (1979a); - Arch. Biochem. Biophys. 193, 346-354 (1979b); - Plant Physiol. 65, 489-494 (1980). - RATHNAM, C.K.M., EDWARDS, G.E.: Plant Cell Physiol. 17, 177-186 (1976). - RAY, T.B., BLACK, C.C.: The C_4 pathway and its regulation, 77-101. In: Encyclopedia of Plant Physiology, N.S., Vol. 6, eds. M. GIBBS, E. LATZKO. Berlin-Heidelberg-New York: Springer 1979. - REED, M.L.: Plant Physiol. 63, 216-217 (1979). - REED, M.L., GRAHAM, D.: Australian J. Plant Physiol. 4, 87-98 (1977). - REGER, B.J., YATES, I.E.: Plant Physiol. 63, 209-212 (1979). - RHODES, D., SIMS, A.P., FOLKES, B.F.: Phytochemistry 19, 357-365 (1980). - ROBINSON, J.M., SMITH, M.G., GIBBS, M.: Plant Physiol. 65, 755-759 (1980). - ROBINSON, S.P., McNEIL, P.H., WALKER, D.A.: FEBS Lett. 97, 296-300 (1979). - ROBINSON, S.P. WALKER, D.A.: Biochim. Biophys. Acta 545, 528-536 (1979a); - Arch. Biochem. Biophys. 196, 319-323 (1979b); - FEBS Lett. 107, 295-299 (1979c). - ROBISON, P.D., TABITA, F.R.: Biochem. Biophys. Res. Commun. 88, 85-91 (1979). - ROBISON, P.D., MARTIN, M.N., TABITA, F.R.: Biochemistry 18, 4453-4458 (1979). - ROUGHAN, P.G., SLACK, C.R., HOLLAND, R.: Biochem. J. 158, 593-601 (1976). - ROUGHAN, P.G., HOLLAND, R., SLACK, C.R.: Biochem. J. 184, 193-202 (1979).

SALUJA, A.K., McFADDEN, B.A.: FEBS Lett. 96, 361-363 (1978). - SAMEJIMA, M., MIYACHI, S.: Plant Cell Physiol. 19, 907-916 (1978). - SAWHNEY, S.K., NAIK, M.S., NICHOLAS, D.J.D.: Nature 272, 647-648 (1978a); - Biochem. Biophys. Res. Commun. 81, 1209-1216 (1978b). - SAYRE, R.T., KENNEDY, R.A., PRINGNITZ, D.J.: Plant Physiol. 64, 293-299 (1979). - SCHEIBE, R., BECK, E.: Plant Physiol. 64, 744-748 (1979). - SCHILLING, N., DITTRICH, P.: Planta 147, 210-215 (1979). - SCHMIDT, A.: Photosynthetic assimilation of sulfur compounds, 481-496. In: Encyclopedia of Plant Physiology, N.S., Vol. 6, eds. M. GIBBS, E. LATZKO. Berlin-Heidelberg-New York: Springer 1979. - SCHNARRENBERGER, C., GROSS, D., BURKHARD, C., HERBERT, M.: Planta 147, 477-484 (1980). - SCHÜRMANN, P., JACQUOT, J.-P.: Biochim. Biophys. Acta 569, 309-312 (1979). - SHIRAHASHI, K., HAYAKAWA, S., SUGIYAMA, T.: Plant Physiol. 62, 826-830, (1978). - SHIRAIWA, Y., MIYACHI, S.: FEBS Lett. 95, 207-210 (1978). - SHOMER-ILAN, A., NEUMANN-GANMORE, R., WAISEL, Y.: Plant Physiol. 64, 963-965 (1979). - SICHER, R.C., JENSEN, R.G.: Plant Physiol. 64, 880-883 (1979). - SIEGELMAN, H.W., HIND, G. (eds.): Photosynthetic Carbon Assimilation. pp. 445. New York: Plenum 1978. - SILVIUS, J.E., KREMER, D.F., LEE, D.R.: Plant Physiol. 62, 54-58 (1978). - SILVIUS, J.E., CHATTERTON, N.J., KREMER, D.F.: Plant Physiol. 64, 872-875 (1979). - SOMERVILLE, C.R., OGREN, W.L.: Nature 280, 833-836 (1979). - SPALDING, M.H., SCHMITT, M.R., KU, S.B., EDWARDS, G.E.: Plant Physiol. 63, 738-743 (1979a). - SPALDING, M.H., STUMPF, D.K., KU, M.S.B., BURRIS, R.H., EDWARDS, G.E.: Australian J. Plant Physiol. 6, 557-567 (1979b). - SPALDING, M.H., ARRON, G.P., EDWARDS, G.E.: Arch. Biochem. Biophys. 199, 448-456 (1980). - STACEY, G., VAN BAALEN, C., TABITA, F.R.: Arch. Biochem. Biophys. 194, 457-467 (1979). - STEUP, M., LATZKO, E.: Planta 145, 69-75 (1979). - STEUP, M., SCHÄCHTELE, CH., LATZKO, E.: Z. Pflanzenphysiol. 96, 365-374 (1979); - Planta 148, 168-173 (1980). - STITT, M., apREES, T.: Phytochemistry 17, 1251-1256 (1978); - ibid. 18, 1905-1911 (1979); - Biochim. Biophys. Acta 627, 131-143 (1980). - STITT, M., BULPIN, P.V., apREES, T.: Biochim. Biophys. Acta 544, 200-214 (1978). - SUGIYAMA, T., SCHMITT, M.R., KU, S.B., EDWARDS, G.E.: Plant Cell Physiol. 20, 965-971 (1979). - SUSKE, G., WAGNER, W., FOLLMANN, H.: Z. Naturforsch. 34c, 214-221 (1979).

TABITA, F.R., COLLETTI, C.: J. Bacteriol. 140, 452-458 (1979). - TAKABE, T., ISHIKAWA, H., MIYAKAWA, M., TAKENAKA, K., NIKAI, S.: J. Biochem. 85, 203-208 (1979). - THOMAS, D.: Physiol. Vég. 14, 843-848 (1976). - THOMAS, S.M., HALL, N.P., MERRETT, M.J.: J. Exp. Bot. 29, 1161-1168 (1978). - THORPE, N., BRADY, C.J., MILTHORPE, F.L.: Australian J. Plant Physiol. 5, 485-493 (1978). - TOLBERT, N.E.: Glycolate metabolism by higher plants and algae, 338-352. In: Encyclopedia of Plant Physiology, N.S., Vol. 6, eds. M. GIBBS, E. LATZKO. Berlin-Heidelberg-New York: Springer 1979. - TOMATI, U., GALLI, E.: J. Exp. Bot. 30, 557-563 (1979). - TORRE, DE LA, A., LARA, C., WOLOSIUK, R.A., BUCHANAN, B.B.: FEBS Lett. 107, 141-145 (1979).

USUDA, H.: Plant Cell Physiol. 21, 211-217 (1980).

VATER, J., SALNIKOW, J.: Arch. Biochem. Biophys. 194, 190-197 (1979). - VENKATARAMANA, S., DAS, V.S.R.: Z. Naturforsch. 34c, 210-213 (1979). - VIDAL, J., JACQUOT, J.P., GADAL, P., VIDAL, D.: Physiol. Plant. 42, 307-314 (1978a). - VIDAL, J., JACQUOT, J.P., MEMBRE, H., GADAL, P.: Plant Sci. Lett. 11, 305-310 (1978b).

WALKER, D.A., ROBINSON, S.P.: Ber. Dtsch. Bot. Ges. 91, 513-526 (1978). - WALLSGROVE, R.M. HAREL, E., LEA, P.J., MIFLIN, B.J.: J. Exp. Bot. 28, 588-596 (1977). - WALLSGROVE, R.M., LEA, P.J., MIFLIN, B.J.: Plant Physiol. 63, 232-236 (1979). - WERDAN, K., HELDT, H.W.: Biochim. Biophys. Acta 283, 430-441 (1972). - WHITMAN, W.B., MARTIN, M.N., TABITA, F.R.:

J. Biol. Chem. 254, 10184-10189 (1979). - WILDMAN, S.G.: Arch. Biochem. Biophys. 196, 598-610 (1979). - WILDNER, G.F., HENKEL, J.: FEBS Lett. 91, 99-103 (1978); - Planta 146, 223-228 (1979a); - FEBS Lett. 103, 246-249 (1979b). - VON WILLERT, D.J., VON WILLERT, K.: Z. Pflanzenphysiol. 95, 43-49 (1979). - VON WILLERT, D.J., BRINCKMANN, E., SCHEITLER, B., THOMAS, D.A., TREICHEL, S.: Planta 147, 31-36 (1979). - WILLIAMS, L.E., KENNEDY, R.A.: Planta 142, 269-274 (1978). - WILLIAMS, M., RANDALL, D.D.: Plant Physiol. 64, 1099-1103 (1979). - WINKLER, F.J., WIRTH, E., LATZKO, E., SCHMIDT, H.-L., HOPPE, W., WIMMER, P.: Z. Pflanzenphysiol. 87, 255-263 (1978). - WINTER, K.: J. Exp. Bot. 29, 1169-1172 (1978). - WOLOSIUK, R.A., BUCHANAN, B.B.: Arch. Biochem. Biophys. 189, 97-101 (1978a); - Plant Physiol. 61, 669-671 (1978b). - WOLOSIUK, R.A., CRAWFORD, N.A., YEE, B.C., BUCHANAN, B.B.: J. Biol. Chem. 254, 1627-1632 (1979). - WOLOSIUK, R.A., PERELMUTER, M.E., CHEHEBAR, C.: FEBS Lett. 109, 289-293 (1980). - WONG, W.W., BENEDICT, C.R., KOHEL, R.J.: Plant Physiol. 63, 852-856 (1979). - WOO, K.C.: Plant Physiol. 63, 783-787 (1979). - WOO, K.C., CANVIN, D.T.: Can. J. Bot. 58, 511-516 (1980a); - ibid. 58, 517-521 (1980b). - WOO, K.C., JOKINEN, M., CANVIN, D.T.: Plant Physiol. 65, 433-436 (1980). - WOODROW, S., O'BRIEN, M.J., EASTERBY, J.S., POWLS, R.: Eur. J. Biochem. 98, 425-430 (1979).

YOKOTA, A., NAKANO, Y., KITAOKA, S.: Agric. Biol. Chem. 42, 115-120 (1978).

ZELITCH, I.: Photorespiration: studies with whole tissues, 353-367. In: Encyclopedia of Plant Physiology, N.S., Vol. 6, eds. M. GIBBS, E. LATZKO. Berlin-Heidelberg-New York: Springer 1979. - ZINDLER-FRANK, E.: Z. Pflanzenphysiol. 80, 1-13 (1976). - ZSCHOCHE, W.C., TING, I.P.: Plant Sci. Lett. 9, 103-106 (1977).

Professor Dr. ERWIN LATZKO
Botanisches Institut der
Universität Münster
Schloßgarten 3
D 4400 Münster

Dr. GRAHAME J. KELLY
Department of Biochemistry
and Nutrition
University of New England
Armidale, N.S.W. 2351
Australia

Photosynthesis

Carbon Metabolism: The Profound Effects of Illumination on the Metabolism of Photosynthetic Cells*

By GRAHAME J. KELLY and ERWIN LATZKO

"The first man I saw was of a meagre aspect, with sooty hands and face, his hair and beard long, ragged and singed in several places. His clothes, shirt, and skin were all of the same colour. He had been eight years upon a project for extracting sunbeams out of cucumbers..."

Jonathan Swift in *Gulliver's Travels*, ca. 1727

1. Introduction

The first investigations of photosynthesis became possible after Jan Ingenhousz demonstrated, sometime during the 1770's, that light was necessary for the "purification of air" by plants. Naturally, no distinction was made at that time between "light reactions" and "dark reactions". This distinction did not develop until after the discovery of the Hill reaction (late 1930's), and photophosphorylation and the Calvin cycle (early 1950's). Once developed, the concept of "light reactions" and "dark reactions" took firm hold, and for a period well into the 1970's, the role of light in photosynthesis was viewed to be more-or-less exclusively restricted to the generation of ATP and NADPH for the Calvin cycle.

However, in recent years other effects of illumination on photosynthetic cells have been discovered, and it is fair to say that at least some of these are as essential as the generation of NADPH and ATP for photosynthesis to continue. It has been found that light causes an increase in both the pH and the Mg^{2+} concentration of the chloroplast stroma which houses the Calvin cycle. Light induces increases in the activities of several Calvin cycle enzymes. Light leads to a decrease in (or cessation of) mitochondrial respiration, but supports another respiration, termed photorespiration, that uses reactions in both microbodies and mitochondria. Finally, besides leading to reduction of $NADP^+$, light also leads to reduction of O_2, an event which can result in the inactivation of Calvin cycle enzymes.

It is clear that the effects of illumination on photosynthetic carbon metabolism are more profound and far-reaching than the term

---

* Abbreviations: CAM, crassulacean acid metabolism; CoA, coenzyme A; FBP, fructose-1,6-P_2; GAP, glyceraldehyde-3-P; OAA, oxaloacetate; PEP, P-enolpyruvate; P_i, inorganic phosphate; RuBP, ribulose-1,5-P_2; SBP, sedoheptulose-1,7-P_2.

G.J. Kelly and E. Latzko:
Thirty Years of Photosynthesis

Originally published in Progress in Botany 44

"supply of ATP and NADPH" implies. This review, which covers the interval since our last review in Volume 42, attempts to give special emphasis to this many-faceted influence of light on the carbon metabolism of photosynthesis.

2. Prologue: Advances in Methodology

The minor revolution in research that followed the introduction of techniques for isolating protoplasts (GALUN, 1981) of photosynthetic cells has continued unabated. A new enzyme, pectolyase Y23, has been recommended by NAGATA and ISHII (1979) to complement cellulase in the isolation of protoplasts in less than 30 minutes. Extensive analyses of the distribution of metabolites between the chloroplasts and cytoplasm have been made possible by utilising photosynthesising protoplasts (GIERSCH et al., 1980; ROBINSON and WALKER, 1980a; STITT et al., 1980; WIRTZ et al., 1980). The combined results have confirmed that photosynthetically fixed carbon can be very quickly channelled into sucrose in the cytoplasm. HAMPP (1980) has developed a more elaborate procedure so that three fractions, viz. chloroplasts, mitochondria, and cytoplasm can be rapidly separated from protoplasts. Experiments with isolated wheat protoplasts have confirmed the suspicions of ROBINSON and WALKER (1980b) that the induction phase of photosynthesis by isolated chloroplasts (i.e. the several-minute lag that occurs between the beginning of illumination and the achievement of a steady rate of CO_2 fixation) reflects predominately the time required to build up the levels of Calvin cycle intermediates, and is not directly related to the light-mediated activation of certain Calvin cycle enzymes (LEEGOOD and WALKER, 1981a; see Sects. 4b and 5).

Several innovative techniques have opened new avenues for investigating algal metabolism. The first reports of the isolation of functional algal protoplasts (from *Chlorella*) have come from WEBB et al. (1980) and YAMADA and SAKAGUCHI (1981). Algal chloroplasts have been isolated from the red alga *Griffithsia monilis* (and termed "rhodoplasts"; LILLEY and LARKUM, 1981) and the green alga *Chlamydomonas reinhardii* (BELKNAP and TOGASAKI, 1981) using two quite different procedures. The effect of Pi on CO_2 fixation by each of these algal chloroplast preparations was similar to the effect observed with the well-characterised spinach chloroplast. An unusual "cytoplast", i.e. chloroplast surrounded by a thin layer of cytoplasm, was obtained from the green alga *Caulerpa simpliciuscula* by GRANT and HOWARD (1980). Unlike "normal" chloroplasts, these cytoplasts formed large quantities of sucrose during CO_2 fixation, presumably because the thin layer of cytoplasm contained all the enzymes necessary for the synthesis of sucrose from triose-P.

Further progress has also been made with higher plants. DAY et al. (1981) prepared functional mesophyll protoplasts from the C_4 plant *Zea mays* and obtained chloroplasts from these. The first isolation of protoplasts from CAM plants was described by DODDS (1980) and SPALDING and EDWARDS (1980). In the latter case, chloroplasts were isolated from the protoplasts and found to be quantitatively different in several respects from spinach chloroplasts.

There are abundant photomicrographs of chloroplasts in the literature, almost all of which show a longitudinal profile that best reveals the stacking of the internal lamellar membranes (JENSEN, 1980). It was refreshing, therefore, to be offered a three-dimensional view of the chloroplast (WILDMAN et al., 1980) that suggests a string of grana arranged into a slight helix, with starch grains sitting in the cavity (Fig. 1).

A word of caution concerning metabolic studies with isolated cells, protoplasts, etc. has come from MADORE and WEBB (1982). These authors found that isolated mesophyll cells of *Cucurbita pepo* would not synthesise stachyose from CO_2, although intact leaves of this plant form 50% of their neutral sugars as stachyose, However, there was no effect of cell isolation on sucrose synthesis.

3. The Uptake of Inorganic Carbon by Photosynthetic Cells

New viewpoints concerning the entry of inorganic carbon into photosynthetic cells are emerging. These have been triggered by studies with unicellular algae. Most investigators (but see FERRIER, 1980) believe that algal cells take up inorganic carbon in the form of HCO_3^-, and that this uptake is mediated by some transfer mechanism that can concentrate the inorganic-carbon pool inside the cell to levels manyfold the external concentration (FINDENEGG, 1980; COLEMAN and COLMAN, 1981; KAPLAN, 1981; KAPLAN et al., 1980; ZENVIRTH and KAPLAN, 1981). The concentrating mechanism was light-dependent and inhibited by electron transport and photophosphorylation inhibitors (BADGER et al., 1980a; MILLER and COLMAN, 1980; CANVIN and SHELP, 1981). The dehydration of this HCO_3^- to generate CO_2 for RuBP carboxylase would be ensured by the presence of carbonic anhydrase (TSUZUKI et al., 1980; REISSER and BENSELER, 1981).

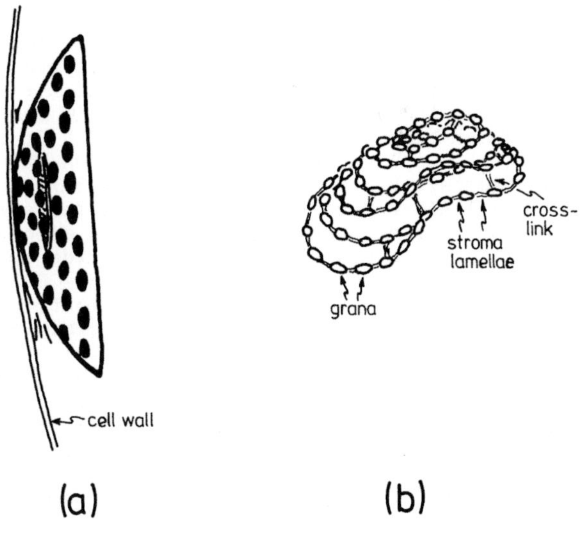

Fig. 1a,b. Three-dimensional view of a chloroplast. (*a*) Side view showing grana (*black dots*) and starch grain (*cross-hatched internal disc*). (*b*) Conception of the arrangement of the thylakoid system. Redrawn from Botanical Gazette 141, pp. 28 & 34, by WILDMAN et al. (1980) by permission of the University of Chicago Press. © by the University of Chicago

VOLOKITA et al. (1981) now have evidence that a similar situation may apply to higher plants. Traditionally, it has been assumed that inorganic carbon passes from the environment to the carboxylation site in chloroplasts of higher plants by simple diffusion of CO_2. VOLOKITA et al. have tested this hypothesis using isolated pea mesophyll protoplasts and obtained results inconsistent with diffusion of CO_2, but which support a light-stimulated transfer of HCO_3^- across the plasmalemma and into the cell.

At this point it is worth noting that, in experiments with wheat and tobacco protoplasts, the exported product of carbon fixation (sucrose) was transferred to the outside of the cell by a plasmalemma-located, energy-requiring mechanism that involved the co-export of 1.5 - 2.5 K^+ per molecule of sucrose exported (HUBER and MORELAND, 1981).

4. The Calvin Cycle: Enzymes and Their Regulation

a) RuBP Carboxylase

Apart from a group of photosynthetic bacteria that achieves net fixation of CO_2 via a reductive tricarboxylic acid cycle (FUCHS et al., 1980a,b), all photosynthetic organisms contain RuBP carboxylase. This enzyme is still one of the most intensely investigated enzymes from plants. Its purification and crystallisation from eight species has been achieved (JOHAL et al., 1980), and from inhibitor experiments a theoretical specific activity of 2.8 μmoles CO_2 fixed per mg protein · min has been calculated (HALL et al., 1981). Mechanistic aspects of the catalysed reaction have been studied (LORIMER, 1981) and these are likely to continue with the discovery of an inhibitor that acts at the CO_2 subsite (BADGER et al., 1980b).

Renewed interest in the affinity of RuBP carboxylase for CO_2 followed a report by GARRETT (1978) that the apparent $Km(CO_2)$ for the enzyme from a tetraploid cultivar of *Lolium perenne* (ryegrass) was less than half that for the enzyme from a diploid cultivar. This difference could not be confirmed by either McNEIL et al. (1981) (who used plant material supplied by GARRETT), or REJDA et al. (1981). However, extensive surveys by YEOH et al. (1980, 1981) show clearly that variation in the $Km(CO_2)$ values of RuBP carboxylases among *different* species of grasses, and between different groups of plants (C_3, C_4, CAM, aquatic), does occur. In addition, the polypeptide composition of the ryegrass enzyme was somewhat unusual (JOHAL and CHOLLET, 1981). Another report (JOSEPH et al., 1981) has indicated that, within a species, ploidy can influence the enzyme activity by determining the contents: in *Festuca arundinacea* (tall fescue) increased ploidy led to an increase in both the content of RuBP carboxylase and net photosynthesis.

The content of RuBP carboxylase in leaves is a major determinant of the rate of photosynthesis. It appears to determine the CO_2 assimilation rate at low pCO_2; equally critical is the supply of RuBP which is reported to limit CO_2 fixation under higher pCO_2 (VON CAEMMERER and FARQUHAR, 1981) or in dim light (PERCHOROWICZ et al., 1981), and in this respect it is interesting that GAUDILLERE and GAUDILLERE (1981) concluded that ATP (which is essential for RuBP regeneration) is the first factor to become limiting under low light conditions.

The activation of RuBP carboxylase by CO_2 and Mg^{2+} to form the physiologically active species enzyme·CO_2·Mg^{2+} (LORIMER, 1981) is a phenomenon now widely recognised. It has explained how the carboxylase in vivo can have an affinity for CO_2 commensurate with air levels of CO_2, and more recently it has simplified explanations concerning the action of effectors such as NADPH and gluconate-6-P. Both McCURRY et al. (1981) and BADGER and LORIMER (1981) concluded that positive effectors (e.g., NADPH and gluconate-6-P) exerted their effect by promoting the activation of the enzyme by $CO_2 + Mg^{2+}$. These authors, and also DAILEY and CRIDDLE (1980), believe that only one effector-binding site, probably on (or overlapping with) the catalytic site, is required to explain the influence of the effectors. Observations by VATER et al. (1980) might also agree with this interpretation. HATCH and JENSEN (1980) measured the effect of NADPH and gluconate-6-P on the concentration of CO_2 required to achieve half-maximal activation of the carboxylase, and found that each effector (at 0.1 mM) lowered this value from 43 μM to ca. 5 μM. The importance of CO_2 for activation was highlighted by the observation of SICHER et al. (1981) that RuBP carboxylase is deactivated in vivo when chloroplasts are illuminated in the absence of CO_2.

Increased attention has been given recently to environmental effects on RuBP carboxylase in vivo. KAISER et al. (1981a,b) have reported that the osmotic stress that may occur in the leaves of wilting plants can reversibly inhibit the activity of RuBP carboxylase, although such stress causes more severe damage to chloroplast membranes. Two views of temperature influences have been provided by HUNER et al. (1981) and WEIS (1981a). The former authors found that the RuBP carboxylases from *Solanum commersonii* (coldhardy potato) and *Solanum tuberosum* (nonhardy potato) were identical with respect to molecular weight, subunit composition, affinity for CO_2, and number of sulphydryl groups. But the enzyme from the hardy plant had its sulphydryl groups less exposed, and a Vmax that was slightly higher when measured at 5°C but much lower when measured at 25°C. WEIS detected a considerable, but reversible, heat inactivation of the RuBP carboxylase in spinach leaves exposed to temperatures between 25°C and 40°C. Further experiments by WEIS (1981b) have related temperature to light-mediated activation of the carboxylase, a concept for which the popularity and persistence have been inversely proportional in recent years. WEIS reported that light-mediated activation is not seen with chloroplasts held at 5°C because, at this temperature, dark-mediated inactivation does not occur. However, at 25°C, an approximately 60% decrease in activity occurred after 10 min in darkness, and this activity was recovered upon illumination. Earlier experiments (MÄCHLER and NÖSBERGER, 1980) showing increased RuBP carboxylase activity in intact wheat leaves under higher light intensities or lower temperatures support a relationship between temperature and light-mediated activation.

Prokaryotic RuBP carboxylases have received most recent attention relating to the enzyme structure. New evidence confirms that the blue-green algal enzyme has the L_8S_8 structure (eight large and eight small subunits) typical of the eukaryotic enzyme (OKABE and CODD, 1980; ANDREWS and ABEL, 1981). The latter authors have speculated that the small subunits are required to maintain a more active conformation of the enzyme as a whole, a cumbersome mechanism which *Rhodospirillum rubrum* RuBP carboxylase (with an L_2 structure) appears to have avoided, thereby making it conceivably the more advanced form of the enzyme.

Not all photosynthetic bacteria have RuBP carboxylase with an L_8S_8 structure (see our preceding review). *Alcaligenes eutrophus* does, but the arrangement of the subunits appeared different to that in the plant enzyme (BOWIEN et al., 1980). Microscopic examination of photosynthetic bacteria has often revealed polyhedral inclusion bodies that have RuBP carboxylase activity; the latest case was for *Nitrosomonas* (HARMS et al., 1981). The inclusion bodies in *Thioba-*

cillus neapolitanus have been described as "functionally enigmatic" (BEUDEKER et al., 1980, 1981). The inclusion bodies recently have been shown to contain many other enzymes, including most of the enzymes of the Calvin cycle (BEUDEKER and KUENEN, 1981).

A final and intriguing aspect of RuBP carboxylase is that an auxin-binding protein purified from bean leaves has been tentatively identified as RuBP carboxylase (WARDROP and POLYA, 1980). The affinity for auxin was very high (dissociation constant ca. 1 μM). No soluble auxin-binding protein was found in roots from the same plants.

b) The Bisphosphatases

Two bisphosphatases are required by the Calvin cycle: one to hydrolyse FBP and the other to hydrolyse SBP. The latter (SBPase) has survived for 25 years despite the fact that an activity sufficient to accomodate even low rates of CO_2 fixation had never been reported. It was a relief, therefore, when WOODROW and WALKER (1980) measured an SBPase activity in wheat chloroplasts approaching 60 μmoles/mg chlorophyll · hr. These authors demonstrated that previous failures to detect high activities could be attributed to (a) the requirement for a light-mediated, reductive activation of the enzyme, and (b) the presence of the substrate SBP for this activation to proceed at an appreciable rate. Confirmatory results with the spinach and maize enzymes were reported shortly afterwards by LAING et al. (1981) and NISHIZAWA and BUCHANAN (1981).

Light-mediated activation of the other bisphosphatase (FBPase) also has been shown to benefit from the presence of its substrate FBP (HERTIG and WOLOSIUK, 1980; CHEHEBAR and WOLOSIUK, 1980; LAING et al., 1981; NISHIZAWA and BUCHANAN, 1981; ROSA, 1981). PRADEL et al. (1981) have further evidence that the activation involves reduction of disulphides to sulphydryls. The proposal that thioredoxin mediates this reduction has been extended to cover the FBPases from the blue-green alga *Synechococcus* (SCHMIDT, 1980, 1981a) and maize leaves (NISHIZAWA and BUCHANAN, 1981). Indeed, PLA and LOPEZ-GORGE (1981) have shown that FBPase binds to an affinity chromatography system that has thioredoxin attached to the support gel.

More concerning the light-mediated control of FBPase activity is given below (Sect. 5).

The higher pH and Mg^{2+} concentration in the stroma developed upon illumination of the chloroplast is known to favour FBPase activity. Artificial reversal of the alkalisation of the stroma was observed to lower FBPase activity (ENSER and HEBER, 1980; FLÜGGE et al., 1980a). Ca^{2+} (which competes with Mg^{2+}) also inhibits the enzyme activity (CHARLES and HALLIWELL, 1980b). However, HERTIG and WOLOSIUK (1980) have shown that it may substitute for Mg^{2+} during reductive activation, although ROSA (1981) has replied that this is not an essential role for Ca^{2+}. The existence of large amounts of Ca^{2+} in plants, mostly in a membrane-bound form (CHEVALIER and PARIS, 1981; YAMAGISHI et al., 1981) will no doubt receive more attention now that the Ca^{2+}-binding protein calmodulin, known to be of central importance in animal cells, has been clearly demonstrated in plants (ANDERSON et al., 1980; WATTERSON et al., 1980).

c) Other Enzymes of the Calvin Cycle

Two other enzymes that undergo light-mediated activation are GAP dehydrogenase and ribulose-5-P kinase. In both cases the dissociation of low-activity oligomers to high-activity protomers may be involved in the activation (WARA-ASWAPATI et al., 1980). GAP dehydrogenase sulphydryl groups are more prone to oxidation when $NADP^+$ is bound to the enzyme (FERRI et al., 1981). The enzyme is extremely susceptible to inactivation by the chemical iodoacetol-P (USUDA and EDWARDS, 1981). The increase in enzyme activity during greening of maize leaves was attributed, at least in part, to the synthesis of new protein in the cytosol (LIN and STOCKING, 1980). Purified preparations of ribulose-5-P kinase from photosynthetic bacteria have been obtained and the earlier demonstrated activation by NADH confirmed (TABITA, 1980; SIEBERT et al., 1981).

Further evidence that the FBP-aldolase and SBP-aldolase activities in chloroplasts are carried by the same enzyme molecule has been obtained by MURPHY and WALKER (1981). However, ANDERSON and HEINRIKSON (1979) have shown that this chloroplast enzyme is distinct from its cytosolic counterpart as far as amino acid composition is concerned.

A thought-provoking report by KNOWLES et al. (1980) presents evidence that P-ribose isomerase from spinach and *Chromatium* (but not the enzyme from yeast or muscle) catalyses the dehydration of ribulose-5-P to a product that is unstable and loses Pi to form a hydroxymethylfuranone that is a potent inhibitor of transketolase.

5. On the Light-Mediated Control of Enzyme Activities

In a recent review, BUCHANAN (1980) indicates that five Calvin cycle enzymes viz. RuBP carboxylase, GAP dehydrogenase, FBPase, SBPase, and ribulose-5-P kinase are claimed to have activities regulated by light-mediated processes. Several enzymes not in the Calvin cycle are also reported to undergo light-mediated regulation, including NADP-linked malate dehydrogenase, glucose-6-P dehydrogenase (BUCHANAN, 1980), transaldolase (ANDERSON, 1981), and *Chlorella* glutamine synthetase (SCHMIDT, 1981b).

It is still not clear how many mechanisms of light-mediated regulation exist. Most workers submit to the involvement of a photosynthetically generated reductant. One line of thought is that electrons flow from ferredoxin, via thioredoxin, to the enzyme molecule. Chloroplast thioredoxins have been found not only in green leaves, but also in etiolated leaves, and even in seeds (CRAWFORD et al., 1981). Analogous thioredoxins occur in the prokaryotic blue-green alga *Nostoc muscorum* (YEE et al., 1981). Two more small proteins, purothionin and ferralterin, thought to function like thioredoxin in light-mediated enzyme regulation, have been reported (LARA et al., 1980; WADA and BUCHANAN, 1981), while a possible third alternative, glutaredoxin (TSANG, 1981), also exists. SCHEIBE (1981) has clearly demonstrated a light-dependent reduction of thioredoxin in pea chloroplasts.

An alternative proposal involves enzyme reduction by a membrane-bound, "light effect mediator". Indications that photosystem I is the electron donor to this system have been reported by MVÉ AKAMBA and ANDERSON (1981). The mediator has been identified as a surface-exposed protein rather tightly bound to the thylakoid membrane (MOHAMED and

ANDERSON, 1981). However, this protein appeared to require the presence of a small soluble stromal protein to be effective, and although this soluble protein was thought to be distinct from thioredoxin (ASHTON and ANDERSON, 1981), the involvement of thioredoxin in the regulation of glucose-6-P dehydrogenase was also noted (ASHTON et al., 1980; SCHEIBE and ANDERSON, 1981).

The "black sheep" among the group of enzymes regulated by light is pyruvate Pi dikinase. Light-meditated regulation of this C_4 plant enzyme does not involve the interconversion of dithiol and disulphide forms of the enzyme, but rather involves an ADP-induced transformation to an inactive form (upon leaf darkening), then a Pi-mediated reactivation (upon illumination) (CHAPMAN and HATCH, 1981; SUGIYAMA and HATCH, 1981). Hence the effect of light on this enzyme appears to be mediated through changes in the relative concentrations of adenine nucleotides and Pi. Furthermore, this process was influenced by temperature when the light intensity was fairly low (EDWARDS et al., 1980). Other C_4 plant enzymes (FBPase, SBPase, and NADP-linked malate dehydrogenase) are activated by the more familiar reductive processes (NISHIZAWA and BUCHANAN, 1981; JACQUOT et al., 1981).

FBPase has been at the centre of investigations that foreshadow a probable radical change of ideas regarding the role of light-mediated enzyme activation. Activation involves reduction of the enzyme, and the reduced enzyme can be re-oxidised (and thereby inactivated) by H_2O_2 which is a product of pseudocyclic electron transport. This type of electron transport is now considered essential for photosynthesis (ZIEM-HANCK and HEBER, 1980; McVETTY and CANVIN, 1981; STEIGER and BECK, 1981). The inactivation by H_2O_2 of chloroplast FBPase from spinach and pea has been well established by CHARLES and HALLIWELL (1980a, 1981a,b). Similar results with glucose-6-P dehydrogenase have been reported (BRENNAN and ANDERSON, 1980). These observations have led LEEGOOD and WALKER (1980a) and CHARLES and HALLIWELL (1980a) to conclude that one purpose of light-mediated activation is to protect enzymes against photosynthetically generated oxidants such as H_2O_2. Experiments with uncouplers of photophosphorylation have provided a different type of evidence that supports this conclusion (ROSA and WHATLEY, 1981). Hence light-mediated activation would appear to be important not only at sunrise, but also during the entire period of illumination.

The dark-mediated inactivation of FBPase is not, however, considered to involve H_2O_2 to a significant degree. Rather, LEEGOOD and WALKER (1980b,c, 1981b) have shown that this inactivation most probably involves a simple reversal of the activation process, i.e. oxidation of the FBPase molecule by oxidised thioredoxin. Moreover, these authors propose that the *extent* of activation during illumination depends on a balance between the supply of photosynthetically generated electrons and availability of alternative electron acceptors (e.g., $NADP^+$, O_2; see Fig. 2). Indeed, in the absence of O_2, the FBPase in chloroplasts could be experimentally forced to become "light activated" in darkness by generating the physiological electron donor NADPH.

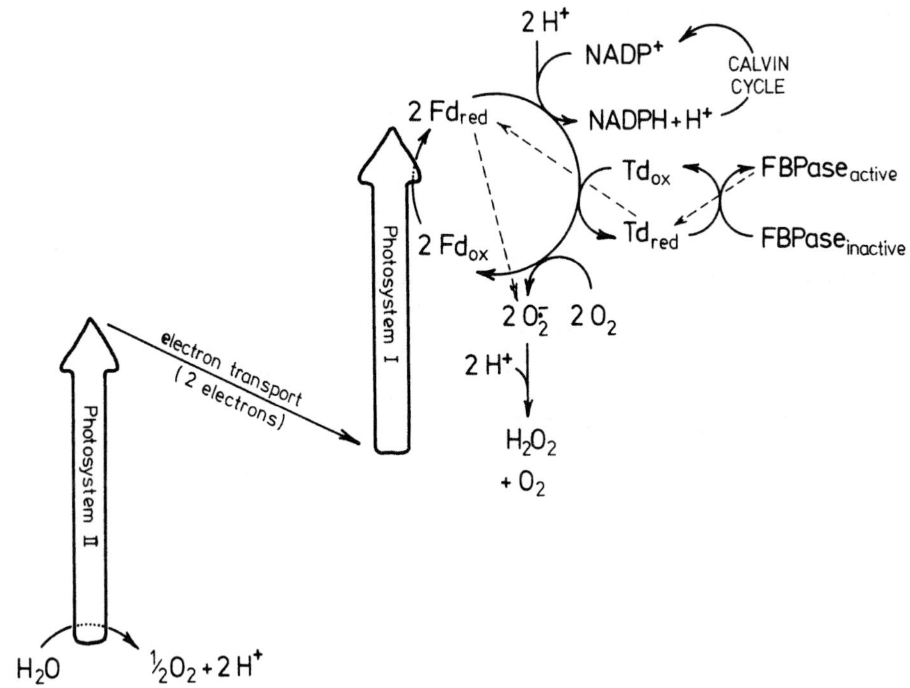

Fig. 2. The extent of light-mediated activation of chloroplast FBPase depends on how many photosynthetically generated electrons flow from ferredoxin to FBPase. Electrons may alternatively flow to O_2 (pseudocyclic electron transport), or to $NADP^+$ and thence to CO_2 fixed by the Calvin cycle. Dark-mediated inactivation of FBPase may involve electron flow (*dashed line*) from the enzyme to O_2, via thioredoxin and ferredoxin. Adapted from LEEGOOD and WALKER (1980b). Fd: ferredoxin; Td: thioredoxin

6. Transport by the Chloroplast Envelope

The inner membrane of the chloroplast envelope possesses several types of translocators which catalyse specific transfer of metabolites into and out of the chloroplast (cf. HEBER and HELDT, 1981). The most powerful is the now well-known phosphate translocator that facilitates the export of fixed carbon (in the form of triose-P or glycerate-3-P) in stoichiometric exchange for Pi. The activity of this translocator must be carefully controlled: there must be sufficient Pi import to permit the export of all the triose-P that represents net carbon fixed, but not excessive Pi import or else the chloroplast's pool of triose-P needed to regenerate the CO_2 acceptor (RuBP) will be depleted (FLÜGGE et al., 1980b). This balance is so critical that it has led HEROLD (1980) to suggest that the real boundary between the "source" and "sink" of photosynthesis is the chloroplast envelope. Another effect of Pi is that, at higher concentrations in the stroma, it can inhibit FBPase and SBPase activities (FURBANK and LILLEY, 1980). Further information on the action of Pi in cells may be expected now that individual pools can be viewed using ^{31}P nuclear magnetic resonance (FOYER et al., 1982).

The carrier protein of the phosphate translocator has been solubilised, purified, and reconstituted by incorporation into liposomes (FLÜGGE and HELDT, 1981). The observation that the phosphate translocator is also present on another type of plastid (chromoplast of the daffodil flower; LIEDVOGEL and KLEINIG, 1980) is important in that it increases the probability that another non-photosynthetic plastid, viz. the starch-storing amyloplast, might also possess phosphate translocators, a phenomenon which would be of central significance to the regulation of carbohydrate metabolism in non-photosynthetic tissues.

Other metabolite transporters are (a) one or more translocators for malate, OAA, α-ketoglutarate, aspartate, and glutamate, (b) an adenylate carrier which appears to have a high capacity only in C_4 plants, (c) a glucose carrier, and (d) a pyruvate carrier that may be only in C_4 plants (HEBER and HELDT, 1981).

Some recent attention has been given to the movement of cations across the chloroplast envelope. MAURY et al. (1981) suggest that two mechanisms can lead to the movement of protons and K^+ across the envelope (see Fig. 3). Entry of K^+ could help to screen negative charges on thylakoid membranes and thus free Mg^{2+} (KAISER et al., 1980), the concentration of which is low in the dark (PORTIS, 1981).

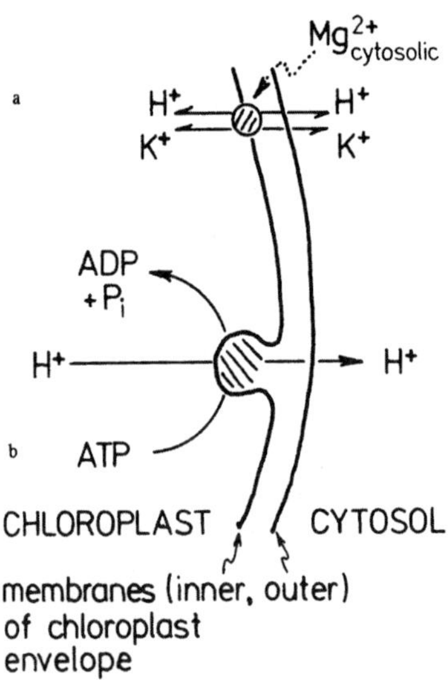

Fig. 3a,b. Two mechanisms postulated to be involved in fluxes of H^+ and K^+ between the chloroplast and cytosol: (*a*) passive, reversible K^+/H^+ exchange activated by cytosolic Mg^{2+}, and (*b*) an ATPase that promotes the unidirectional movement of H^+, i.e., exit from the chloroplast. Adapted from MAURY et al. (1981)

7. Products of CO_2 Fixation

a) Starch

Starch synthesis has received little attention in recent years. One interesting study showed that the rate of starch accumulation in leaves increases dramatically when plants are moved from 14-hour to 7-hour photoperiods (CHATTERTON and SILVIUS, 1980). The principal regulatory enzyme involved in starch synthesis (ADP-glucose pyrophosphorylase) has been purified using a new procedure (COPELAND and PREISS, 1981).

Much more interest has been shown in the biochemistry of starch degradation. The initial step in this process may involve either hydrolytic enzymes, ultimately producing glucose (OKITA and PREISS, 1980), or phosphorylase, producing glucose-1-P (STITT and HELDT, 1981a). Leaves contain hexokinases able to phosphorylate any glucose produced (BALDUS et al., 1981). There is also evidence that maltose can be a photosynthetic product, particularly when the Pi supply is restricted (HEROLD et al., 1980, 1981); utilisation of this maltose may also require hexokinase activity.

Chloroplast phosphorylase has been purified to homogeneity (STEUP, 1981). It has a higher affinity for amylose, but lower affinity for amylopectin, compared to the enigmatic non-chloroplast enzyme (PREISS et al., 1980; STEUP and SCHÄCHTELE, 1981). The latter authors reported the intriguing observation that maltotetraose (i.e. linear polymer of four glucose moieties) is a powerful inhibitor of the chloroplast enzyme. Several forms of phosphorylase were found in the green algae *Dunaliella marina* and *Eremosphaera viridis*, although in the former case all forms were shown to be located in the chloroplast (KOMBRINK and WÖBER, 1980; STEUP and MELKONIAN, 1981).

Subsequent metabolism of glucose-1-P could occur via either glycolysis or the oxidative pentose-P pathway. Experiments with reconstituted chloroplast systems have demonstrated that the latter pathway is possible (LENDZIAN, 1980; FURBANK and LILLEY, 1981), and evidence that as much as 30% of the metabolism takes place by this pathway has been obtained (STITT and AP REES, 1980; STITT and HELDT, 1981a). This apparently occurs despite the fact that glycolysis is far more beneficial to the carbon economy of the plant (LENDZIAN, 1980; see Fig. 4).

It usually has been assumed that chloroplast starch is synthesised during the day and degraded at night. This assumption has been cautiously questioned by STITT and HELDT (1981b), who concluded that synthesis and degradation could occur simultaneously in isolated spinach chloroplasts. Supporting studies were reported by NAKAMURA and MIYACHI (1982), who found that a rise in temperature (from 20°C to 38°C) caused photosynthesising *Chlorella vulgaris* to degrade starch and mobilise the products of this degradation for the synthesis of extra sucrose.

Control mechanisms linking chloroplast starch and cytosolic sucrose clearly exist. In barley leaves, starch mobilisation was not observed (even at night) until after most of the sucrose had been exported (GORDON et al., 1980). On the other hand, the ratio of the amounts of these two major products formed varies widely according to plant species (HUBER, 1981a). A final possibility, previously not even remotely considered, is that starch may occur in the cytosol; REPKA et al. (1981) observed the release of starch granules from chloroplasts in the flag leaf of wheat, but only during the period of maximum grain filling.

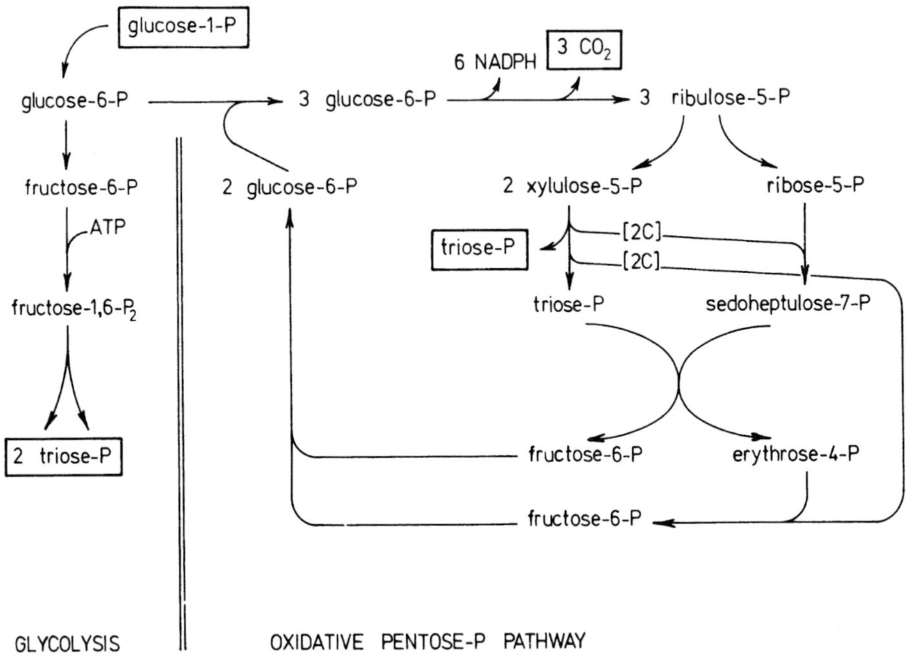

Fig. 4. Alternative pathways for the metabolism of glucose-1-P (product of chloroplast starch degradation) to triose-P (which can be exported from the chloroplast). Only one triose-P becomes available per glucose-1-P metabolised when the oxidative pentose-P pathway is utilised (the other half of the carbon is lost as CO_2), but two triose-P become available when glycolysis is utilised

b) Sucrose

Sucrose is synthesised in the cytosol, using triose-P exported from the chloroplast (KELLY and LATZKO, 1980a; USUDA and EDWARDS, 1980a). The amount synthesised has been observed to decrease when NH_4^+ is supplied because a greater proportion of the triose-P is metabolised to form C-skeletons for amino acid biosynthesis (LARSEN et al., 1981). LAWYER et al. (1981) have found that this compensatory mechanism is not operative when photorespiration is suppressed and net carbon fixation is consequently higher.

Several enzymes have been involved in considerations of how sucrose biosynthesis is controlled. One possibility is the cytosolic FBPase, which is inhibited by AMP, UDP, and fructose-6-P (HARBRON et al., 1981). However, UDP inhibition is unlikely since leaves contain UDPase activity (HUBER and PHARR, 1981). Also, regulation of this enzyme by pH (as occurs with the FBPase in chloroplasts) is difficult to envisage since variation in the cytosolic pH does not occur (ESPIE and COLMAN, 1981; KALLAS and DAHLQUIST, 1981); in some leaves this pH stability may be attributed to lactate dehydrogenase (BETSCHE, 1981).

Other possible controls of sucrose synthesis may involve P-hexose isomerase, for which erythrose-4-P is a potent competitive inhibitor (KELLY and LATZKO, 1980b), or sucrose-P synthetase, which is inhibited by FBP, Pi, UDP (HARBRON et al., 1981), and, in some plants but not others, sucrose (HUBER, 1981b). A continuous spectrophotometric assay for this enzyme was developed recently (HARBRON et al., 1980).

c) Amino Acids and Fatty Acids

The generation of NH_4^+ for amino acid biosynthesis in leaves involves the cytosolic reduction of NO_3^- to NO_2^-, and then the reduction of NO_2^- to NH_4^+ in the chloroplast. The first step utilises NADH generated in the mitochondrion (KADAM et al., 1980; RAMARAO et al., 1981), but the mechanism by which this NADH is indirectly shuttled into the cytosol is still being debated (NAIK and NICHOLAS, 1981). Nitrate reduction by CAM plants, as with other plants, occurs only during the day (CHANG et al., 1981). The NO_3^- in barley leaves has been found to occur in two pools: a small metabolic pool and a large storage pool in the vacuole (MARTINOIA et al., 1981).

Once NH_4^+ is formed, it is incorporated into amino acids via glutamine and glutamate (see KELLY and LATZKO, 1980a). Glutamine synthetase from the blue-green alga *Anabaena* was not subject to control by adenylylation/deadenylylation, as is typical of the animal and bacterial enzymes (ORR and HASELKORN, 1981) but the enzyme from the photosynthetic bacterium *Rhodopseudomonas palustris* was (ALEF and ZUMFT, 1981). α-Ketoglutarate (the C-skeleton for the subsequent glutamine-linked formation of glutamate) apparently can be synthesised from isocitrate in chloroplasts (RANDALL and GIVAN, 1981), although this is not essential since chloroplasts can import α-ketoglutarate using the envelope dicarboxylate carrier (HEBER and HELDT, 1981; PROUDLOVE and THURMAN, 1981). Light-stimulated synthesis of isoleucine, lysine, methionine, and threonine in pea chloroplasts was observed by MILLS et al. (1980). While some activity of at least one threonine-synthesising enzyme was in the cytosol (SAINIS et al., 1981), all activity of the regulatory enzyme of lysine synthesis (dihydrodipicolinate synthase) was found in the chloroplast (WALLSGROVE and MAZELIS, 1980).

The biosynthesis of fatty acids in plant leaves is a particularly active and interesting area of research at the present time. It is generally accepted that all synthesis up to stearate and oleate occurs in the chloroplast. The required starting material (acetate) had been proposed to be generated in the chloroplast from imported pyruvate, but KUHN et al. (1981) and MURPHY and STUMPF (1981) have vigorously promoted an alternative possibility, i.e. that acetate originates from the mitochondrial pool and diffuses into the chloroplast where it is converted to acetyl-CoA by acetyl-CoA synthetase, all activity of which is in the chloroplast. Leaf mitochondria contain an active and highly regulated pyruvate dehydrogenase complex for the formation of acetyl-CoA (which can then be hydrolysed to acetate) from pyruvate (RANDALL et al., 1981).

Fatty acid synthesis from acetyl-CoA by isolated intact chloroplasts is light dependent: the rate in the dark is only 2% of that in the light. The reason for this light dependency is unknown, although regulation of acetyl-CoA carboxylase (which catalyses the next step in fatty acid synthesis) is a favoured possibility (ROUGHAN et al., 1980a). Recent studies of this enzyme have been reported (MOHAN and KEKWICK, 1980; NIKOLAU et al., 1981).

Chloroplasts contain transferases able to link fatty acids onto glycerol-3-P to form 1,2-diacylglycerol (ROUGHAN et al., 1980b; BERTRAMS and HEINZ, 1981). Moreover, the CTP-mediated addition of a second glycerol moiety to the C-3 of the first glycerol to form phosphatidyl glycerol has been demonstrated (MUDD and DEZACKS, 1981). However, the chloroplast is not autonomous if the polyunsaturated fatty acids linoleate and linolenate, which are so abundant in plant lipids, are to be included in any lipid (MURPHY and LEECH, 1981). To obtain these two fatty acids, oleate is specifically exported out of the chloroplast (JOYARD and STUMPF, 1980; DRAPIER et al., 1982) and moves to the endoplasmic reticulum where further desaturation to linoleate occurs (TREMOLIERES et al., 1980a,b). The synthesis of the lipids phosphatidylcholine and phosphatidylethanolamine also occurs on the endoplasmic reticulum (ROUGHAN et al., 1980b; TREMOLIERES et al., 1980b) and there is evidence that oleate is not desaturated to linoleate until after its incorporation into one of these lipids (MURPHY and STUMPF, 1980a). The final desaturation of linoleate to linolenate (which is the most prevalent fatty acid in nature) is poorly understood, although it seems certain that the desaturation occurs only with linoleate present in either phosphatidylcholine (STOBART et al., 1980) or monogalacto syldiacylglycerol (HAWKE and STUMPF, 1980a; JONES and HARWOOD, 1980). In the latter case, lipids containing linoleate formed at the endoplasmic reticulum must first move back to the chloroplast, at least as far as the envelope, because this is where galactosylation to form the mono- and di-galactosyl diacylglycerols takes place (ROUGHAN et al., 1980b; TREMOLIERES et al., 1980b). There is some evidence that these desaturation processes require light in cucumber and barley leaves (MURPHY and STUMPF, 1980b; STOBART et al., 1980), but not in maize leaves (HAWKE and STUMPF, 1980b).

Almost no attention has been given to fatty acid degradation in leaves. GERHARDT (1981) has now reported that the β-oxidation of fatty acids in leaves occurs in microbodies, as is the case with germinating fatty seeds, and not in mitochondria as is the case in animals.

8. C_4 Metabolism

When the carboxylation of PEP to OAA plays an important part in carbon metabolism, C_4 metabolism is said to be involved. Four principal cases are reviewed below.

a) C_4 Plants

At its simplest, carbon flow in C_4 plants involves (a) CO_2 fixation by PEP carboxylase, forming OAA which is reduced to malate (mesophyll cells), then (b) movement of this malate to bundle sheath cells where it is decarboxylated by malic enzyme, producing CO_2 and NADPH (for the Calvin cycle) and pyruvate (which is returned to the mesophyll and used by pyruvate Pi dikinase to regenerate PEP). This carbon flow is called the C_4 pathway. The arrangement of mesophyll and bundle sheath cells is termed Kranz anatomy. A provocative report by PAN and TAN (1981) has resurrected the alternative possibility of a direct transcarboxylation from the carboxyl group of a C_4-compound (e.g., OAA) to a pentose-P.

Some CO_2 fixation by young leaf bases and by older leaves of C_4 plants evidently bypasses PEP carboxylase and is directly catered for by the Calvin cycle (PERCHOROWICZ and GIBBS, 1980; RAGHAVENDRA, 1980a). In mature leaves, the C_4 pathway is thought to concentrate CO_2 at the site of RuBP carboxylase and this appears to permit C_4 plants to require less of this enzyme (KU et al., 1979).

The required metabolite transport across the mesophyll chloroplast envelope of maize has been demonstrated by DAY and HATCH (1981a,b): OAA and malate are exchanged across a dicarboxylate carrier, while a phosphate translocator moves all four of PEP, Pi, glycerate-3-P, and dihydroxyacetone-P. The glycerate-3-P originates from bundle sheath cells and, by competing with OAA for NADPH in mesophyll chloroplasts, may in effect assist the regulation of mesophyll carbon metabolism

(DAY and HATCH, 1981b; HALLBERG and LARSSON, 1981). A complementary regulatory mechanism may originate from the transfer of malate into the bundle sheath. Decarboxylation of this malate yields NADPH which, to be used in the Calvin cycle, needs a corresponding quota of ATP. This seems to be generated during cyclic electron transport (CHAPMAN et al., 1980), and in fact the electrons needed to trigger this cyclic flow appear to originate from some of the NADPH (LEEGOOD et al., 1981).

More non-conventional C_4 plants have been uncovered. Most C_4 plants are of tropical origin, but another temperate example, *Cyperus longus*, has been identified by JONES et al. (1981a). An interesting species is *Portulaca grandiflora*, a CAM-like C_4 plant which has bundle sheath cells containing not only the usual enzymes (for malate decarboxylation and the Calvin cycle) but also a duplicate set of the three enzymes usually restricted to the mesophyll cell (viz. pyruvate Pi dikinase, PEP carboxylase, and NADP-linked malate dehydrogenase; KU et al., 1981; see Fig. 5). Hence the entire C_4 pathway could potentially occur within one cell (a bundle sheath cell) of this plant. The situation with respect to C_3-C_4 intermediate plants has received further attention (BROWN and MORGAN, 1980; RAGHAVENDRA, 1980b) and two new species (crucifers of the genus *Moricandia*) have been discovered (APEL, 1980; HOLADAY et al., 1981). Finally, *Parthenium hysterophorus* has been noted as an example of a C_3 plant with Kranz anatomy (HEDGE and PATIL, 1981).

Three enzymes of the C_4 pathway have received recent attention. The cytosolic (PERROT et al., 1981) PEP carboxylase has been purified from three C_4 grasses (MAREŠ and LEBLOVÁ, 1980). The increase in activity that accompanies greening of etiolated leaves is attributed to de novo synthesis (HAGUE and SIMS, 1980; HAYAKAWA et al., 1981). The enzyme from three C_3-C_4 intermediate *Panicum* species was more similar to the "C_3-enzyme" (from *Panicum laxum*) than to the "C_4-enzyme" (from *Panicum prionitis*; HOLADAY and BLACK, 1981). Regulatory properties of the bundle sheath's NADP-linked malic enzyme have been presented by ASAMI et al. (1979) and the unusual mechanism for light-mediated regulation of pyruvate Pi dikinase has been further studied (EDWARDS et al., 1980; CHAPMAN and HATCH, 1981; SUGIYAMA and HATCH, 1981; see Sect. 5). In plants, this enzyme was once thought to be restricted to CAM plants and the mesophyll cells of C_4 plants, but it is now clear that it also occurs in bundle-sheath cells of some C_4 plants (KU et al., 1981) and in the aleurone layer of developing C_3-cereal grains (MEYER et al., 1982).

b) CAM Plants

In CAM plants, PEP carboxylation occurs nocturnally. The resultant OAA is reduced to malate, which accumulates in the vacuole. Next day, the malate is returned to the cytosol and decarboxylated, the CO_2 being donated into the Calvin cycle so that light-dependent net CO_2 fixation can occur behind closed stomata. Three sub-phases of daytime metabolism have been identified (KENYON et al., 1981). The diurnal rhythm of CAM is resistant to short-term interruption by sudden changes in the dark/light regime (KLUGE et al., 1981a), but field experiments have shown aspects of adaptability, e.g., switch to daytime CO_2 fixation during cool, foggy periods in winter (VON WILLERT et al., 1980). Pseudo-CAM plants which "idle" during water stress by cycling CO_2 through a pool of malate, but which do not display nocturnal CO_2 uptake, have been identified. Examples are the primitive Cactaceae members *Pereskia grandifolia* and *P. aculeata* (RAYDER and TING, 1981).

Malate synthesised in CAM is actively accumulated in vacuoles using a tonoplast-located ATPase (LÜTTGE et al., 1981). It accumulates as

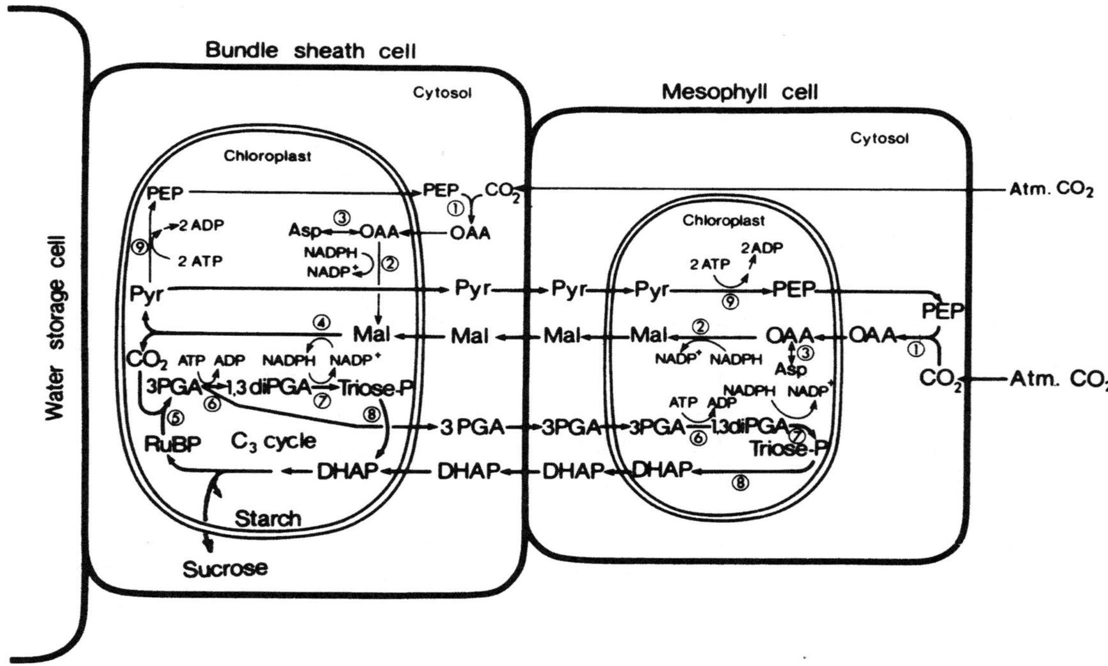

Fig. 5. A proposed scheme for C_4 photosynthesis in *Portulaca grandiflora*. Note the duplication of the mesophyll cell metabolism in the bundle sheath cell. Asp, aspartate; DHAP, dihydroxyacetone-P; Mal, malate; PGA, glycerate-3-P; Pyr, pyruvate. From KU et al. (1981). Reprinted with permission

malic acid (i.e., equivalent amounts of malate anions and protons); a claim that the protons exchange in part with K^+ and Na^+ of epidermal cells in *Mesambryanthemum crystallinum* has been discounted (WINTER et al., 1981). MOREL et al. (1980) have evidence that polyamines might counter the tendency for acidification of the cytosol during CAM, and other influences of the acidification level have been suggested (PHILIPPS, 1980). The daytime decarboxylation of malate is dependent on the Calvin cycle accepting the CO_2 (NISHIDA and HAYASHI, 1980). When decarboxylation is catalysed by malic enzyme, pyruvate Pi dikinase is present to convert the pyruvate to PEP, but when malate dehydrogenase plus PEP carboxykinase are utilised, pyruvate Pi dikinase is not required and in fact is not found (HOLTUM and OSMOND, 1981).

One regulator of CAM is feedback inhibition of PEP carboxylase by malate. The sensitivity of the enzyme to malate appears to vary with temperature and with the diurnal cycle (KLUGE et al., 1980; WINTER, 1980; BUCHANAN-BOLLIG and KLUGE, 1981; KLUGE et al., 1981b). Much greater sensitivity is seen with enzyme in freshly-prepared crude extracts, indicating a clear danger that artifacts are easily obtained in studies of this enzyme (BUCHANAN-BOLLIG et al., 1980; JONES et al., 1981b; WINTER, 1981). At least some similarity between the PEP carboxylases from CAM-members and a C_3-member of the genus *Sedum* is evident (MÜLLER et al., 1980). Difficulties in investigations of another CAM enzyme (NAD-linked malic enzyme from *Crassula argentea*) also have been reported (WEDDING et al., 1981).

c) Aquatic Plants

The possibility that some aquatic plants may be C_4 plants is no longer seriously considered. However, a significant contribution of PEP carboxylase to the carbon metabolism of many aquatic plants has been detected. Examples include the blue-green alga *Coccochloris peniocystis* (COLEMAN and COLMAN, 1980b), several brown algae (KREMER, 1981), aquatic angiosperms (BEER and WETZEL, 1981; SALVUCCI and BOWES, 1981), and seagrasses (BEER et al., 1980). However, one criterion for detecting C_4 metabolism (discrimination of ^{13}C) is equivocal (SMITH and WALKER, 1980).

d) Stomata

The opening of a leaf stoma is a consequence of the swelling of the associated guard cells. This swelling is the result of solute accumulation; in most plants the accumulated solute is K-malate. During opening, starch in guard cells is degraded to PEP which is carboxylated to OAA; this OAA is reduced to malic acid, and while the malate anion remains in the cell, the proton is exported to surrounding cells in exchange for an imported K^+ ion. Hence K-malate accumulates. A notable exception is *Allium cepa*, which does not have starch in guard cells and does not synthesise malate. Instead, all of the solute is imported as KCl (SCHNABL and ZIEGLER, 1977). These features have been recently elegantly demonstrated using isolated guard-cell protoplasts (SCHNABL, 1980a,b; SCHNABL and RASCHKE, 1980). One problem plant is an orchid of the genus *Paphiopedilum* in which K^+ appears not to be the major osmoticum causing stoma opening (OUTLAW et al., 1982).

Malate is presumably used to replenish the starch reserve when stomata close. The pathway by which this occurs is unclear: initial malic enzyme-catalysed decarboxylation to pyruvate is probably involved (OUTLAW) et al., 1981), but the presence of pyruvate Pi dikinase necessary to convert this pyruvate to PEP (SCHNABL, 1981) is in doubt (OUTLAW et al., 1981).

There are obvious similarities between stomatal guard cells and CAM plants with respect to malate accumulation (SCHNABL, 1981; TING, 1981) and after reflecting upon this, and upon features of C_4 plants, COCKBURN (1981) expressed the opinion that CAM, and maybe C_4 photosynthesis, result from the expression in photosynthetic cells of genetic information which is expressed only in guard cells of C_3 plants.

9. Photorespiration

a) Wir wissen, was passiert ...

Photorespiration involves the oxidative cleavage of RuBP into glycerate-3-P and glycolate-P, and the subsequent conversion of two glycolate-P to one glycerate-3-P, one CO_2, and one Pi via a series of seven reactions in four cellular compartments (chloroplast, mitochondrion, peroxisome, and cytosol; see our two preceding reviews). The cleavage of RuBP is catalysed by the oxygenase activity of RuBP carboxylase. The proposed 1:1:1:1 stoichiometry of the reaction : RuBP + O_2 → glycerate-3-P + glycolate-2-P was recently demonstrated by JORDAN and OGREN (1981a). Indications that carbon reserves can be used to supply RuBP for this reaction have been obtained by KAMIŃSKA and MALESZEWSKI (1980) and LAWLOR and PEARLMAN (1981).

Attempts to reduce specifically the oxygenase activity of the enzyme (and thereby reduce photorespiration) have had little success, and a previous claim of manipulation using hydroxylamine has been refuted (BROWN et al., 1980; CHRISTELLER, 1981; JORDAN and OGREN, 1981a; RAGHAVENDRA et al., 1981). However, an observation that O_2 (probably as O_2^-, the superoxide anion) is required for both activities of the enzyme (HENKEL et al., 1980; WILDNER et al., 1980) has been followed by the interesting demonstration that deactivation by removal of O_2 applies only to the oxygenase activity (and not to the carboxylase activity) in the presence of certain N-heterocyclic compounds such as atebrin (WILDNER and HENKEL, 1980). Nevertheless, confidence that an oxygenase-less RuBP carboxylase can be obtained has been low recently, hence it was encouraging to find that evolutionary pressures have increased the proclivity towards carboxylase activity (JORDAN and OGREN, 1981b), thus raising hopes that further increases may be possible on a shorter time scale in the laboratory.

Photosynthetic bacteria can also produce glycolate from an RuBP oxygenase reaction; some of this glycolate may be excreted (KING and ANDERSEN, 1980; STORRØ AND McFADDEN, 1981). An unusual problem has been created by YOKOTA and KITAOKA (1981), who found that mutant bleached cells of *Euglena gracilis* that lack RuBP carboxylase activity still presumably produce glycolate. Some consideration is still given to the alternative of glycolate formation by oxidation of a 2C-thiamin-P_2 complex (attached to transketolase), possibly by O_2^- (CHRISTEN and GASSER, 1980; TAKABE et al., 1980).

The view that photorespiration is virtually absent in C_4 plants has been tempered to include the possibility that some photorespiration occurs in young leaves of maize (MOROT-GAUDRY et al., 1980) and in mature leaves of *Amaranthus graecizans* which have a lower diffusive resistance to O_2 and CO_2 (KU and EDWARDS, 1980). Confusion still reigns in the realm of photorespiration by green algae. *Chlorella* do not seem to photorespire when grown in air levels of CO_2, but this property could not always be attributed solely to a mechanism for concentrating CO_2 in the cell (SHELP and CANVIN, 1980, 1981; cf. Sect. 3). Glycolate formation by *Chlorella* was detected by NAKAMURA and MIYACHI (1980). These authors reported that formation was favoured by a lower growth temperature. However, COLEMAN and COLMAN (1980a) obtained data consistent with some photorespiration occurring in *Chlamydomonas reinhardii* and found that this is favoured by higher temperatures. SNELGAR and GREEN (1981) found that the algal component of a lichen photorespired less when the lichen was moistened. In contrast to green algae, blue-green algae are more firmly believed to lack photorespiration (COLEMAN and COLMAN, 1980a).

Glycolate diffuses out of chloroplasts and into peroxisomes (TAKABE and AKAZAWA, 1981; TOLBERT, 1981) where it is oxidised to glyoxylate and then aminated to form glycine. Support for the theory that some photorespiratory CO_2 may be produced by peroxisomal reactions is diminishing (CHANG and HUANG, 1981; OLIVER, 1981), although a new possibility has arisen (WALTON and BUTT, 1981). Studies with peroxisomes are hampered by the action during isolation of a destructive, presumably vacuolar enzyme (TOMOMATSU and ASAHI, 1980).

Most photorespiratory CO_2 is believed to be released when glycine permeates into the mitochondrion (DAY and WISKICH, 1980) and undergoes the reaction: 2 glycine + NAD^+ → serine + CO_2 + NH_4^+ + NADH. [This reaction also has been found in a photosynthetic bacterium, *Chromatium vinosum* (SADO et al., 1980)]. The development of this mitochondrial capacity is light-dependent (ARRON and EDWARDS, 1980). Reassimilation of the released NH_4^+ is most probably catalysed by glutamine synthetase, two forms of which are present in leaves: one in the chloroplast, the other in the cytosol (WALLSGROVE et al., 1980; BERGMAN et al., 1981; KRETOVICH et al., 1981; HIREL and GADAL, 1982). The NADH formed may be reoxidised by the mitochondrial electron transport chain, or shuttled into the peroxisome to support the reduction of hydroxypyruvate to glycerate (DAY and WISKICH, 1981; JOURNET et al., 1981). The movement of nitrogen and reducing power during the photorespiratory conversion of two glycolate to one glycerate is summarised schematically in Fig. 6.

In our preceding review, we theoretically deduced that, for a plant to survive, all glycerate produced from glycolate would need to be returned to the chloroplast *before* phosphorylation to form glycerate-3-P. This deduction has been substantiated by USUDA and EDWARDS (1980a,b), who found that all glycerate kinase occurs in the chloroplast of both C_3 and C_4 plants, and concluded that all flow of carbon to sucrose, whether direct through the Calvin cycle or indirect through the glycolate pathway, is linked to the phosphate translocator of the chloroplast envelope. Supporting evidence has been obtained by VIIL and PÄRNIK (1979) and LAWLOR and PEARLMAN (1981).

Artificial inhibitors of various reactions of the glycolate pathway have been found, but their application to photosynthesising cells did not increase net CO_2 fixation (DORAVARI and CANVIN, 1980; USUDA et al., 1980). A natural inhibitor (glyoxylate) has shown more promise (OLIVER, 1980), particularly with blue-green algae (BERGMAN, 1980, 1981). Sustained efforts by SOMERVILLE and OGREN (1980a,b, 1981, 1982) have produced four mutants of *Arabidopsis thaliana*, each lacking one of the enzymes involved in photorespiration. All of the mutations were lethal to plants when growing in normal air, but not when glycolate formation was suppressed by growing under a high CO_2 or low O_2 atmosphere. This indicated that the enzymes had no essential function unrelated to

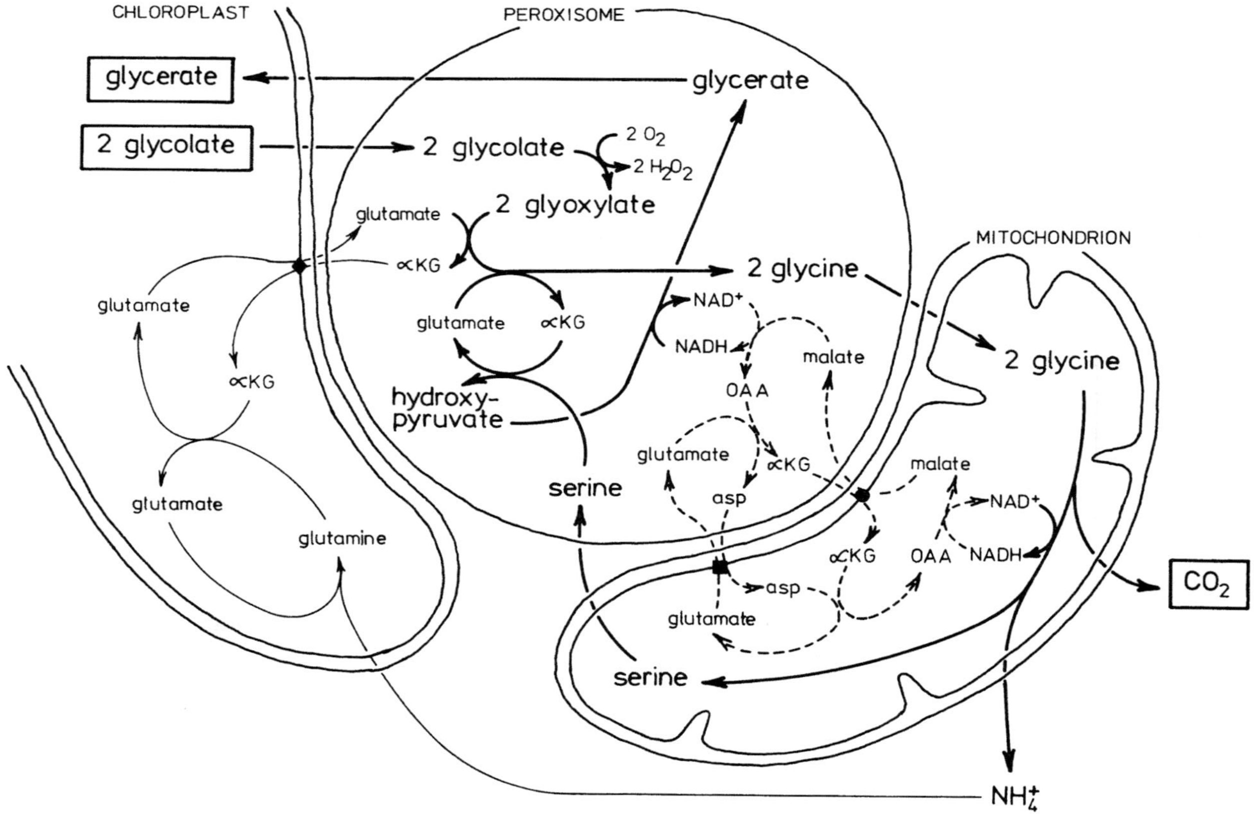

Fig. 6. Cycling of nitrogen and reducing power during the conversion of glycolate to glycerate in photorespiration. The carbon balance of the reactions shown is: 2 glycolate → glycerate + CO_2. *Bold arrows*, glycolate pathway; *thin arrows*, photorespiratory N-cycle; *dashed arrows*, reducing power shuttle; ●, ■, ◆, membrane translocators. Asp, aspartate; αKG, α-ketoglutarate. Adapted from JOURNET et al. (1981)

photorespiration. A tantalizing question is whether or not efforts to remove photorespiration from plants will be commercially successful before the current enrichment of atmospheric CO_2 reaches a point (in 30-50 years?) where it is sufficient to suppress photorespiration (OSMOND, 1981).

The status of mitochondrial "dark" respiration in illuminated plant leaves is still unclear. Current indications are that the tricarboxylic acid cycle and mitochondrial electron transport persist during photosynthesis, but at a reduced rate (CANVIN et al., 1980; GERBAUD and ANDRÉ, 1980; PEISKER and APEL, 1980; AZCÓN-BIETO et al., 1981).

b) ... aber wissen wir, warum?

Most terrestrial plants partially or completely close their stomata during warm, relatively dry periods. While this successfully prevents water loss from the leaf, it also prevents entry of CO_2. Sunlight, of course, continues to reach the chloroplast, and an extremely attractive suggestion is that photorespiration deals with this light energy by cycling CO_2 through a pathway (Fig. 7A) that achieves nothing other than the utilisation of the light-generated ATP and NADPH (OSMOND, 1981). The required oxygen can enter by diffusion through the leaf cuticle. It had been earlier shown that illuminated leaves deprived of this O_2 (in the absence of CO_2), and therefore unable to photorespire, suffered "photoinhibition": reduced photosynthesis upon return to a normal atmosphere, apparently due to damage to one or more components of the photosynthetic apparatus. Similar photoinhibition was seen in normal air when shade plants (e.g., low-light-grown *Phaseolus vulgaris* and the deep-shade fern *Lastreopsis microsora*) were placed in full sunlight (POWLES and CRITCHLEY, 1980; POWLES and THORNE, 1981). However, photoinhibition of C_4 plant leaves required that only CO_2, and not O_2, be absent during illumination, suggesting that C_4 plants have protective mechanisms other than photorespiration (POWLES et al., 1980). The Mehler reaction (in which O_2 acts as electron acceptor, ultimately producing H_2O_2; see Fig. $7B_1$) was also discounted as a dissipator of energy under these conditions. This reaction is believed to occur in all leaves to some extent during photosynthesis to assist in supplying, either directly or indirectly, sufficient ATP for CO_2 fixation (VIIL and PÄRNIK, 1979; ZIEM-HANCK and HEBER, 1980; McVETTY and CANVIN, 1981; STEIGER and BECK, 1981). The H_2O_2 produced could be removed by the system shown in Fig. $7B_1$ (JABLONSKI and ANDERSON, 1981; NAKANO and ASADA, 1981). However, this system, in itself, would not be expected to act as an "energy dissipator" for photosynthesis because coupled electron transport would quickly become restricted by the lack of any reaction regenerating ADP (OSMOND, 1981). On the other hand, if it occurred in conjunction with some cyclic mechanism that achieves nothing other than ATP hydrolysis, the combination might be a successful energy dissipator. A possible ATP-utilising system, based on the "idling" seen with a primitive cactus (RAYDER and TING, 1981), is outlined in Fig. $7B_2$, and in fact some type of CO_2 metabolism was implicated in the protection of C_4 plant leaves against photoinhibition (POWLES et al., 1980). An even simpler ATP disposal system is the "futile cycle" created by the simultaneous activity of FBPase and P-fructokinase [as occurs in bumble-bee flight muscle (NEWSHOLME et al., 1972)], but it is not known whether the P-fructokinase activity in leaves would be sufficient to match the rate of ATP synthesis by the chloroplasts.

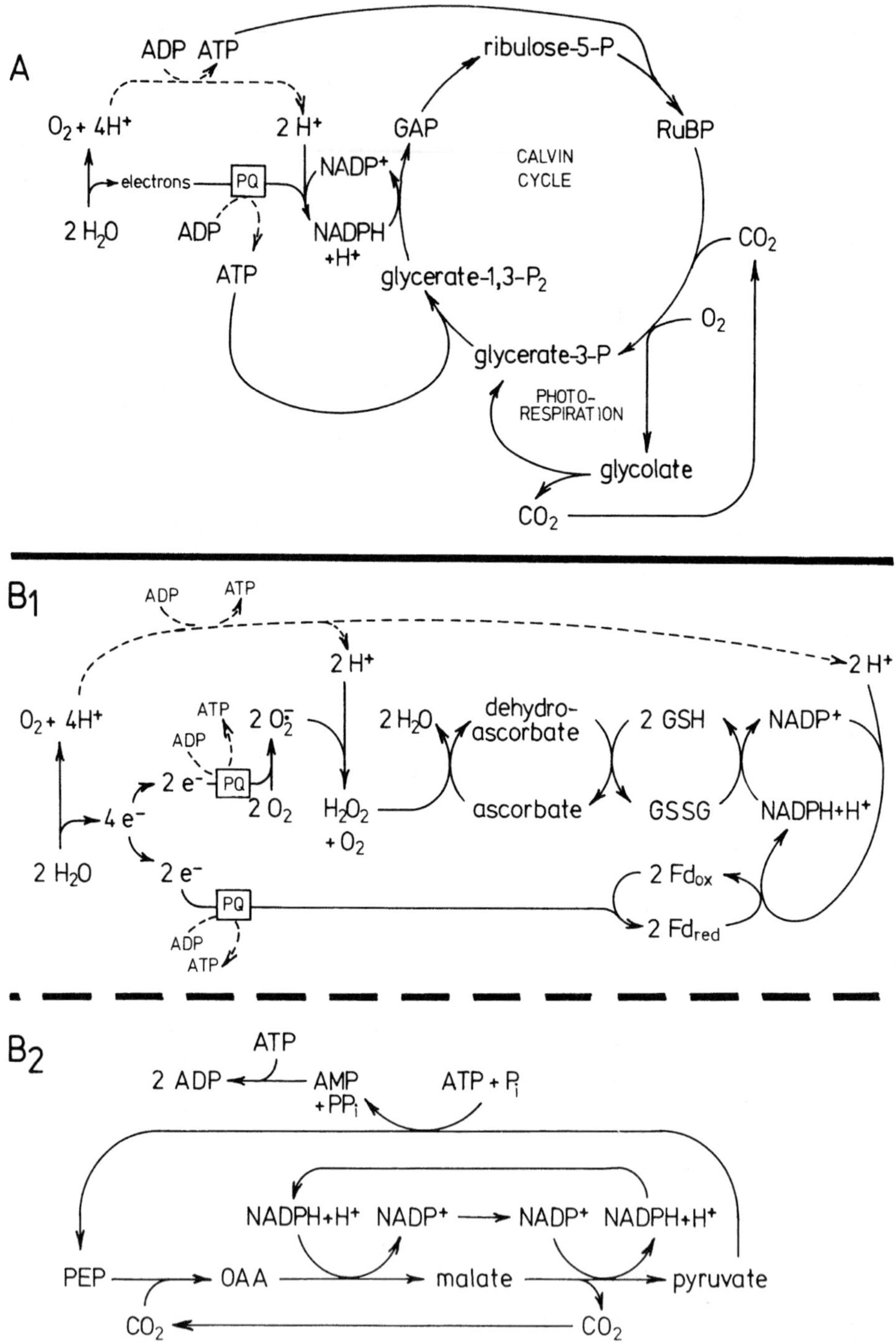
A
ADP ATP
ribulose-5-P
O_2 + 4H⁺
2 H⁺
GAP
RuBP
CALVIN CYCLE
electrons
PQ
NADP⁺
2 H_2O
ADP
NADPH +H⁺
CO_2
ATP
glycerate-1,3-P_2
O_2
glycerate-3-P
PHOTO-RESPIRATION
glycolate
CO_2
B1
ADP
ATP
2 H⁺
2 H⁺
O_2 + 4H⁺
ATP
ADP
2 O_2^-
2 H_2O
dehydro-ascorbate
2 GSH
NADP⁺
2 e⁻
PQ
4 e⁻
2 O_2
H_2O_2 + O_2
ascorbate
GSSG
NADPH+H⁺
2 H_2O
2 e⁻
2 Fd_{ox}
PQ
2 Fd_{red}
ADP
ATP
B2
ATP
2 ADP
AMP +PP_i
ATP + P_i
NADPH+H⁺
NADP⁺
NADP⁺
NADPH+H⁺
PEP
OAA
malate
pyruvate
CO_2
CO_2

Photorespiration still remains a puzzle. If a process other than photorespiration protects C_4 plant leaves against photoinhibition (POWLES et al., 1980), then it seems unfortunate that C_3 plant leaves have chosen instead to photorespire. Even more strange is their persistence in this regard: photorespiration operates continually, losing (as CO_2) 30% - 50% of the carbon fixed and reduced to the level of triose-P by photosynthesis, irrespective of whether or not the chloroplast is momentarily in a photoinhibitory microenvironment. This seems unnecessary, particularly in view of evidence that a C_3 plant lacking several of the reactions of photorespiration would survive quite well under nonphotoinhibitory conditions (SOMERVILLE and OGREN, 1980a,b, 1981, 1982). One could argue that C_3 plants would have gained much by having RuBP oxygenase activity switched on only when needed (e.g., when the CO_2 concentration fell to the compensation point level [1.5 μM]), and by having it switched off at other times (e.g., through more-or-less complete inhibition by air levels of CO_2 [10 μM]). Similar control mechanisms operate at many other branch points in biochemical pathways. Perhaps it is not possible here because the two relevant activities, RuBP carboxylase and RuBP oxygenase, evolved on the one protein. Either plants have been most unfortunate in developing these two activities on this one protein (a protein that is not only an extraordinarily complex enzyme, but also a very slow one), or else there are other aspects of photorespiration about which we are currently in total ignorance.

References

ALEF, K., ZUMFT, W.G.: Z. Naturforsch. 36c, 784-789 (1981). - ANDERSON, J.M., CHARBONNEAU, H., JONES, H.P., McCANN, R.O., CORMIER, M.J.: Biochemistry 19, 3113-3120 (1980). - ANDERSON, L.E.: Biochem. Biophys. Res. Commun. 99, 1199-1202 (1981). - ANDERSON, L.E., HEINRIKSON, R.L.: Plant Physiol. 64, 404-405 (1979). - ANDREWS, T.J., ABEL, K.M.: J. Biol. Chem. 256, 8445-8451 (1981). - APEL, P.: Biochem. Physiol. Pflanzen 175, 386-388 (1980). - ARRON, C.P., EDWARDS, G.E.: Plant Sci. Lett. 18, 229-235 (1980). - ASAMI, S., INOUE, K., AKAZAWA, T.: Arch. Biochem. Biophys. 196, 581-587 (1979). - ASHTON, A.R., ANDERSON, L.E.: Biochim. Biophys. Acta 638, 242-249 (1981). - ASHTON, A.R., BRENNAN, T., ANDERSON, L.E.: Plant Physiol. 66, 605-608 (1980). - AZCÓN-BIETO, J., FARQUHAR, G.D., CABALLERO, A.: Planta 152, 497-504 (1981).

BADGER, M.R., LORIMER, G.H.: Biochemistry 20, 2219-2225 (1981). - BADGER, M.R., KAPLAN, A., BERRY, J.A.: Plant Physiol. 66, 407-413 (1980a). - BADGER, M.R., ANDREWS, T.J., CANVIN, D.T., LORIMER, G.H.: J. Biol. Chem. 255, 7870-7875 (1980b). - BALDUS, B., KELLY, G.J., LATZKO, E.: Phytochemistry 20, 1811-1814 (1981). - BEER, S., WETZEL, R.G.: Plant Sci. Lett. 21, 199-207 (1981). - BEER, S., SHOMER-ILAN,

←

Fig. 7A,B. Pathways that dissipate photosynthetic energy in leaves. (*A*) Use of reducing power and ATP during the simultaneous operation of the Calvin cycle and photorespiration. (B_1) Use of reducing power to produce H_2O_2 which is scavenged by ascorbate peroxidase. Regeneration of ascorbate uses more reducing power. (B_2) Use of ATP during "idling". *Dotted lines* designate photophosphorylation in *A* and B_1. Adapted from NAKANO and ASADA (1981), OSMOND (1981), RAYDER and TING (1981), and Fig. 2 in our review in Vol. 40 of Progress in Botany. Fd_{ox} and Fd_{red}, ferredoxin (oxidised and reduced); GSH and GSSG, glutathione (oxidised and reduced); PQ, plastoquinone

A., WAISEL, Y.: J. Exp. Bot. 31, 1019-1026 (1980). - BELKNAP, W.R., TOGASAKI, R.K.: Proc. Natl. Acad. Sci. USA 78, 2310-2314 (1981). - BERGMAN, A., GARDESTRÖM, P., ERICSON, I.: Physiol. Plant 53, 528-532 (1981). - BERGMAN, B.: Physiol. Plant 49, 398-404 (1980); - Planta 152, 302-306 (1981). - BERTRAMS, M., HEINZ, E.: Plant Physiol. 68, 653-657 (1981). - BETSCHE, T.: Biochem. J. 195, 615-622 (1981). - BEUDEKER, R.F., KUENEN, J.G.: FEBS Lett. 131, 269-274 (1981). - BEUDEKER, R.F., CANNON, G.C., KUENEN, J.G., SHIVELY, J.M.: Arch. Microbiol. 124, 185-189 (1980). - BEUDEKER, R.F., CODD, G.A., KUENEN, J.G.: Arch. Microbiol. 129, 361-367 (1981). - BOWIEN, B., MAYER, F., SPIESS, E., PÄHLER, A., ENGLISCH, U., SAENGER, W.: Europ. J. Biochem. 106, 405-410 (1980). - BRENNAN, T., ANDERSON, L.E.: Plant Physiol. 66, 815-817 (1980). - BROWN, H.M., REJDA, J.M., CHOLLET, R.: Biochim. Biophys. Acta 614, 545-552 (1980). - BROWN, R.H., MORGAN, J.A.: Plant Physiol. 66, 541-544 (1980). - BUCHANAN, B.B.: Annu. Rev. Plant Physiol. 31, 341-374 (1980). - BUCHANAN-BOLLIG, I.C., KLUGE, M.: Planta 152, 181-188 (1981). - BUCHANAN-BOLLIG, I.C., KLUGE, M., LÜTTGE, U.: Z. Pflanzenphysiol. 97, 457-470 (1980).

CAEMMERER, S. von, FARQUHAR, G.D.: Planta 153, 376-387 (1981). - CANVIN, D.T., SHELP, B.J.: Abst. XIII Int. Bot. Congr. Sydney, 15 (1981). - CANVIN, D.T., BERRY, J.A., BADGER, M.R., FOCK, H., OSMOND, C.B.: Plant Physiol. 66, 302-307 (1980). - CHANG, C.-C., HUANG, A.H.C.: Plant Physiol. 67, 1003-1006 (1981). - CHANG, N.K., VINES, H.M., BLACK, C.C.: Plant Physiol. 68, 464-468 (1981). - CHAPMAN, K.S.R., HATCH, M.D.: Arch. Biochem. Biophys. 210, 82-89 (1981). - CHAPMAN, K.S.R., BERRY, J.A., HATCH, M.D.: Arch. Biochem. Biophys. 202, 330-341 (1980). - CHARLES, S.A., HALLIWELL, B.: Biochem. J. 189, 373-376 (1980a); - Biochem. J. 188, 775-779 (1980b); - Planta 151, 242-246 (1981a); - Biochem. J. 200, 357-363 (1981b). - CHATTERTON, N.J., SILVIUS, J.E.: Physiol. Plant 49, 141-144 (1980). - CHEHEBAR, C., WOLOSIUK, R.A.: Biochim. Biophys. Acta 613, 429-438 (1980). - CHEVALIER, S., PARIS, N.: Physiol. Vég. 19, 23-31 (1981). - CHRISTELLER, J.T.: Biochem. J. 193, 839-844 (1981). - CHRISTEN, P., GASSER, A.: Europ. J. Biochem. 107, 73-77 (1980). - COCKBURN, W.: Plant Cell Environ. 4, 417-418 (1981). - COLEMAN, J.R., COLMAN, B.: Plant Physiol. 65, 980-983 (1980a); - Planta 149, 318-320 (1980b); - Plant Physiol. 67, 917-921 (1981). - COPELAND, L., PREISS, J.: Plant Physiol. 68, 996-1001 (1981). - CRAWFORD, N.A., YEE, B.C., BUCHANAN, B.B.: Plant Sci. Lett. 22, 317-326 (1981).

DAILEY, F., CRIDDLE, R.S.: Arch. Biochem. Biophys. 201, 594-602 (1980). - DAY, D.A., HATCH, M.D.: Arch. Biochem. Biophys. 211, 738-742 (1981a); - ibid. 211, 743-749 (1981b). - DAY, D.A., WISKICH, J.T.: FEBS Lett. 112, 191-194 (1980); - Plant Physiol. 68, 425-429 (1981). - DAY, D.A., JENKINS, C.L.D., HATCH, M.D.: Aust. J. Plant Physiol. 8, 21-29 (1981). - DODDS, J.H.: Z. Pflanzenphysiol. 96, 177-179 (1980). - DORAVARI, S., CANVIN, D.T.: Plant Physiol. 66, 628-631 (1980). - DRAPIER, D., DUBACQ, J.-P., TRÉMOLIÈRES, A., MAZLIAK, P.: Plant Cell Physiol. 23, 125-135 (1982).

EDWARDS, G.E., UJIHIRA, M., SUGIYAMA, T.: Photosyn. Res. 1, 199-207 (1980). - ENSER, U., HEBER, U.: Biochim. Biophys. Acta 592, 577-591 (1980). - ESPIE, G.S., COLMAN, B.: Planta 153, 210-216 (1981).

FERRI, G., IADAROLA, P., ZAPPONI, M.C.: Biochim. Biophys. Acta 660, 325-332 (1981). - FERRIER, J.M.: Plant Physiol. 66, 1198-1199 (1980). - FINDENEGG, G.R.: Plant Sci. Lett. 18, 289-297 (1980). - FLÜGGE, U.I., HELDT, H.W.: Biochim. Biophys. Acta 638, 296-304 (1981). - FLÜGGE, U.I., FREISL, M., HELDT, H.W.: Planta 149, 48-51 (1980a); - Plant Physiol. 65, 574-577 (1980b). - FOYER, C., WALKER, D., SPENCER, C.,

MANN, B.: Biochem. J. 202, 429-434 (1982). - FUCHS, G., STUPPERICH, E., JAENCHEN, R.: Arch. Microbiol. 128, 56-63 (1980a). - FUCHS, G., STUPPERICH, E., EDEN, G.: Arch. Microbiol. 128, 64-71 (1980b). - FURBANK, R.T., LILLEY, R.McC.: Biochim. Biophys. Acta 592, 65-75 (1980); - Plant Physiol. 67, 1036-1041 (1981).

GALUN, E.: Annu. Rev. Plant Physiol. 32, 237-266 (1981). - GARRETT, M.K.: Nature 274, 913-915 (1978). - GAUDILLERE, J.P.: GAUDILLERE, M.: Photosyn. Res. 2, 167-173 (1981). - GERBAUD, A., ANDRÉ, M.: Plant Physiol. 66, 1032-1036 (1980). - GERHARDT, B.: FEBS Lett. 126, 71-73 (1981). - GIERSCH, C., HEBER, U., KAISER, G., WALKER, D.A., ROBINSON, S.P.: Arch. Biochem. Biophys. 205, 246-259 (1980). - GORDON, A.J., RYLE, G.J.A., WEBB, G.: J. Exp. Bot. 31, 845-850 (1980). - GRANT, B.R., HOWARD, R.J.: Plant Physiol. 66, 29-33 (1980).

HAGUE, D.R., SIMS, T.L.: Plant Physiol. 66, 505-509 (1980). - HALL, N.P., PIERCE, J., TOLBERT, N.E.: Arch. Biochem. Biophys. 212, 115-119 (1981). - HALLBERG, M., LARSSON, C.: Arch. Biochem. Biophys. 208, 121-130 (1981). - HAMPP, R.: Planta 150, 291-298 (1980). - HARBRON, S., WOODROW, I.E., KELLY, G.J., ROBINSON, S.P., LATZKO, E., WALKER, D.A.: Anal. Biochem. 107, 56-59 (1980). - HARBRON, S., FOYER, C., WALKER, D.: Arch. Biochem. Biophys. 212, 237-246 (1981). - HARMS, H., KOOPS, H.-P., MARTINY, H., WULLENWEBER, M.: Arch. Microbiol. 128, 280-281 (1981). - HATCH, A.L., JENSEN, R.G.: Arch. Biochem. Biophys. 205, 587-594 (1980). - HAWKE, J.C., STUMPF, P.K.: Arch. Biochem. Biophys. 203, 296-306 (1980a); - Plant Physiol. 65, 1027-1030 (1980b). - HAYAKAWA, S., MATSUNAGA, K., SUGIYAMA, T.: Plant Physiol. 67, 133-138 (1981). - HEBER, U., HELDT, H.W.: Annu. Rev. Plant Physiol. 32, 139-168 (1981). - HEDGE, B.A., PATIL, T.M.: Photosynthetica 15, 1-4 (1981). - HENKEL, J., SANE, P.V., WILDNER, G.F.: Biochim. Biophys. Acta 614, 564-576 (1980). - HEROLD, A.: New Phytol. 86, 131-144 (1980). - HEROLD, A., McGEE, E.E.M., LEWIS, D.H.: New Phytol. 85, 1-13 (1980). - HEROLD, A., LEEGOOD, R.C., McNEIL, P.H., ROBINSON, S.P.: Plant Physiol. 67, 85-88 (1981). - HERTIG, C., WOLOSIUK, R.A.: Biochem. Biophys. Res. Commun. 97, 325-333 (1980). - HIREL, B., GADAL, P.: Physiol. Plant 54, 69-74 (1982). - HOLADAY, A.S., BLACK, C.C.: Plant Physiol. 67, 330-334 (1981). - HOLADAY, A.S., SHIEH, Y.-J., LEE, K.W., CHOLLET, R.: Biochim. Biophys. Acta 637, 334-341 (1981). - HOLTUM, J.A.M., OSMOND, C.B.: Aust. J. Plant Physiol. 8, 31-44 (1981). - HUBER, S.C.: Z. Pflanzenphysiol. 101, 49-54 (1981a); - ibid. 102, 443-450 (1981b). - HUBER, S.C., MORELAND, D.E.: Plant Physiol. 67, 163-169 (1981). - HUBER, S.C., PHARR, D.M.: Plant Physiol. 68, 1294-1298 (1981). - HUNER, N.P.A., PALTA, J.P., LI, P.H., CARTER, J.V.: Can. J. Biochem. 59, 280-289 (1981).

JABLONSKI, P.P., ANDERSON, J.W.: Plant Physiol. 67, 1239-1244 (1981). - JACQUOT, J.-P.P., BUCHANAN, B.B., MARTIN, F., VIDAL, J.: Plant Physiol. 68, 300-304 (1981). - JENSEN, R.G.: pp. 273-313 in: The Biochemistry of Plants, Vol. 1, ed. N.E. TOLBERT. New York: Academic Press 1980. - JOHAL, S., CHOLLET, R.: FEBS Lett. 124, 75-78 (1981). - JOHAL, S., BOURQUE, D.P., SMITH, W.W., SUH, S.W., EISENBERG, D.: J. Biol. Chem. 255, 8873-8880 (1980). - JONES, A.V.M., HARWOOD, J.L.: Biochem. J. 190, 851-854 (1980). - JONES, M.B., HANNON, G.E., COFFEY, M.D.: Plant Cell Environ. 4, 161-168 (1981a). - JONES, R., BUCHANAN, I.C., WILKINS, M.B., FEWSON, C.A., MALCOLM, A.D.B.: J. Exp. Bot. 32, 427-441 (1981b). - JORDAN, D.B., OGREN, W.L.: Plant Physiol. 67, 237-245 (1981a); - Nature 291, 513-515 (1981b). - JOSEPH, M.C., RANDALL, D.D., NELSON, C.J.: Plant Physiol. 68, 894-898 (1981). - JOURNET, E.-P., NEUBURGER, M., DOUCE, R.: Plant Physiol. 67, 467-469 (1981). - JOYARD, J., STUMPF, P.K.: Plant Physiol. 65, 1039-1043 (1980).

KADAM, S.S., JOHARI, R.P., RAMARAO, C.S., SRINIVASAN: Phytochemistry 19, 2095-2097 (1980). - KAISER, W.M., URBACH, W., GIMMLER, H.: Planta 149, 170-175 (1980). - KAISER, W.M., KAISER, G., PRACHUAB, P.K., WILDMAN, S.G., HEBER, U.: Planta 153, 416-422 (1981a). - KAISER, W.M., KAISER, G., SCHÖNER, S., NEIMANIS, S.: Planta 153, 430-435 (1981b). - KALLAS, T., DAHLQUIST, F.W.: Biochemistry 20, 5900-5907 (1981). - KAMIŃSKA, Z., MALESZEWSKI, S.: Photosyn. Res. 1, 45-51 (1980). - KAPLAN, A.: J. Exp. Bot. 32, 669-677 (1981). - KAPLAN, A., BADGER, M.R., BERRY, J.A.: Planta 149, 219-226 (1980). - KELLY, G.J., LATZKO, E.: pp. 183-208 in: The Biochemistry of Plants, Vol. 1, ed. N.E. TOLBERT. New York: Academic Press 1980a; - Photosyn. Res. 1, 181-187 (1980b). - KENYON, W.H., HOLADAY, A.S., BLACK, C.C.: Plant Physiol. 68, 1002-1007 (1981). - KING, W.R., ANDERSEN, K.: Arch. Microbiol. 128, 84-90 (1980). - KLUGE, M., BÖCHER, M., JUNGNICKEL, G.: Z. Pflanzenphysiol. 97, 197-204 (1980). - KLUGE, M., BÖHLKE, C., QUEIROZ, O.: Planta 152, 87-92 (1981a). - KLUGE, M., BRULFERT, J., QUEIROZ, O.: Plant Cell Environ. 4, 251-256 (1981b). - KNOWLES, F.C., CHANLEY, J.D., PON, N.G.: Arch. Biochem. Biophys. 202, 106-115 (1980). - KOMBRINK, E., WÖBER, G.: Planta 149, 130-137 (1980). - KREMER, B.P.: Z. Naturforsch. 36c, 840-847 (1981). - KRETOVICH, W.L., EVSTIGNEEVA, Z.G., PUSHKIN, A.V., DZHOKHARIDZE, T.Z.: Phytochemistry 20, 625-629 (1981). - KU, S.B., EDWARDS, G.E.: Planta 147, 277-282 (1980). - KU, S.B., SCHMITT, M.R., EDWARDS, G.E.: J. Exp. Bot. 30, 89-98 (1979). - KU, S.B., SHIEH, Y.-J., REGER, B.J., BLACK, C.C.: Plant Physiol. 68, 1073-1080 (1981). - KUHN, D.N., KNAUF, M., STUMPF, P.K.: Arch. Biochem. Biophys. 209, 441-450 (1981).

LAING, W.A., STITT, M., HELDT, H.W.: Biochim. Biophys. Acta 637, 348-359 (1981). - LARA, C., de la TORRE, A., BUCHANAN, B.B.: Biochem. Biophys. Res. Commun. 94, 1337-1344 (1980). - LARSEN, P.O., CORNWELL, K.L., GEE, S.L., BASSHAM, J.A.: Plant Physiol. 68, 292-299 (1981). - LAWLOR, D.W., PEARLMAN, J.G.: Plant Cell Environ. 4, 37-52 (1981). - LAWYER, A.L., CORNWELL, K.L., LARSEN, P.O., BASSHAM, J.A.: Plant Physiol. 68, 1231-1236 (1981). - LEEGOOD, R.C., WALKER, D.A.: Arch. Biochem. Biophys. 200, 575-582 (1980a); - Biochim. Biophys. Acta 593, 362-370 (1980b); - FEBS Lett. 116, 21-24 (1980c); - Plant Cell Environ. 4, 59-66 (1981a); - Arch. Biochem. Biophys. 212, 644-650 (1981b). - LEEGOOD, R.C., CROWTHER, D., WALKER, D.A., HIND, G.: FEBS Lett. 126, 89-92 (1981). - LENDZIAN, K.J.: Plant Physiol. 66, 8-12 (1980). - LIEDVOGEL, B., KLEINIG, H.: Planta 150, 170-173 (1980). - LILLEY, R. McC., LARKUM, A.W.D.: Plant Physiol. 67, 5-8 (1981). - LIN, C.H., STOCKING, C.R.: Plant Physiol. 65, 897-901 (1980). - LORIMER, G.H., Annu. Rev. Plant Physiol. 32, 349-383 (1981). - LÜTTGE, U., SMITH, J.A.C., MARIGO, G., OSMOND, C.B.: FEBS Lett. 126, 81-84 (1981).

MÄCHLER, F., NÖSBERGER, J.: J. Exp. Bot. 31, 1485-1491 (1980). - MADORE, M., WEBB, J.A.: Can. J. Bot. 60, 126-130 (1982). - MAREŠ, J., LEBLOVÁ, S.: Photosynthetica 14, 25-31 (1980). - MARTINOIA, E., HECK, U., WIEMKEN, A.: Nature 289, 292-294 (1981). - MAURY, W.J., HUBER, S.C., MORELAND, D.E.: Plant Physiol. 68, 1257-1263 (1981). - McCURRY, S.D., PIERCE, J., TOLBERT, N.E., ORME-JOHNSON, W.H.: J. Biol. Chem. 256, 6623-6628 (1981). - McNEIL, P.H., FOYER, C.H., WALKER, D.A., BIRD, I.F., CORNELIUS, M.J., KEYS, A.J.: Plant Physiol. 67, 530-534 (1981). - McVETTY, P.B.E., CANVIN, D.T.: Can. J. Bot. 59, 721-725 (1981). - MEYER, A.O., KELLY, G.J., LATZKO, E.: Plant Physiol. 69, 7-10 (1982). - MILLER, A.G., COLMAN, B.: J. Bacteriol. 143, 1253-1259 (1980). - MILLS, W.R., LEA, P.J., MIFLIN, B.J.: Plant Physiol. 65, 1166-1172 (1980). - MOHAMED, A.H., ANDERSON, L.E.: Arch. Biochem. Biophys. 209, 606-612 (1981). - MOHAN, S.B., KEKWICK, R.G.O.: Biochem. J. 187, 667-676 (1980). - MOREL, C., VILLANUEVA, V.R., QUEIROZ, O.: Planta 149, 440-444 (1980). - MOROT-GAUDRY, J.F., FARINEAU, J.P.,

HUET, J.C.: Plant Physiol. 66, 1079-1084 (1980). - MUDD, J.B., DEZACKS, R.: Arch. Biochem. Biophys. 209, 584-591 (1981). - MÜLLER, D., KLUGE, M., GRÖSCHEL-STEWART, U.: Plant Cell Environ. 3, 249-253 (1980). - MURPHY, D.J., LEECH, R.M.: Plant Physiol. 68, 762-765 (1981). - MURPHY, D.J., STUMPF, P.K.: Plant Physiol. 66, 660-665 (1980a); - ibid. 66, 666-671 (1980b); - Arch. Biochem. Biophys. 212, 730-739 (1981). - MURPHY, D.J., WALKER, D.A.: FEBS Lett. 134, 163-166 (1981). - MVÉ AKAMBA, L., ANDERSON, L.E.: Plant Physiol. 67, 197-200 (1981).

NAGATA, T., ISHII, S.: Can. J. Bot. 57, 1820-1823 (1979). - NAIK, M.S., NICHOLAS, D.J.D.: Aust. J. Plant Physiol. 8, 515-524 (1981). - NAKAMURA, Y., MIYACHI, S.: Plant Cell Physiol. 21, 765-774 (1980); - ibid. 23, 333-341 (1982). - NAKANO, Y., ASADA, K.: Plant Cell Physiol. 22, 867-880 (1981). - NEWSHOLME, E.A., CRABTREE, B., HIGGENS, S.J., THORNTON, S.D., START, C.: Biochem. J. 128, 89-97 (1972). - NIKOLAU, B.J., HAWKE, J.C., SLACK, C.R.: Arch. Biochem. Biophys. 211, 605-612 (1981). - NISHIDA, K., HAYASHI, Y.: Plant Sci. Lett. 19, 271-276 (1980). - NISHIZAWA, A.N., BUCHANAN, B.B.: J. Biol. Chem. 256, 6119-6126 (1981).

OKABE, K., CODD, G.A.: Plant Cell Physiol. 21, 1117-1127 (1980). - OKITA, T.W., PREISS, J.: Plant Physiol. 66, 870-876 (1980). - OLIVER, D.J.: Plant Physiol. 65, 888-892 (1980); - ibid. 68, 1031-1034 (1981). - ORR, J., HASELKORN, R.: J. Biol. Chem. 256, 13099-13104 (1981). - OSMOND, C.B.: Biochim. Biophys. Acta 639, 77-98 (1981). - OUTLAW, W.H., MANCHESTER, J., BROWN, P.H.: Plant Physiol. 68, 1047-1051 (1981). - OUTLAW, W.H., MANCHESTER, J., ZENGER, V.E.: Can. J. Bot. 60, 240-244 (1982).

PAN, D., TAN, K.H.: Z. Naturforsch. 36c, 688-691 (1981). - PEISKER, M., APEL, P.: Z. Pflanzenphysiol. 100, 389-395 (1980). - PERCHOROWICZ, J.T., GIBBS, M.: Plant Physiol. 65, 802-809 (1980). - PERCHOROWICZ, J.T., RAYNES, D.A., JENSEN, R.G.: Proc. Natl. Acad. Sci. USA 78, 2985-2989 (1981). - PERROT, C., VIDAL, J., BURLET, A., GADAL, P.: Planta 151, 226-231 (1981). - PHILLIPS, R.: Nature 287, 727-728 (1980). - PLA, A., LOPEZ-GORGE, J.: Biochim. Biophys. Acta 636, 113-118 (1981). - PORTIS, A.R.: Plant Physiol. 67, 985-989 (1981). - POWLES, S.B., CRITCHLEY, C.: Plant Physiol. 65, 1181-1187 (1980). - POWLES, S.B., THORNE, S.W.: Planta 152, 471-477 (1981). - POWLES, S.B., CHAPMAN, K.S.R., OSMOND, C.B.: Aust. J. Plant Physiol. 7, 737-747 (1980). - PRADEL, J., SOULIÉ, J.-M., BUC, J., MEUNIER, J.C., RICARD, J.: Europ. J. Biochem. 113, 507-511 (1981). - PREISS, J., OKITA, T.W., GREENBERG, E.: Plant Physiol. 66, 864-869 (1980). - PROUDLOVE, M.O., THURMAN, D.A.: New Phytol. 88, 255-264 (1981).

RAGHAVENDRA, A.S.: Physiol. Plant 49, 405-409 (1980a); - Photosynthetica 14, 271-283 (1980b). - RAGHAVENDRA, A.S., CARRILLO, N.J., VALLEJOS, R.H.: Plant Cell Physiol. 22, 1113-1117 (1981). - RAMARAO, C.S., SRINIVASAN, NAIK, M.S.: New Phytol. 87, 517-525 (1981). - RANDALL, D.D., GIVAN, C.V.: Plant Physiol. 68, 70-73 (1981). - RANDALL, D.D., WILLIAMS, M., RAPP, B.J.: Arch. Biochem. Biophys. 207, 437-444 (1981). - RAYDER, L., TING, I.P.: Plant Physiol. 68, 139-142 (1981). - REISSER, W., BENSELER, W.: Arch. Microbiol. 129, 178-180 (1981). - REJDA, J.M., JOHAL, S., CHOLLET, R.: Arch. Biochem. Biophys. 210, 617-624 (1981). - REPKA, J., MAREK, J., HRAŠKA, Š., BEŽO, M.: Photosynthetica 15, 256-257 (1981). - ROBINSON, S.P., WALKER, D.A.: Plant Physiol. 65, 902-905 (1980a); - Arch. Biochem. Biophys. 202, 617-623 (1980b). - ROSA, L.: FEBS Lett. 134, 151-154 (1981). - ROSA, L., WHATLEY, F.R.: Plant Physiol. 68, 364-370 (1981). - ROUGHAN, P.G., KAGAWA, T., BEEVERS, H.: Plant Sci. Lett. 18, 221-228 (1980a). - ROUGHAN, P.G., HOLLAND, R., SLACK, C.R.: Biochem. J. 188, 17-24 (1980b).

SADO, M., ASAMI, S., NISHIMURA, M., AKAZAWA, T.: Plant Cell Physiol. 21, 1077-1084 (1980). - SAINIS, J.K., MAYNE, R.G., WALLSGROVE, R.M. LEA, P.J., MIFLIN, B.J.: Planta 152, 491-496 (1981). - SALVUCCI, M.E., BOWES, G.: Plant Physiol. 67, 335-340 (1981). - SCHEIBE, R.: FEBS Lett. 133, 301-304 (1981). - SCHEIBE, R., ANDERSON, L.E.: Biochim. Biophys. Acta 636, 58-64 (1981). - SCHMIDT, A.: Arch. Microbiol. 127, 259-265 (1980); - Planta 152, 101-104 (1981a); - Z. Naturforsch. 36c, 396-399 (1981b). - SCHNABL, H.: Z. Naturforsch. 35c, 621-626 (1980a); - Planta 149, 52-58 (1980b); - Planta 152, 307-313 (1981). - SCHNABL, H., RASCHKE, K.: Plant Physiol. 65, 88-93 (1980). - SCHNABL, H., ZIEGLER, H.: Planta 136, 37-43 (1977). - SHELP, B.J., CANVIN, D.T.: Plant Physiol. 65, 780-784 (1980); - Plant Physiol. 68, 1500-1503 (1981). - SICHER, R.C., HATCH, A.L., STUMPF, D.K., JENSEN, R.G.: Plant Physiol. 68, 252-255 (1981). - SIEBERT, K., SCHOBERT, P., BOWIEN, B.: Biochim. Biophys. Acta 658, 35-44 (1981). - SMITH, F.A., WALKER, N.A.: New Phytol. 86, 245-259 (1980). - SNELGAR, W.P., GREEN, T.G.A.: J. Exp. Bot. 32, 661-668 (1981). - SOMERVILLE, C.R., OGREN, W.L.: Nature 286, 257-259 (1980a); - Proc. Natl. Acad. Sci. USA 77, 2684-2687 (1980b); - Plant Physiol. 67, 666-671 (1981); - Biochem. J. 202, 373-380 (1982). - SPALDING, M.H., EDWARDS, G.E.: Plant Physiol. 65, 1044-1048 (1980). - STEIGER, H.-M., BECK, E.: Plant Cell Physiol. 22, 561-576 (1981). - STEUP, M.: Biochim. Biophys. Acta 659, 123-131 (1981). - STEUP, M., MELKONIAN, M.: Physiol. Plant 51, 343-348 (1981). - STEUP, M., SCHÄCHTELE, C.: Planta 153, 351-361 (1981). - STITT, M., ap REES, T.: Phytochemistry 19, 1583-1585 (1980). - STITT, M., HELDT, H.W.: Plant Physiol. 68, 755-761 (1981a); - Biochim. Biophys. Acta 638, 1-11 (1981b). - STITT, M., WIRTZ, W., HELDT, H.W.: Biochim. Biophys. Acta 593, 85-102 (1980). - STOBART, A.K., STYMNE, S., APPELQVIST, L.-Å.: Phytochemistry 19, 1397-1402 (1980). - STORRØ, I., McFADDEN, B.A.: Arch. Microbiol. 129, 317-320 (1981). - SUGIYAMA, T., HATCH, M.D.: Plant Cell Physiol. 22, 115-126 (1981).

TABITA, F.R.: J. Bacteriol. 143, 1275-1280 (1980). - TAKABE, T., AKAZAWA, T.: Plant Physiol. 68, 1093-1097 (1981). - TAKABE, T., ASAMI, S., AKAZAWA, T.: Biochemistry 19, 3985-3989 (1980). - TING, I.P.: Plant Sci. Lett. 21, 215-221 (1981). - TOLBERT, N.E.: Annu. Rev. Biochem. 50, 133-157 (1981). - TOMOMATSU, A., ASAHI, T.: Plant Cell Physiol. 21, 689-698 (1980). - TREMOLIERES, A., DRAPIER, D., DUBACQ, J.P., MAZLIAK, P.: Plant Sci. Lett. 18, 257-269 (1980a). - TREMOLIERES, A., DUBACQ, J.P., DRAPIER, D., MULLER, M., MAZLIAK, P.: FEBS Lett. 114, 135-138 (1980b). - TSANG, M.L.-S.: J. Bacteriol. 146, 1059-1066 (1981). - TSUZUKI, M., SHIRAIWA, Y., MIYACHI, S.: Plant Cell Physiol. 21, 677-688 (1980).

USUDA, H., EDWARDS, G.E.: Plant Physiol. 65, 1017-1022 (1980a); - Aust. J. Plant Physiol. 7, 655-662 (1980b); - Plant Physiol. 67, 854-858 (1981). - USUDA, H., ARRON, G.P., EDWARDS, G.E.: J. Exp. Bot. 31, 1477-1483 (1980).

VATER, J., GAUDSZUN, T., SCHARNOW, H., SALNIKOW, J.: Z. Naturforsch. 35c, 416-422 (1980). - VIIL, J., PÄRNIK, T.: Z. Pflanzenphysiol. 95, 213-225 (1979). - VOLOKITA, M., KAPLAN, A., REINHOLD, L.: Plant Physiol. 67, 1119-1123 (1981).

WADA, K., BUCHANAN, B.B.: FEBS Lett. 124, 237-240 (1981). - WALLSGROVE, R.M., MAZELIS, M.: FEBS Lett. 116, 189-192 (1980). - WALLSGROVE, R.M., KEYS, A.J., BIRD, I.F., CORNELIUS, M.J., LEA, P.J., MIFLIN, B.J.: J. Exp. Bot. 31, 1005-1017 (1980). - WALTON, N.J., BUTT, V.S.: Planta 153, 225-231 (1981). - WARA-ASWAPATI, O., KEMBLE, R.J., BRADBEER, J.W.: Plant Physiol. 66, 34-39 (1980). - WARDROP, A.J., POLYA, G.M.: Plant Physiol. 66, 105-111 (1980). - WATTERSON, D.M., IVERSON, D.B., van ELDIK, L.J.: Biochemistry 19, 5762-5768 (1980). - WEBB, D.T., BERLINER, M.D., CARLSON, I.: Z. Pflanzenphysiol. 96, 325-329 (1980). - WEDDING, R.T., CANELLAS, P.F., BLACK, M.K.: Plant Physiol. 68, 1416-1423 (1981). - WEIS, E.: Planta 151, 33-39 (1981a); - FEBS Lett. 129, 197-200 (1981b). - WILDMAN, S.G., JOPE, C.A., ATCHISON, B.A.: Bot. Gaz. 141, 24-36 (1980). - WILDNER, G.F., HENKEL, J.: FEBS Lett. 113, 81-84 (1980). - WILDNER, G.F., SANE, P.V., HENKEL, J.: Z. Naturforsch. 35c, 239-248 (1980). - WILLERT, D.J. von, BRINCKMANN, E., SCHEITLER, B., SCHULZE, E.-D., THOMAS, D.A., TREICHEL, S.: Naturwissenschaften 67, 21-28 (1980). - WINTER, K.: Plant Physiol. 65, 792-796 (1980); - Aust. J. Plant Physiol. 8, 115-119 (1981). - WINTER, K., EDWARDS, G.E., HOLTUM, J.A.M.: Plant Physiol. 68, 355-357 (1981). - WIRTZ, W., STITT, M., HELDT, H.W.: Plant Physiol. 66, 187-193 (1980). - WOODROW, I.E., WALKER, D.A.: Biochem. J. 191, 845-849 (1980).

YAMADA, T., SAKAGUCHI, K.: Agric. Biol. Chem. 45, 1905-1909 (1981). - YAMAGISHI, A., SATOH, K., KATOH, S.: Biochim. Biophys. Acta 637, 252-263 (1981). - YEE, B.C., de la TORRE, A., CRAWFORD, N.A., LARA, C., CARLSON, D.E., BUCHANAN, B.B.: Arch. Microbiol. 130, 14-18 (1981). - YEOH, H.-H., BADGER, M.R., WATSON, L.: Plant Physiol. 66, 1110-1112 (1980); - Plant Physiol. 67, 1151-1155 (1981). - YOKOTA, A., KITAOKA, S.: Agric. Biol. Chem. 45, 15-22 (1981).

ZENVIRTH, D., KAPLAN, A.: Planta 152, 8-12 (1981). - ZIEM-HANCK, U., HEBER, U.: Biochim. Biophys. Acta 591, 266-274 (1980).

Professor Dr. ERWIN LATZKO
Botanisches Institut der
Universität Münster
Schloßgarten 3
D 4400 Münster

Dr. GRAHAME J. KELLY
Department of Biochemistry
and Nutrition
University of New England
AUS Armidale, N.S.W. 2351

Photosynthesis. Carbon Metabolism: On Land and at Sea[1]

By Grahame J. Kelly and Erwin Latzko

"If you were to row out from the coast on a summer's day and haul in a bucket of water, it might surprise you to learn that it contained several million plants."

E. STEMANN NIELSEN in
The Galathea Deep Sea Expedition
(1950-52)

1. Introduction

This contribution celebrates, for our part, a decade of reviewing photosynthetic carbon metabolism in *Progress in Botany*. Each 2 years an attempt has been made to highlight a particular aspect of this topic, and one would expect that by now the supply of aspects would be exhausted. Indeed, this may have appeared likely, had not one of us moved to a new Institute. This move gave birth to a realization that we have, to a considerable extent, ignored about 30% of the world's CO_2 fixation over the past decade of reviewing. The greater portion of this 30% belongs to the layer of unicellular algae (commonly termed marine phytoplankton) that occurs in the world's oceans. Suspended at a depth where light intensity is low, light quality is blue-greenish, and in a medium of pH about 8 and containing 2 mM HCO_3^-, these marine photosynthetic CO_2 fixers face problems quite unlike those that have made CAM, the C_4 pathway, and photorespiration so predominant in their terrestrial counterparts. The remainder of the 30% is attributed to macrophytic algae and sea grasses, mainly in shallow waters near land masses. Carbon metabolism by the macrophytic algae has been recently thoroughly reviewed by KREMER (1981a).

Consequently, this 1984 review endeavors to look into the ocean, as well as to consider investigations with terrestrial plants reported since our last review in Vol. 44. The great majority of publications relate to the latter, but a glance at the available reports on marine algae should at least promote a more balanced outlook on global photosynthetic carbon metabolism.

[1]Abbreviations: CAM, crassulacean acid metabolism; CoA, coenzyme A; FBP, fructose-1,6-P_2; GAP, glyceraldehyde-3-P; OAA, oxaloacetate; PEP, P-enolpyruvate; P_i, inorganic phosphate; RuBP, ribulose-1,5-P_2; SBP, sedoheptulose-1,7-P_2

G.J. Kelly and E. Latzko:
Thirty Years of Photosynthesis

Originally published in Progress in Botany 46

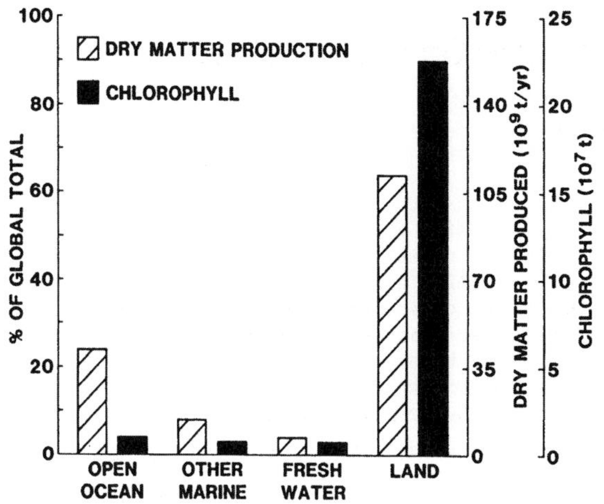

Fig. 1. Distribution of dry matter production and chlorophyll between land and sea. Adapted from WHITTAKER and LIKENS (1975)

2. The Superior Performance of Marine Phytoplankton in Photosynthetic Carbon Assimilation

It is estimated that the open ocean phytoplankton carry out 24% of the world's carbon assimilation, using only 4% of its chlorophyll (WHITTAKER and LIKENS 1975, Fig. 1). Therefore, the marine algae appear to be eight times more effective than terrestrial plants as far as CO_2-fixation expressed on a chlorophyll basis is concerned: the respective rates, derived from Figure 1 on the assumption that dry matter is 45% carbon, are (in µmol CO_2-fixed/mg chlorophyll·h) 36 for marine algae and 4.3 for terrestrial plants. The superior performance by the marine algae is all the more impressive when it is noted that the irradiance at the average depth where they occur is below that required to saturate their photosynthesis. Productivity experiments with samples of seawater exposed to saturating or near-saturating irradiance have yielded rates of 33 to over 1000 µmol/mg chl·h, with a mean value of about 350 µmol/mg chl·h (RYTHER and YENTSCH 1957, STRICKLAND et al. 1969, VENRICK et al. 1977, FURNAS 1983, WILLIAMS et al. 1983). A similarly high mean value was obtained for 12 unialgal cultures in the laboratory, although the distribution of values was not so great, the highest being 480 µmol/mg chl·h (KREMER and BERKS 1978, GLOVER 1980). One exception noted by GLOVER (1980) was *Coccolithus pelagicus* for which a rate of 1730 µmol/mg chl·h was obtained, and attributed to a low content of chlorophyll. However, the high CO_2 fixation rates for marine algae cannot generally be explained by low chlorophyll concentrations (and the "missing" chlorophyll being replaced by other light-harvesting accessory pigments), since, in fact, most measurements point to chlorophyll contents similar to those of green algae and higher plant cells (PRÉZELIN and ALBERTE 1978, FALKOWSKI 1981). One tends to conclude, then, that the open ocean phytoplankton consists mostly of healthy algal cells capable of photosynthesizing at maximal rates, and suffering neither water nor CO_2 limitation (seawater contains 2 mM HCO_3^-). The main factor that reduces their in situ CO_2 fixation from the potential 350 to actual 36 µmol/mg chl·h is the low irradiance, and this is a consequence of the depth at which the algae choose to grow.

The viewpoint that the phytoplankton in seawater samples photosynthesize at maximum possible rates when exposed to saturating light is supported by the fact that the value of 350 µmol/mg ch·h is equal to the best rates of photosynthetic electron transport measured (albeit with higher plants; DELANEY and WALKER 1978). Electron transport (specifically, the rate of oxidation of reduced plastoquinone) is widely believed to limit the rate of light-saturated and CO_2-saturated photosynthesis (RICHARDSON et al. 1983).

The above discussion does not pretend that the photosynthetic cells of land plants are not capable of photosynthesizing at a rate approaching that of phytoplankton. Indeed, suspensions of such cells isolated from tobacco leaves fixed 120 to 150 µmol CO_2/mg chl·h (CATALDO and BERLYN 1974), and intact alfalfa leaflets can fix CO_2 at 90 µmol/mg chl·h (PLATT et al. 1976). However, the rigors of land-based photosynthesis (periodic water and/or CO_2 deficits, shading, aged cells) result in actual rates well below the potential rates, and only one-eighth of the actual rates for oceanic phytoplankton.

3. The Uptake of Inorganic Carbon, and Evidence of a Role for Carbonic Anhydrase

Terrestrial plants obtain all their inorganic carbon as CO_2 from the air, and although some of this CO_2 may be converted to HCO_3^- (see below) in an aqueous film surrounding the plasmalemma, earlier evidence that HCO_3^- may be transported across the plasmalemma has been reinvestigated and it now seems that CO_2 is the principal permeating species of inorganic carbon (ESPIE and COLMAN 1982, VOLOKITA et al. 1983).

Aquatic plants, however, are exposed to both CO_2 and HCO_3^-, because CO_2 from the air dissolves in water to a concentration of 10 µM, and this dissolved CO_2 can be converted to HCO_3^- [Eq. (1)], the equilibrium concentration of which is dependent on the H^+ concentration, i.e., on the pH

$$CO_2 + H_2O \rightleftharpoons H_2CO_3 \rightleftharpoons H^+ + HCO_3^-. \quad (1)$$

It appears that algae and cyanobacteria, both freshwater and marine, utilize the option of taking up HCO_3^- (MARCUS et al. 1982, MERRETT and ARMITAGE 1982, COLMAN and GEHL 1983, IMAMURA et al. 1983). The uptake by marine algae is of particular interest because the seawater pH of 8.2 results in a HCO_3^- concentration 200 times that of CO_2. Uptake of HCO_3^- appears to be an active process, involving an enzyme-mediated transport of HCO_3^- across the plasmalemma, driven by ATP hydrolysis and capable of concentrating HCO_3^- inside the cell (BADGER and ANDREWS 1982, KAPLAN et al. 1982, OGAWA and INOUE 1983, SPALDING and OGREN 1983, SPALDING et al. 1983a). Once inside the cell CO_2 can be obtained from the HCO_3^- by a reversal of Eq. (1), and may be held in the cell as protein carbamate (YOKOTA et al. 1983a). The reaction producing CO_2 by H_2CO_3 dehydration is catalyzed by carbonic anhydrase. Studies supporting the importance of this enzyme (NILSEN and JOHNSEN 1982, SPALDING and OGREN 1982, SHIRAIWA and MIYACHI 1983) have been reinforced by a description from SPALDING et al. (1983b) of a *Chlamydomonas reinhardii* mutant with low carbonic anhydrase activity. The mutant grew poorly, yet it accumulated HCO_3^- to an internal concentration six times that of wild-type cells. The authors concluded that the accumulated HCO_3^-, in the absence of sufficient carbonic anhydrase activity, was not dehydrated to CO_2 rapidly enough to support normal rates of photosynthesis.

Nevertheless, there is still much to be learnt about carbonic anhydrase. Reports from Tokyo indicate that some algae have carbonic anhydrase on the cell surface, outside the plasmalemma, and also that certain algae take up CO_2 instead of, or as well as, HCO_3^- (KIMPEL et al. 1983, MIYACHI et al. 1983, TSUZUKI 1983). In higher plants, where the enzyme makes up 1% - 2% of all protein (OKABE et al. 1984), its function is unclear since HCO_3^- uptake seems not to be important (see above). Thus it is interesting that PORTER and GRODZINSKI (1984) quite recently observed a much greater decline in carbonic anhydrase activity in bean leaves maturing under 1200 µl/l CO_2 than under air-level CO_2 (300 µl/l). This response is reminiscent of the very low enzyme activity in algae that are grown under a high level of CO_2 (SPALDING and OGREN 1982, SHIRAIWA and MIYACHI 1983).

4. Chloroplast Carbon Metabolism

A comprehensive and thorough review of chloroplast metabolism, emphasizing not only enzyme properties, but also metabolite concentrations and the considerably artificial environments in which isolated enzymes and chloroplasts are usually studied, has been prepared recently by LILLEY (1983).

a) RuBP Carboxylase

The aerial roots of leafy orchids have become the latest addition to the list of photosynthetic tissues with the demonstration by KWOK-KI et al. (1983) that they contain substantial levels of RuBP carboxylase activity. Studies with rice leaves have provided further evidence that this enzyme serves an additional purpose to that of being the Calvin cycle's CO_2 fixer, viz. it is a store of nitrogen which is remobilized for the growth of young tissues (MAE et al. 1983). Its level in rice leaves can be a limiting factor to photosynthesis by these leaves (MAKINO et al. 1983).

The reaction catalyzed by RuBP carboxylase has received special attention during the last 2 years. Students being introduced to photosynthesis in introductory biochemistry courses have inevitably been told that "carboxylation of RuBP produces an unstable six-carbon intermediate that breaks down to two molecules of glycerate-3-P". It now appears that this six-carbon intermediate finally has been trapped and identified by SCHLOSS and LORIMER (1982) as 2-carboxy-3-keto-D-arabinitol-1,5-P_2. Combined with direct evidence for the intermediacy of the enediol between carbons 2 and 3 (SAVER and KNOWLES 1982), details

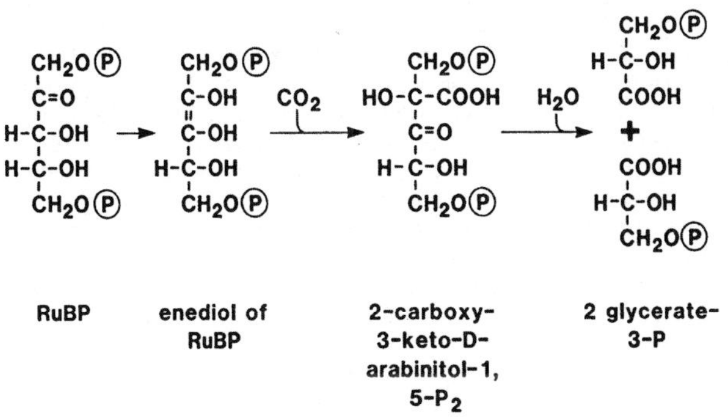

Fig. 2. Details of the reaction catalyzed by RuBP carboxylase. Based on SCHLOSS and LORIMER (1982) and SAVER and KNOWLES (1982)

of the CO_2 fixation reaction can now be written with some considerable degree of confidence (Fig. 2).

The true CO_2-acceptor substrate of the higher plant enzyme also has been investigated, and found to be $RuBP^{4-}$ (rather than $Mg \cdot RuBP^{2-}$; ROACH and McFADDEN 1983). The high affinity reported for this substrate (Km = 2.2 μM) concurs with the conclusion of BOURQUIN and FOCK (1983) that its concentration will normally not restrict CO_2 assimilation.

Activation of RuBP carboxylase via formation of the enzyme$\cdot CO_2 \cdot Mg^{2+}$ complex has been studied further. It is claimed that Ca^{2+} can replace Mg^{2+} in this complex (BÁRCENA 1983, PARRY et al. 1983). The activation may be more involved than previously thought, in that GUTTERIDGE et al. (1982) have found two forms of wheat RuBP carboxylase that differ in their rates of activation by CO_2 and Mg^{2+}, the half-times for activation being 8 s and 120 min, respectively. These authors, and also JORDAN and CHOLLET (1983), JORDAN et al. (1983), and VATER et al. (1983) have attributed the action of RuBP carboxylase effectors, such as gluconate-6-P, FBP, NADPH, and RuBP to somewhat complex influences on the CO_2 + Mg^{2+} activation process. There is also renewed interest in the effect of light on RuBP carboxylase activity in vivo, and recent strong evidence that the activity doubles upon illumination (SICHER 1982, VU et al. 1983) has been followed by a suggestion that light leads to conditions more favorable for activation of the carboxylase by CO_2 and Mg^{2+} (PERCHOROWICZ and JENSEN 1983). It seems that more attention should be devoted to the in vivo enzyme (STUMPF and JENSEN 1982), the properties of which may be altered upon purification (BRAUN et al. 1983).

The RuBP carboxylase enzyme is composed of eight large and eight small subunits. Each large subunit contains one active site, but the function of the small subunit has been something of a mystery. Working with the enzyme from blue-green algae, ANDREWS and BALLMENT (1983) and TAKABE et al. (1984) have now concluded that large subunits are not catalytically competent unless they have small subunits attached to them. It will be important to ascertain the extent to which this conclusion applies to the enzyme from other organisms.

RuBP carboxylase has been identified in a number of marine algae, but reported levels of activity usually have been below observed rates of photosynthesis. Best agreement has been obtained with the diatom *Phaeodactylum tricornutum* (BEARDALL and MORRIS 1976, HOLDSWORTH and COLBECK 1976), but with several other species (GLOVER and MORRIS 1979, RIVKIN et al. 1982) and with natural populations of phytoplankton (PRISCU and GOLDMAN 1983, SMITH et al. 1983) photosynthesis was more than three times the measured RuBP carboxylase activity. Some authors are tempted to attribute the balance of carbon fixation to enzymes using PEP as CO_2 acceptor, but it should be remembered that none of these is connected to an autocatalytic sequence, such as the Calvin cycle, that permits growth. The demonstration by GLOVER and MORRIS (1979) of a positive correlation between RuBP carboxylase activity (but not PEP carboxylase activity) and rate of photosynthesis attests to the requirement for more effort being directed toward demonstrating RuBP carboxylase activities in marine algae commensurate with their capabilities for CO_2 fixation.

b) Other Calvin Cycle Enzymes

The struggle to understand chloroplast GAP dehydrogenase is continuing. Much of the confusion appears to be due to the presence of homotetrameric (identical subunits) and heterotetrameric (two different subunits) forms of the enzyme, the ability of light in vivo and NAD^+ (during extraction) to influence the ratio of these two forms, and the proclivity toward aggregation (IADAROLA et al. 1983, LOOZE and WAGNER 1983a,b). Sophisticated experiments by CERFF and KLOPPSTECH (1982), using messenger RNA translation by wheat germ ribosomes, have confirmed that the chloroplast enzyme is distinct from its cytosolic counterpart.

Chloroplast and cytosolic forms of triose-P isomerase (PICHERSKY and GOTTLIEB 1984) and FBP-aldolase (KRÜGER and SCHNARRENBERGER 1983) have been purified. In the latter case, a separate experiment with corn leaves indicated only one form of aldolase, although GASPERINI and PUPILLO (1982) did find two. Wheat, spinach, pea and rye leaves all have been shown to contain transketolase in both chloroplasts and the cytosol (MURPHY and WALKER 1982a, FEIERABEND and GRINGEL 1983).

The pivotal regulatory function of the Calvin cycle's FBPase has continued to attract attention. The light-mediated reductive activation of this enzyme (LEEGOOD and WALKER 1982) has been shown to be very dependent on pH and the presence of Mg^{2+} and FBP (LEEGOOD et al. 1982, MINOT et al. 1982, HERTIG and WOLOSIUK 1983). An unusual observation is that certain polyamines interfere with this activation (CORLEY et al. 1983). Chloroplast FBPase now has been purified using affinity chromatography (PLÁ et al. 1981). In the prokaryote *Anacystis nidulans* two FBPases have been identified (GERBLING et al. 1984). One of these was found to have properties similar to those of the higher plant chloroplast enzyme (UDVARDY et al. 1982). As with chloroplast FBPase, light-mediated activation of SBPase also benefits from the presence of the substrate (SBP) and Mg^{2+} (WOODROW and WALKER 1982, WOLOSIUK et al. 1982). Specifically, SBP increases the rate of reductive activation (WOODROW and WALKER 1983), while Mg^{2+} influences the dissociation constant for protonation, and thus modulates the effect of pH on the activity (WOODROW et al. 1984). A continuous spectrophotometric assay for SBPase has been developed by WOODROW and WALKER (1982).

Intensive efforts in Göttingen (GARDEMANN et al. 1982, 1983, FLÜGGE et al. 1982) have produced indications that the rapid cessation of CO_2 fixation upon darkening of leaves is largely due to ribulose-5-P kinase being rapidly inactivated by the effects of rapid changes in stromal metabolite levels superimposed on dark-mediated oxidation. Simulation experiments suggest that the enzyme achieves only 4% of its potential activity during illumination, and is completely inactive in darkened leaves. An affinity label that binds only to the reduced form of the enzyme was discovered recently (ASHTON 1984).

Other experiments on light-mediated enzyme regulation have been reported. WIRTZ et al. (1982) managed to measure the activities of enzymes from illuminated pea leaves, within 5 to 10 s after extraction, by developing a violent squashing technique. They detected light-mediated activations of FBPase, SBPase, and ribulose-5-P kinase, but not of RuBP carboxylase or chloroplast GAP dehydrogenase. The presence of thioredoxins that can mediate light activation, and activations of FBPase and SBPase, have been confirmed in a CAM plant (HUTCHESON and BUCHANAN 1979). Ferralterin, an alternative to the ferrodoxin/thioredoxin system, has been discovered (DE LA TORRE et al. 1982).

Certain enzymes outside the Calvin cycle are also light-regulated. Recent information indicates that chloroplast NADP-linked malate dehydrogenase activation is inhibited by $NADP^+$ (SCHEIBE and JACQUOT 1983), that the light-mediated regulation of glucose-6-P dehydrogenase is not mechanistically comparable to reduction of the enzyme by dithiothreitol

(SRIVASTAVA and ANDERSON 1983), and that even a pig-heart enzyme (in this case aconitase) can be activated by the chloroplast's light-activation machinery (MOHAMED and ANDERSON 1983). Evidence for some light-mediated activation of PEP carboxylase also has appeared (KARABOURNIOTIS et al. 1983, RAO and ANDERSON 1983).

c) Starch Metabolism

A transitory reserve of reduced carbon, in the form of starch, is deposited in chloroplasts during the day, and dissimilated to soluble sugar-Ps at night (KRUGER et al. 1983a). The evidence is as strong as ever (CAM plants and stomatal guard cells included) that the ratio of the concentrations of glycerate-3-P to P_i regulates leaf starch synthesis through its effect on ADP-glucose pyrophosphorylase activity (PREISS 1982, PORTIS 1982, OUTLAW and TARCZYNSKI 1984, SINGH et al. 1984). In *Chlorella*, synthesis was concentrated into the first half of the cell cycle (HIROKAWA et al. 1982) and was affected by temperature: growth at 38°C (rather than 20°C) produced a different quality starch, more amylose-like (NAKAMURA and MIYACHI 1982, NAKAMURA and IMAMURA 1983).

Concerted efforts continue to probe the relative contributions of hydrolysis (via amylases) and phosphorolysis (via phosphorylase) to starch degradation. Leaf phosphorylases, including the enigmatic extra-chloroplast enzyme, have been further investigated (SHIMOMURA et al. 1982, HAMMOND and PREISS 1983, KRUGER and AP REES 1983a). The properties of amylase are less well understood, perhaps because so many forms have been identified in leaves (STEUP and GERBLING 1983). A chloroplast debranching enzyme has been recently purified (LUDWIG et al. 1984). Two reports imply that hydrolysis is the primary degradation reaction (CHANG 1982, STEUP et al. 1983) and the suggestion has been made that phosphorolysis is most likely restricted to a pool of soluble glucan intermediates (STEUP et al. 1983). However, other reports give phosphorolysis at least an equal status to hydrolysis (HAMMOND and BURTON 1983, HAMPP and GOLLER 1983, LEVI and GIBBS 1984). Interestingly, maltose (and higher maltodextrins) are formed in chloroplasts both in the light and in darkness, but enzymes other than amylases are involved (SCHILLING 1982, KRUGER and AP REES 1983b, LINDEN and SCHILLING 1984).

d) "Chloroplast Respiration"

Conversion of the immediate products of starch degradation to triose-P may be regulated at the reaction catalyzed by chloroplast phosphofructokinase [now also reported in a green alga (KOMBRINK and WÖBER 1982)]. The leaf enzyme is reportedly inhibited by NADPH (CSÉKE et al. 1982a) and reduced in activity by a light-mediated mechanism (HEUER et al. 1982). In the dark, ATP for the phosphofructokinase could be generated if one of the two triose-P (formed from the product FBP) is converted further to glycerate-3-P. However, this also produces an NADPH, and KOW et al. (1982a) have coined the novel term "chloroplast respiration" to accommodate evidence for the oxidation of this NADPH, with oxygen as the final electron acceptor, to produce H_2O_2 (Fig. 3). The ascorbate-mediated reduction of this H_2O_2 to water is quite possibly linked to the oxidation of another NADPH through glutathione (LATZKO and KELLY 1983a). A similar type of chloroplast respiration has been described by BENNOUN (1982).

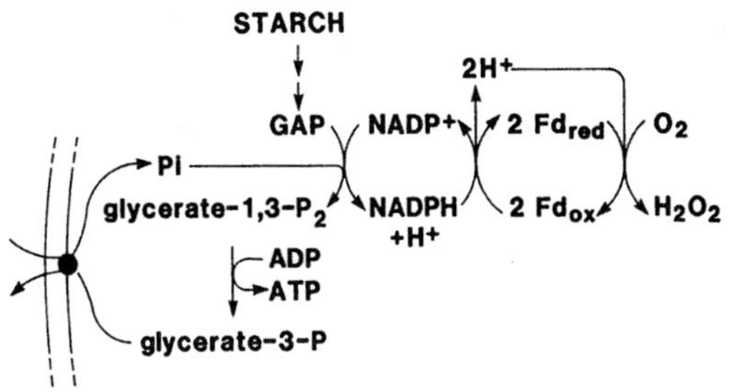

Fig. 3. "Chloroplast respiration". Adapted from KOW et al. (1982a). *Fd* ferredoxin (oxidized or reduced); • phosphate translocator

e) Potassium, Calcium, and pH

Several new regulatory aspects of chloroplast carbon metabolism have appeared. An intriguing study by DEMMIG and GIMMLER (1983) showed that CO_2 fixation by isolated spinach chloroplasts could be increased from 10 to 110 μmol/mg chl·h by boosting the external KCl concentration from 10 to 100 mM. Calcium may also regulate carbon metabolism, possibly by activating NAD kinase. It does this in the form of a complex with the small regulatory protein calmodulin (MUTO 1983), which is receiving much attention in the world of mammalian biochemistry. A small but definite percentage (0.6% to 2%) of the cell's calmodulin was found to be in chloroplasts (JARRETT et al. 1982, MUTO 1982), and uptake of Ca^{2+} by chloroplasts has been demonstrated (MUTO et al. 1982). On the other hand, NAD kinase has been found in both the stroma and envelope of pea chloroplasts, and in the outer mitochondrial membrane of dark-grown corn seedlings, but only the membrane-bound enzymes underwent Ca^{2+}-calmodulin activation (DIETER and MARMÉ 1984, SIMON et al. 1984). It will be interesting to find out whether Ca^{2+}-calmodulin-dependent protein kinases, recently found in wheat germ and dark-grown corn coleoptiles (POLYA and MICUCCI 1984, VELUTHAMBI and POOVAIAH 1984), also occur in photosynthetic cells.

Experiments with salt (NaCl)-stressed plants have revealed a remarkable ability of chloroplasts to maintain their NaCl content (at about 70 μM) when the total leaf content varied 60-fold (KAISER et al. 1983). The chloroplasts in water-stressed plants have carbon fixation inhibited by stromal acidification which is somehow induced by the increased osmolarity, and results in reduced FBPase activity (BERKOWITZ and GIBBS 1983, BERKOWITZ et al. 1983).

5. Cytosolic Synthesis of Sucrose from Carbon Exported from the Chloroplast

a) Movement of Metabolites Across the Chloroplast Envelope

The outer membrane of the chloroplast envelope poses no problem for metabolite movement: it has pores of 2.5 - 3 nm diameter, and thus acts as a molecular sieve, excluding only those molecules with molecular weight over 7000 (FLÜGGE and BENZ 1984). It is the inner membrane that controls the import/export of most small molecules, usually via specific transporting proteins. Reduced carbon is exported as triose-P or glycerate-3-P via the well-established phosphate translocator. Evidence for the presence of this translocator in chloroplasts from two algae, *Chlamydomonas reinhardii* (BELKNAP 1983, KLEIN et al. 1983) and *Codium fragile* (RUTTER and COBB 1983), has been obtained, but SCHÜRMANN and ORTIZ (1982) could not detect evidence for a translocator in *Euglena gracilis* chloroplasts. The phosphate translocator of C_4-plant mesophyll chloroplasts has acquired the principal role of exporting PEP (HALLBERG and LARSSON 1983) generated in the chloroplast from imported pyruvate. Chloroplasts from CAM plants also appear to possess phosphate translocators, but somehow seem able to shield Calvin cycle intermediates from export, so that only products of starch degradation move into the cytosol (PIAZZA et al. 1982, MONSON et al. 1983).

The phosphate translocator clearly plays a central role in metabolic regulation, as evidenced by its sensitivity to P_i concentration and to the ΔpH across the chloroplast envelope (USUDA and EDWARDS 1982, FLÜGGE et al. 1983, PORTIS 1983). Fortunately for research, the development by KOW and GIBBS (1982) of a particulate reconstituted chloroplast system allows the sensitive translocator to be circumvented where desired during investigations.

Other chloroplast envelope translocators have also received attention. The capacity of the equally well-established dicarboxylate translocator to indirectly export reducing equivalents to the cytosol has been seriously questioned (GIERSCH 1982, SOMERVILLE and OGREN 1983), and the probability that there is more than one type of dicarboxylate translocator has been raised (DRY and WISKICH 1983, SOMERVILLE and OGREN 1983, WOO 1983, PROUDLOVE et al. 1984). In addition, a light-stimulated glycerate carrier has been identified (ROBINSON 1982a,b), and a translocator specific to C_4 and CAM plants has been proposed from experiments with isolated envelope membranes (FOSTER and EDWARDS 1983).

b) The Regulation of Sucrose Biosynthesis

Until recently (see following paragraph), the regulation of sucrose biosynthesis was elusive to the point where its nonexistence was contemplated. Feedback inhibition of any enzyme by sucrose was effectively ruled out by FOYER et al. (1983a), a conclusion consistent with the partitioning of sucrose into the vacuole in isolated protoplasts (KAISER et al. 1982). However, the possibility of some regulatory phenomena was kept alive by the persistent efforts of HUBER (1983) and co-workers who established positive correlations between the synthesis and/or content of sucrose, and the activity of sucrose-P synthase (HUBER and ISRAEL 1982, RUFTY and HUBER 1983, RUFTY et al. 1983). Supporting results were obtained with *Solanum melongena* (CLAUSSEN and LENZ 1983) and with wheat leaf protoplasts (STITT et al. 1983a). However, great sensitivity of the synthase to known effector molecules has not been found (AMIR and PREISS 1982, DOEHLERT and HUBER 1983), and this no doubt caused some workers to consider an alternative enzyme for regulation, i.e., the cytosolic FBPase (FOYER et al. 1982, STITT et al. 1983a). Again, however, apart from the interesting property of cold lability (WEEDEN and BUCHANAN 1983), great sensitivity to effectors had not been recorded for this enzyme.

This changed, and a completely new perspective to the regulation of sucrose biosynthesis was introduced, when CSÉKE et al. (1982b) and STITT et al. (1982a) reported that a sugar bisphosphate recently discovered in animal tissues, fructose-2,6-P_2, also occurs in the cytosol of leaf cells and is a potent inhibitor of the cytosolic FBPase. Subsequent work correlated decreased sucrose formation with increased fructose-2,6-P_2 levels (STITT et al. 1983b, HUBER and BICKETT 1984). However, this neat correlation overshadows the obvious complexity of the regulatory system involved. Fructose-2,6-P_2 is not a metabolic intermediate; it occurs in (and is effective at) nanomolar concentrations. [In fact, it markedly influenced experiments when present as a contaminant of commercial fructose-6-P at a level of only 0.0002% (KRUGER et al. 1983b)!] A special kinase exists for its synthesis

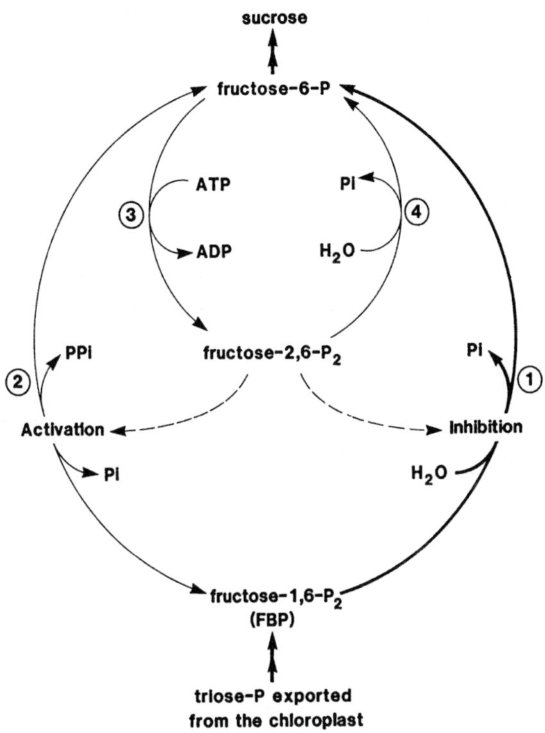

Fig. 4. Regulation of cytosolic sucrose biosynthesis at the stage of FBP hydrolysis (*reaction 1*; enzyme FBPase). An apparently small amount of the product of FBP hydrolysis, fructose-6-P, is used to generate (*reaction 3*; enzyme fructose-6-P,2 kinase) a different fructose bisphosphate, i.e., fructose-2,6-P_2, that inhibitis FBPase and thus slows down the hydrolysis of further FBP. The fructose-2,6-P_2 can be eliminated subsequently by hydrolysis (*reaction 4*; enzyme fructose-2,6-bisphosphatase). The significance of *reaction 2* (enzyme: PP_i-linked P-fructokinase), which is activated by fructose-2,6-P_2, is not yet clear. Adapted from several references (see text)

(CSÉKE and BUCHANAN 1983) and a special phosphatase hydrolyzes it (SMYTH et al. 1984). Obviously, regulation of the activities of these two enzymes will also have to be considered. In addition, fructose-2,6-P_2 is a powerful effector (in this case, activator) of a second cytosolic enzyme, the PP_i-linked P-fructokinase that has been found so far only in plants and certain microorganisms (PFLEIDERER and KLEMME 1980, SABULARSE and ANDERSON 1981, CSÉKE et al. 1982b, STITT et al. 1982a, CARNAL and BLACK 1983, SMYTH et al. 1984). Since the reaction catalyzed by this unusual P-fructokinase is reversible, its role vis-a-vis that of FBPase remains in question. The current status of these new discoveries is outlined in Figure 4. Interestingly, there is essentially no evidence for fructose-2,6-P_2, or for enzymes regulated by it, in the chloroplast (CSÉKE et al. 1982b).

c) Soluble Low-Molecular-Weight Carbohydrates Other than Sucrose

While it must be acknowledged that sucrose is the major soluble carbohydrate in a wide variety of plants, it should be emphasized that other soluble carbohydrates occur at comparable concentrations in many species. In higher plants alditols, such as mannitol and sorbitol, are particularly prevalent. RUMPHO et al. (1983) have recently demonstrated that celery leaves, which form as much mannitol as sucrose, contain in the cytosol activities of the three enzymes required for mannitol synthesis. They propose that the reduction of mannose-6-P to mannitol-1-P is driven by photosynthetically generated NADPH indirectly transferred from the chloroplast by the triose-P/glycerate-3-P shuttle that utilises nonreversible GAP dehydrogenase in the cytosol. In the leaves of apple, and most probably other members of the Rosaceae, NADPH is used to reduce glucose-6-P to sorbitol-6-P for the production of sorbitol, which is the major translocated form of carbon (LOESCHER et al. 1982), and if this reaction is also cytosolic, then operation of the same shuttle could be envisaged here.

Marine algae are noted for the variety of soluble carbohydrates accumulated (KREMER 1981a). An extensive analysis by KREMER and KIRST (1982) led to the identification of four main groups: (1) those that accumulate free monosaccharides, mainly glucose and fructose (e.g., cryptomonads), (2) those that accumulate disaccharides (mainly sucrose) as well as some glucose and fructose (e.g., *Ulva lactuca* and certain other green algae), (3) those that accumulate alditols (e.g., brown algae, renowned for mannitol accumulation, and certain green algae – a variety of C_3 to C_7 alditols are encountered), and (4) those that accumulate heterosides, i.e., a monosaccharide (usually galactose or mannose) attached to an aglycone (typically glycerol or glycerate); many examples belong to the red algae. In almost all cases it would seem feasible that these carbohydrates could be synthesized from triose-P exported from chloroplasts, in a manner analogous to sucrose synthesis. Even volemitol, a C_7 alditol derived from sedoheptulose-7-P, could be generated by partial operation of the oxidative pentose-P pathway with a loss (as CO_2) of only one-sixth of the entering carbon.

6. Nitrogen Assimilation, and Lipid Biosynthesis

a) Nitrogen Assimilation

Plants acquire most nitrogen either as NH_4^+, NO_3^- (which is then reduced to NH_4^+), or through nitrogen fixation (which produces NH_4^+). The NH_4^+ is then incorporated into the amino acid pool by glutamine synthetase. Higher plant leaves (McNALLY et al. 1983) and green algae (FLORENCIO and VEGA 1983, SUMAR et al. 1984) contain two forms of this enzyme. With higher plants one of these is in the chloroplast while the other is cytosolic, but while many species have both forms, some lack the cytosolic enzyme (HIREL et al. 1982a, 1983, McNALLY et al. 1983). Activity of the chloroplast enzyme markedly increases when dark-grown leaves are illuminated (HIREL et al. 1982b, NISHIMURA et al. 1982a). The next enzyme of NH_4^+ assimilation, glutamate synthase, has been purified from rice leaves (SUZUKI and GADAL 1982). Evidence for light-mediated activation of this enzyme in *Chlorella sorokiniana* has been reported by TISCHNER and SCHMIDT (1982).

Carbon skeletons for the net synthesis of amino acids must be derived ultimately from the Calvin cycle. The conversion of triose-P, exported from the chloroplast, to pyruvate, oxaloacetate, and α-oxoglutarate must therefore be an integral component of nitrogen assimilation (LATZKO and KELLY 1983b). This conversion involves PEP as an intermediate, which then may be either carboxylated by PEP carboxylase or converted to pyruvate by pyruvate kinase. There are conflicting reports as to whether NH_4^+ uptake leads to increased PEP carboxylase activity (KANAZAWA et al. 1983, MARQUES et al. 1983). Two forms of pyruvate kinase have been found in leaves, and the regulatory properties of one of these seemed particularly suited to NH_4^+ assimilation (RUYTERS and MIYACHI 1983, BAYSDORFER and BASSHAM 1984). Growth under blue light, which favors protein synthesis, appears to stimulate pyruvate kinase activity (RUYTERS 1982, RUYTERS and MIYACHI 1983).

Marine phytoplankton obtain nitrogen as NH_4^+, NO_3^- or urea (REES and BEKHEET 1982, ROSSLENBROICH and DÖHLER 1982). In the latter case, Ni becomes an essential nutrient where urease, a Ni-containing enzyme, is used to hydrolyze the urea. Recent evidence suggests that marine phytoplankton take up NH_4^+, using an active transport system with very high affinity, presumably to compensate for the low background concentration (less than 30 mM) of NH_4^+ in the ocean (KOIKE et al. 1983).

b) Lipid Biosynthesis

The active research into lipid biosynthesis mentioned in our last review has continued unabated. Controversy has emerged concerning the source of acetate for fatty acid synthesis in the chloroplast. The pathway proposed for spinach leaves, i.e., hydrolysis of acetyl-CoA in the mitochondrion, and movement of the acetate to the chloroplast where resynthesis of acetyl-CoA occurs (LIEDVOGEL and STUMPF 1982, MURPHY and WALKER 1982b), does not seem applicable to pea leaves, in which direct generation of acetyl-CoA by pyruvate dehydrogenase appears more likely (GIVAN 1983, GIVAN and HODGSON 1983). The distribution of these alternative approaches among other plants remains to be established.

The synthesis of fatty acids from acetyl-CoA proceeds in the stroma of plastids, catalyzed by a fatty acid synthetase complex of the non-associated type, i.e., similar to that in the prokaryote *Escherichia coli* (SHIMAKATA and STUMPF 1982). In chloroplasts, synthesis appears to be favored during illumination (SAUER and HEISE 1983). The synthetase produces palmitate, which may be incorporated into lipids, or used for the production of other fatty acids. In the plastid stroma it may be elongated to stearate, which may in turn be desaturated to oleate (in both reactions the fatty acid is attached to acyl carrier protein), but further desaturation of oleate occurs only after the oleate is exported from plastids and esterified into a lipid, probably phosphatidylcholine in the endoplasmic reticulum (ROUGHAN and SLACK 1982). Enzymes for the esterification of oleate into phosphatidylcholine and for its desaturation to linoleate have been demonstrated in microsomal preparations from pea leaves (MURPHY et al. 1983a,b), and evidence for the channeling of newly synthesized dioleoyl-phosphatidylcholine directly to the desaturase enzyme has been obtained by MURPHY et al. (1984). Details of the further desaturation of linoleate to linolenate have not been worked out. Current indications are that it occurs on the chloroplast envelope, where galactose units are also added to produce the envelope's galactolipids (DORNE et al. 1982, OHNISHI and YAMADA 1982, ROUGHAN and SLACK 1982, BLOCK et al. 1983, CLINE and KEEGSTRA 1983, FRENTZEN et al. 1983), but desaturation on the endoplasmic reticulum has not been ruled out (DUBACQ et al. 1983).

Oceanic algae have some unique features to their fatty acid composition. Linoleic (18 : 2, i.e., 18 carbons, 2 double bonds) and linolenic (18 : 3) acids are less prevalent than in land plants and, in fact, linoleic acid is essentially absent from diatoms and dinoflagellates, two principal algal groups in marine phytoplankton. On the other hand, longer chain polyunsaturated fatty acids, especially eicosapentaenoic (20 : 5) and docosahexaenoic (22 : 6), seldom encountered in land plants, are particularly predominant in marine algae (KATES and VOLCANI 1966, ACKMAN et al. 1968, VOLKMAN et al. 1981, JONES et al. 1983). Nothing is known about the biosynthesis of these marine fatty acids, but the incentive for research will no doubt be stimulated by burgeoning evidence that these fatty acids, which persist through the oceanic food web into fish, are particularly effective at reducing the incidence of cardiovascular disease in man (SIESS et al. 1980, GOODNIGHT et al. 1982, COREY et al. 1983).

7. C_4 Metabolism

a) C_4 Plants

Some species of *Cymbopogon* and *Vetiveria*, which are aromatic tropical grasses, and *Microstegium vimineum*, a shade-adapted grass, have been added to the list of plants with the C_4 pathway of photosynthesis (RAJENDRUDU and DAS 1981, WINTER et al. 1982a). The pathway can make use of aspartate as a reservoir of dicarboxylic acid in some species (CHAPMAN and HATCH 1981, SHIEH et al. 1982). Decarboxylation of malate in the bundle-sheath cells is catalyzed, in one subgroup of C_4 plants, by a mitochondrial NAD-linked malic enzyme. GARDESTRÖM and EDWARDS (1983) have now reported the interesting observation that the mitochondria from these cells (but not those from mesophyll cells) oxidize malate without respiratory control, apparently via the alternative oxidase (cyanide-resistant) pathway, thus ensuring that the C_4 pathway in these plants may operate without the constraints that coupled electron flow might impose.

The remaining research with C_4 plants has concentrated on the enzymes involved. A report by SUGIYAMA and HIRAYAMA (1983) that the biomass of maize seedlings correlated with the PEP carboxylase (but not RuBP carboxylase) activity agreed with figures from BASSI and PASSERA (1982), showing that higher light intensity during growth was associated with a much higher level of PEP carboxylase, but only slightly higher level of RuBP carboxylase. There are indications for more than one form of PEP carboxylase in leaves (PAN and TAN 1981, STIBOROVÁ and LEBLOVÁ 1983, VIDAL and GADAL 1983), and also for light-mediated activation (KARABOURNIOTIS et al. 1983). Immunochemical techniques have confirmed that the main form is cytosolic, while the subsequent C_4 pathway enzyme, NADP-linked malate dehydrogenase, is in the chloroplast (PERROT-RECHENMANN et al. 1982, 1983). This latter enzyme is synthesized de novo during greening of leaves (VIDAL and GADAL 1981), and undergoes reductive light-mediated activation, the extent of which depends on the redox state of ferredoxin and associated electron carriers (LEEGOOD and WALKER 1983). An affinity label that binds only to the reduced, active enzyme has been reported (ASHTON 1983). The location of PEP carboxykinase in those C_4 plants that use this enzyme for decarboxylation in bundle-sheath cells has been finally established as the cytosol (CHAPMAN and HATCH 1983, WATANABE et al. 1984).

The unique mechanism of light/dark-mediated regulation of pyruvate P_i dikinase has been consistently probed (NAKAMOTO and SUGIYAMA 1982, NAKAMOTO and EDWARDS 1983a,b,c, HATCH and BURNELL 1983). Inactivation was finally traced to a protein factor (enzyme?)-mediated phosphorylation of the enzyme, with ADP as the phosphate donor (ASHTON and HATCH 1983). This is the first known protein phosphorylation by ADP. It only occurred with enzyme already phosphorylated "conventionally", i.e., by ATP (BURNELL and HATCH 1984). Activation by P_i appears to involve the same protein factor (BURNELL and HATCH 1983). A further report that pyruvate P_i dikinase occurs in C_3 plants, at levels lower than in C_4 plants, has appeared (AOYAGI and BASSHAM 1983).

b) C_3-C_4 Intermediate Plants

Evidence for the existence of C_3-C_4 intermediate plants has become tenuous. Properties of the enzymes PEP carboxylase and glutamine synthetase from species claimed to be C_3-C_4 were in some respects intermediate between those of the enzymes from closely related C_3 and C_4 plants (NAKAMOTO et al. 1983, HIREL et al. 1983). However, from the three main examples of C_3-C_4 intermediates, viz. *Panicum milioides*, *Moricandia arvensis* (HOLADAY et al. 1982), and several species of *Flaveria* (APEL and MAASS 1981, KU et al. 1983), only the latter remains unchallenged. A consensus has emerged that *M. arvensis* is a C_3 plant (WINTER et al. 1982b, HOLADAY and CHOLLET 1983), but more interesting is the erosion of evidence for *P. milioides* being C_3-C_4, since its reputation has been more long-standing. In particular, previous claims that *P. milioides* leaves contain pyruvate P_i dikinase activity have been squelched (MEYER et al. 1982, EDWARDS et al. 1982, HOLADAY and CHOLLET 1983). The report by KU et al. (1983) that this enzyme is in C_3-C_4 *Flaveria* species will therefore bear further investigation.

c) CAM

Interest in CAM has not waned. Rapid publication of the proceedings of a symposium on CAM (TING and GIBBS 1982) attests to this. The conclusions by these editors that "CAM is an adaptation to dry environments" is reflected in an odd way by the findings of WINTER et al. (1983) that Australian CAM plants, although found mainly in rain forests, are epiphytic species (e.g., many Orchidaceae) and therefore liable to short but frequent periods of water stress. Among more conventional CAM plants, a range of types is emerging, including C_3 plants with some CAM characteristics (MARTIN and ZEE 1983, TING and BURK 1983), *Cissus quadrangularis* which has C_3 leaves but CAM stems (TING et al. 1983), and further examples of species with CAM when well watered, but CAM-idling when water-stressed (RAYDER and TING 1983a,b).

Malate formed from nocturnal PEP carboxylation during CAM is thought to be actively transported into vacuoles (SMITH et al. 1982) in which the pH falls (STIDHAM et al. 1983), although a permease capable of mediating an exchange diffusion of malate across the vacuole membrane also exists (BUSER-SUTER et al. 1982). Cytosolic pH in CAM cells is held relatively constant during CAM (MARIGO et al. 1983). Malate decarboxylation the following day proceeds more rapidly than photosynthesis, thus CO_2 accumulates behind closed stomata (BARROW and COCKBURN 1982).

With respect to the enzymology of CAM, NAD-linked malic enzyme has been purified and shown to be hysteretic (WEDDING and BLACK 1983), and pyruvate P_i dikinase has been shown to have a wide distribution among CAM plants (SANADA and NISHIDA 1982). Adenylate kinase was found to be 4 - 20 times more active in CAM plants utilizing malic enzyme for decarboxylation, compared to those utilizing PEP carboxykinase (MANETAS et al. 1983). Most other studies have centered on PEP carboxylase. This enzyme was more prominent in older leaves (AMAGASA 1982, WINTER et al. 1982c). Its regulatory properties appear to change diurnally (BRULFERT and QUEIROZ 1982, BRULFERT et al. 1982a, MANETAS 1982, WINTER 1982), and are affected by temperature in vitro (BUCHANAN-BOLLIG et al. 1984). Discussion continues as to whether more than one form of enzyme is present: in most cases only one form is seen (MÜLLER et al. 1982, NOTT and OSMOND 1982), but BRULFERT et al. (1982b) claim that a second form appeared when CAM was induced in *Kalanchoe blossfeldiana*. Marked increase in activity due to synthesis (but not necessarily of a new enzyme form) also occurred with another species in which CAM can be induced, viz. *Mesembryanthemum crystallinum* (FOSTER et al. 1982, HOLTUM and WINTER 1982). Chloroplasts have been obtained from protoplasts of this plant and, rather remarkably, showed distinctly different properties depending on whether the parent tissue was performing CAM or C_3 photosynthesis (WINTER et al. 1982d, DEMMIG and WINTER 1983).

d) Freshwater Macrophytes, and Stomata

C_4 metabolism is a notable feature in submersed freshwater plants. It is most conspicuous under summer-like growth conditions, and is distinguished by high PEP carboxylase activity (BEER and WETZEL 1982, ASCENCIO and BOWES 1983). Low photorespiration was also observed, and attributed to CO_2-concentration via either a C_4-like mechanism or active uptake of inorganic carbon (SALVUCCI and BOWES 1983a,b).

Stomatal guard cells accumulate malate via carboxylation of PEP (THORPE 1983) generated from starch. Activation of P-fructokinase (SCHNABL et al. 1982) and a high capacity for glycolysis (HAMPP et al. 1982a) must benefit this process. These cells also possess a high potential for respiration (SHIMAZAKI et al. 1983). However, the absence of RuBP carboxylase and a Calvin cycle has been confirmed (HAMPP et al. 1982a, OUTLAW et al. 1982), except perhaps for CAM plants (MADHAVAN and SMITH 1982).

e) Marine Algae

C_4 metabolism appears to have assumed great importance in the ocean, particularly for the unicellular algae. This is because high growth rates, coupled to high rates of cell division, result in these plants concentrating on protein synthesis, rather than synthesis of carbohydrates. Protein synthesis requires, in turn, a supply of amino acids, the majority of which have carbon skeletons derived from the tricarboxylic acid cycle. Replenishment of intermediates taken from this cycle is a principal function of PEP carboxylation (see LATZKO and KELLY 1983b), hence it is not surprising that short-term feeding of $[^{14}C]O_2$ to marine algae has resulted in labeling of C_4 compounds approaching, or even surpassing, that of C_3 compounds (BEARDALL et al. 1976, MUKERJI et al. 1978, ROSSLENBROICH and DÖHLER 1982, KULKARNI and NIMBALKAR 1983). Labeling becomes almost exclusively C_4 type in cultures approaching or at stationary phase (MUKERJI et al. 1978) when PEP carboxylation persists after RuBP carboxylation has significantly decreased (GLOVER and MORRIS 1979). Initially, results such as these led to speculation that, in some algae, PEP carboxylase had achieved an importance comparable to that of RuBP carboxylase, and although such thoughts still appear (PRISCU and GOLDMAN 1983), they have in general been checked by reminders from KREMER (1981a,b) that all the PEP provided is derived from the Calvin cycle, either directly from glycerate-3-P in the light, or (in the dark) from stored carbohydrates synthesized from Calvin cycle intermediates during a preceding light period. Figure 5 shows that such metabolic activity can lead to a spectrum of C_3/C_4 labeling ratios.

An interesting aspect of marine PEP carboxylation is that it is most predominant in brown algae (KREMER 1981a), including diatoms, in which the carboxylating enzyme is PEP carboxykinase; PEP carboxylase appears to be absent (AKAGAWA et al. 1972, HOLDSWORTH and BRUCK 1977, ROSSLENBROICH and DÖHLER 1982, KERBY and EVANS 1983).

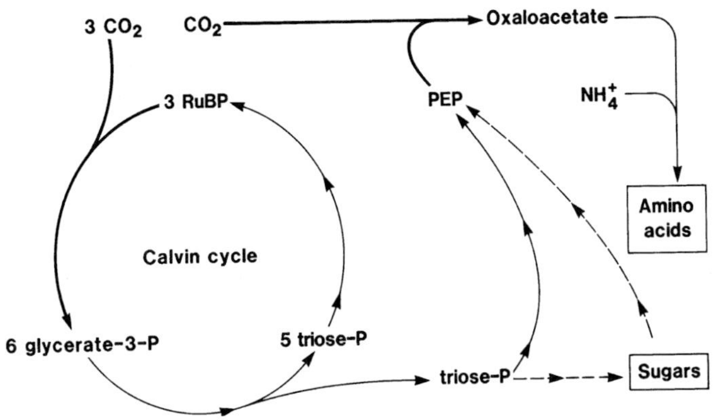

Fig. 5. The net fixation of three CO_2 by the Calvin cycle produces one triose-P. If each triose-P is used to synthesize a C_4 amino acid (in a metabolic sequence that includes fixation of a fourth CO_2), a short-term labeling pattern of "75% C_3/25% C_4" will ensue *(heavy lines)*. The proportion of label in C_4 compounds would, however, be expected to be less since the triose-P is also required for other biosyntheses, e.g., sugars *(broken lines)*. On the other hand, at later growth stages these sugars may be remobilized to generate additional PEP *(broken lines)*, and if this occurs when the Calvin cycle activity is low, label may appear predominantly in C_4 compounds

8. Respiration

a) Mitochondrial (Dark) Respiration

The popular belief that mitochondrial respiration in photosynthetic cells is largely inhibited during photosynthesis (presumably due to a lack of ADP which is fully utilized by the chloroplast) has received support from some quarters, but has been challenged in others. In green and blue-green algae, inhibition seemed to vary according to growth rate and species (SCHERER and BÖGER 1982, HARRIS and PICCININ 1983). In higher plants, illumination led to a marked increase in the cytosolic dihydroxyacetone-P/glycerate-3-P ratio, which could act through the adenylate system to inhibit respiration (HEBER et al. 1982). However, two groups of workers, using new techniques for the rapid fractionation of protoplasts, could find little or no change in the cytosolic and mitochondrial ATP/ADP ratios when dark-kept protoplasts were illuminated (GOLLER et al. 1982, HAMPP et al. 1982b, LILLEY et al. 1982, STITT et al. 1982b) and it has been concluded that any inhibition of mitochondrial respiration in the light cannot be due to changes in the cytosolic ATP/ADP ratio (DRY and WISKICH 1982, STITT et al. 1982b). Adenylate pools are equilibrated in the chloroplasts (by a stromal adenylate kinase) and cytosol (by an adenylate kinase in the mitochondrial intermembrane space), but not in the mitochondrial matrix (BIRKENHEAD et al. 1982, HAMPP et al. 1982b, STITT et al. 1982b).

Studies with intact wheat leaves have produced data to support the inhibition of dark respiration by light in one case (PEISKER and APEL 1981), but not in another (AZCÓN-BIETO and OSMOND 1983). If any respiration proceeds in the light, a new question will arise, viz. what proportion can be attributed to the alternative (cyanide-resistant) pathway recently demonstrated to be present in leaves, particularly when the supply of respiratory substrate is abundant (AZCÓN-BIETO et al. 1983a,b). These and other questions should receive answers soon, now that high-quality preparations of leaf mitochondria are available (NEUBURGER et al. 1982, NISHIMURA et al. 1982b, NASH and WISKICH 1983).

b) Photorespiration

Oxygenation of the Calvin cycle's CO_2 acceptor (RuBP), catalyzed by RuBP carboxylase/oxygenase, and subsequent metabolism of one of the products (glycolate-P) to glycerate-P and CO_2, via the glycolate pathway, form the basis of photorespiration. From one viewpoint it is a consequence of an unfavorable ratio: that of the concentrations of CO_2 and O_2 in air. These two gases compete for the active site of the enzyme, but only when CO_2 is incorporated does plant growth result. By increasing the CO_2/O_2 ratio, WULFF and STRAIN (1982) increased the dry matter production of *Desmodium paniculatum* twofold, despite a decline in photosynthesis/unit leaf area, and KAMIŃSKA and MALESZEWSKI (1982) and SOMERVILLE and SOMERVILLE (1983) found a decreased flow of assimilated carbon through the glycolate pathway in bean and *Arabidopsis* leaves when the CO_2/O_2 ratio was similarly increased. The complementary dream of researchers, i.e., to develop a mutant RuBP carboxylase with lowered affinity for O_2 (i.e., low oxygenase activity in air) is still being pursued (SPREITZER et al. 1982).

Improved methods of estimating photorespiration have been developed (PETERSON 1983, VINES et al. 1983). The apparent increase of photorespiration with temperature (MONSON et al. 1982, PETERSON 1983) has been explained by HALL and KEYS (1983) to be due to changes in the relative solubilities of CO_2 and O_2 with temperature. In C_4 plant leaves, O_2 uptake by RuBP carboxylase/oxygenase seemed to be insensitive to CO_2 levels, but some evidence indicated that increased CO_2 did, as expected, decrease this O_2 uptake, but that the decrease was masked by a concomitant increase in O_2 uptake by the Mehler reaction (i.e., oxygen reduction leading to H_2O_2 formation) (FURBANK and BADGER 1982, 1983, JOLIVET-TOURNIER and GERSTER 1984).

Extensive investigation of the glycolate pathway has occurred. Six enzymes of the pathway have received detailed attention (BIEKMANN and FEIERABEND 1982, WINKLER et al. 1982, KLECZKOWSKI and RANDALL 1983, NISHIMURA et al. 1983a,b, SAROJINI and OLIVER 1983, TITUS et al. 1983, HUSIC and TOLBERT 1984). A transporter that facilitates glycolate export from the chloroplast has been identified (HOWITZ and McCARTY 1983); the glycolate can then easily enter peroxisomes (LIANG and HUANG 1983). Quality preparations of peroxisomes can now be obtained from leaves (LIANG et al. 1982, SCHMITT and EDWARDS 1982). Peroxisomal oxidation of glycolate produces glyoxylate, for which four fates have been considered: (1) reduction back to glycolate (PETERSON 1982), (2) condensation with succinate to produce isocitrate (SINGH and NAIK 1983), (3) nonenzymic decarboxylation by H_2O_2 [no longer considered physiologically important in leaves (WALTON 1982, WALTON and WOOLHOUSE 1983, KENDALL et al. 1983), but is significant in the unusual case of *Euglena* (YOKOTA and KITAOKA 1982, YOKOTA et al. 1983b)], and (4) amination to form glycine (the most favored possibility). New evidence suggests that alanine may be an important amine donor to glyoxylate (BETSCHE 1983, NAKAMURA and TOLBERT 1983, WALTON 1983). Glycine moves into mitochondria, either by diffusion or via an active uptake system (PROUDLOVE and MOORE 1982, WALKER et al. 1982a, YU et al. 1983). In the mitochondrion, glycine is oxidized in preference to tricarboxylic acid cycle intermediates (WALKER et al. 1982b, BERGMAN and ERICSON 1983, DRY et al. 1983). An indication that glycine is generated by a glycolate pathway in C_4 plant leaves (MAREK and STEWART 1983) has been supplemented by the observation that only bundle-sheath mitochondria are capable of oxidizing glycine; mesophyll mitochondria have lost this ability (OHNISHI and KANAI 1983). Glycine oxidation produces CO_2, NH_3, and a one-carbon unit that condenses with a second glycine to form serine which returns to the peroxisome and is deaminated to hydroxypyruvate. The latter is reduced to glycerate, but the source of reducing power is unclear now that a malate/aspartate shuttle has been effectively ruled out (SCHMITT and EDWARDS 1983). Finally, glycerate returns to the chloroplast via a light-stimulated chloroplast transporter (ROBINSON 1982a,b).

Reassimilation of the NH_3 produced during glycine oxidation occurs in the chloroplast, utilizing the enzymes normally employed for nitrogen assimilation, viz. glutamine synthetase and glutamate synthase. The first of these enzymes is absent from mitochondria (NISHIMURA et al. 1982b). The chloroplast system has been demonstrated, using inhibitors, in C_3 and C_4 plant leaves and green algae (LARSSON et al. 1982, PLATT and RAND 1982, MARTIN et al. 1983, PELTIER and THIBAULT 1983), and further confirmed using [^{15}N] (WOO et al. 1982, BERGER and FOCK 1983). It has been shown operative with isolated chloroplasts (WOO and OSMOND 1982) and a reconstituted chloroplast system (ANDERSON and WALKER 1983). The principal role of the chloroplast envelope's dicarboxylate transporter is now considered to be the import of α-oxoglutarate and export of glutamate during this reassimilation of NH_3 (SOMERVILLE and OGREN 1983).

The role of photorespiration in the disposal of excess photosynthetically generated energy (10 ATP and 6 NADPH are utilized for each CO_2 released and subsequently refixed: see LATZKO and KELLY 1978) has been discussed less vigorously in the last 2 years. It had been shown that intact leaves suffered photoinhibition (i.e., damage to photosystem II) in the light if not allowed to regenerate $NADP^+$ through either photorespiration or photosynthesis, but it was not possible to imitate this situation with isolated chloroplasts (CORNIC et al. 1982). This result means that, while the prevention of photoinhibition by photorespiration is still a real possibility, the mechanism is not completely clear. Photoinhibition can be potentiated by other stresses, e.g., low temperature (POWLES et al. 1983). It is thought to be initiated by the Mehler reaction, i.e., electrons being passed from reduced ferredoxin to oxygen (in the absence of $NADP^+$, the "usual" electron acceptor), with consequent accumulation of toxic products such as superoxide ions and H_2O_2. However, it is clear that the Mehler reaction proceeds to some extent even under more-or-less normal conditions (BEHRENS et al. 1982, FURBANK et al. 1982), concomitantly with $NADP^+$ reduction (ROBINSON and GIBBS 1982), and is coupled to ATP synthesis (i.e., "pseudocyclic" photophosphorylation) in both C_3 and C_4 plant leaves, thus enabling chloroplasts to balance their ATP to NADPH requirements (FURBANK et al. 1982, 1983). The H_2O_2 produced by this Mehler reaction is presumably reduced to H_2O by NADPH through the ascorbate/glutathione system shown in Fig. 7B_1 of our last review (KELLY and LATZKO 1982). This system has been demonstrated with both intact and ruptured chloroplasts (JABLONSKI and ANDERSON 1982, ANDERSON et al. 1983a,b), and another of the constituent enzymes characterized (HOSSAIN and ASADA 1984). Uptake of ascorbate by spinach chloroplasts has been observed (ANDERSON et al. 1983b, BECK et al. 1983), and considerable concentrations of both ascorbate and glutathione, each maintained predominantly in the reduced form in both light and darkness, have been demonstrated in chloroplasts (ANDERSON et al. 1983a, FOYER et al. 1983b, LAW et al. 1983). In darkness, NADPH generated during mobilization of starch to glycerate-3-P is believed to be oxidized by a similar ascorbate-based system (KOW et al. 1982b).

Further clues to the mechanism of photoinhibition, and the proposed protective role of photorespiration, may be expected from studies with phytoplankton algae. These algae, especially those in the ocean, grow at depths where the light intensity is between 1% and 10% of full sunlight. Many species become photoinhibited at higher light intensities, even in their natural environment where photosynthesis and photorespiration would seem possible. For example, photoinhibition of the dinoflagellate *Amphidinium carterae* occurs at about 1/20th of full sunlight (SAMUELSSON and RICHARDSON 1982). The reason for such sensitivity remains to be explained, but two hints are available. Firstly, in brown algae the Mehler reaction appears to be quite active (POPOVIC et al. 1983). Secondly, it is likely that photorespiration does not occur in the ocean (KELLY 1984). Although there is evidence for the production of glycolate by phytoplankton (FOGG 1983), and for enzymes of the glycolate pathway in the diatom *Cylindrotheca fusiformis* (PAUL and VOLCANI 1976), studies with the diatom *Navicula pelliculosa* and the red unicellular alga *Porphyridium cruentum* suggest that photorespiration is largely repressed by a high internal CO_2 concentration consequent upon the ocean's abundance of inorganic carbon (as 2 mM HCO_3^-) and the proclivity for these algae to actively transport HCO_3^- into the cell (BIRMINGHAM et al. 1982, COLMAN and GEHL 1983). Similar conclusions may apply to freshwater algae, since MERRETT and ARMITAGE (1982) found that the CO_2-concentrating mechanism of *Chlamydomonas reinhardii* apparently repressed glycolate formation, and VINCENT et al. (1984) found that the bright light experienced by phytoplankton in the high-altitude Lake Titicaca (Peru) led to marked photoinhibition, an indication that the level (if any) of photorespiration in these algae was too low to prevent them from experiencing this photoinhibition.

References

ACKMAN, R.G., TOCHER, C.S., McLACHLAN, J.: J. Fish. Res. Board Can. 25, 1603-1620 (1968). - AKAGAWA, H., IKAWA, T., NISIZAWA, K.: Plant Cell Physiol. 13, 999-1016 (1972). - AMAGASA, T.: Plant Cell Physiol. 23, 1471-1474 (1982). - AMIR, F., PREISS, J.: Plant Physiol. 69, 1027-1030 (1982). - ANDERSON, J.W., WALKER, D.A.: Planta 159, 247-253 (1983). - ANDERSON, J.W., FOYER, C.H., WALKER, D.A.: Biochim. Biophys. Acta 724, 69-74 (1983a); - Planta 158, 442-450 (1983b). - ANDREWS, T.J., BALLMENT, B.: J. Biol. Chem. 258, 7514-7518 (1983). - AOYAGI, K., BASSHAM, J.A.: Plant Physiol. 73, 853-854 (1983). - APEL, P., MAASS, I.: Biochem. Physiol. Pflanzen 197, 396-399 (1981). - ASCENCIO, J., BOWES, G.: Photosynth. Res. 4, 151-170 (1983). - ASHTON, A.R.: Biochim. Biophys. Acta 749, 24-31 (1983); - Biochem. J. 217, 79-84 (1984). - ASHTON, A.R., HATCH, M.D.: Biochem. Biophys. Res. Commun. 115, 53-60 (1983). - AZCÓN-BIETO, J., OSMOND, C.B.: Plant Physiol. 71, 574-581 (1983). - AZCÓN-BIETO, J., LAMBERS, H., DAY, D.A.: Aust. J. Plant Physiol. 10, 237-245 (1983a); - Plant Physiol. 72, 598-603 (1983b).

BADGER, M.R., ANDREWS, T.J.: Plant Physiol. 70, 517-523 (1982). - BÁRCENA, J.A.: Biochem. Int. 7, 755-760 (1983). - BARROW, S.R., COCKBURN, W.: Plant Physiol. 69, 568-571 (1982). - BASSI, R., PASSERA, C.: Photosynth. Res. 3, 53-58 (1982). - BAYSDORFER, C., BASSHAM, J.A.: Plant Physiol. 74, 374-379 (1984). - BEARDALL, J., MORRIS, I.: Mar. Biol. 37, 377-387 (1976). - BEARDALL, J., MUKERJI, D., GLOVER, H.E., MORRIS, I.: J. Phycol. 12, 409-417 (1976). - BECK, E., BURKERT, A., HOFMANN, M.: Plant Physiol. 73, 41-45 (1983). - BEER, S., WETZEL, R.G.: Plant Physiol. 70, 488-492 (1982). - BEHRENS, P.W., MARSHO, T.V., RADMER, R.J.: Plant Physiol. 70, 179-185 (1982). - BELKNAP, W.R.:

Plant Physiol. 72, 1130-1132 (1983). - BENNOUN, P.: Proc. Natl. Acad. Sci. USA 79, 4352-4356 (1982). - BERGER, M.G., FOCK, H.P.: Photosynth. Res. 4, 3-7 (1983). - BERGMAN, A., ERICSON, I.: Physiol. Plant. 59, 421-427 (1983). - BERKOWITZ, G.A., GIBBS, M.: Plant Physiol. 72, 1100-1109 (1983). - BERKOWITZ, G.A., CHEN, C., GIBBS, M.: Plant Physiol. 72, 1123-1126 (1983). - BETSCHE, T.: Plant Physiol. 71, 961-965 (1983). - BIEKMANN, S., FEIERABEND, J.: Biochim. Biophys. Acta 721, 268-279 (1982). - BIRKENHEAD, K., WALKER, D., FOYER, C.: Planta 156, 171-175 (1982). - BIRMINGHAM, B.C., COLEMAN, J.R., COLMAN, B.: Plant Physiol. 69, 259-262 (1982). - BLOCK, M.A., DORNE, A.-J., JOYARD, J., DOUCE, R.: J. Biol. Chem. 258, 13281-13286 (1983). - BOURQUIN, P.W., FOCK, H.P.: Photosynthetica 17, 182-188 (1983). - BRAUN, P., BODE, J., WILD, A.: Z. Naturforsch. 38c, 243-246 (1983). - BRULFERT, J., QUEIROZ, O.: Planta 154, 339-343 (1982). - BRULFERT, J., VIDAL, J., GADAL, P., QUEIROZ, O.: Planta 156, 92-94 (1982a). - BRULFERT, J., MÜLLER, D., KLUGE, M., QUEIROZ, O.: Planta 154, 326-331 (1982b). - BUCHANAN-BOLLIG, I.C., KLUGE, M., MÜLLER, D.: Plant Cell Environ. 7, 63-70 (1984). - BURNELL, J.N., HATCH, M.D.: Biochem. Biophys. Res. Commun. 111, 288-293 (1983); - ibid. 118, 65-72 (1984). - BUSER-SUTER, C., WIEMKEN, A., MATILE, P.: Plant Physiol. 69, 456-459 (1982).

CARNAL, N.W., BLACK, C.C.: Plant Physiol. 71, 150-155 (1983). - CATALDO, D.A., BERLYN, G.P.: Am. J. Bot. 61, 957-963 (1974). - CERFF, R., KLOPPSTECH, K.: Proc. Natl. Acad. Sci. USA 79, 7624-7628 (1982). - CHANG, C.W.: Phytochemistry 21, 1263-1269 (1982). - CHAPMAN, K.S.R., HATCH, M.D.: Aust. J. Plant Physiol. 8, 237-248 (1981); - Plant Sci. Lett. 29, 145-154 (1983). - CLAUSSEN, W., LENZ, F.: Z. Pflanzenphysiol. 109, 459-468 (1983). - CLINE, K., KEEGSTRA, K.: Plant Physiol. 71, 366-372 (1983). - COLMAN, B., GEHL, K.A.: J. Phycol. 19, 216-219 (1983). - COREY, E.J., SHIH, C., CASHMAN, J.R.: Proc. Natl. Acad. Sci. USA 80, 3581-3584 (1983). - CORLEY, E., WOLOSIUK, R.A., HERTIG, C.M.: Biochem. Biophys. Res. Commun. 115, 707-714 (1983). - CORNIC, G., WOO, K.C., OSMOND, C.B.: Plant Physiol. 70, 1310-1315 (1982). - CSÉKE, C., BUCHANAN, B.B.: FEBS Lett. 155, 139-142 (1983). - CSÉKE, C., NISHIZAWA, A.N., BUCHANAN, B.B.: Plant Physiol. 70, 658-661 (1982a). - CSÉKE, C., WEEDEN, N.F., BUCHANAN, B.B., UYEDA, K.: Proc. Natl. Acad. Sci. USA 79, 4322-4326 (1982b).

DELANEY, M.E., WALKER, D.A.: Biochem. J. 171, 477-482 (1978). - DE LA TORRE, A., LARA, C., YEE, B.C., MALKIN, R., BUCHANAN, B.B.: Arch. Biochem. Biophys. 213, 545-550 (1982). - DEMMIG, B., GIMMLER, H.: Plant Physiol. 73, 169-174 (1983). - DEMMIG, B., WINTER, K.: Planta 159, 66-76 (1983). - DIETER, P., MARMÉ, D.: J. Biol. Chem. 259, 184-189 (1984). - DOEHLERT, D.C., HUBER, S.C.: Plant Physiol. 73, 989-994 (1983). - DORNE, A.-J., BLOCK, M.A., JOYARD, J., DOUCE, R.: FEBS Lett. 145, 30-34 (1982). - DRY, I.B., WISKICH, J.T.: Arch. Biochem. Biophys. 217, 72-79 (1982); - Plant Physiol. 72, 291-296 (1983). - DRY, I.B., DAY, D.A., WISKICH, J.T.: FEBS Lett. 158, 154-158 (1983). - DUBACQ, J.-P., DRAPIER, D., TRÉMOLIÈRES, A.: Plant Cell Physiol. 24, 1-9 (1983).

EDWARDS, G.E., KU, M.S.B., HATCH, M.D.: Plant Cell Physiol. 23, 1185-1195 (1982). - ESPIE, G.S., COLMAN, B.: Plant Physiol. 70, 649-654 (1982).

FALKOWSKI, P.G.: J. Plankton Res. 3, 203-216 (1981). - FEIERABEND, J., GRINGEL, G.: Z. Pflanzenphysiol. 110, 247-258 (1983). - FLORENCIO, F.J., VEGA, J.M.: Z. Naturforsch. 38c, 531-538 (1983). - FLÜGGE, U.I., BENZ, R.: FEBS Lett. 169, 85-89 (1984). - FLÜGGE, U.I., STITT, M., FREISL, M., HELDT, H.W.: Plant Physiol. 69, 263-267 (1982). - FLÜGGE, U.I., GERBER, J., HELDT, H.W.: Biochim. Biophys. Acta 725, 229-237 (1983). - FOGG, G.E.: Bot. Mar. 26, 3-14 (1983). - FOSTER, J.G., EDWARDS, G.E.: Plant Cell Physiol. 24, 1191-1202 (1983). - FOSTER, J.G., EDWARDS, G.E.,

WINTER, K.: Plant Cell Physiol. 23, 585-594 (1982). - FOYER, C., WALKER, D., LATZKO, E.: Z. Pflanzenphysiol. 107, 457-465 (1982). - FOYER, C., ROWELL, J., WALKER, D.: Arch. Biochem. Biophys. 220, 232-238 (1983a); - Planta 157, 239-244 (1983b). - FRENTZEN, M., HEINZ, E., McKEON, T.A., STUMPF, P.K.: Eur. J. Biochem. 129, 629-636 (1983). - FURBANK, R.T., BADGER, M.R.: Aust. J. Plant Physiol. 9, 553-558 (1982); - ibid. 10, 451-458 (1983). - FURBANK, R.T., BADGER, M.R., OSMOND, C.B.: Plant Physiol. 70, 927-931 (1982); - ibid. 73, 1038-1041 (1983). - FURNAS, M.J.: J. Plankton Res. 5, 637-655 (1983).

GARDEMANN, A., STITT, M., HELDT, H.W.: FEBS Lett. 137, 213-216 (1982); - Biochim. Biophys. Acta 722, 51-60 (1983). - GARDESTRÖM, P., EDWARDS, G.E.: Plant Physiol. 71, 24-29 (1983). - GASPERINI, C., PUPILLO, P.: Plant Sci. Lett. 28, 163-171 (1982). - GERBLING, K.-P., STEUP, M., LATZKO, E.: Arch. Microbiol. 137, 109-114 (1984). - GIERSCH, C.: Arch. Biochem. Biophys. 219, 379-387 (1982). - GIVAN, C.V.: Physiol. Plant. 57, 311-316 (1983). - GIVAN, C.V., HODGSON, J.M.: Plant Sci. Lett. 32, 233-242 (1983). - GLOVER, H.E.: J. Plankton Res. 2, 69-79 (1980). - GLOVER, H.E., MORRIS, I.: Limnol. Oceanogr. 24, 510-519 (1979). - GOLLER, M., HAMPP, R., ZIEGLER, H.: Planta 156, 255-263 (1982). - GOODNIGHT, S.H., HARRIS, W.S., CONNER, W.E., ILLINGWORTH, D.R.: Arteriosclerosis 2, 87-113 (1982). - GUTTERIDGE, S., PARRY, M.A.J., SCHMIDT, C.N.G.: Eur. J. Biochem. 126, 597-602 (1982).

HALL, N.P., KEYS, A.J.: Plant Physiol. 72, 945-948 (1983). - HALLBERG, M., LARSSON, C.: Physiol. Plant. 57, 330-338 (1983). - HAMMOND, J.B.W., BURTON, K.S.: Plant Physiol. 73, 61-65 (1983). - HAMMOND, J.B.W., PREISS, J.: Plant Physiol. 73, 709-712 (1983). - HAMPP, R., GOLLER, M.: Planta 159, 314-321 (1983). - HAMPP, R., OUTLAW, W.H., TARCZYNSKI, M.C.: Plant Physiol. 70, 1582-1585 (1982a). - HAMPP, R., GOLLER, M., ZIEGLER, H.: Plant Physiol. 69, 448-455 (1982b). - HARRIS, G.P., PICCININ, B.B.: J. Phycol. 19, 185-192 (1983). - HATCH, M.D., BURNELL, J.N.: Aust. J. Plant Physiol. 10, 179-186 (1983). - HEBER, U., TAKAHAMA, U., NEIMANIS, S., SHIMIZU-TAKAHAMA, M.: Biochim. Biophys. Acta 679, 287-299 (1982). - HERTIG, C.M., WOLOSIUK, R.A.: J. Biol. Chem. 258, 984-989 (1983). - HEUER, B., HANSEN, M.J., ANDERSON, L.E.: Plant Physiol. 69, 1404-1406 (1982). - HIREL, B., PERROT-RECHENMANN, C., SUZUKI, A., VIDAL, J., GADAL, P.: Plant Physiol. 69, 983-987 (1982a). - HIREL, B., VIDAL, J., GADAL, P.: Planta 155, 17-23 (1982b). - HIREL, B., LAYZELL, D.B., McCASHIN, B., McNALLY, S.F., CANVIN, D.T.: Can. J. Bot. 61, 2257-2259 (1983). - HIROKAWA, T., HATA, M., TAKEDA, H.: Plant Cell Physiol. 23, 813-820 (1982). - HOLADAY, A.S., CHOLLET, R.: Plant Physiol. 73, 740-745 (1983). - HOLADAY, A.S., HARRISON, A.T., CHOLLET, R.: Plant Sci. Lett. 27, 181-189 (1982). - HOLDSWORTH, E.S., BRUCK, K.: Arch. Biochem. Biophys. 182, 87-94 (1977). - HOLDSWORTH, E.S., COLBECK, J.: Mar. Biol. 38, 189-199 (1976). - HOLTUM, J.A.M., WINTER, K.: Planta 155, 8-16 (1982). - HOSSAIN, M.A., ASADA, K.: Plant Cell Physiol. 25, 85-92 (1984). - HOWITZ, K.T., McCARTY, R.E.: FEBS Lett. 154, 339-342 (1983). - HUBER, S.C.: Plant Physiol. 71, 818-821 (1983). - HUBER, S.C., BICKETT, D.M.: Plant Physiol. 74, 445-447 (1984). - HUBER, S.C., ISRAEL, D.W.: Plant Physiol. 69, 691-696 (1982). - HUSIC, H.D., TOLBERT, N.E.: Arch. Biochem. Biophys. 229, 64-72 (1984). - HUTCHESON, S.W., BUCHANAN, B.B.: Plant Physiol. 72, 877-885 (1979).

IADAROLA, P., ZAPPONI, M.C., FERRI, G.: Experentia 39, 50-52 (1983). - IMAMURA, M., TSUZUKI, M., SHIRAIWA, Y., MIYACHI, S.: Plant Cell Physiol. 24, 533-540 (1983).

JABLONSKI, P.P., ANDERSON, J.W.: Plant Physiol. 69, 1407-1413 (1982). - JARRETT, H.W., BROWN, C.J., BLACK, C.C., CORMIER, M.J.: J. Biol. Chem. 257, 13795-13804 (1982). - JOLIVET-TOURNIER, P., GERSTER, R.:

Plant Physiol. 74, 108-111 (1984). - JONES, G.J., NICHOLS, P.D., JOHNS, R.B.: J. Phycol. 19, 416-420 (1983). - JORDAN, D.B., CHOLLET, R.: J. Biol. Chem. 258, 13752-13758 (1983). - JORDAN, D.B., CHOLLET, R., OGREN, W.L.: Biochemistry 22, 3410-3418 (1983).

KAISER, G., MARTINOIA, E., WIEMKEN, A.: Z. Pflanzenphysiol. 107, 103-113 (1982). - KAISER, W.M., WEBER, H., SAUER, M.: Z. Pflanzenphysiol. 113, 15-27 (1983). - KAMIŃSKA, Z., MALESZEWSKI, S.: Z. Pflanzenphysiol. 108, 201-206 (1982). - KANAZAWA, T., DISTEFANO, M., BASSHAM, J.A.: Plant Cell Physiol. 24, 979-986 (1983). - KAPLAN, A., ZENVIRTH, D., REINHOLD, L., BERRY, J.A.: Plant Physiol. 69, 978-982 (1982). - KARABOURNIOTIS, G., MANETAS, Y., GAVALAS, N.A.: Plant Physiol. 73, 735-739 (1983). - KATES, M., VOLCANI, B.E.: Biochim. Biophys. Acta 116, 264-278 (1966). - KELLY, G.J.: Trends Biochem. Sci. 9, 255-256 (1984). - KELLY, G.J., LATZKO, E.: Prog. Bot. 44, 103-131 (1982). - KENDALL, A.C., KEYS, A.J., TURNER, J.C., LEA, P.J., MIFLIN, B.J.: Planta 159, 505-511 (1983). - KERBY, N.W., EVANS, L.V.: J. Phycol. 19, 1-3 (1983). - KIMPEL, D.L., TOGASAKI, R.K., MIYACHI, S.: Plant Cell Physiol. 24, 255-259 (1983). - KLECZKOWSKI, L.A., RANDALL, D.D.: FEBS Lett. 158, 313-316 (1983). - KLEIN, U., CHEN, C., GIBBS, M.: Plant Physiol. 72, 488-491 (1983). - KOIKE, I., REDALJE, D.G., AMMERMAN, J.W., HOLM-HANSEN, O.: Mar. Biol. 74, 161-168 (1983). - KOMBRINK, E., WÖBER, G.: Arch. Biochem. Biophys. 213, 602-619 (1982). - KOW, Y.W., GIBBS, M.: Plant Physiol. 69, 179-186 (1982). - KOW, Y.W., ERBES, D.L., GIBBS, M.: Plant Physiol. 69, 442-447 (1982a). - KOW, Y.W., SMYTH, D.A., GIBBS, M.: Plant Physiol. 69, 740-741 (1982b). - KREMER, B.P.: Oceanogr. Mar. Biol. Annu. Rev. 19, 41-94 (1981a); - Phycologia 20, 242-250 (1981b). - KREMER, B.P., BERKS, R.: Z. Pflanzenphysiol. 87, 149-165 (1978). - KREMER, B.P., KIRST, G.: Z. Naturforsch. 37c, 761-771 (1982). - KRÜGER, I., SCHNARRENBERGER, C.: Eur. J. Biochem. 136, 101-106 (1983). - KRUGER, N.J., ap REES, T.: Phytochemistry 22, 1891-1898 (1983a); - Planta 158, 179-184 (1983b). - KRUGER, N.J., BULPIN, P.V., ap REES, T.: Planta 157, 271-273 (1983a). - KRUGER, N.J., KOMBRINK, E., BEEVERS, H:: Biochem. Biophys. Res. Commun. 117, 37-42 (1983b). - KU, M.S.B., MONSON, R.K., LITTLEJOHN, R.O., NAKAMOTO, H., FISHER, D.B., EDWARDS, G.E.: Plant Physiol. 71, 944-948 (1983). - KULKARNI, A.H., NIMBALKAR, J.D.: Photosynthetica 17, 189-192 (1983). - KWOK-KI, H., HOCK-HIN, Y., CHOY-SIN, H.: Plant Cell Physiol. 24, 1317-1321 (1983).

LARSSON, M., LARSSON, C.-M., ULLRICH, W.R.: Plant Physiol. 70, 1637-1640 (1982). - LATZKO, E., KELLY, G.J.: Prog. Bot. 40, 99-118 (1978); - pp. 59-69 in: The New Frontiers in Plant Biochemistry, ed. T. AKAZAWA, T. ASAHI, H. IMASEKI. Tokyo: Japan Scientific Societies Press 1983a; - Physiol. Vég. 21, 805-815 (1983b). - LAW, M.Y., CHARLES, S.A., HALLIWELL, B.: Biochem. J. 210, 899-903 (1983). - LEEGOOD, R.C., WALKER, D.A.: Planta 156, 449-456 (1982); - Plant Physiol. 71, 513-518 (1983). - LEEGOOD, R.C., KOBAYASHI, Y., NEIMANIS, S., WALKER, D.A., HEBER, U.: Biochim. Biophys. Acta 682, 168-178 (1982). - LEVI, C., GIBBS, M.: Plant Physiol. 74, 459-463 (1984). - LIANG, Z., HUANG, A.H.C.: Plant Physiol. 73, 147-152 (1983). - LIANG, Z., YU, C., HUANG, A.H.C.: Plant Physiol. 70, 1210-1212 (1982). - LIEDVOGEL, B., STUMPF, P.K.: Plant Physiol. 69, 897-903 (1982). - LILLEY, R.McC.: Plant Cell Environ. 6, 329-343 (1983). - LILLEY, R.McC., STITT, M., MADER, G., HELDT, H.W.: Plant Physiol. 70, 965-970 (1982). - LINDEN, J.C., SCHILLING, N.: Plant Physiol. 74, 795-799 (1984). - LOESCHER, W.H., MARLOW, G.C., KENNEDY, R.A.: Plant Physiol. 70, 335-339 (1982). - LOOZE, S. de, WAGNER, E.: Physiol. Plant. 57, 231-237 (1983a); - ibid. 57, 238-242 (1983b). - LUDWIG, I., ZIEGLER, P., BECK, E.: Plant Physiol. 74, 856-861 (1984).

MADHAVAN, S., SMITH, B.N.: Plant Physiol. 69, 273-277 (1982). - MAE, T., MAKINO, A., OHIRA, K.: Plant Cell Physiol. 24, 1079-1086 (1983). -

MAKINO, A., MAE, T., OHIRA, K.: Plant Physiol. 73, 1002-1007 (1983). - MANETAS, Y.: Photosynth. Res. 3, 321-333 (1982). - MANETAS, Y., SAMARAS, Y., STAMATAKIS, K.: Z. Pflanzenphysiol. 111, 365-368 (1983). - MARCUS, Y., ZENVIRTH, D., HAREL, E., KAPLAN, A.: Plant Physiol. 69, 1008-1012 (1982). - MAREK, L.F., STEWART, C.R.: Plant Physiol. 73, 118-120 (1983). - MARIGO, G., LÜTTGE, U., SMITH, J.A.C.: Z. Pflanzenphysiol. 109, 405-413 (1983). - MARQUES, I.A., OBERHOLZER, M.J., ERISMANN, K.H.: Plant Physiol. 71, 555-561 (1983). - MARTIN, C.E., ZEE, A.K.: Plant Physiol. 73, 718-723 (1983). - MARTIN, F., WINSPEAR, M.J., MacFARLANE, J.D., OAKS, A.: Plant Physiol. 71, 177-181 (1983). - McNALLY, S.F., HIREL, B., GADAL, P., MANN, A.F., STEWART, G.R.: Plant Physiol. 72, 22-25 (1983). - MERRETT, M.J., ARMITAGE, T.L.: Planta 155, 95-96 (1982). - MEYER, A.O., KELLY, G.J., LATZKO, E.: Plant Physiol. 69, 7-10 (1982). - MINOT, R., MEUNIER, J.-C., BUC, L., RICARD, J.: FEBS Lett. 142, 118-120 (1982). - MIYACHI, S., TSUZUKI, M., AVRAMOVA, S.T.: Plant Cell Physiol. 24, 441-451 (1983). - MOHAMED, A.H., ANDERSON, L.E.: Plant Physiol. 71, 248-250 (1983). - MONSON, R.K., STIDHAM, M.A., WILLIAMS, G.J., EDWARDS, G.E., URIBE, E.G.: Plant Physiol. 69, 921-928 (1982). - MONSON, R.K., RUMPHO, M.E., EDWARDS, G.E.: Planta 159, 97-104 (1983). - MUKERJI, D., GLOVER, H.E., MORRIS, I.: J. Phycol. 14, 137-142 (1978). - MÜLLER, D., KLUGE, M., GRÖSCHEL-STEWART, U.: Plant Cell Environ. 5, 223-230 (1982). - MURPHY, D.J., WALKER, D.A.: Planta 155, 316-320 (1982a); - ibid. 156, 84-88 (1982b). - MURPHY, D.J., MUKHERJEE, K.D., LATZKO, E.: Biochem. J. 213, 249-252 (1983a). - MURPHY, D.J., WOODROW, I.E., LATZKO, E., MUKHERJEE, K.D.: FEBS Lett. 162, 442-446 (1983b). - MURPHY, D.J., MUKHERJEE, K.D., WOODROW, I.E.: Eur. J. Biochem. 139, 373-379 (1984). - MUTO, S.: FEBS Lett. 147, 161-164 (1982); - Z. Pflanzenphysiol. 109, 385-393 (1983). - MUTO, S., IZAWA, S., MIYACHI, S.: FEBS Lett. 139, 250-254 (1982).

NAKAMOTO, H., EDWARDS, G.E.: Plant Physiol. 71, 568-573 (1983a); - Aust. J. Plant Physiol. 10, 279-289 (1983b); - Biochem. Biophys. Res. Commun. 115, 673-679 (1983c). - NAKAMOTO, H., SUGIYAMA, T.: Plant Physiol. 69, 749-753 (1982). - NAKAMOTO, H., KU, M.S.B., EDWARDS, G.E.: Plant Cell Physiol. 24, 1387-1393 (1983). - NAKAMURA, Y., IMAMURA, M.: Plant Sci. Lett. 31, 123-131 (1983). - NAKAMURA, Y., MIYACHI, S.: Plant Sci. Lett. 27, 1-6 (1982). - NAKAMURA, Y., TOLBERT, N.E.: J. Biol. Chem. 258, 7631-7638 (1983). - NASH, D., WISKICH, J.T.: Plant Physiol. 71, 627-634 (1983). - NEUBURGER, M., JOURNET, E.-P., BLIGNY, R., CARDE, J.-P., DOUCE, R.: Arch. Biochem. Biophys. 217, 312-323 (1982). - NILSEN, S., JOHNSON, Ø.: Physiol. Plant. 56, 273-280 (1982). - NISHIMURA, M., BHUSAWANG, P., STRZALKA, K., AKAZAWA, T.: Plant Physiol. 70, 353-356 (1982a). - NISHIMURA, M., DOUCE, R., AKAZAWA, T.: Plant Physiol. 69, 916-920 (1982b). - NISHIMURA, M., AKHMEDOV, Y.D., STRZALKA, K., AKAZAWA, T.: Arch. Biochem. Biophys. 222, 397-402 (1983a). - NISHIMURA, M., AKHMEDOV, Y.D., AKAZAWA, T.: Photosynth. Res. 4, 99-109 (1983b). - NOTT, D.L., OSMOND, C.B.: Aust. J. Plant Physiol. 9, 409-422 (1982).

OGAWA, T., INOUE, Y.: Biochim. Biophys. Acta 724, 490-493 (1983). - OHNISHI, J., KANAI, R.: Plant Cell Physiol. 24, 1411-1420 (1983). - OHNISHI, J., YAMADA, M.: Plant Cell Physiol. 23, 767-773 (1982). - OKABE, K., YANG, S.-Y., TSUZUKI, M., MIYACHI, S.: Plant Sci. Lett. 33, 145-153 (1984). - OUTLAW, W.H., TARCZYNSKI, M.C.: Plant Physiol. 74, 424-429 (1984). - OUTLAW, W.H., TARCZYNSKI, M.C., ANDERSON, L.C.: Plant Physiol. 70, 1218-1220 (1982).

PAN, D., TAN, K.H.: Photosynthetica 15, 511-517 (1981). - PARRY, M.A.J., SCHMIDT, C.N.G., KEYS, A.J., GUTTERIDGE, S.: FEBS Lett. 159, 107-111 (1983). - PAUL, J.S., VOLCANI, B.E.: Arch. Microbiol. 110, 247-252 (1976). - PEISKER, M., APEL, P.: Photosynthetica 15, 435-441 (1981). -

PELTIER, G., THIBAULT, P.: Plant Physiol. 71, 888-892 (1983). - PERCHOROWICZ, J.T., JENSEN, R.G.: Plant Physiol. 71, 955-960 (1983). - PERROT-RECHENMANN, C., VIDAL, J., BRULFERT, J., BURLET, A., GADAL, P.: Planta 155, 24-30 (1982). - PERROT-RECHENMANN, C., JACQUOT, J.P., GADAL, P., WEEDEN, N.F., CSÉKE, C., BUCHANAN, B.B.: Plant Sci. Lett. 30, 219-226 (1983). - PETERSON, R.B.: Plant Physiol. 69, 192-197 (1982); - ibid. 73, 983-988 (1983). - PFLEIDERER, C., KLEMME, J.-H.: Z. Naturforsch. 35c, 229-238 (1980). - PIAZZA, G.J., SMITH, M.G., GIBBS, M.: Plant Physiol. 70, 1748-1758 (1982). - PICHERSKY, E., GOTTLIEB, L.D.: Plant Physiol. 74, 340-347 (1984). - PLÁ, A., CHUECA, A., LÓPEZ-GORGÉ, J.: Photosynth. Res. 2, 291-296 (1981). - PLATT, S.G., RAND, L.: Plant Cell Physiol. 23, 917-921 (1982). - PLATT, S.G., PLAUT, Z., BASSHAM, J.A.: Plant Physiol. 57, 69-73 (1976). - POLYA, G.M., MICUCCI, V.: Biochim. Biophys. Acta 785, 68-74 (1984). - POPOVIC, R., COLBOW, K., VIDAVER, W., BRUCE, D.: Plant Physiol. 73, 889-892 (1983). - PORTER, M.A., GRODZINSKI, B.: Plant Physiol. 74, 413-416 (1984). - PORTIS, A.R.: Plant Physiol. 70, 393-396 (1982); - ibid. 71, 936-943 (1983). - POWLES, S.B., BERRY, J.A., BJÖRKMAN, O.: Plant Cell Environ. 6, 117-123 (1983). - PREISS, J.: Annu. Rev. Plant Physiol. 33, 431-454 (1982). - PRÉZELIN, B.B., ALBERTE, R.S.: Proc. Natl. Acad. Sci. USA 75, 1801-1804 (1978). - PRISCU, J.C., GOLDMAN, C.R.: Limnol. Oceanogr. 28, 1168-1181 (1983). - PROUDLOVE, M.O., MOORE, A.L.: FEBS Lett. 147, 26-30 (1982). - PROUDLOVE, M.O., THURMAN, D.A., SALISBURY, J.: New Phytol. 96, 1-5 (1984).

RAJENDRUDU, G., DAS, V.S.R.: Photosynth. Res. 2, 225-233 (1981). - RAO, I.M., ANDERSON, L.E.: Plant Physiol. 71, 451-455 (1983). - RAYDER, L., TING, I.P.: Photosynth. Res. 4, 203-211 (1983a); - Plant Physiol. 72, 606-610 (1983b). - REES, T.A.V., BEKHEET, I.A.: Planta 156, 385-387 (1982). - RICHARDSON, K., BEARDALL, J., RAVEN, J.A.: New Phytol. 93, 157-191 (1983). - RIVKIN, R.B., SELIGER, H.H., SWIFT, E., BIGGLEY, W.H.: Mar. Biol. 68, 181-191 (1982). - ROACH, D.J.W., McFADDEN, B.A.: Photosynth. Res. 4, 111-117 (1983). - ROBINSON, J.M., GIBBS, M.: Plant Physiol. 70, 1249-1254 (1982). - ROBINSON, S.P.: Biochem. Biophys. Res. Commun. 106, 1027-1034 (1982a); - Plant Physiol. 70, 1032-1038 (1982b). - ROSSLENBROICH, H.-J., DÖHLER, G.: Biochem. Physiol. Pflanzen 177, 363-373 (1982). - ROUGHAN, P.G., SLACK, C.R.: Annu. Rev. Plant Physiol. 33, 97-132 (1982). - RUFTY, T.W., HUBER, S.C.: Plant Physiol. 72, 474-480 (1983). - RUFTY, T.W., KERR, P.S., HUBER, S.C.: Plant Physiol. 73, 428-433 (1983). - RUMPHO, M.E., EDWARDS, G.E., LOESCHER, W.H.: Plant Physiol. 73, 869-873 (1983). - RUTTER, J.C., COBB, A.H.: New Phytol. 95, 559-568 (1983). - RUYTERS, G.: Z. Pflanzenphysiol. 108, 207-214 (1982). - RUYTERS, G., MIYACHI, S.: Plant Cell Physiol. 24, 863-871 (1983). - RYTHER, J.H., YENTSCH, C.S.: Limnol. Oceanogr. 2, 281-286 (1957).

SABULARSE, D.C., ANDERSON, R.L.: Biochem. Biophys. Res. Commun. 103, 848-855 (1981). - SALVUCCI, M.E., BOWES, G.: Plant Physiol. 73, 488-496 (1983a); - Planta 158, 27-34 (1983b). - SAMUELSSON, G., RICHARDSON, K.: Mar. Biol. 70, 21-26 (1982). - SANADA, Y., NISHIDA, K.: Z. Pflanzenphysiol. 105, 189-192 (1982). - SAROJINI, G., OLIVER, D.J.: Plant Physiol. 72, 194-199 (1983). - SAUER, A., HEISE, K.-P.: Plant Physiol. 73, 11-15 (1983). - SAVER, B.G., KNOWLES, J.R.: Biochemistry 21, 5398-5403 (1982). - SCHEIBE, R., JACQUOT, J.-P.: Planta 157, 548-553 (1983). - SCHERER, S., BÖGER, P.: Arch. Microbiol. 132, 329-332 (1982). - SCHILLING, N.: Planta 154, 87-93 (1982). - SCHLOSS, J.V., LORIMER, G.H.: J. Biol. Chem. 257, 4691-4694 (1982). - SCHMITT, M.R., EDWARDS, G.E.: Plant Physiol. 70, 1213-1217 (1982); - ibid. 72, 728-734 (1983). - SCHNABL, H., ELBERT, C., KRÄMER, G.: J. Exp. Bot. 33, 996-1003 (1982). - SCHÜRMANN, P., ORTIZ, W.: Planta 154, 70-75 (1982). - SHIEH, Y.-J., KU, M.S.B., BLACK, C.C.: Plant Physiol. 69, 776-780 (1982). - SHIMAKATA, T., STUMPF, P.K.: Plant Physiol. 69, 1257-1262 (1982). - SHIMAZAKI, K.,

GOTOW, K., SAKAKI, T., KONDO, N.: Plant Cell Physiol. 24, 1049-1056 (1983). - SHIMOMURA, S., NAGAI, M., FUKUI, T.: J. Biochem. 91, 703-717 (1982). - SHIRAIWA, Y., MIYACHI, S.: Plant Cell Physiol. 24, 919-923 (1983). - SICHER, R.C.: Plant Physiol. 70, 366-369 (1982). - SIESS, W., SCHERER, B., BÖHLIG, B., ROTH, P., KURZMANN, I., WEBER, P.C.: Lancet 1, 441-444 (1980). - SIMON, P., BONZON, M., GREPPIN, H., MARME, D.: FEBS Lett. 167, 332-338 (1984). - SINGH, B.K., GREENBERG, E, PREISS, J.: Plant Physiol. 74, 711-716 (1984). - SINGH, P., NAIK, M.S.: Plant Sci. Lett. 30, 9-16 (1983). - SMITH, J.A.C., MARIGO, G., LÜTTGE, U., BALL, E.: Plant Sci. Lett. 26, 13-21 (1982). - SMITH, J.C., PLATT, T., HARRISON, W.G.: Prog. Oceanogr. 12, 425-549 (1983). - SMYTH, D.A., WU, M.-X., BLACK, C.C.: Plant Sci. Lett. 33, 61-70 (1984). - SOMERVILLE, S.C., OGREN, W.L.: Proc. Natl. Acad. Sci. USA 80, 1290-1294 (1983). - SOMERVILLE, S.C., SOMERVILLE, C.R.: J. Exp. Bot. 34, 415-424 (1983). - SPALDING, M.H., OGREN, W.L.: FEBS Lett. 145, 41-44 (1982); - ibid. 154, 335-338 (1983). - SPALDING, M.H., SPREITZER, R.J., OGREN, W.L.: Plant Physiol. 73, 273-276 (1983a); - ibid. 73, 268-272 (1983b). - SPREITZER, R.J., JORDAN, D.B., OGREN, W.L.: FEBS Lett. 148, 117-121 (1982). - SRIVASTAVA, D.K., ANDERSON, L.E.: Biochim. Biophys. Acta 724, 359-369 (1983). - STEUP, M., GERBLING, K.-P.: Anal. Biochem. 134, 96-100 (1983). - STEUP, M., ROBENEK, H., MELKONIAN, M.: Planta 158, 428-436 (1983). - STIBOROVÁ, M., LEBLOVÁ, S.: Photosynthetica 17, 379-385 (1983). - STIDHAM, M.A., MORELAND, D.E., SIEDOW, J.N.: Plant Physiol. 73, 517-520 (1983). - STITT, M., MIESKES, G., SOLING, H.-D., HELDT, H.W.: FEBS Lett. 145, 217-222 (1982a). - STITT, M., LILLEY, R.McC., HELDT, H.W.: Plant Physiol. 70, 971-977 (1982b). - STITT, M., WIRTZ, W., HELDT, H.W.: Plant Physiol. 72, 767-774 (1983a). - STITT, M., GERHARDT, R., KÜRZEL, B., HELDT, H.W.: Plant Physiol. 72, 1139-1141 (1983b). - STRICKLAND, J.D.H., EPPLEY, R.W., ROJAS de MENDIOLA, B.: Bol. Inst. Mar Peru (Callao) 2, 1-45 (1969). - STUMPF, D.K., JENSEN, R.G.: Plant Physiol. 69, 1263-1267 (1982). - SUGIYAMA, T., HIRAYAMA, Y.: Plant Cell Physiol. 24, 783-787 (1983). - SUMAR, N., CASSELTON, P.J., McNALLY, S.F., STEWART, G.R.: Plant Physiol. 74, 204-207 (1984). - SUZUKI, A., GADAL, P.: Plant Physiol. 69, 848-852 (1982).

TAKABE, T., RAI, A.K., AKAZAWA, T.: Arch. Biochem. Biophys. 229, 202-211 (1984). - THORPE, N.: Plant Sci. Lett. 30, 331-338 (1983). - TING, I.P., BURK, J.H.: Plant Sci. Lett. 32, 279-285 (1983). - TING, I.P., GIBBS, M. (eds.). pp. 316 in: Crassulacean Acid Metabolism. Baltimore: Waverly Press 1982. - TING, I.P., STERNBERG, L.O., DeNIRO, M.J.: Plant Physiol. 71, 677-679 (1983). - TISCHNER, R., SCHMIDT, A.: Plant Physiol. 70, 113-116 (1982). - TITUS, D.E., HONDRED, D., BECKER, W.M.: Plant Physiol. 72, 402-408 (1983). - TSUZUKI, M.: Z. Pflanzenphysiol. 110, 29-37 (1983).

UDVARDY, J., GODEH, M.M., FARKAS, G.L.: J. Bacteriol. 151, 203-208 (1982). - USUDA, H., EDWARDS, G.E.: Plant Physiol. 69, 469-473 (1982).

VATER, J., GAUDSZUN, T., LANGE, B., ERDIN, N., SALNIKOW, J.: Z. Naturforsch. 38c, 418-427 (1983). - VELUTHAMBI, K., POOVAIAH, B.W.: Science 223, 167-169 (1984). - VENRICK, E.L., BEERS, J.R., HEINBOKEL, J.F.: J. Exp. Mar. Biol. Ecol. 26, 55-76 (1977). - VIDAL, J., GADAL, P.: Physiol. Vég. 19, 483-489 (1981); - Physiol. Plant. 57, 119-123 (1983). - VINCENT, W.F., NEALE, P.J., RICHERSON, P.J.: J. Phycol. 20, 201-211 (1984). - VINES, H.M., TU, Z.-P., ARMITAGE, A.M., CHEN, S.-S., BLACK, C.C.: Plant Physiol. 73, 25-30 (1983). - VOLKMAN, J.K., SMITH, D.J., EGLINTON, G., FORSBERG, T.E.V., CORNER, E.D.S.: J. Mar. Biol. Ass. U.K. 61, 509-527 (1981). - VOLOKITA, M., KAPLAN, A., REINHOLD, L.: Plant Physiol. 72, 886-890 (1983). - VU, C.V., ALLEN, L.H., BOWES, G.: Plant Physiol. 73, 729-734 (1983).

WALKER, G.H., SAROJINI, G., OLIVER, D.J.: Biochem. Biophys. Res. Commun. 107, 856-861 (1982a). - WALKER, G.H., OLIVER, D.J., SAROJINI. G.: Plant Physiol. 70, 1465-1469 (1982b). - WALTON, N.J.: Planta 155, 218-224 (1982); - Plant Sci. Lett. 30, 203-209 (1983). - WALTON, N.J., WOOLHOUSE, H.W.: Planta 158, 469-471 (1983). - WATANABE, M., OHNISHI, J., KANAI, R.: Plant Cell Physiol. 25, 69-76 (1984). - WEDDING, R.T., BLACK, M.K.: Plant Physiol. 72, 1021-1028 (1983). - WEEDEN, N.F., BUCHANAN, B.B.: Plant Physiol. 72, 259-261 (1983). - WHITTAKER, R.H., LIKENS, G.E.: pp. 305-328 in: Primary Productivity of the Biosphere, ed. H. LIETH, R.H. WHITTAKER. Berlin,Heidelberg, New York: Springer 1975. - WILLIAMS, P.J.leB., HEINEMANN, K.R., MARRA, J., PURDIE, D.A.: Nature 305, 49-50 (1983). - WINKLER, U., SÄFTEL, W., STABENAU, H.: Plant Physiol. 70, 340-343 (1982). - WINTER, K.: Planta 154, 298-308 (1982). - WINTER, K., SCHMITT, M.R., EDWARDS, G.E.: Plant Sci. Lett. 24, 311-318 (1982a). - WINTER, K., USUDA, H., TSUZUKI, M., SCHMITT, M., EDWARDS, G.E., THOMAS, R.J., EVERT, R.F.: Plant Physiol. 70, 616-625 (1982b). - WINTER, K., FOSTER, J.G., SCHMITT, M.R., EDWARDS, G.E.: Planta 154, 309-317 (1982c). - WINTER, K., FOSTER, J.G., EDWARDS, G.E., HOLTUM, J.A.M.: Plant Physiol. 69, 300-307 (1982d). - WINTER, K., WALLACE, B.J., STOCKER, G.C., ROKSANDIC, Z.: Oecologia 57, 129-141 (1983). - WIRTZ, W., STITT, M., HELDT, H.W.: FEBS Lett. 142, 223-226 (1982). - WOLOSIUK, R.A., HERTIG, C.M., NISHIZAWA, A.N., BUCHANAN, B.B.: FEBS Lett. 140, 31-35 (1982). - WOO, K.C.: Plant Physiol. 71, 112-117 (1983). - WOO, K.C., OSMOND, C.B.: Plant Physiol. 69, 591-596 (1982). - WOO, K.C., MOROT-GAUDRY, J.F., SUMMONS, R.E., OSMOND, C.B.: Plant Physiol. 70, 1514-1517 (1982). - WOODROW, I.E., WALKER, D.A.: Arch. Biochem. Biophys. 216, 416-422 (1982); - Biochim. Biophys. Acta 722, 508-516 (1983). - WOODROW, I.E., MURPHY, D.J., LATZKO, E.: J. Biol. Chem. 259, 3791-3795 (1984). - WULFF, R.D., STRAIN, B.R.: Can. J. Bot. 60, 1084-1091 (1982).

YOKOTA, A., KITAOKA, S.: Plant Physiol. 70, 760-764 (1982). - YOKOTA, A., KOMURA, H., KITAOKA, S.: Biochem. Biophys. Res. Commun. 111, 544-550 (1983a). - YOKOTA, A., KAWABATA, A., KITAOKA, S.: Plant Physiol. 71, 772-776 (1983b). - YU, C., CLAYBROOK, D.L., HUANG, A.H.C.: Arch. Biochem. Biophys. 227, 180-187 (1983).

Dr. GRAHAME J. KELLY
CSIRO
Marine Laboratories
7000 Hobart, Tasmania
Australia

Professor Dr. ERWIN LATZKO
Botanisches Institut der
Universität Münster
Schloßgarten 3
D-4400 Münster, FRG

Notes added in proof (2005):
Marine photosynthesis (Sect. 1) is now estimated to represent over 40 % of the global total.
The concentration of NH_4^+ in the ocean is 30nM, not 30 mM (Sect. 6a, last line).

Photosynthesis: Carbon Metabolism: By Day and by Night

By JOSEPH A. M. HOLTUM, GRAHAME J. KELLY, and ERWIN LATZKO

"Therefore, potatoes take up much nutrient from the air and little from the soil, and the notion of growing potatoes in cellars on a little soil could well be a good idea".

J.A. Uilkens in *Handboek van Vaderlandschen Landhuishoudkunde* (1819)

1. Introduction

The extent and complexity of metabolic activity in photosynthetic cells during darkness has been overshadowed by those more novel portions of metabolism that are unique to photosynthetic cells in the light. This review attempts to correct some of this imbalance by emphasizing recent advances in our understanding of the nighttime activity of photosynthetic cells, including the role of the newly discovered regulatory metabolite fructose-2,6-P_2, and a detailed review of that variant of photosynthesis which has a cellar-like attribute to it, i.e. CAM, in which CO_2 is taken up nocturnally. The recently discovered finer details of daytime photosynthetic carbon metabolism are also covered for the period since our last review in Volume 46 of Progress in Botany.

2. Inorganic Carbon Uptake

Real photosynthetic carbon metabolism appears to begin with an earlier step for algae when compared with terrestial plants. In the latter, inorganic carbon simply diffuses into cells in the form of CO_2, but for algae there is now evidence that inorganic carbon is actively accumulated (COLMAN 1984; LARSSON et al. 1985). Much effort has been devoted to studies with blue-green algae which can develop an internal inorganic carbon concentration over 100 times the external concentration (MILLER et al. 1984; SHELP and CANVIN 1984). It has been argued that CO_2 is the species of inorganic carbon actually taken up (SHIRAIWA and MIYACHI 1985), although VOLOKITA et al. (1984) suggest that this CO_2 is converted to HCO_3^- by a carbonic anhydrase-like subunit of the transport apparatus, and that it is ultimately a HCO_3^- ion that is moved into the cell through a membrane porter. This HCO_3^- movement is stimulated by Na^+ (KAPLAN et al. 1984; MILLER and CANVIN 1985), and the requirement of both ATP and electron transport through photosystem I for inorganic carbon uptake (OGAWA et al. 1985) is thought to reflect the energy requirements for a Na^+/H^+ antiport running in conjunction with HCO_3^- import, so that intracellular charge and pH are maintained during the generation of CO_2 via $HCO_3^- + H^+ \rightarrow H_2CO_3 \rightarrow H_2O + CO_2$.

G.J. Kelly and E. Latzko:
Thirty Years of Photosynthesis

Originally published in Progress in Botany 48

However, experiments with unicellular green algae indicate that CO_2 is the species taken up, and that this CO_2 is generated from HCO_3^- when necessary (i.e., when the external CO_2 concentration is low but HCO_3^- concentration is high) using carbonic anhydrase located on the outside surface of the cell (AIZAWA and MIYACHI 1984; MARCUS et al. 1985; MORONEY et al. 1984; AIZAWA et al. 1986). Practically all the carbonic anhydrase of *Chlamydomonas reinhardtii* was found in this extra-cellular location (COLEMAN et al. 1984; YAGAWA et al. 1986), having been exported across the plasmalemma as a glycoprotein (COLEMAN and GROSSMAN 1984).

There is still a possibility that terrestial plants have an active inorganic carbon uptake mechanism (MÄCHLER et al. 1985), but any role for the abundant carbonic anhydrase activity found in the leaves of these plants is still obscure (SEETHAMBARAM et al. 1985; TSUZUKI et al. 1985).

3. A Pot-Pourri of Observations with Isolated Intact Chloroplasts

Two new procedures for the aquisition of quality intact chloroplasts have been described (CEROVIĆ and PLESNIČAR 1984; LIN et al. 1985). The ever-present concern that such chloroplast preparations may contain a heterogenous population of organelles has been reinforced by TERASHIMA and INOUE (1985a,b), who showed that both CO_2 fixation rate and RuBP carboxylase content in spinach palisade tissue chloroplasts were at least double the value in spongy tissue chloroplasts. These authors demonstrated that a gradual but continuous adaptation from sun to shade type chloroplast occurs with depth through the leaf.

Past findings have shown that chloroplasts are not only the location of carbohydrate synthesis from CO_2, but also the site of nitrogen assimilation, biosynthesis of many amino acids, and most lipid biosynthesis. Some of these processes no doubt continue by night. DOREMUS and JAGENDORF (1985) and SHIBITA et al. (1986) have confirmed that four of the five enzymes which convert carbamoyl-P plus aspartate to UMP during pyrimidine nucleotide biosynthesis are located in the chloroplast. An unusual observation from RAMASWAMY and NAIR (1984) is that chloroplasts isolated from green potato tubers can photoreduce CO_2 directly to formate and use the latter to synthesize the steroid solanidine.

The selective and controlled permeability of chloroplast envelopes to metabolites and ions has received less attention than in past years. An interesting point made by ROBINSON (1985) is that the selectivity must be viewed in relative terms, in that chloroplasts osmotically stressed by transfer from 0.33 to 0.72 M sorbitol recover within several min by taking up this solute to a final concentration of 0.5 M, even though it does not normally penetrate the chloroplast envelope.

The importance of cytosolic Pi concentration in determining the activity of the phosphate translocator (which normally imports Pi in exchange for the immediate product of photosynthesis, i.e., triose-P) has led to investigations of how, in turn, the cytosolic Pi concentration is controlled. The indications are that, in dark and in light, any drop in the level of cytosolic Pi is followed by its replenishment from reserve Pi in the vacuole, but only relatively slowly, so that the initial drop has time to retard triose-P export and consequently direct assimilated carbon toward intrachloroplastic starch synthesis (WATERTON et al. 1983; WOODROW et al. 1984a). It is not yet known how Pi export from the vacuole is controlled.

Fig. 1. An impression of the subunit arrangement of spinach RuBP carboxylase. The eight small subunits are arranged equatorially around a two-layered core of eight large subunits (4 per layer) containing a large central hole (After BARCENA and SHAW 1985)

Two further chloroplast metabolite transporters have been characterized: one mediating a light-dependent, active uptake of pyruvate by maize mesophyll chloroplasts (FLÜGGE et al. 1985), obviously a component of the C_4 pathway of photosynthesis, and a second mediating oxaloacetate uptake by chloroplasts from either maize or spinach (HATCH et al. 1984). In addition, an electrogenic uniport mediating Ca^{2+} uptake by chloroplasts, and ultimately driven by photosynthetic electron transport, has been described by KREIMER et al. (1985a,b). Ca^{2+} can directly influence the activity of some enzymes such as FBPase. Support for any regulatory action of the Ca^{2+}/calmodulin couple in the stroma has decreased, although it seems likely that Ca^{2+}/calmodulin may regulate envelope enzymes, including NAD kinase and an ATPase (SIMON et al. 1984; NGUYEN and SIEGENTHALER 1985).

4. RuBP Carboxylase

Continuing to hold its place as the most popular Calvin cycle enzyme for research, RuBP carboxylase from spinach leaves has been crystallized (BARCENA and SHAW 1985; PAGLIARO and JOHAL 1985), and in the former case electron microscopy was used to obtain a good idea of what the enzyme molecule looks like (Fig. 1). Immuno-gold identification of the enzyme in vivo indicated that it is uniformly distributed through the stromal contents of barley mesophyll chloroplasts (SHAW and HENWOOD 1985), although some of this enzyme may be loosely associated with thylakoid membranes (MORI et al. 1984). In more primitive plants, distribution is not uniform in that the enzyme is a predominant constituent of green algal pyrenoids and blue-green carboxysomes (SATOH et al. 1984; TURPIN et al. 1984). Finally, there is a report of what might be termed "slipped synthesis of the small subunit" in the crysophyte alga *Ochromonas danica*, in which this subunit is found in the mitochondria (as well as, of course, in the chloroplast; LACOSTE-ROYAL and GIBBS 1985).

The role of the small subunits in their proper place, i.e., associated with large subunits into holoenzyme, has been studied further. With a blue-green algal enzyme, the small subunits were not found necessary for the activators CO_2 and Mg^{2+} to bind to the large subunits, but they were found essential for the activation to be expressed and the actual catalytic reaction to occur (ANDREWS and BALLMENT 1984; ANDREWS et al. 1984; INCHAROENSAKDI et al. 1985). Using spinach enzyme, VOORDOUW et al. (1984) concluded that there is a tightening of the interactions between the large and small subunits upon CO_2-plus-Mg^{2+} activation. Even in the photosynthetic bacterium *Chromatium vinosum*, an indispensibility of the small subunit for carboxylase activity has been observed by JORDAN and CHOLLET (1985), although not by TORRES-RUIZ and McFADDEN (1985). It is, perhaps, worth recalling the odd case of the *Rhodospirillum rubrum* carboxylase, which has been understood for many years now to exist as a simple dimer of two large subunits.

There is much evidence from past work (see BJÖRKMAN 1981) and recent studies (DIETZ and HEBER 1984; USUDA et al. 1984a; MAKINO et al. 1985) that the rate of light-saturated photosynthesis by leaves in air is closely correlated to the content of RuBP carboxylase, hence there is an active interest in measuring this carboxylase content. YOKOTA and CANVIN (1985) have introduced an improved procedure (utilizing measurements of the binding of a ^{14}C-labelled reaction intermediate analogue), and shown that algae contain less enzyme than C_3 and C_4 plants. The moss *Ceratodon purpureus*, and willow leaves were also shown to contain comparable levels of enzyme (RINTAMÄKI and ARO 1984; VAPAAVUORI 1986). The relatively lower content of enzyme in algae may be compensated for by a higher activity per unit weight of enzyme protein: the specific activity of the carboxylase from *Chlamydomonas reinhardtii* (and also from C_4 plant leaves) was about double that of the enzyme from C_3 plant leaves (SEEMAN et al. 1984; SERVAITES 1985a). On the other hand, the studies by SEEMAN et al. (1984) showed that the affinities of the C_4 and algal carboxylases for CO_2 were only about half that of the C_3 enzyme, but this difference may not affect comparison of in vivo activities since algae (see Sect. 2) and C_4 plants concentrate CO_2 in the vicinity of the carboxylase. While on the subject of affinity for CO_2, a valid cautionary note has been made by YOKOTA and KITAOKA (1985), who point out that the pK value for the dissociation of H_2CO_3 is significantly influenced by ionic strength, a point not considered by most authors, thereby leading to their conclusion that "arguments on the diversity of the K_m (of the carboxylase for CO_2) among plants and algae between different laboratories do not hold water..".

While it is generally accepted that light-saturated photosynthesis in air is limited by the supply of one of the RuBP carboxylase substrates (i.e., CO_2), it appears that the rate of regeneration of the other substrate (RuBP) is such that, under normal conditions, the concentration of this RuBP is not limiting (MAKINO et al. 1985). However, perturbations in irradiance or CO_2 supply could momentarily produce RuBP limitation, but most evidence points to the subsequent intervention of control processes designed to re-establish the original RuBP level (BADGER et al. 1984; MOTT et al. 1984; DIETZ and HEBER 1986). It seems likely that chloroplast FBPase is centrally involved in these control processes (WOODROW et al. 1985).

Investigations of the CO_2-plus-Mg^{2+} activation of RuBP carboxylase have quietened down in recent years. The previous claim that wheat contains two enzyme forms with vastly different activation rates has been modified into a more conservative interpretation involving interconversion of the two "forms" (SCHMIDT et al. 1984). Pi at 10 mM interacted somewhat with the activation of this wheat enzyme (PARRY et al. 1985). Crystallization of spinach RuBP carboxylase produced differently shaped crystals according to whether or not the enzyme was activated by CO_2-plus-Mg^{2+} (PAGLIARO and JOHAL 1985), an observation which may be relevant to the in vivo crystallization of enzyme and simultaneous increase in activity when willow leaves were water-stressed (VAPAAVUORI 1986). However, severe water stress inhibits photosynthesis, due at least in part to Pi inhibition of RuBP carboxylase (KAISER et al. 1986) and inhibition of light-mediated activation of FBPase and SBPase (BOAG and PORTIS 1984, 1985).

After many years of controversy, the true story relating to a light-mediated activation of RuBP carboxylase seems to be emerging. A convincing description of light-mediated activation in white clover leaves (SCHNYDER et al. 1984) has been accompanied by the observation by VU et al. (1984) that this phenomenon is restricted to certain plants such as soybean, but is clearly absent from others such as wheat. Shortly afterwards, a first indication of the biochemical mechanism involved was obtained by SERVAITES (1985b), who concluded that, at night, about 50% of the enzyme in tobacco leaves was inactive because of the tight binding of a small molecular weight, phosphorylated inhibitor at the active site. Evidence for the presence of an inhibitor on the maize enzyme in the dark has also been presented (USUDA 1985a). Finally, SEEMAN et al. (1985) re-emphasized the division of plants into those that experience light-mediated activation of the carboxylase (particularly legumes) and those that do not, and independently concluded that dark-mediated inactivation involved binding of a P-ester to the active site. They showed that treatment of the enzyme-inhibitor complex with alkaline phosphatase restored the full activity, and hypothesized that chloroplasts might contain a similar phosphatase with activity controlled in some fashion by a product of photosynthetic electron transport. The big question is: What is the nature of this P-ester that helps keep RuBP carboxylase resting at night? Is it a metabolite as novel as fructose-2,6-P_2?

5. Enzymes by Day and by Night

The great majority of recent research on other enzymes of photosynthetic carbon metabolism has concentrated on the mechanisms whereby certain of these enzymes are kept active by day, but more or less inactive by night.

Chloroplast glyceraldehyde-3-P dehydrogenase has received some attention: The *Euglena gracilis* enzyme was purified by THEISS-SEUBERLING (1984) and shown to contain two types of subunits, as demonstrated previously for the higher plant enzyme. The controversy surrounding association/dissociation of the higher plant enzyme has reappeared with a claim by LEVY and BETTS (1985) that properly prepared enzyme free of RNA behaves quite respectably, being uninfluenced by NAD(P) and displaying straightforward Michaelis-Menten kinetics.

Light-mediated activation of chloroplast FBPase (MARQUES and ANDERSON 1985) has been further shown to involve reduction of disulfide bonds of the enzyme in the presence of its substrate (FBP) and Mg^{2+}; in fact, reduction in the absence of these lead to inactivation (ROSA and WHATLEY 1984). Media pH is also intimately involved, as shown with the spinach enzyme. Following pH changes between 7 and 8, ionization (or protonation) of strategic groups led to slow conformational changes of the reduced (i.e., activated) spinach FBPase that could be viewed using a fluorescent technique (GONTERO et al. 1984a). The presence of FBP and Mg^{2+} modified these pH-induced changes such that critical sulfhydryl groups became deeply embedded in the protein at pH 7, but were exposed at pH 8, thereby making them more susceptible to oxidation at this higher pH (ARAGNOL et al. 1985). The reversible dissociation of tetramer into dimer was also demonstrated (GONTERO et al. 1985). This dissociation may be more important in blue-green algae, where pH changes between light and dark are less than in chloroplasts. GERBLING et al. (1985) have shown that a FBPase from *Synechococcus leopoliensis* undergoes oxidation/reduction as an inactive dimer, and that active enzyme is generated by an FBP-plus-Mg^{2+}-induced association of reduced dimers into reduced tetramers. A surprising report is that from GONTERO et al. (1984b), claiming that chloroplasts contain a second FBPase that will also act as a SBPase. Meanwhile, further work on the specific SBPase of chloroplasts has related the interactions between enzyme, SBP, Mg^{2+}, and pH, and led to calculations showing that the enzyme is maintained with less than 1% of its activity by night (WOODROW et al. 1984b).

Wheat leaf ribulose-5-P kinase has been purified to homogeneity (SUREK et al. 1985). Work with the spinach enzyme has shown that, of two reactive sulfhydryl groups on the active enzyme, only one (which is at or near the ATP-substrate binding site) is involved in light-mediated activation (OMNAAS et al. 1985). Similarly, light-mediated activation of NADP malate dehydrogenase from C_3 plants appears to be connected to the formation of two (out of a total of three) sulfhydryl groups per enzyme subunit (FERTÉ et al. 1984; SCHEIBE 1984). An unusual observation is that the oxidized (i.e., inactive) enzyme recovers some of its activity in the presence of 0.2 M guanidine-HCl, presumably due to the induction of an appropriate conformational change (SCHEIBE and FICKENSCHER 1985). The C_4 NADP malate dehydrogenase is also activated through the generation of sulfhydryl groups by day (JACQUOT et al. 1984), or even by night to some extent, when respiration reduces the cellular oxygen concentration (NAKAMOTO and EDWARDS 1984).

First reports of other enzymes that appear to undergo light-mediated reductive activations have appeared for the photorespiratory enzyme glycerate kinase from maize (but not that from spinach; KLECZKOWSKI and RANDALL 1985), the spinach chloroplast acetyl-CoA carboxylase involved in fatty acid synthesis (SAUER and HEISE 1984), and the mustard chloroplast glutamine synthetase (MANDERSCHEID and WILD 1985). In addition, light-mediated activation of maize sucrose-P synthase via an unknown mechanism has been detected by SICHER and KREMER (1985). Further information on the novel light-mediated activation of the C_4 enzyme pyruvate Pi dikinase via an enzyme-catalyzed, ADP-linked phosphorylation has also been obtained (see Section 11b).

The collection of small proteins called thioredoxins that transfer photosynthetically generated reducing power to many of the enzymes that undergo light-mediated reductive activation have received much attention (SOULIÉ et al. 1985; SCHEIBE et al. 1986). The specificity of different thioredoxins toward these enzymes is somewhat loose (SCHÜRMANN et al. 1985; CRAWFORD et al. 1986), whereas there seems to be a greater specificity of thioredoxins between the plant species than was previously recognized (HÄBERLEIN et al. 1985). Thioredoxin activation proceeds more effectively in a medium less polar than water, e.g., in 10% propanol (CORLEY and WOLOSIUK 1985; WOLOSIUK et al. 1985).

Far less attention has been given to the reverse of light-mediated activation, i.e., to dark-mediated inactivation, which is generally thought to involve a flow of reducing power from enzymes via thioredoxin to oxygen. UDVARDY et al. (1984) found that ferric-chelates could also mediate the oxidative inactivations.

6. Respiration by Night and by Day

The dual respiratory metabolism of photosynthetic cells is now well established. Energy-yielding mitochondrial respiration occurs by night, but its extent of operation by day is still something of a mystery. Measurable rates of energy-consuming photorespiration occur by day (and by day only) but are possibly absent from C_4 plants and algae.

Further refinements of techniques for isolating mitochondria from chlorophyll-containing cells have been developed (SCHWITZGUEBEL and SIEGENTHALER 1984; DAY et al. 1985; HRUBEC et al. 1985). In the former case, peroxisomes were also obtained.

Arguments favouring the inhibition of mitochondrial respiration during photosynthesis have been presented for oat leaf protoplasts by HAMPP et al. (1985). A reduction of oxidative phosphorylation by triose-P

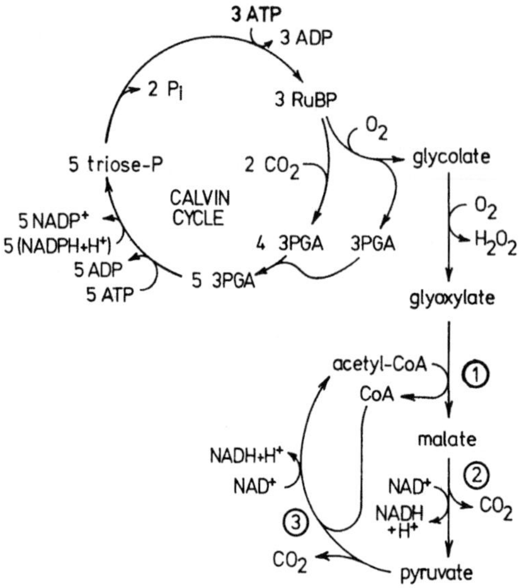

Fig. 2. An alternative to the "usual" photorespiratory glycolate pathway that may operate in NH_4^+-grown *Chlorella sorokiniana*. The situation depicted represents the CO_2 compensation point, where the two CO_2 fixed by the Calvin cycle are balanced by the two CO_2 released when glyoxylate is oxidized by the enzymes malate synthase (1), NAD malic enzyme (2), and the pyruvate dehydrogenase complex (3). (Based on BEUDEKER and TABITA 1984)

exported from chloroplasts appeared to be important (HAMPP 1985). However, measurements by WALKER et al. (1984a) and BROOKS and FARQUHAR (1985) point to the persistance of at least some mitochondrial respiration in illuminated leaves, while experiments with the green algae *Scenedesmus* and *Chlamydomonas* also indicated that this respiration was present in the light, although in the former case it was shown that the evolved CO_2 was completely reassimilated by photosynthesis without leaving the cell (SCHERER et al. 1984; PELTIER and THIBAULT 1985a).

Evidence that all plants do not perform conventional mitochondrial respiration by night has also appeared. Leaves from *Lactuca sativa* and *Capsella bursa-pastoris* formed lactate, even under anaerobic conditions (BETSCHE 1983). The fate of the lactate was not clear. Even more intriguing is the report from KOWALCZYK and MASLOWSKI (1984) that maize shoot mitochondria possess a variant of oxidative phosphorylation that generates pyrophosphate instead of ATP. If confirmed, this would have fundamental repercussions on the regulation of carbohydrate metabolism, given that a PPi-linked P-fructokinase occurs in the cytosol of plant cells.

Turning to photorespiration, the extent of this process in C_3 leaves in air has been estimated by HANSON (1984) using glycerate labelled in appropriate positions with both ^{14}C and ^{3}H. It was found that 39% of the RuBP was oxygenated, 32% of the fixed CO_2 was released again as CO_2, and that photorespiration was therefore equal to 46% of the net photosynthetic CO_2 fixation. This real situation is thus very close to that illustrated in Fig. 2 of our earlier review (LATZKO and KELLY 1980). However, these values change dramatically with changes in the concentration of either CO_2 or O_2 (HANSON and PETERSON 1985), as expected since these two gases compete for the active site of RuBP carboxylase/oxygenase. Temperature also influences the relative rates of photorespiration and photosynthesis, not only due to the differential effects of temperature on the solubilities of CO_2 and O_2, but also due to its differential effects on the respective K_m and V_{max} values of the carboxylase and oxygenase activities (JORDAN and OGREN 1984; BROOKS and FARQUHAR 1985). The altered relative rates affect, in turn, the distribution of assimilated carbon (VIIL et al. 1985).

Final proof has now been obtained that an enediol of RuBP is formed on RuBP carboxylase prior to reaction with either CO_2 or O_2 (JAWOROWSKI et al. 1984). A claim that the actual reaction with CO_2 or O_2 occurs more or less nonenzymically, with no "formal" enzyme binding site for this gaseous substrate (GUTTERIDGE et al. 1984) could quelch all hopes of selecting for a low-oxygenase enzyme. There are still, however, hints that some glycolate may originate from other sources (JOLIVET et al. 1985).

In most plants, glycolate is metabolized in peroxisomes via glyoxylate to glycine, two molecules of which are then decarboxylated and deaminated in mitochondria to generate the respiratory CO_2, NH_4^+, and one molecule of serine. The enzyme system involved has an access to NAD different to that of malate-oxidizing enzymes (DRY and WISKICH 1985). A somewhat different approach to glycolate metabolism is used by certain algae. In *Chlorella sorokiniana* growing with NH_4^+ as a N-source, the glyoxylate generated from glycolate oxidation reacted with acetyl-CoA, to form malate (BEUDEKER and TABITA 1984). In this case, a cycle of three reactions for the complete oxidation of the glyoxylate to CO_2 can be formulated (Fig. 2). One wonders why this much simpler, one-compartment (mitochondrion) glycolate pathway has not been adopted by other photosynthetic cells, unless (a) the conventional glycolate pathway has important functions additional to that of disposing of excess photosynthetic energy, or (b) the restriction that no more than 33% of the RuBP can be oxygenated without growth ceasing is too severe (Fig. 2). In *Euglena gracilis*, glycolate goes directly to mitochondria, where it is oxidized to glyoxylate, but in this case most of the glyoxylate is decarboxylated to CO_2 and formate. The latter is activated to a so-called C_1 unit and combined with a glycine to form serine (YOKOTA et al. 1985a,b), which is then presumably metabolized, as in higher plants, to glycerate for return to the Calvin cycle (EDWARDS and WALKER 1984). Entry of glycerate into chloroplasts is via a light-stimulated transporter on the envelope; it seems that the light-generated increase in stromal pH, rather than light-driven ATP synthesis, is involved (ROBINSON 1984). An alternative fate proposed for glycerate, viz. oxidation to hydroxypyruvate, which is in turn decarboxylated to glycolate (GAUDILLERE et al. 1983), also needs evaluation in terms of the stoichiometry of carbon flow from glycolate back to the Calvin cycle.

The NH_4^+ released during glycine decarboxylation in mitochondria is reassimilated in chloroplasts by glutamine synthetase. Experimental inhibition of this enzyme disturbs the balanced flow of intermediates, leading to an inhibition of photosynthesis in wheat leaves (WALKER et al. 1984b), and to straightforward excretion of NH_4^+ and glycolate in *Chlamydomonas* (PELTIER and THIBAULT 1985b). Nevertheless, the closed cyclic process of NH_4^+ formation in mitochondria and reassimilation in chloroplasts may not be absolute, as evidenced by the entry of asparagine-N, transported from other parts of pea plants to leaves, into the intermediates of photorespiration (TA et al. 1985).

Photorespiration is often looked upon as a mechanism for the disposal of excess photosynthetically generated reducing power, for example when CO_2 supply is restricted. Interestingly, the habitat of algae seems to offer an alternative approach, viz. use of the excess energy to reduce NO_3^- to NH_4^+ which, in the absence of a sufficient carbohydrate supply, is simply excreted (AZUARA and APARICIO 1984).

Another alternative energy disposal system is the Mehler reaction (SAMUELSSON et al. 1985; BRECHIGNAC and ANDRE 1985) in which O_2 is reduced to O_2^-, and the latter is reacted with protons (catalyzed by

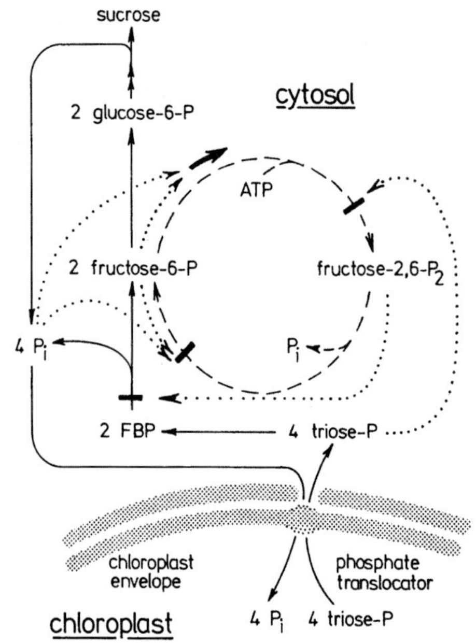

Fig. 3. Regulation of sucrose biosynthesis. A principal regulatory mechanism is the inhibition of FBP hydrolysis by fructose-2,6-P_2; the concentration of the latter is controlled in turn by the relative rates of the enzymes synthesizing and degrading it *(dashed lines)*, as determined by the levels of various metabolites *(dotted lines)*. The thick arrow represents activation; the thick bars inhibitions. The overall scheme *(solid lines)* illustrates the net synthesis of one sucrose from four triose-P. (Based on STITT et al. 1984c)

superoxide dismutase) to form H_2O_2. The H_2O_2 is subsequently scavenged in chloroplasts by the ascorbate/glutathione system (ASADA and BADGER 1984; HOSSAIN et al. 1984; TEL-OR et al. 1985; GILLHAM and DODGE 1986), which can be rapidly inactivated if H_2O_2 is added to dark-kept chloroplasts (ASADA and BADGER 1984; HOSSAIN and ASADA 1984). Renewed claims for an alternate system of H_2O_2 removal in plants based on glutathione peroxidase (DROTAR et al. 1985; OVERBAUGH and FALL 1985) will need further scrutiny (JABLONSKI and ANDERSON 1984).

H_2O_2 is also generated during oxidation of glycolate to glyoxylate in peroxisomes. This H_2O_2 is presumably disposed of by catalase, which is also in the peroxisomes. It is of interest, therefore, that glutathione synthesis was increased threefold when catalase was inhibited or deficient; all this extra glutathione was in the oxidized form (SMITH 1985; SMITH et al. 1985).

7. Fructose-2,6-P_2 Regulation of Cytosolic Carbohydrate Metabolism

As pointed out in the preceding review, a new outlook on how cytosolic carbohydrate metabolism is regulated, particularly from the viewpoint of sucrose biosynthesis, was initiated by the discovery of fructose-2,6-P_2 in the cytosol and the great sensitivity of the cytosolic enzymes FBPase and PPi-linked P-fructokinase to it. A wave of experiments by STITT et al. (1984a,b,c, 1985) and HERZOG et al. (1984) has now provided details of the regulatory network involved, including the inhibition of cytosolic FBPase by fructose-2,6-P_2, the potentiation of this inhibition by AMP, the inhibition of the fructose-2,6-P_2 hydrolyzing enzyme by fructose-6-P and Pi, the regulation of the fructose-2,6-P_2 synthesizing enzyme (activation by fructose-6-P and Pi; inhibition by glycerate-3-P and dihydroxyacetone-P), and various mass action effects (Fig. 3).

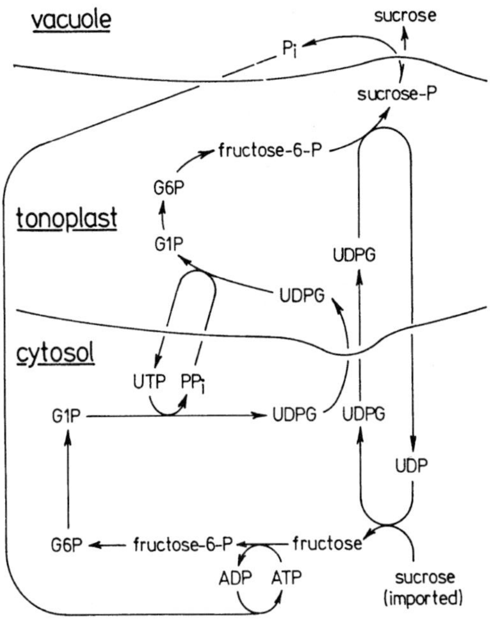

Fig. 4. Metabolism of sucrose by sugar cane cells. Imported sucrose is converted to UDP-glucose in the cytosol. This UDP-glucose is specifically taken up by the tonoplast that contains the enzymes for converting it back to sucrose. These enzymes are aligned so that the newly synthesized sucrose is directed into the vacuole. (Based on THOM and MARETZKI 1985). Additional abbreviations: *G1P* and *G6P* glucose-1-P and glucose-6-P; *UDPG* UDP-glucose

Nevertheless, other aspects of sucrose metabolism have not been ignored. The inhibition of sucrose-P synthase by Pi was reported by DOEHLERT and HUBER (1984), and more fundamentally, the equilibrium constant for the reaction catalyzed by this enzyme was re-evaluated and found to be approximately 2, and not 3250 as has appeared in text books for over 20 years (BARBER 1985). New insights into sucrose metabolism may follow the development of a procedure for isolating a "cytosolic fraction" from clover protoplasts (CANUT et al. 1984), and of particular interest, the evidence that sugar cane cells accumulate sucrose in vacuoles via the uptake of UDP-glucose by the outside of the vacuole membrane (tonoplast) and the subsequent "directional" metabolism of this UDP-glucose by five tonoplast-bound enzymes such that sucrose is released into the vacuole (THOM and MARETZKI 1985, see Fig. 4). It is not known whether the pathway in Figure 4 operates in leaves. Sucrose has been found in barley leaf vacuoles (KAISER et al. 1982) but not in spinach leaf vacuoles (ASAMI et al. 1985).

Returning to the fructose-2,6-P_2 saga: Little more is understood concerning the role of the PPi-linked P-fructokinase, whose regulation is also known to be a primary target of the sugar bisphosphate. The enzyme is widespread in plants (CARNAL and BLACK 1979; KOWALCZYK et al. 1984), and may occur in two related forms (WU et al. 1983, 1984; YAN and TAO 1984). A shemozzle was created by claims from BALOGH et al. (1984) and WONG et al. (1984) that interconversion between this PPi-linked P-fructokinase and the classical ATP-linked P-fructokinase was possible. The provocative claims were challenged by GANCEDO (1984) and KRUGER and DENNIS (1985a,b), who provided more conservative interpretations of the data in question.

Three more enzymes have been found to be sensitive to fructose-2,6-P_2. These are P-glucomutase (activation; GALLOWAY et al. 1985), 6-P-gluconate dehydrogenase (inhibition; MIERNYK et al. 1984), and trehalose phosphorylase of *Euglena* (inhibition; MIYATAKE et al. 1984). In addition, fructose-2,6-P_2 appears to play a central role in determining the direction of carbohydrate metabolism in guard cells (HEDRICH et al. 1985), and in initiating metabolic responses to flooding (PAZ et al. 1985). The rapidity (within 15 min) with which fructose-2,6-P_2 levels responded to environmental changes in these experiments (see also BAYSDORFER and ROBINSON 1985a) is particularly noteworthy.

8. Starch Metabolism

During photosynthesis, most assimilated carbon is directed toward either the cytosolic synthesis of sucrose, or the formation of a transitory reserve of chloroplast starch. The balance between these two fates is influenced by fructose-2,6-P_2, in that an increase in the concentration of this regulator molecule will inhibit sucrose synthesis, thereby diverting assimilated carbon to starch (STITT et al. 1984b). However, there is clearly no *direct* effect of fructose-2,6-P_2 on starch synthesis: BAYSDORFER and ROBINSON (1985b) observed a fourfold increase in the rate of starch synthesis when the photosynthetic period of spinach plants was reduced from 12 h to 7 h, even though the fructose-2,6-P_2 level fell to almost half. However, this fall did result, as expected, in an increase in the rate of sucrose synthesis. A threefold increase in the concentration of CO_2 around cotton leaves also stimulated starch accumulation in chloroplasts, to the point where the chloroplasts themselves were damaged (SASEK et al. 1985).

Although the metabolite-regulated activity of ADP-glucose pyrophosphorylase is well established as the "day on/night off" switch for starch synthesis, recent experiments by DIETZ (1985) indicate that hexose-P isomerase may limit the rate of starch synthesis during the day. This author proposed that inhibition by glycerate-3-P may determine the enzyme's rate, but the far more potent inhibition by erythrose-4-P might also have a role (KELLY and LATZKO 1980).

By night, starch and sucrose metabolism continue. Starch is degraded via glycolysis to triose-P and/or glycerate-3-P in chloroplasts, and these intermediates are then used to synthesize sucrose in the cytosol. In effect, leaves continue to be the whole plant's primary source of carbon by night (KERR et al. 1985; FARRER and FARRER 1985). In algae, a similar partitioning of starch degradation between chloroplast and cytosol appears to exist (KLEIN 1986). The rate of degradation increases under nitrogen, indicating the presence of a Pasteur effect (PEAVEY et al. 1983). In *Chlamydomonas reinhardtii*, degradation may begin using either phosphorylase or α-amylase (LEVI and GIBBS 1984), and where glycerate-3P is the end-product in chloroplasts, NADH generated at the glyceraldehyde-3-P dehydrogenase reaction is reoxidized by a NADH-linked plastoquinone reductase. The reduced plastoquinone then reduces O_2 to water (GFELLER and GIBBS 1985).

9. Lipid Biosynthesis

It remains unclear whether acetyl-CoA is generated from acetate or pyruvate in spinach chloroplasts, since the measured concentrations of both these intermediates were low (LIEDVOGEL 1985; TREEDE and HEISE 1985), but in pea chloroplasts synthesis from pyruvate is favored (CAMP and RANDALL 1985). There is evidence that a principal regulatory enzyme of fatty acid synthesis (acetyl-CoA carboxylase) undergoes light-mediated activation (SAUER and HEISE 1984). Fatty acid synthesis is located in the chloroplast stroma (NIKOLAU et al. 1984a; WALKER and

HARWOOD 1985). The assembly of phosphatidylglycerol is also restricted to the chloroplast (BISHOP et al. 1985; ROUGHAN 1985), but participation of the endoplasmic reticulum is required for the desaturation of oleate to linoleate (which occurs on the 2-position of phosphatidylcholine; MURPHY et al. 1985), and for the assembly of the galactolipids and sulfolipids (BISHOP et al. 1985). An intriguing report from VIGH et al. (1985) indicates that the leaves of certain varieties of wheat contain considerable amounts of the highly unsaturated fatty acids, arachidonic, eicosapentaenoic, and docosahexaenic acid.

10. Nitrogen Assimilation

Experiments with algae, at least some of which contain nitrate reductase in pyrenoids (LOPEZ-RUIZ et al. 1985), have confirmed that NH_4^+ is the actual co-repressor of nitrate reductase synthesis in NH_4^+ grown cells, although NH_4^+-derivatives can influence NO_3^- utilisation by inhibiting its transport into the cell (FRANCO et al. 1984; ROMERO et al. 1985). An important source of reductant for nitrate reductase is the pair of reactions catalyzed by the glucose-6-P and 6-P-gluconate dehydrogenase (HIPKIN and CANNONS 1985).

NH_4^+ is usually assimilated by the glutamine synthetase/glutamate synthase (so-called "GS/GOGAT") sequence. The former enzyme, which occurs in at least four forms in pea plants (PUSHKIN et al. 1985), occurs in the chloroplast as a glycoprotein (NATO et al. 1984). The latter enzyme occurs in forms using either reduced ferredoxin or NAD(P)H as reductant (AVILA et al. 1984; MARQUÉZ et al. 1984). However, the GS/GOGAT pathway is not ubiquitous; certain algae utilize instead a NADPH-linked glutamate dehydrogenase with unusually high affinity for NH_4^+ (EVEREST and SYRETT 1983; TISCHNER 1984; AHMAD and HELLEBUST 1985).

11. C_4 Pathway

a) PEP Carboxylase

PEP carboxylase is the primary carboxylating enzyme in both C_4 leaves and in CAM tissues. In the former, the enzyme must be active during the light, while in the latter it must function during the dark. It is perhaps not surprising, then, that PEP carboxylase extracted from illuminated C_4 leaves exhibits a greater activity than that extracted from darkened leaves (IGLESIAS and ANDREO 1984; GONZALES et al. 1984). The dark activity approaches the light activity following incubation with dithiothreitol: two disulfide bonds are reduced in the process. STIBOROVÀ and LEBLOVÀ (1985) have shown that maize leaves contain two PEP carboxylase forms, one of which is strongly activated by glucose-6-P and by glycine. Both substances instigate conformational changes in the protein.

b) Pyruvate Orthophosphate Dikinase

Pyruvate orthophosphate dikinase (PPDK) is currently the topic of considerable study, much of which is summarized in two recent reviews (BURNELL and HATCH 1985b; EDWARDS et al. 1985). It is now clear that PPDK is present in the chlorophyllous cells of C_4 plants, some CAM

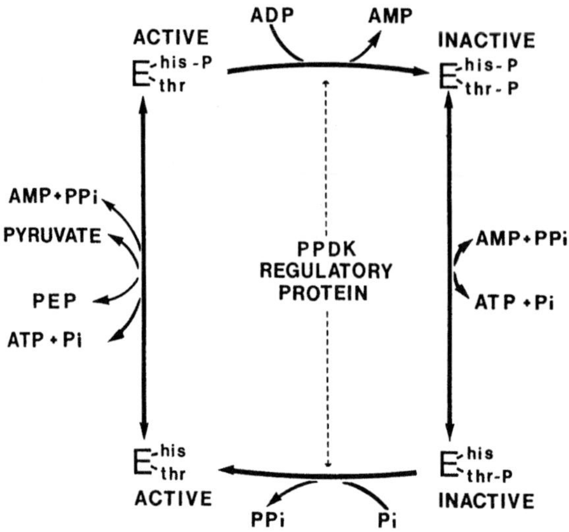

Fig. 5. The roles proposed for the dikinase regulatory protein, adenylates and Pi in the light/dark activation/ inactivation of PPDK. *his* and *thr* represent the enzyme histidine and threonine residues which can be reversibly phosphorylated. (After EDWARDS et al. 1985)

plants, many C_3 plants and in some C_3 and C_4 grass caryopses (AOYAGI and BASSHAM 1984). Although bundle-sheath preparations from many C_4 plants lack PPDK, perhaps 10% of the PPDK activity in maize is associated with the bundle-sheath cells (AOYAGI and NAKAMOTO 1985).

Our understanding of the mechanism of the regulation of leaf PPDK has shown sufficient recent progress for the scheme in Fig. 5 to be proposed. Active PPDK is inactivated by an ADP-dependent phosphorylation of a threonine residue not directly associated with the catalytic site (ASHTON and HATCH 1983a; ASHTON et al. 1984; BURNELL 1984a). A phosphorylation by ATP of a catalytic site histidine residue is a prerequisite for the threonine phosphorylation, and thus for inactivation (BURNELL and HATCH 1983). The histidine phosphate is not lost during this process (BURNELL and HATCH 1984). The phosphorylytic reactivation is most rapid with the enzyme phosphorylated on the threonine residue only, which is formed following the removal of the histidine phosphate in the presence of AMP and PPi (BURNELL 1983, 1984b; BURNELL and HATCH 1985a).

Both the ADP-plus-ATP inactivation and the Pi-dependent reactivation are catalyzed by a heat-labile protein, the so-called PPDK regulatory protein, which is located in the chloroplast. This protein, with a subunit size of 45 kD, is a dimer at pH 7.5 and a tetramer at pH 8.3 (BURNELL and HATCH 1983). Although earlier workers found it to be unstable, this problem has now been surmounted in both crude and purified extracts (BAER and SCHRADER 1985; EDWARDS et al. 1985). It has been suggested that the light/dark activation/inactivation of PPDK may be controlled by changes in adenylate ratios or, perhaps, through covalent or allosteric modification of the regulatory protein (BURNELL and HATCH 1985b).

c) Other Enzymes

EDWARDS et al. (1985) have reviewed the regulation and dark/light inactivation/activation of NADP malate dehydrogenase by processes associated with either the redox state of the chloroplast electron transport system or by NADPH/NADP ratios. In addition to activation by reduced thioredoxin and inactivation by oxidized thioredoxin, the enzyme is sensitive to NADPH/NADP ratios such that at any fixed ratio of reduced to oxidized thioredoxin, high rates of oxaloacetate reduction can only occur when the NADPH/NADP ratios are also large (ASHTON and HATCH 1983b,c).

Carbonic anhydrase and enzymes involved in the biosynthesis of starch, fatty acids, and sulfur-containing amino acids have received brief attention in C_4 plants. Carbonic anhydrase from C_4 dicotyledons, but not from C_4 monocotyledons, showed cross-reactivity with antibody raised against spinach leaf carbonic anhydrase (OKABE et al. 1984). Distinct starch phosphorylase forms have been found in bundle-sheath and mesophyll cells of maize (MATEYKA and SCHNARRENBERGER 1984).

Acetyl-CoA carboxylase (see Sect. 9) is present in epidermal and mesophyll protoplasts and in bundle-sheath strands from both sorghum and maize (NIKOLAU et al. 1984b). Both bundle-sheath and mesophyll chloroplasts can synthesize fatty acids from acetate.

Over 90% of ATP sulfurylase and adenosine-5'-phosphosulfate sulfotransferase are located in the bundle-sheath in maize (BURNELL 1984c; SCHMUTZ and BRUNOLD 1984). Sulfate reductase, O-acetyl-L-serine-sulfohydrylase, cysteine synthase and cystathione β-lyase are roughly equally distributed between mesophyll and bundle-sheath cells. These locations support earlier proposals that sulfate is activated in bundle-sheaths and may, at least in part, be transferred to mesophyll cells as sulfite or as adenosine-5'-phosphosulfate.

d) Inter- and Intracellular Distinctions

Investigations into the physiology of the C_4 pathway have long been hampered by the difficulty in obtaining physiologically intact and active organelles from bundle-sheath cells. Methods have been reported for the isolation of bundle-sheath protoplasts and chloroplasts from *Panicum maximum*, a PEP carboxykinase C_4 species (CHAPMAN and HATCH 1983), *Zea mays*, a NADP malic enzyme species (JENKINS and BOAG 1985), and for several C_4 dicotyledons (MOORE et al. 1984). A method for the large-scale preparation of mesophyll chloroplasts from several C_4 species has also been described (JENKINS and RUSS 1984).

In vivo metabolite concentrations have been estimated in maize bundle-sheath and mesophyll cells in attempts to describe, in more quantitative terms, the flow of carbon through the C_4 pathway (LEEGOOD 1985; STITT and HELDT 1985a,b). Between 70 and 95% of the malate, triose phosphates and PEP are restricted to the mesophyll cells. The bulk of the glycerate-3-P is in the bundle-sheath cells. The approximately 10 mM gradients of dihydroxyacetone-P and glycerate-3-P between the two cell types are sufficient to power their diffusive transport. Although pyruvate is evenly distributed between the cell types, it is unevenly distributed within the mesophyll cells (STITT and HELDT 1985b).

Active light-dependent uptake of pyruvate by mesophyll cell chloroplasts (FLÜGGE et al. 1985) ensures a low cytoplasmic pyruvate concentration in the mesophyll and probably creates a substantial pyruvate gradient between the mesophyll and bundle-sheath cells. In *Spartina anglica*, a PEP carboxykinase plant, SMITH and WOOLHOUSE (1983) found that the transport of pyruvate or glycerate-3-P from the bundle-sheath to the mesophyll cells, and the rates of PEP formation in the mesophyll cells, were too slow to account for the observed rates of CO_2 uptake. They therefore postulated that either PEP is transported between the bundle-sheath and mesophyll cells, or that more than one pathway of PEP regeneration is involved.

In maize, PEP is predominantly found in the mesophyll cytoplasm although it is produced in the chloroplast by PPDK. This unequal PEP distribution is probably effected by a special Pi translocator similar to that found in *Digitaria sanguinalis*. The translocator in *Digitaria* exhibits a higher affinity for PEP than its C_3 equivalent, and transports PEP as well as glycerate-3-P and Pi (RUMPHO and EDWARDS 1984, 1985; HALLBERG and LARSSON 1983). Maize mesophyll chloroplasts also possess an ATP/PEP transporter (WOLDEGIORGIS et al. 1983).

Malate decarboxylation in maize bundle-sheath chloroplasts requires, in addition to malate and NADP, aspartate and an external carbon source, preferably glycerate-3-P (JENKINS and BOAG 1985). Aspartate decreases the K_m (malate) for decarboxylation 10-fold and increases the V_{max} 2-3 fold (BOAG and JENKINS 1985). Maize bundle-sheath chloroplasts may possess a specific malate-carrier that requires aspartate.

In common with the majority of other plant leaves, many C_4 leaves form sucrose as a major product of CO_2 fixation. In maize, sucrose synthesis and the bulk of the sucrose-P synthase and cytoplasmic FBPase are restricted to the mesophyll cytoplasm (FURBANK et al. 1985). In contrast, chloroplastic FBPase is largely restricted to bundle-sheath chloroplasts (USUDA et al. 1984b). In mesophyll cells, one might expect a competition for cytoplasmic triose-P between sucrose synthesis and the transport requirements of the cell, since high triose-P concentrations are required to drive the diffusion of triose-P from the mesophyll to the bundle-sheath cells. Apparently, this is prevented by a low affinity of the maize cytoplasmic FBPase for its substrate (STITT and HELDT 1985a). In protoplast extracts; the affinity of the maize enzyme for FBP is 5-15-fold lower than that of the equivalent spinach enzyme. As for the spinach enzyme, there is a threshold triose-P level below which the FBPase activity will be very low. It appears likely that this threshold concentration will, under conditions prevailing in the cytoplasm, be high enough to enable both sucrose synthesis and gradient-powered diffusion. In maize, sucrose-P synthase displays daily fluctuations of the maximal activity extracted from young and old leaves (HUBER et al. 1985). Activity is greatest during the first few hours of the light period and lowest at the end of the light period. The affinity of the enzyme for its substrates UDP-glucose and fructose-6-P also varies during a 24-h period (SICHER and KREMER 1985).

Fructose-2,6-P_2, a compound involved in the regulation of some enzyme-catalyzed interconversions of FBP and fructose-6-P (see Sect. 7), is present in maize and is restricted to the mesophyll cell cytoplasm, as are fructose-6-P 2-kinase and fructose-2,6-P_2ase, the two enzymes which synthesize and degrade it (SOLL et al. 1985). Both enzymes are inhibited by PEP. The fructose-2,6-P_2ase is also, unlike its spinach counterpart, inhibited by glycerate-3-P and dihydroxyacetone-P. As in other plants and an animal tissues, the two activities have not been separated from each other and are thought to be properties of a single protein. Despite this, oxaloacetate selectively inhibits the kinase activity, and pyruvate, UDP-glucose and fructose-6-P selectively inhibit the phosphatase activity.

In C_4 leaves, the induction of photosynthetic carbon metabolism following the transition from darkness to light requires not only the activation of certain Calvin cycle enzymes, but also the establishment of metabolite gradients between the mesophyll cells and the bundle-sheath cells. Light rapidly activates NADP malate dehydrogenase, PPDK and chloroplastic FBPase during the induction period in maize (USUDA et al. 1984b). After 2.5 min illumination, these three enzymes were fully activated, although the rate of CO_2 uptake had only reached half that of the steady-state rate. Obviously other processes were rate-limiting. The biphasic CO_2 uptake kinetics that are observed during induction are most likely a direct consequence of the time required to fill metabolite pools and to create the intercellular metabolite gradients that are necessary for the integrated operation of the C_4 pathway (FURBANK and WALKER 1985; LEEGOOD and FURBANK 1984; USUDA and EDWARDS 1984). As carbon is accumulated in the various metabolites (particularly in triose-P) more rapidly than it is assimilated from the atmosphere, a transfer of carbon from previously formed metabolites must be invoked. The source of carbon for the triose-P pools is probably pyruvate, which accumulates during the dark. Upon illumination, this pyruvate is converted to PEP via PPDK (FURBANK and LEEGOOD 1984, but see USUDA 1985b). Under steady-state conditions, the rate of CO_2 uptake is strongly correlated with the extractable activities of RuBP carboxylase and PPDK, but not with FBPase nor the malate dehydrogenase (USUDA 1984; USUDA et al. 1984a, 1985b).

Headway has been made in the study of the puzzling specific requirement of C_4 tissues for micronutrient amounts of Na^+. Sodium-deficiency symptoms in two C_4 species were alleviated in the presence of 1500 µl CO_2 l^{-1} and were accentuated at 15 µl CO_2 l^{-1} (JOHNSTON et al. 1984). It was suggested that the kinetics of carbon transport between the mesophyll and bundle-sheath cells are altered under Na-deficient conditions.

In a review of oxygen exchange in photosynthesis, BADGER (1985) conclude that two types of O_2 uptake are possible in C_4 leaves, one CO_2-sensitive and the other CO_2-insensitive. The CO_2-sensitive uptake is probably due to the oxygenase activity of RuBP carboxylase and is predominant when the external CO_2 tensions are low. When the external CO_2 tensions are high and photosynthetic rates are increased, the oxygenase activity is reduced and pseudocyclic O_2 photoreduction occurs. In mesophyll cells, which lack RuBP carboxylase, the rates of O_2 uptake are related to the energy requirements of the substrate used to accept electrons during electron transport (FURBANK et al. 1983). The theoretical O_2 requirements of the different C_4 types are, in general, greater than those observed. One explanation of the difference is that cyclic photophosphorylation occurs in the mesophyll chloroplasts and various metabolites, and so reducing equivalents are transported between the cell types.

The debate on the capacity and the extent of photorespiration in C_4 plants continues. Isolated bundle-sheath strands can convert glycolate →glycine→serine (FARINEAU et al. 1984). Although it is possible that these intermediates can be transferred out of the bundle-sheath, extensive transport is unlikely since bundle-sheath mitochondria can oxidize glycine at rates approaching those of C_3 plants while mesophyll mitochondria either lack or have low activities of serine hydroxymethyltransferase and glycine decarboxylase (GARDSTRÖM et al. 1985; OHNISHI and KANAI 1983). Although peroxisomes isolated from *Panicum miliaceum* mesophyll and bundle-sheath cells contain all photorespiratory glycolate pathway enzymes, the bundle-sheath activities are, with the exception of hydroxypyruvate reductase, only 20-30% of the mesophyll activities (OHNISHI et al. 1985).

12. C_3 - C_4 Intermediates

A timely review by MONSON et al. (1984) on C_3 - C_4 intermediates defines "intermediate" metabolism on the basis of photosynthetic strategy.

C_3 - C_4 intermediates are divided into two general groups. In the first, which includes various *Panicum* species and the crucifer *Moricandia arvensis*, plants exhibit C_3 patterns of $^{14}CO_2$ uptake, reduced apparent photorespiration, and often have a relatively large number of mitochondria and peroxisomes in their bundle-sheath cells (BROWN et al. 1985; HOLBROOK et al. 1985). In some species which lack bundle-sheath cells, the mitochondria and peroxisomes tend to inhabit centripetal positions. The close physical association between chloroplasts and the photorespiratory organelles presumably results in a more efficient refixation of photorespiratory CO_2 and, consequently, reduced apparent photorespiration. In *Panicum milioides*, glycolate produced in the mesophyll cells may be shunted into the bundle-sheath cells, where the ensuing CO_2 has a greater chance of being refixed. Enzyme compartmentation can vary. In *P. decipens* and in *M. arvensis*, PEP carboxylase and RuBP carboxylase are found in all chlorophyllous, nonepidermal cells, but in *P. milioides* PEP carboxylase is restricted to the mesophyll cytoplasm, while RuBP carboxylase is present in both mesophyll and bundle-sheath chloroplasts (PERROT-RECHENMANN et al. 1984; HOLBROOK et al. 1985).

In the second group, which includes a number of *Flaveria* species, the initial products of $^{14}CO_2$ uptake include both 3-C and 4-C compounds (MONSON et al. 1984). Plants in this group are also characterized by Kranz anatomy, relatively high activities of C_4 pathway enzymes, reduced apparent photorespiration, and high concentrations of organelles in bundle-sheath cells (KU et al. 1983; NAKAMOTO et al. 1983; RUMPHO et al. 1984). These plants are capable of, at least to some extent, concentrating CO_2 in the bundle-sheath cells in a manner similar to true C_4 plants. Unlike PEP carboxylase from *P. milioides* and *M. arvensis* which exhibit C_3-like kinetics, PEP carboxylase from *Flaveria* intermediates exhibits C_4-like properties (NAKAMOTO et al. 1983). Nevertheless, in C_3, C_4, and C_3 - C_4 intermediate species, both carboxylases are found throughout the leaf chlorenchyma (REED and CHOLLET 1985). HOLADAY et al. (1985) have successfully bred a $C_3 \times C_4$ *Flaveria* F1 hybrid which exhibits reduced photorespiration. Crosses between C_4 and C_3 - C_4 intermediate *Flaveria* plants yielded progeny with carbon isotope contents similar to their maternal parents (SMITH and POWELL 1984).

HATTERSLEY and ROKSANDIC (1983) suggest, on the basis of $\delta^{13}C$ composition and anatomical features, that the grass genus *Neurachne* may contain C_3 - C_4 intermediates, one possible species being *N. minor*.

13. Crassulacean Acid Metabolism (CAM)

Three reviews (COCKBURN 1985a; TING 1985; WINTER 1985) and a proceedings volume (MEDINA 1984) devoted to CAM have appeared since our last review.

a) PEP Carboxylase

The contents of a recent proceedings volume dedicated to PEP carboxylase illustrate how the regulatory properties exhibited by this enzyme in a number of tissues are attuned to its physiological role in each tissue both by day and by night (BRULFERT et al. 1983). PEP carboxylase from CAM tissues is no exception. The enzyme from darkened CAM tissues displays a high affinity for PEP and a low sensitivity to inhibition by malate and is physiologically active. In contrast, the enzyme from illuminated tissue exhibits a low affinity for PEP, is sensitive to malate, and is, at least during deacidification, physiologically inactive (cf. WINTER 1985). In *Crassula argentea*, the dark form of PEP carboxylase is a tetramer and the light form a dimer (WU and WEDDING 1985a,b). The darkened form from leaves of *Kalanchoe daigremontiana*, *K. blossfeldiana*, and *Bryophyllum tubiflorum* contains a covalently bound phosphate (NIMMO et al. 1984; BRULFERT et al. 1986). This phosphate is absent from the enzyme extracted during the light. When the dark form is incubated in vitro with alkaline phosphatase, the phosphate is lost and the kinetic characteristics of the enzmye change to those of the light form. The stimulus(i) in vivo for the interconversion of these two forms is not known, but pH and the cytoplasmic concentrations of Pi, PEP, malate, and Mg^{2+} are obvious candidates. Long-term changes in the properties of PEP carboxylase occur when pennywort *(Umbilicus rupestris)* switches from C_3 photosynthesis to CAM-idling following water stress (DANIEL et al. 1984). The physiological significance of three PEP carboxylase forms, each with a similar molecular weight, in *K. daigremontiana* extracts is not known (AMAGASA 1984).

In several obligate CAM species the in vivo rates of CO_2 hydration in the dark, although more rapid than the rates of carboxylation by PEP carboxylase, are slow enough to be partially rate-limiting and may affect carbon isotope fractionation calculations (HOLTUM et al. 1984). Carbon isotope fractionation is subject to direct and indirect temperature effects (DELEENS et al. 1985). The indirect effects, which show considerable species to species variation, include effects on stomatal aperture. The direct effects include temperature effects on gaseous diffusion and on the carboxylation reactions.

b) Malate Fluctuations

A proposal that the accumulation of malic acid in CAM vacuoles is an active process which involves the pumping of protons into the vacuole by a tonoplast ATPase followed by the passive influx of malate has received solid support with the discovery of an ATPase on vacuolar membranes (tonoplasts) isolated from *K. daigremontiana* (AOKI and NISHIDA 1984; SMITH et al. 1984a,b) and from CAM *Mesembryanthemum crystallinum* (STRUVE et al. 1985). The in vitro ATPase activity of 46 nmol m^{-2} s^{-1} is close to the 62 nmol m^{-2} s^{-1} required in vivo. The ATPase appears to possess a proton pumping ability (JOCHEM et al. 1984).

In contrast to uptake, malate efflux from the vacuoles is dominated by passive diffusion of H_2 malate (LÜTTGE and SMITH 1984). When the vacuolar pH is greater than 4.5, some efflux of H malate$^-$ and malate^{2-} is also observed. Efflux of the charged species may be facilitated by a malate permease. The malic acid efflux against which the tonoplast ATPase must operate may, in concert with thermodynamic restrictions, limit both the rate and extent of malic acid accumulation. Isotope fractionation studies indicate that the rate of diffusion of malic acid from the vacuole is slow compared to the rate of malate decarboxylation (DELEENS et al. 1985). Energy demand in leaves of *B. pinnatum* is greatest during the phase of malate decarboxylation, when photosynthesis is operating under conditions of high CO_2 supply (KÖSTER and WINTER 1985).

c) Other Enzymes

The characterization of CAM plant enzymes other than PEP carboxylase continues to receive sporadic attention. In contrast to PEP carboxylase, no differences have been observed in the properties of NAD malic enzyme extracted from *M. crystallinum* in the dark or in the light (ARTUS and EDWARDS 1985a,b). NAD malic enzyme from C_3 and CAM *M. crystallinum* differ only in that the CAM enzyme has a higher V_{max}. ADP-glucose pyrophosphorylase from two CAM species, in common with the enzyme from a number of C_3 and C_4 species, is most active when glycerate-3-P/Pi ratios are high (SINGH et al. 1984). The affinity for glucose-1-P is strongly increased in the presence of glycerate-3-P.

d) Phases and Rhythms

CAM cells rarely operate under steady-state conditions. It has been confirmed that during the burst of CO_2 assimilation which is often present following the daily transition from dark to light, uptake by PEP carboxylase is initially dominant, but gradually decreases in favour of uptake by RuBP carboxylase (LITTLEJOHN and KU 1984; FISCHER and KLUGE 1984). In the presence of $^{13}CO_2$, multiply labeled malate is formed during the late afternoon, but not during the dark (RITZ et al. 1986). The simplest explanation for multiply labeled malate is that two carboxylations are involved. The first, by RuBP carboxylase, produces ^{13}C-labeled 3-C compounds which must then be exported to the cytoplasm where they form PEP, the substrate for β-carboxylation. The unexpectedly high proportion of singly labelled malate probably reflected the production of PEP from precursors not in equilibrium with Calvin cycle intermediates. If true, this latter proposal poses interesting questions about the compartmentation of cellular carbon flow.

Plants which display fluctuations in titratable acidity without a concomitant net dark CO_2 uptake have been designated as either CAM-idlers or as CAM-cyclers (TING 1985). The CAM-idlers exhibit acid fluctuations in the absence of net gas exchange during the light or during the dark. Additions to the list of CAM-idlers include both facultative and obligate CAM species (RAYDER and TING 1983a,b; SIPES and TING 1985) and, possibly, at least one C_3 plant (DANIEL et al. 1984). In CAM-cyclers (called weak-CAM plants by some), some nocturnal acid accumulation occurs, but gas exchange is restricted, for at least the most part, to the daylight hours. CAM-cycling has been observed recently in two non-CAM halophytes (MARTIN and ZEE 1983; NEALES et al. 1983), the gymnosperm *Welwitschia mirabilis* (TING and BURK 1983, but see von WILLERT 1985), and in young leaves of *Peperomia camptotricha* (SIPES and TING 1985). The source of CO_2 for nocturnal malate synthesis in CAM-idlers and -cyclers is presumed to be respiratory CO_2, and such metabolism is usually expressed as a response to water stress. Under extreme water stress, some CAM plants display neither acid fluctuations nor net gas exchange (von WILLERT et al. 1985), while some C_3 plants accumulate organic acids during the light (von WILLERT et al. 1984).

Circadian rhythmic CO_2 exchange is sometimes observed when CAM tissues are subjected to continuous light or darkness. In *Kalanchoe*, this CO_2 exchange is not under stomatal control, but is determined by a metabolic rhythm in the leaf mesophyll (BUCHANAN-BOLLIG and SMITH 1984). It is suspected that this rhythm is the result of changes in PEP carboxylase activity, possibly induced by the periodic influx and efflux of malate into and out of the vacuole (WILKINS 1983a,b; BUCHANAN-BOLLIG 1984; BUCHANAN-BOLLIG et al. 1984a,b).

A computor model which simulates many of the dark/light rhythms typical of CAM has been constructed (NUNGESSER et al. 1985). Although the model has an empirical rather than a true kinetic basis, its ability to simulate CAM behaviour successfully under various conditions suggests that the major factors which control the dominant carbon fluxes of CAM have been described correctly.

e) CAM in Rainforests, Bogs and Under Water

Of the 15,000 to 20,000 species that exhibit CAM, a number that exceeds that of C_4 plants by close to an order of magnitude, perhaps 70% are rainforest epiphytes which inhabit microhabitats subject to periodic water stress.

A group that has received considerable recent attention is the orchids. Most CAM orchids are epiphytic and have thick leaves (WINTER et al. 1983) but not all thick-leaved orchids exhibit CAM (DONOVAN et al. 1984). Although nonepiphytic orchids with CAM occur (CASTRILLO et al. 1984), it is not clear if they originally developed from an epiphytic habitat. Some orchids and ferns progressively fix a greater proportion of their total CO_2 during the night when subjected to continued water stress (FU and HEW 1982; SINCLAIR 1984). Many green, aerial orchid roots exhibit CAM although they lack stomata (HO et al. 1983; HEW et al. 1984; GOH et al. 1983). Presumably CO_2 enters the roots by simple diffusion through the spongy velamen. An emerging pattern is that, when CAM is present, green roots from leafy orchid species perform a weak CAM although they exhibit a net negative carbon balance, whereas in orchid species which lack chlorophyllous shoots, the green roots exhibit typical CAM CO_2 exchange and malic acid accumulation patterns (WINTER et al. 1984; COCKBURN et al. 1985). In Trinidadian bromeliads, the distribution of C_3 and CAM is influenced by both habitat and phylogeny (GRIFFITHS and SMITH 1983).

The occurrence of CAM in submerged *Isoetes* species and in some aquatic angiosperms (KEELEY and MORTON 1982; HELDER and van HARMELEN 1982; RICHARDSON et al. 1984; FARMER and SPENCE 1985) is unlikely to be an adaptation to counter water stress! Aquatic CAM plants inhabit oligotrophic lakes or seasonal freshwater pools in which the levels of dissolved inorganic carbon are either permanently very low or are reduced during the light period due to photosynthetic activity (KEELEY et al. 1983). The possession of CAM enables plants to assimilate carbon more efficiently in these carbon-depleted waters. In *Isoetes*, submerged species either lack or have nonfunctional stomata. Often, a substantial proportion of the CO_2 that is fixed in the submerged leaves is initially absorbed from the inorganic carbon-rich sediments via the roots.

The carbon isotope compositions of aquatic C_3 and CAM plants are strongly dependent upon the inorganic carbon source and content of their habitat. Consequently, carbon isotope contents provide a rather unreliable estimate of the relative contributions of light and dark CO_2 uptake to the carbon mass of aquatic plants. The combined estimation of C, H, and O isotope ratios appears to be a more sensitive method (STERNBERG et al. 1984a,b).

Stylites andicola, a recently discovered primitive relative of *Isoetes*, which grows in peat bogs and mud in the Peruvian Andes, exhibits CAM (KEELEY et al. 1984). The sporophylls lack stomata and are covered in a thick, CO_2- and H_2O-impervious cuticle. The bulk of the external CO_2 assimilated in the sporophylls diffuses there from the roots.

Debate continues over the theory of COCKBURN (1981), that CAM originally developed in stomatal guard cells. KEELEY et al. (1984, 1985) consider that the presence of CAM in astomatal tissues, both leaf and root, argues against the hypothesis, while COCKBURN (1985a,b) and THOMAS (1985) have countered that it is probable that at least some of the astomatal species may have developed from predecessors which contained stomata. At present, the evidence both for and against the theory is tenuous.

References

AHMAD, I., HELLEBUST, J.A.: Plant Physiol. 79, 1125-1126 (1985). - AIZAWA, K., MIYACHI, S.: FEBS Lett. 173, 41-44 (1984). - AIZAWA, K., TSUZUKI, M., MIYACHI, S.: Plant Cell Physiol. 27, 37-47 (1986). - AMAGASA, T.: Plant Cell Physiol. 25, 625-633 (1984). - ANDREWS, T.J., BALLMENT, B.: Proc. Natl. Acad. Sci. USA 81, 3660-3664 (1984). - ANDREWS, T.J., GREENWOOD, D.M., YELLOWLEES, D.: Arch. Biochem. Biophys. 234, 313-317 (1984). - AOKI, K., NISHIDA, K.: Physiol. Plant. 60, 21-25 (1984). - AOYAGI, K., BASSHAM, J.A.: Plant Physiol. 75, 387-392 (1984). - AOYAGI, K., NAKAMOTO, H.: Plant Physiol. 78, 661-664 (1985). - ARAGNOL, D., PRADEL, J., CECCHINI, J.-P.: Biochim. Biophys. Acta 829, 275-281 (1985). - ARTUS, N.N., EDWARDS, G.E.: FEBS Lett. 182, 225-233 (1985a); - Plant Cell Physiol. 26, 341-350 (1985b). - ASADA, K., BADGER, M.R.: Plant Cell Physiol. 25, 1169-1179 (1984). - ASAMI, S., HARA-NISHIMURA, I., NISHIMURA, M., AKAZAWA, T.: Plant Physiol. 77, 963-968 (1985). - ASHTON, A.R., HATCH, M.D.: Biochem. Biophys. Res. Commun. 115, 53-60 (1983a); - Arch. Biochem. Biophys. 227, 406-415 (1983b); - ibid. 227, 416-424 (1983c). - ASHTON, A.R., BURNELL, J.N., HATCH, M.D.: Arch. Biochem. Biophys. 230, 492-503 (1984). - AVILA, C., CÁNOVAS, F., NÚÑEZ DE CASTRO, I., VALPUESTA, V.: Biochem. Biophys. Res. Commun. 122, 1125-1130 (1984). - AZUARA, M.P., APARICIO, P.J.: Photosynth. Res. 5, 97-103 (1984).

BADGER, M.R.: Annu. Rev. Plant Physiol. 36, 27-53 (1985). - BADGER, M.R., SHARKEY, T.D., von CAEMMERER, S.: Planta 160, 305-313 (1984). - BAER, G.R., SCHRADER, L.E.: Plant Physiol. 77, 608-611 (1985). - BALOGH, A., WONG, J.H., WÖTZEL, C., SOLL, J., CSÉKE, C., BUCHANAN, B.B.: FEBS Lett. 169, 287-292 (1984). - BARBER, G.A.: Plant Physiol. 79, 1127-1128 (1985). - BARCENA, J.A., SHAW, P.J.: Planta 163, 141-144 (1985). - BAYSDORFER, C., ROBINSON, J.M.: Plant Physiol. 79, 911-913 (1985a); - ibid. 79, 838-842 (1985b). - BETSCHE, T.: Phytochemistry 22, 1341-1344 (1983). - BEUDEKER, R.F., TABITA, F.R.: Plant Physiol. 75, 516-520 (1984). - BISHOP, D.G., SPARASE, S.A., MUDD, J.B.: Arch. Biochem. Biophys. 240, 851-858 (1985). - BJÖRKMAN, O.: Responses to different quantum flux densities, 57-107. In: Encyclopedia of Plant Physiology, N.S. Vol. 12A, Physiological Plant Ecology I, eds. O.L. LANGE, P.S. NOBEL, C.B. OSMOND, H. ZIEGLER. Berlin-Heidelberg-New York: Springer 1981. - BOAG, S., JENKINS, C.L.D.: Plant Physiol. 79, 165-170 (1985). - BOAG, S., PORTIS, A.R.: Planta 160, 33-40 (1984); - ibid. 165, 416-423 (1985). - BRECHIGNAC, F., ANDRÉ, M.: Plant Physiol. 78, 545-550 (1985). - BROOKS, A., FARQUHAR, G.D.: Planta 165, 397-406 (1985). - BROWN, R.H., BOUTON, J.H., EVANS, P.T., MALTER, H.E., RIGSBY, L.L.: Plant Physiol. 77, 653-658 (1985). - BRULFERT, J., KLUGE, M., QUEIROZ, O.: Physiol. Vég. 21, 803-804 (1983). - BRULFERT, J., VIDAL, J., Le MARECHAL, P., GADAL, P., QUEIROZ, O., KLUGE, M., KRUGER, I.: Biochem. Biophys. Res. Commun. 136, 151-159 (1986). - BUCHANAN-BOLLIG, I.C.: Planta 160, 264-271 (1984). - BUCHANAN-BOLLIG, I.C., SMITH, J.A.C.: Planta 161, 314-319 (1984). - BUCHANAN-BOLLIG, I.C., FISCHER, A., KLUGE, M.: Planta 161, 71-80 (1984a). - BUCHANAN-BOLLIG, I.C., KLUGE, M., MÜLLER, D.: Plant Cell Environ. 7, 63-70 (1984b). - BURNELL, J.N.: Biochem. Int. 9, 683-689 (1984a); - Biochem. Biophys. Res. Commun. 120, 559-565 (1984b); - Plant Physiol. 75, 873-875 (1984c). - BURNELL, J.N., HATCH, M.D.: Biochem. Biophys. Res. Commun. 111, 288-293 (1983); - Arch. Biochem. Biophys. 231, 175-182 (1984); - ibid. 237, 490-503 (1985a); - Trends Biochem. Sci. 10, 288-291 (1985b).

CAMP, P.J., RANDALL, D.D.: Plant Physiol. 77, 571-577 (1985). - CANUT, H., ALIBERT, G., BOUDET, A.M.: Z. Pflanzenphysiol. 114, 247-253 (1984). - CARNAL, N.W., BLACK, C.C.: Plant Physiol. 71, 150-155 (1983). -

CASTRILLO, M., GUARIGUATA, M., CALCAGNO, A.M., GARCIA, D.: 83-97 in: Physiological Ecology of CAM Plants, ed. E. MEDINA. Caracas: CIET (IVIC-UNESCO) 1984. - CEROVIĆ, Z.G., PLESNIČAR, M.: Biochem. J. 223, 543-545 (1984). - CHAPMAN, K.S.R., HATCH, M.D.: Plant Sci. Lett. 29, 145-154 (1983). - COCKBURN, W.: Plant Cell Environ. 4, 417-418 (1981); - New Phytol. 101, 3-24 (1985a); - Nature 314, 200 (1985b). - COCKBURN, W., GOH, C.J., AVADHANI, P.N.: Plant Physiol. 77, 83-86 (1985). - COLEMAN, J.R., GROSSMAN, A.R.: Proc. Natl. Acad. Sci. USA 81, 6049-6053 (1984). - COLEMAN, J.R., BERRY, J.A., TOGASAKI, R.K., GROSSMAN, A.R.: Plant Physiol. 76, 472-477 (1984). - COLMAN, B.: Plant Cell Environ. 7, 619-621 (1984). - CORLEY, E., WOLOSIUK, R.A.: J. Biol. Chem. 260, 3978-3983 (1985). - CRAWFORD, N.A., YEE, B.C., HUTCHESON, S.W., WOLOSIUK, R.A., BUCHANAN, B.B.: Arch. Biochem. Biophys. 244, 1-15 (1986).

DANIEL, P.P., BRYANT, J.A., WOODWARD, F.I.: Biochem. J. 218, 387-394 (1984). - DAY, D.A., NEUBERGER, M., DOUCE, R.: Aust. J. Plant Physiol. 12, 219-228 (1985). - DELEENS, E., TREICHEL, I., O'LEARY, M.H.: Plant Physiol. 79, 202-206 (1985). - DIETZ, K.-J.: Biochim. Biophys. Acta 839, 240-248 (1985). - DIETZ, K.-J., HEBER, U.: Biochim. Biophys. Acta 767, 432-443 (1984); - ibid. 848, 392-401 (1986). - DOEHLERT, D.C., HUBER, S.C.: Plant Physiol. 76, 250-253 (1984). - DONOVAN, D.C., ARDITTI, J., TING, I.P.: Ann. Bot. 54, 583-586 (1984). - DOREMUS, H.D., JAGENDORF, A.T.: Plant Physiol. 79, 856-861 (1985). - DROTAR, A., PHELPS, P., FALL, R.: Plant Sci. 42, 35-40 (1985). - DRY, I.B., WISKICH, J.T.: Aust. J. Plant Physiol. 12, 329-339 (1985).

EDWARDS, G.E., WALKER, D.A.: Arch. Biochem. Biophys. 231, 124-135 (1984). - EDWARDS, G.E., NAKAMOTO, H., BURNELL, J.N., HATCH, M.D.: Annu. Rev. Plant Physiol. 36, 255-286 (1985). - EVEREST, S.A., SYRETT, P.J.: New Phytol. 93, 581-589 (1983).

FARINEAU, J., LELANDAIS, M., MOROT-GAUDRY, J.-F.: Physiol. Plant. 60, 208-214 (1984). - FARMER, A.M., SPENCE, D.H.N.: Ann. Bot. 56, 347-350 (1985). - FARRAR, S.C., FARRAR, J.F.: New Phytol. 100, 271-283 (1985). - FERTÉ, N., MEUNIER, J.-C., SAUVE, P., RICARD, J.: Plant Sci. Lett. 35, 175-179 (1984). - FISCHER, A., KLUGE, M.: Planta 160, 121-128 (1984). - FLÜGGE, U.I., STITT, M., HELDT, H.W.: FEBS Lett. 183, 335-339 (1985). - FRANCO, A.R., CÁRDENAS, J., FERNÁNDEZ, E.: FEBS Lett. 176, 453-456 (1984). - FU, C.F., HEW, C.S.: Bot. Gaz. 143, 294-297 (1982). - FURBANK, R.T., LEEGOOD, R.C.: Planta 162, 457-462 (1984). - FURBANK, R.T., WALKER, D.A.: Planta 163, 75-83 (1985). - FURBANK, R.T., BADGER, M.R., OSMOND, C.B.: Plant Physiol. 73, 1038-1041 (1983). - FURBANK, R.T., STITT, M., FOYER, C.H.: Planta 164, 172-178 (1985).

GALLOWAY, C.M., DUGGER, W.M., BLACK, C.C.: Plant Physiol. 79, 920-922 (1985). - GANCEDO, J.M.: FEBS Lett. 175, 369-370 (1984). - GARDSTRÖM, P., EDWARDS, G.E., HENRICSON, D., ERICSON, I.: Physiol. Plant. 64, 29-33 (1985). - GAUDILLÈRE, M., JOLIVET-TOURNIER, P., COSTES, C.: Physiol. Vég. 21, 1115-1124 (1983). - GERBLING, K.-P., STEUP, M., LATZKO, E.: Europ. J. Biochem. 147, 207-215 (1985). - GFELLER, R.P., GIBBS, M.: Plant Physiol. 77, 509-511 (1985). - GILLHAM, D.J., DODGE, A.D.: Planta 167, 246-251 (1986). - GOH, C.J., ARDITTI, J., AVADHANI, P.N.: New Phytol. 95, 367-374 (1983). - GONTERO, B., MEUNIER, J.-C., BUC, J., RICARD, J.: Europ. J. Biochem. 145, 485-488 (1984a). - GONTERO, B., MEUNIER, J.-C., RICARD, J.: Plant Sci. Lett. 36, 137-142 (1984b). - GONTERO, B., MEUNIER, J.-C., SAUVE, P., RICARD, J.: Plant Sci. Lett. 38, 17-22 (1985). - GONZALES, D.H., IGLESIAS, A.A., ANDREO, C.S.: Plant Physiol. 116, 425-434 (1984). - GRIFFITHS, H., SMITH, J.A.C.: Oecologia 60, 176-184 (1983). - GUTTERIDGE, S., PARRY, M.A.J., SCHMIDT, C.N.G., FEENEY, J.: FEBS Lett. 170, 355-359 (1984).

HÄBERLEIN, I., SCHIMPFF-WEILAND, G., FOLLMANN, H.: Biochem. Biophys. Res. Commun. 127, 401-406 (1985). - HALLBERG, M., LARSSON, C.: Physiol. Plant. 57, 330-338 (1983). - HAMPP, R.: Plant Physiol. 79, 690-694 (1985). - HAMPP, R., GOLLER, M., FÜLLGRAF, H., EBERLE, I.: Plant Cell Physiol. 26, 99-108 (1985). - HANSON, K.R.: Arch. Biochem. Biophys. 232, 58-75 (1984). - HANSON, K.R., PETERSON, R.B.: Arch. Biochem. Biophys. 237, 300-313 (1985). - HATCH, M.D., DRÖSCHER, L., FLÜGGE, U.I., HELDT, H.W.: FEBS Lett. 178, 15-19 (1984). - HATTERSLEY, P.W., ROKSANDIC, Z.: Aust. J. Bot. 31, 317-321 (1983). - HEDRICH, R., RASCHKE, K., STITT, M.: Plant Physiol. 75, 977-982 (1985). - HELDER, R.J., van HARMELEN, M.: Acta Bot. Neerl. 31, 281-295 (1982). - HERZOG, B., STITT, M., HELDT, H.W.: Plant Physiol. 75, 561-565 (1984). - HEW, C.S., NG, Y.W., WONG, S.C., YEOH, H.H., HO, K.K.: Physiol. Plant. 60, 154-158 (1984). - HIPKIN, C.R., CANNONS, A.C.: Plant Sci. 41, 155-160 (1985). - HO, K.K., YEOH, H.H., HEW, C.S.: Plant Cell Physiol. 24, 1317-1321 (1983). - HOLADAY, A.S., TALKMITT, S., DOOHAN, M.E.: Plant Sci. 41, 31-39 (1985). - HOLBROOK, G.P., JORDAN, D.B., CHOLLET, R.: Plant Physiol. 77, 578-583 (1985). - HOLTUM, J.A.M., SUMMONS, R.E., ROESKE, C.A., COMINS, H.N., O'LEARY, M.H.: J. Biol. Chem. 259, 6870-6881 (1984). - HOSSAIN, M.A., ASADA, K.: Plant Cell Physiol. 25, 1285-1295 (1984). - HOSSAIN, M.A., NAKANO, Y., ASADA, K.: Plant Cell Physiol. 25, 385-395 (1984). - HRUBEC, T.C., ROBINSON, J.M., DONALDSON, R.P.: Plant Physiol. 77, 1010-1012 (1985). - HUBER, S.C., KERR, P.S., RUFTY, T.W.: Physiol. Plant. 64, 81-87 (1985).

IGLESIAS, A.A., ANDREO, C.S.: Plant Physiol. 75, 983-987 (1984). - INCHAROENSAKDI, A., TAKABE, T., AKAZAWA, T.: Arch. Biochem. Biophys. 237, 445-453 (1985).

JABLONSKI, P.P., ANDERSON, J.W.: Phytochemistry 23, 1865-1869 (1984). - JACQUOT, J.-P., GADAL, P., NISHIZAWA, A.N., YEE, B.C., CRAWFORD, N.A., BUCHANAN, B.B.: Arch. Biochem. Biophys. 228, 170-178 (1984). - JAWOROWSKI, A., HARTMAN, F.C., ROSE, I.A.: J. Biol. Chem. 259, 6783-6789 (1984). - JENKINS, C.L.D., BOAG, S.: Plant Physiol. 79, 84-89 (1985). - JENKINS, C.L.D., RUSS, V.J.: Plant Sci. Lett. 35, 19-24 (1984). - JOCHEM, P., RONA, J.-P., SMITH, J.A.C., LÜTTGE, U.: Physiol. Plant. 62, 410-415 (1984). - JOHNSTON, M., GRÖF, C.P.L., BROWNELL, P.F.B.: Aust. J. Plant Physiol. 11, 137-141 (1984). - JOLIVET, P., GANS, P., TRIANTAPHYLIDES, C.: Anal. Biochem. 147, 86-91 (1985). - JORDAN, D.B., CHOLLET, R.: Arch. Biochem. Biophys. 236, 487-496 (1985). - JORDAN, D.B., OGREN, W.L.: Planta 161, 308-313 (1984).

KAISER, G., MARTINOIA, E., WEIMKEN, A.: Z. Pflanzenphysiol. 107, 103-113 (1982). - KAISER, W.M., SCHROPPEL-MEIER, G., WIRTH, E.: Planta 167, 292-299 (1986). - KAPLAN, A., VOLOKITA, M., ZENVIRTH, D., REINHOLD, L.: FEBS Lett. 176, 166-168 (1984). - KEELEY, J.E., MORTON, B.A.: Photosynthetica 16, 546-553 (1982). - KEELEY, J.E., WALKER, C.M., MATHEWS, R.P.: Oecologia 58, 63-69 (1983). - KEELEY, J.E., OSMOND, C.B., RAVEN, J.A.: Nature 310, 694-695 (1984); - ibid. 314, 200 (1985). - KELLY, G.J., LATZKO, E.: Photosynth. Res. 1, 181-187 (1980). - KERR, P.S., RUFTY, T.W., HUBER, S.C.: Plant Physiol. 78, 576-581 (1985). - KLECZKOWSKI, L.A., RANDALL, D.D.: Plant Physiol. 79, 274-277 (1985). - KLEIN, U.: Planta 167, 81-86 (1986). - KÖSTER, S., WINTER, K.: Plant Physiol. 79, 520-524 (1985). - KOWALCZYK, S., MASLOWSKI, P.: Biochim. Biophys. Acta 766, 570-575 (1984). - KOWALCZYK, S., JANUSZEWSKA, B., CYMERSKA, E., MASLOWSKI, P.: Physiol. Plant. 60, 31-37 (1984). - KREIMER, G., MELKONIAN, M., LATZKO, E.: FEBS Lett. 180, 253-258 (1985a). - KREIMER, G., MELKONIAN, M., HOLTUM, J.A.M., LATZKO, E.: Planta 166, 515-523 (1985b). - KRUGER, N.J., DENNIS, D.T.: Plant Physiol. 78, 645-648 (1985a); - Biochem. Biophys. Res. Commun. 126, 320-326 (1985b). - KU, M.S.B., MONSON, R.K., LITTLEJOHN, R.O.,

NAKAMOTO, H., FISHER, D.B., EDWARDS, G.E.: Plant Physiol. 71, 944-948 (1983).

LACOSTE-ROYAL, G., GIBBS, S.P.: Proc. Natl. Acad. Sci. USA 82, 1456-1459 (1985). - LARSSON, M., LARSSON, C.-M., GUERRERO, M.G.: J. Exp. Bot. 36, 1373-1386 (1985). - LATZKO, E., KELLY, G.J.: Prog. Bot. 42, 58-79 (1980). - LEEGOOD, R.C.: Planta 164, 163-171 (1985). - LEEGOOD, R.C., FURBANK, R.T., PELZÜFER, B.B.: Planta 162, 450-456 (1984). - LEVI, C., GIBBS, M.: Plant Physiol. 74, 459-463 (1984). - LEVY, L.M., BETTS, G.F.: Biochim. Biophys. Acta 832, 186-191 (1985). - LIEDVOGEL, B.: Z. Naturforsch. 40c, 182-188 (1985). - LIN, J.-T., GRIFFITH, O.M., CORSE, J.W.: Anal. Biochem. 148, 10-14 (1985). - LITTLEJOHN, R.O., KU, M.S.B.: Plant Physiol. 74, 1050-1054 (1984). - LOPEZ-RUIZ, A., VERBELEN, J.P., ROLDAN, J.M., DIEZ, J.: Plant Physiol. 79, 1006-1010 (1985). - LÜTTGE, U., SMITH, J.A.C.: J. Membr. Biol. 81, 149-158 (1984).

MÄCHLER, F., LEHNHERR, B., SCHNYDER, H., NÖSBERGER, J.: J. Exp. Bot. 36, 1542-1550 (1985). - MAKINO, A., MAE, T., OHIRA, K.: Planta 166, 414-420 (1985). - MANDERSCHEID, R., WILD, A.: Z. Naturforsch. 40c, 295-296 (1985). - MARCUS, Y., VOLOKITA, M., KAPLAN, J.: J. Exp. Bot. 35, 1136-1144 (1985). - MARQUES, I.A., ANDERSON, L.E.: Plant Physiol. 77, 807-810 (1985). - MÁRQUEZ, A.J., GALVÁN, F., VEGA, J.M.: Plant Sci. Lett. 34, 305-314 (1984). - MARTIN, C.E., ZEE, A.K.: Plant Physiol. 73, 718-723 (1983). - MATEYKA, C., SCHNARRENBERGER, C.: Plant Sci. Lett. 36, 119-123 (1984). - MEDINA, E.: Themas Sobre Eco-fisiologia de Plantas CAM, 174 pp. Caracas: CIET (IVIC-UNESCO). - MIERNYK, J.A., MacDOUGALL, P.S., DENNIS, D.T.: Plant Physiol. 76, 1093-1094 (1984). - MILLER, A.G., CANVIN, D.T.: FEBS Lett. 187, 29-32 (1985). - MILLER, A.G., TURPIN, D.H., CANVIN, D.T.: Plant Physiol. 75, 1064-1070 (1984). - MIYATAKE, K., KURAMOTO, Y., KITAOKA, S.: Biochem. Biophys. Res. Commun. 122, 906-911 (1984). - MONSON, R.K., EDWARDS, G.E., KU, M.B.S.: Bioscience 34, 563-566 (1984). - MOORE, B.D., KU, M.S.B., EDWARDS, G.E.: Plant Sci. Lett. 35, 127-138 (1984). - MORI, H., TAKABE, T., AKAZAWA, T.: Photosynth. Res. 5, 17-28 (1984). - MORONEY, J.V., HUSIC, H.D., TOLBERT, N.E.: Plant Physiol. 76, 968-971 (1984). - MOTT, K.A., JENSEN, R.G., O'LEARY, J.W., BERRY, J.A.: Plant Physiol. 76, 968-971 (1984). - MURPHY, D.J., WOODROW, I.E., MUKHERJEE, K.D.: Biochem. J. 225, 267-270 (1985).

NAKAMOTO, H., EDWARDS, G.E.: Z. Pflanzenphysiol. 114, 315-320 (1984). - NAKAMOTO, H., KU, M.S.B., EDWARDS, G.E.: Plant Cell Physiol. 24, 1387-1393 (1983). - NATO, F., HIREL, B., NATO, B., GADAL, P.: FEBS Lett. 175, 443-446 (1984). - NEALES, T.F., FRASER, M.S., ROKSANDIC, Z.: Aust. J. Plant Physiol. 10, 437-444 (1983). - NGUYEN, T.D., SIEGENTHALER, P.-A.: Biochim. Biophys. Acta 840, 99-106 (1985). - NIKOLAU, B.J., WURTELE, E.S., STUMPF, P.K.: Arch. Biochem. Biophys. 235, 555-561 (1984a); - Plant Physiol. 74, 895-901 (1984b). - NIMMO, G.A., NIMMO, H.G., FEWSON, C.A., WILKINS, M.B.: FEBS Lett. 178, 199-203 (1984). - NUNGESSER, D., KLUGE, M., TOLLE, H., OPPELT, W.: Planta 162, 204-214 (1984).

OGAWA, T., MIYANO, A., INOUE, Y.: Biochim. Biophys. Acta 808, 77-84 (1985). - OHNISHI, J.-I., KANAI, R.: Plant Cell Physiol. 24, 1411-1420 (1983). - OHNISHI, J.-I., YAMAZAKI, M., KANAI, R.: Plant Cell Physiol. 26, 797-803 (1985). - OKABE, K., YANG, S.-Y., TSUZUKI, M., MIYACHI, S.: Plant Sci. Lett. 33, 145-153 (1984). - OMNAAS, J., PORTER, M.A., HARTMAN, F.C.: Arch. Biochem. Biophys. 236, 646-653 (1985). - OVERBAUGH, J.M., FALL, R.: Plant Physiol. 77, 437-442 (1985).

PAGLIARO, L., JOHAL, S.: Biochim. Biophys. Acta 831, 34-39 (1985). - PARRY, M.A.J., SCHMIDT, C.N.G., CORNELIUS, M.J., KEYS, A.J., MILLARD,

B.N., GUTTERIDGE, S.: J. Exp. Bot. 36, 1396-1404 (1985). - PAZ, N., XU, D.-P., BLACK, C.C.: Plant Physiol. 79, 1133-1136 (1985). - PEAVEY, D.G., GFELLER, R.P., GIBBS, M.: Physiol. Vég. 21, 337-342 (1983). - PELTIER, G., THIBAULT, P.: Plant Physiol. 79, 225-230 (1985a); - ibid. 77, 281-284 (1985b). - PERROT-RECHENMANN, C., CHOLLET, R., GADAL, P.: Planta 161, 261-271 (1984). - PUSHKIN, A.V., ANTONIUK, L.P., SOLOVIEVA, N.A., SHUBIN, V.V., EVSTIGNEEVA, Z.G., KRETOVICH, W.L., CHEREDNIKOVA, T.V., TSUPRUN, V.L., ZOGRAF, O.N., KISELEV, N.A.: Biochim. Biophys. Acta 828, 336-350 (1985).

RAMASWAMY, N.K., NAIR, P.M.: Plant Sci. Lett. 34, 261-267 (1984). - RAYDER, L., TING, I.P.: Photosynth. Res. 4, 203-212 (1983a); - Plant Physiol. 72, 606-610 (1983b). - REED, J.E., CHOLLET, R.: Planta 165, 439-445 (1985). - RICHARDSON, K., GRIFFITHS, H., REED, M.L., RAVEN, J.A., GRIFFITHS, N.M.: Oecologia 61, 115-121 (1984). - RINTAMÄKI, E., ARO, E.-M.: Photosynthetica 18, 357-364 (1984). - RITZ, D., KLUGE, M., VEITH, H.J.: Planta 167, 284-291 (1986). - ROBINSON, S.P.: Plant Physiol. 75, 425-430 (1984); - ibid. 79, 996-1002 (1985). - ROMERO, J.M., LARA, C., GUERRERO, M.G.: Arch. Biochem. Biophys. 237, 396-401 (1985). - ROSA, L., WHATLEY, F.R.: Plant Physiol. 75, 131-137 (1984). - ROUGHAN, P.G.: Biochim. Biophys. Acta 835, 527-532 (1985). - RUMPHO, M.E., EDWARDS, G.E.: Plant Physiol. 76, 711-718 (1984); - ibid. 78, 537-544 (1985). - RUMPHO, M.E., KU, M.S.B., CHENG, S.-H., EDWARDS, G.E.: Plant Physiol. 75, 993-996 (1984).

SAMUELSSON, G., LÖNNEBORG, A., ROSENQVIST, E., GUSTAFSSON, P., ÖQUIST, G.: Plant Physiol. 79, 992-995 (1985). - SASEK, T.W., DeLUCIA, E.H., STRAIN, B.R.: Plant Physiol. 78, 619-622 (1985). - SATOH, H., OKADA, M., NAKAYAMA, K., MIYAJI, K.: Plant Cell Physiol. 25, 1205-1214 (1984). - SAUER, A., HEISE, K.-P.: Z. Naturforsch. 39c, 268-275 (1984). - SCHEIBE, R.: Biochim. Biophys. Acta 788, 241-247 (1984). - SCHEIBE, R., FICKENSCHER, K.: FEBS Lett. 180, 317-320 (1985). - SCHEIBE, R., FICKENSCHER, K., ASHTON, A.R.: Biochim. Biophys. Acta 870, 191-197 (1986). - SCHERER, S., STÜRZL, E., BÖGER, P.: Physiol. Plant. 60, 557-560 (1984). - SCHMIDT, C.N.G., GUTTERIDGE, S., PARRY, M.A.J., KEYS, A.J.: Biochem. J. 220, 781-785 (1984). - SCHMUTZ, D., BRUNOLD, C.: Plant Physiol. 74, 866-870 (1984). - SCHNYDER, H., MÄCHLER, F., NÖSBERGER, J.: J. Exp. Bot. 35, 147-156 (1984). - SCHÜRMANN, P., ROUX, J., SALVI, L.: Physiol. Vég. 23, 813-818 (1985). - SCHWITZGUEBEL, J.-P., SIEGENTHALER, P.-A.: Plant Physiol. 75, 670-674 (1984). - SEEMANN, J.R., BADGER, M.R., BERRY, J.A.: Plant Physiol. 74, 791-794 (1984). - SEEMANN, J.R., BERRY, J.A., FREAS, S.M., KRUMP, M.A.: Proc. Natl. Acad. Sci. USA 82, 8024-8028 (1985). - SEETHAMBARAM, Y., RAO, A.N., DAS, V.S.R.: Biochem. Physiol. Pflanzen 180, 107-113 (1985). - SERVAITES, J.C.: Arch. Biochem. Biophys. 238, 154-160 (1985a); - Plant Physiol. 78, 839-843 (1985b). - SHAW, P.J., HENWOOD, J.A.: Planta 165, 333-339 (1985). - SHELP, B.J., CANVIN, D.T.: Can. J. Bot. 62, 1398-1403 (1984). - SHIBATA, H., OCHIAI, H., SAWA, Y., MIYOSHI, S.: Plant Physiol. 80, 126-129 (1986). - SHIRAIWA, Y., MIYACHI, S.: Plant Cell Phsiol. 26, 109-116 (1985). - SICHER, R.C., KREMER, D.F.: Plant Physiol. 79, 695-698 (1985). - SIMON, P., BONZON, M., GREPPIN, H., MARMÉ, D.: FEBS Lett. 167, 332-338 (1984). - SINCLAIR, R.: J. Exp. Bot. 35, 1-7 (1984). - SINGH, B.K., GREENBERG, E., PREISS, J.: Plant Physiol. 74, 711-716 (1984). - SIPES, D., TING, I.P.: Plant Physiol. 77, 59-63 (1985). - SMITH, A.M., WOOLHOUSE, H.W.: Planta 159, 570-578 (1983). - SMITH, B.N., POWELL, A.M.: Naturwissenschaften 71, 217-218 (1984). - SMITH, I.K.: Plant Physiol. 79, 1044-1047 (1985). - SMITH, I.K., KENDALL, A.C., KEYS, A.J., TURNER, J.C., LEA, P.J.: Plant Sci. 41, 11-17 (1985). - SMITH, J.A.C., URIBE, E.G., BALL, E., HEUER, S., LÜTTGE, U.: Europ. J. Biochem. 141, 415-420 (1984a). - SMITH, J.A.C., URIBE, E.G., BALL, E., LÜTTGE, U.: Planta 162, 299-304 (1984b). -

SOLL, J., WÖTZEL, C., BUCHANAN, B.B.: Plant Physiol. 77, 999-1003 (1985). - SOULIÉ, J.-M., BUC, J., RIVIÈRE, M., RICARD, J.: Europ. J. Biochem. 152, 565-568 (1985). - STERNBERG, L., DENIRO, M.J., AJIE, H.: Phytochemistry 23, 2475-2477 (1984a). - STERNBERG, L., DENIRO, M.J., KEELEY, J.E.: Plant Physiol. 76, 68-70 (1984b). - STIBOROVÁ, M., LEBLOVÁ, S.: Phytosynthetica 19, 177-182 (1985). - STITT, M., HELDT, H.W.: Planta 164, 179-188 (1985a); - Biochim. Biophys. Acta 808, 400-414 (1985b). - STITT, M., HERZOG, B., HELDT, H.W.: Plant Physiol. 75, 548-553 (1984a). - STITT, M., KÜRZEL, B., HELDT, H.W.: Plant Physiol. 75, 554-560 (1984b). - STITT, M., CSÉKE, C., BUCHANAN, B.B.: Europ. J. Biochem. 143, 89-93 (1984c). - STITT, M., HERZOG, B., HELDT, H.W.: Plant Physiol. 79, 590-598 (1985). - STRUVE, I., WEBER, A., LÜTTGE, U., BALL, E., SMITH, J.A.C.: J. Plant Physiol. 117, 451-468 (1985). - SUREK, B., HEILBRONN, A., AUSTEN, A., LATZKO, E.: Planta 165, 507-512 (1985).

TA, T.C., JOY, K.W., IRELAND, R.J.: Plant Physiol. 78, 334-337 (1985). - TEL-OR, E., HUFLEJT, M., PACKER, L.: Biochem. Biophys. Res. Commun. 132, 533-539 (1985). - TERASHIMA, I., INOUE, Y.: Plant Cell Physiol. 26, 63-75 (1985a); - ibid. 26, 781-785 (1985b). - THEISS-SEUBERLING, H.-B.: Plant Cell Physiol. 25, 601-609 (1984). - THOM, M., MARETZKI, A.: Proc. Natl. Acad. Sci. USA 82, 4697-4701 (1985). - THOMAS, B.A.: Nature 314, 200 (1985). - TING, I.P.: Annu. Rev. Plant Physiol. 36, 595-622 (1985). - TING, I.P., BURK, J.H.: Plant Sci. Lett. 32, 279-285 (1983). - TISCHNER, R.: Plant Sci. Lett. 34, 73-80 (1984). - TORRES-RUIZ, J.A., McFADDEN, B.A.: Arch. Microbiol. 142, 55-60 (1985). - TREEDE, H.-J., HEISE, K.-P.: Z. Naturforsch. 40c, 496-502 (1985). - TSUZUKI, M., MIYACHI, S., EDWARDS, G.E.: Plant Cell Physiol. 26, 881-891 (1985). - TURPIN, D.H., MILLER, A.G., CANVIN, D.T.: J. Phycol. 20, 249-253 (1984).

UDVARDY, J., BORBÉLY, G., JUHÁSZ, A., FARKAS, G.L.: FEBS Lett. 172, 11-16 (1984). - USUDA, H.: Plant Cell Physiol. 25, 1297-1301 (1984); ibid. 26, 1455-1463 (1985a); - Plant Physiol. 78, 859-864 (1985b). - USUDA, H., EDWARDS, G.E.: Plant Sci. Lett. 37, 41-45 (1984). - USUDA, H., KU, M.S.B., EDWARDS, G.E.: Aust. J. Plant Physiol. 11, 509-517 (1984a); - Plant Physiol. 76, 238-243 (1984b); - Physiol. Plant. 63, 65-70 (1985).

VAPAAVUORI, E.M.: J. Exp. Bot. 37, 89-98 (1986). - VIGH, L., HORVÀTH, I., van HASSELT, P.R., KUIPER, P.J.C.: Plant Physiol. 79, 756-759 (1985). - VIIL, J., IVANOVA, H., PÄRNIK, T.: Photosynthetica 19, 88-97 (1985). - VOLOKITA, M., ZENVIRTH, D., KAPLAN, A., REINHOLD, L.: Plant Physiol. 76, 599-602 (1984). - VOORDOUW, G., van der VIES, S.M., BOUWMEISTER, P.P.: Europ. J. Biochem. 141, 313-318 (1984). - VU, J.C.V., ALLEN, L.H., BOWES, G.: Plant Physiol. 76, 843-845 (1984).

WALKER, K.A., HARWOOD, J.L.: Biochem. J. 226, 551-556 (1985). - WALKER, K.A., GIVAN, C.V., KEYS, A.J.: Plant Physiol. 75, 60-66 (1984a). - WALKER, K.A., KEYS, A.J., GIVAN, C.V.: J. Exp. Bot. 35, 1800-1810 (1984b). - WATERTON, J.C., BRIDGES, I.G., IRVING, M.P.: Biochim. Biophys. Acta 763, 315-320 (1983). - WILKINS, M.B.: Physiol. Vég. 21, 997-1005 (1983); - Planta 161, 381-384 (1984). - WILLERT, D.J. von: 157-200 in: Advances in Botanical Research, Vol. 11, ed. J.A. CALLOW, H.W. WOOLHOUSE. London: Academic Press 1985. - WILLERT, D.J. von, BRINCKMANN, E., ELLER, B.M., SCHEITLER, B.: Oecologia 61, 393-397 (1984). - WILLERT, D.J. von, BRINCKMANN, E., SCHEITLER, B., ELLER, B.M.: Planta 164, 44-55 (1985). - WINTER, K.: 329-388 in: Photosynthetic Mechanisms and the Environment, ed. J. BARBER, N.R. BAKER. London: Elsevier 1985. - WINTER, K., WALLACE, B.J., STOCKER, G.C., ROKSANDIC, Z.: Oecologia 57, 129-141 (1983). - WINTER, K., MEDINA, E.,

GARCIA, V., MAYORAL, M.L., MUNIZ, R.: J. Plant Physiol. 118, 73-78 (1984). - WOLDEGIORGIS, G., VOSS, S., SHRAGO, E., WERNER-WASHBURNE, M., KEEGSTRA, K.: Biochem. Biophys. Res. Commun. 116, 945-951 (1983). - WOLOSIUK, R.A., CORLEY, E., CRAWFORD, N.A., BUCHANAN, B.B.: FEBS Lett. 189, 212-216 (1985). - WONG, J.H., BALOGH, Á., BUCHANAN, B.B.: Biochem. Biophys. Res. Commun. 121, 842-847 (1984). - WOODROW, I.E., ELLIS, J.R., JELLINGS, A., FOYER, C.H.: Planta 161, 525-530 (1984a). - WOODROW, I.E., MURPHY, D.J., LATZKO, E.: J. Biol. Chem. 259, 3791-3795 (1984b). - WOODROW, I.E., FURBANK, R.T., BROOKS, A., MURPHY, D.J.: Biochim. Biophys. Acta 807, 263-271 (1985). - WU, M.-X., WEDDING, R.T.: Plant Physiol. 77, 667-675 (1985a); - Arch. Biochem. Biophys. 240, 655-675 (1985b). - WU, M.-X., SMYTH, D.A., BLACK, C.C.: Plant Physiol. 73, 188-191 (1983); - Proc. Natl. Acad. Sci. USA 81, 5051-5055 (1984).

YAGAWA, Y., AIZAWA, K., YANG, S.-Y., MIYACHI, S.: Plant Cell Physiol. 27, 215-221 (1986). - YAN, T.-F.J., TAO, M.: J. Biol. Chem. 259, 5087-5092 (1984). - YOKOTA, A., CANVIN, D.T.: Plant Physiol. 77, 735-739 (1985). - YOKOTA, A., KITAOKA, S.: Biochem. Biophys. Res. Commun. 131, 1075-1079 (1985). - YOKOTA, A., KOMURA, H., KITAOKA, S.: Agric. Biol. Chem. 49, 3309-3310 (1985a); - Arch. Biochem. Biophys. 242, 498-506 (1985b).

Dr. JOSEPH A.M. HOLTUM
Botanisches Institut der Universität
Schloßgarten 3
D-4400 Münster, FRG

Dr. GRAHAME J. KELLY
CSIRO
Marine Laboratories
AUS-7000 Hobart, Tasmania, Australia

Prof. Dr. ERWIN LATZKO
Botanisches Institut der Universität
Schloßgarten 3
D-4400 Münster, FRG

Notes added in proof (2005):
In the fourth paragraph of Sect. 6, “anaerobic” should read “aerobic”.
The scheme in Fig. 4 has since been challenged (Sect. 5, 1990 review).
Reactions (curved arrows) on the left vertical arrow of Figure 5 are unclear. There are actually two alternatives: either a PEP-to-pyruvate interconversion, or a (ATP + Pi)-to-(AMP + PPi) interconversion.

Photosynthesis. Carbon Metabolism: New Regulators of CO_2 Fixation, the New Importance of Pyrophosphate, and the Old Problem of Oxygen Involvement Revisited

By Grahame J. Kelly, Joseph A.M. Holtum, and Erwin Latzko

"When some organisms developed the ability
to tap sunlight for energy,
photosynthesis was born....
The more advanced forms of photosynthesis
released a corrosive and poisonous gas
into the atmosphere: oxygen ...
bacteria and blue-green algae developed ways
to neutralize gaseous oxygen by combining it
with their own waste products.
The next step was to harness the energy
released by oxidation of these waste compounds.
(If you are going to burn your garbage,
you might as well keep warm by the fire).
This was the beginning of oxidation, or respiration...."

R.E.Dickerson in Scientific American (1972) 226:58-72

1. Introduction

Research into photosynthetic carbon metabolism has experienced a shift of emphasis in recent years. Detailed analyses of C_4 photosynthesis, of photorespiration, of light-mediated regulation of enzyme activities and of photosynthesis by isolated chloroplasts have become less frequent, while the topics of glycolysis and mitochondrial respiration in photosynthetic cells, somewhat neglected during the past two decades, are receiving an increasing amount of attention. In particular, the central role of glycolysis in crassulacean acid metabolism (CAM) is being actively researched, and is helping to create a new awareness of the pivotal role of inorganic pyrophosphate (PP_i) in plant carbohydrate metabolism. But, as emphasized in this review, a central participant in the PP_i-story, the PP_i-linked PFK, is still largely an enigma due to the reversibility of the reaction it catalyzes. This is despite the great attention given to its activator, fructose-2,6-P_2. Meanwhile, a new sugar phosphate, and a new protein, both regulators of ribulose-1,5-bisphosphate (RuBP) carboxylase, have been discovered in chloroplasts and are producing major alterations to our perception of the regulation of CO_2 fixation, and to our evaluation of photosynthetic performance (given the major role being attributed to this enzyme in determining photosynthetic capacity). Last but not least, the diverse effects of photosynthetically generated O_2 on photosynthetic cells are receiving ever-increasing recognition. In a sense, O_2 is probably the great unrecognized "feedback inhibitor" of photosynthesis. These themes are examined in this review, concentrating on the literature that has appeared since our last review in Volume 48 of Progress in Botany.

G.J. Kelly and E. Latzko:
Thirty Years of Photosynthesis

Originally published in Progress in Botany 50

2. Inorganic Carbon Uptake

Terrestrial and aquatic plants differ markedly with respect to the passage of inorganic carbon from the cell exterior to the site of RuBP carboxylation. Terrestrial C_3 plants appear to rely on diffusion (Espie and Colman 1987; but see Mächler et al. 1986); C_4 plants initially depend on diffusion into mesophyll cells, then utilize the C_4 pathway to develop CO_2 concentrations in bundle-sheath cells of between 160 and 990 μM (Furbank and Hatch 1987), sufficient to effectively exclude RuBP oxygenation and thus photorespiration (Hanson and Peterson 1987). Aquatic plants, on the other hand, actively take up inorganic carbon. They possess membrane-located transporter proteins (Omata and Ogawa 1986) to pump inorganic carbon into cells, but the mechanistic details of this process are not entirely clear. In the cyanobacterium *Synechococcus* and in *Chlamydomonas reinhardtii* the transport of both CO_2 and HCO_3^- is believed to occur (Badger and Gallagher 1987; Miller and Canvin 1987; Williams and Turpin 1987; Miller et al. 1988), but work with two other species of cyanobacteria and several species of eukaryotic microalgae has identified HCO_3^- as the only form of inorganic carbon transported (Abe et al. 1987; Dixon et al. 1987; Dixon and Merrett 1988) in a process that is Na^+-dependent and involves an efflux of H^+ (Katz et al. 1986; Ogawa and Kaplan 1987; Scherer et al. 1988). The suggestion that, in some cases, HCO_3^- might be generated from CO_2 by carbonic anhydrase located outside the plasmalemma and close to the HCO_3^- transporter has been questioned (Williams and Turpin 1987) and in any event could not be considered in species where all carbonic anhydrase is intracellular (Dixon et al. 1987; Yagawa et al. 1987). In addition, Moroney et al. (1987) recently concluded that the *C.reinhardtii* HCO_3^- transporter is not on the plasmalemma, but rather is on the chloroplast envelope, in which case plasmalemma-linked carbonic anhydrase would not be closely involved with HCO_3^- transport.

It is widely accepted that HCO_3^- uptake is an energy-dependent process (Sundblad et al. 1986), driven by ATP from cyclic or pseudocyclic electron transport (Sueltemeyer et al. 1986; Brechignac and Lucas 1987; Yokota et al. 1987; Miller et al. 1988). Evidence for a separate light-mediated activation of the process has also been obtained (Kaplan et al. 1987). However, the possibility that this HCO_3^- transport boosts internal CO_2 concentrations sufficiently to inhibit RuBP oxygenation, and thus photorespiration, is not considered great, at least for *C.reinhardtii*, in which there is evidence that RuBP oxygenation continues (Moroney et al. 1986), possibly because the increased CO_2 concentration stimulates photosynthesis which leads, in turn, to an increased concentration of oxygen (Yokota et al. 1987).

3. Investigations with Isolated Plastids

Since the first acquisition of photosynthetically competent chloroplasts more than 20 years ago, experiments have almost invariably employed a suspension of these organelles, the density of which was far below that in vivo. Now Cerovic et al. (1987) have developed a procedure for forming a thin layer of isolated chloroplasts on the solid support of a cellulose nitrate membrane, making studies under more natural conditions possible.

Other adaptations of conventional procedures have been beneficial. Robinson (1986) described how chloroplasts isolated in media containing 0.2 M KCl as osmoticum (rather than 0.33 M sorbitol) photosynthesized faster, presumably because leakage of K^+ from the chloroplasts was prevented. Persistent efforts by Entwistle et al. (1988) have led to a method for isolating starch-rich amyloplasts from wheat endosperm. Results from studies with these organelles will be eagerly awaited – there is already evidence that amyloplasts are chloroplast-like in that triose-P is the preferred form of carbon taken up as substrate for starch synthesis (Echeverria et al. 1988). Returning to chloroplasts: these have now been isolated from *Kalanchoe blossfeldiana* and *Mesembryanthemum crystallinum* performing CAM (Adam and Rustin 1986; Neuhaus et al. 1988) and from photoautotrophically grown *Euglena gracilis* (Suzuki et al. 1987a); in the latter two cases photosynthesis rates exceeding 80 μmol/mg Chl/h were recorded. Finally, by utilizing ethylene glycol, Fu and Gibbs (1987) have been able to study the photosynthesis of isolated chloroplasts at temperatures as low as -15°C.

Envelope-mediated metabolite transport has received minor attention. The well-known phosphate translocator has been identified as a polypeptide of about 30 kDa in spinach, pea, and maize mesophyll chloroplasts (Thompson et al. 1987; Rumpho et al. 1988). A particularly specific inhibitor for this translocator has been identified by Rumpho et al. (1988). Transport of PEP by chloroplasts from *M.crystallinum* doing CAM has been described (Neuhaus et al. 1988). In C_4 plants the light-dependent uptake of pyruvate by mesophyll chloroplasts is now thought to involve pyruvate/Na^+ symport (Ohnishi and Kanai 1987a), an interesting observation in view of the recorded Na^+ requirement of C_4 plants for growth.

Studies of the biosynthesis of metabolites derived from pyruvate and PEP have again raised questions concerning the presence or absence of certain chloroplast enzymes. Formation of phosphoenolpyruvate (PEP) from glycerate-3-P requires P-glyceromutase and enolase. These two enzymes are clearly present in plastids with little or no photosynthetic ability (Liedvogel and Bäuerle 1986; Denyer and Smith 1988), but in chloroplasts the enzymes are either absent, or present in very small amounts (Bagge and Larsson 1986; Schulze-Siebert et al. 1987). With respect to acetyl-CoA formation, roles for both pyruvate dehydrogenase and acetyl-CoA synthetase are being entertained (Treede et al. 1986). Progress in studying the former enzyme, which is a complex, has been made (Treede and Heise 1986; Camp et al. 1988).

4. The Return of RuBP Carboxylase

To all but those working closely in the field, the CO_2-plus-Mg^{2+} activation of RuBP carboxylase appeared adequate for demonstrating enzyme activities commensurate with observed rates of photosynthesis by leaves. Indeed, it is often adequate for in vitro tests. But some workers noted inconsistencies when relating observations to in vivo conditions: the CO_2 concentration required for maximal activation was well in excess of that present in vivo, at least in C_3 plants; low concentrations of the substrate RuBP severely retarded the activation; reports of "light-mediated activation" of the carboxylase persisted.

Questions that were raised induced experiments that have revealed two new regulators of RuBP carboxylase. The first followed indications that in many (but not all) species, a tight-binding inhibitor maintained the enzyme in an inactive form during the night (Servaites et al. 1986; Sharkey et al. 1986; Vu et al. 1986). The inhibitor was soon identified in potato and bean leaves as 2-carboxy-D-arabinitol-1-P (Gutteridge et al. 1986a; Berry et al. 1987), a close analog of the reaction intermediate 2-carboxy-3-keto-D-arabinitol-1,5-P_2. There is no information as to how this inhibitor, which occurs in chloroplasts in the dark at a concentration of 2-4 mM, is synthetisized, but there is information concerning its metabolism: Salvucci et al. (1988) have partially purified a chloroplast protein that catalyses an NADPH-dependent metabolism of the inhibitor. A strong reductant (dithiothreitol) stabilized the protein's activity and this, together with its utilization of NADPH, provides clues as to how the level of the inhibitor might be decreased upon illumination.

The second regulator of RuBP carboxylase was also uncovered during studies of the light-mediated regulation of the enzyme's activity. In this case, a mutant of *Arabidopsis thaliana* in which light activation of the enzyme was absent (Somerville et al. 1982) was further examined and found to lack a soluble chloroplast protein possessed by the wild-type plant – this protein was found to be capable of activating RuBP carboxylase, and has been appropriately termed "rubisco activase" (Salvucci et al. 1985, 1986). Its particular attribute is its ability to activate the carboxylase in vitro in the presence of physiological concentrations of CO_2 and RuBP (Portis et al. 1986). Unlike the carboxyarabinitol-1-P regulatory mechanism described above, the activase mechanism is not restricted to certain species, but appears to be ubiquitous (Salvucci et al. 1987a).

Differences also exist with the type of light-mediated activation (Salvucci and Anderson 1987). While the carboxyarabinitol-1-P operates a "day/night switch", the activase participates in a light-mediated activation that appears to be linked to photosynthetically generated ATP (Salvucci et al. 1987b; Streusand and Portis 1987; Robinson and Portis 1988a). Although the mechanism of this interaction with ATP is yet to be elucidated, recent reports of phosphorylated forms of RuBP carboxylase (Kaul et al. 1986; Guitton and Mache 1987) may not be without relevance. Experiments on the interaction of the carboxylase with its activase, on the other hand, have drawn carboxyarabinitol-1-P into the picture, in that Robinson and Portis (1988b) obtained data suggesting that the activase interacts with the enzyme so that the inhibitor more easily dissociates from the active site.

In other experiments with RuBP carboxylase, further evidence for essential roles of the small subunits has been obtained (Andrews et al. 1986; Incharoensakdi et al. 1986; Voordouw et al. 1987). These subunits, which are encoded on the nuclear genome of plant leaves, are encoded on the chloroplast genome in the chromophyte alga *Olisthodiscus luteus* (Newman and Cattolico 1987). The arrangement of the small and large subunits is such that a keg ("Bierfass") shaped molecule results – detailed descriptions of the enzyme's structure have been presented (Chapman et al. 1987, 1988; Pilz et al. 1987).

Interference in the extraction of active RuBP carboxylase by proteolysis and by shear-induced inactivation have been described (Gutteridge et al. 1986b; Paech and Dybing 1986; Peñarrubia and Moreno 1987). In vivo, operation of the fully active enzyme can be influenced by the concentration of its product glycerate-3-P, which acts as an inhibitor (Foyer et al. 1987). The mechanism of the RuBP carboxylase reaction is still being questioned by some workers, particularly with respect to evidence that the two glycerate-3-P molecules produced are not necessarily treated equally (Tomany and Kent 1986; Fong and Butcher 1988). Finally, there is further evidence that certain primitive photosynthetic bacteria completely lack RuBP carboxylase (Holo and Sirevåg 1986).

5. RuBP Carboxylase and the Limits to Photosynthesis

Considerable effort is being devoted to ascertaining what biophysical and/or biochemical steps limit photosynthesis under various growth conditions, and under conditions where the cell's photosynthetic capacity is achieved (i.e., light- and CO_2-saturated photosynthesis). The menagerie of results are not always in agreement, but at least there is a consensus of opinion that under most conditions photosynthesis is limited by one or more of four factors: (a) light and/or CO_2 supply, (b) rate of photosynthetic electron transport, (c) activity of RuBP carboxylase, and (d) supply of P_i to the chloroplast.

Experiments continue to establish a close correlation between photosynthetic rate and the leaf content of RuBP carboxylase (Zima and Šesták 1985; Evans 1986; Ramachandra Reddy and Das 1987; Seemann et al. 1987; Suzuki et al. 1987b). It has even been possible to correlate the leaf carboxylase content with grain yield (Loza-Tavera et al. 1987). However, attempts to describe exactly how the potential carboxylase activity (which is usually equivalent to several times the rate of light-saturated photosynthesis in air) is reduced to an actual activity similar to the photosynthetic rate have reached no firm conclusions. Certainly the fact that the enzyme is not saturated with CO_2, at least in C_3 plant leaves, must be a contributing factor, but it is unclear whether or not the in vivo concentration of the other substrate (RuBP) is also insufficient to saturate the enzyme (von Caemmerer and Edmondson 1986; Yokota and Canvin 1986), thereby further reducing its activity. In addition, the degrees to which CO_2-plus-Mg^{2+} activation, rubisco activase activation, carboxyarabinitol-1-P inhibition and glycerate-3-P inhibition operate will need to be further examined (Gustafson et al. 1987; Farineau et al. 1988; see preceding section).

It now seems to be less often considered that the rate of photosynthetic electron transport limits photosynthesis, except when CO_2 is increased to a level that saturates RuBP carboxylase, and the light intensity during growth is high (Kobza and Edwards 1987a; Sukenik et al. 1987). The rate of diffusion of plastoquinone within the photosynthetic membranes, as it moves reducing equivalents from the photosystem II complex to the cytochrome f/b_6 complex, contributes to the rate limitation under such conditions. More often, however, it appears that reduced electron transport capacity leads to a lower level of activation of the RuBP carboxylase, so that this enzyme once more bears a major burden in determining the photosynthetic rate (Taylor and Terry 1986; Vu et al. 1987; Mächler et al. 1988). Thus a reasonable balance between electron transport and RuBP carboxylase activity is maintained (Brooks 1986; Heber et al. 1988; Terashima and Evans 1988).

The rate at which phosphate is cycled between sugar-P and P_i is receiving new emphasis as a major determinant of photosynthetic rate. Following observations that P_i supply can limit photosynthesis (Foyer and Spencer 1986; Sivak and Walker 1986), an investigation by Stitt (1986) demonstrated that in fact photosynthesis in the chloroplast is ultimately limited by the capacity for cytosolic sucrose synthesis and return of the released P_i to the chloroplast (Fig.1) – the activity of the phosphate translocator was subsequently identified as being a kinetically rate-limiting step (Gerhardt et al. 1987).

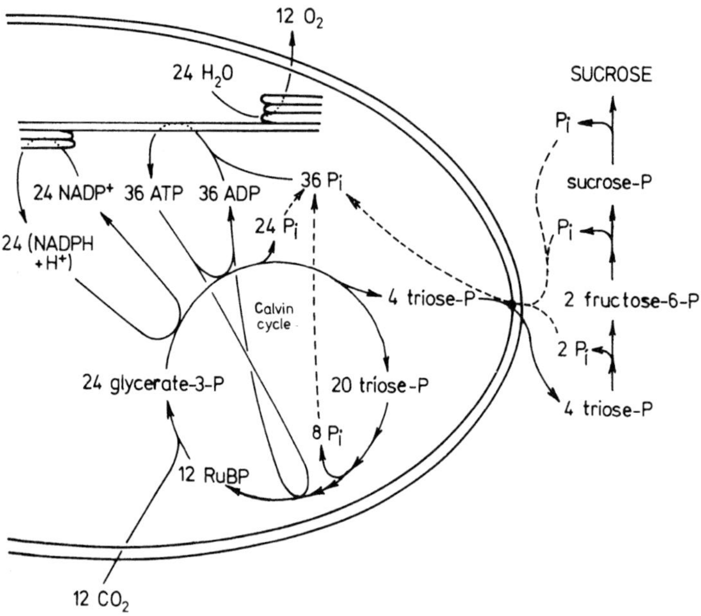

Fig.1. The synthesis of one molecule of sucrose requires the fixation of 12 CO_2 and utilization of 24 NADPH and 36 ATP. During steady-state photosynthesis, the regeneration of the 36 ATP will require 36 P_i. Thirty-two of these are released in the chloroplast by Calvin cycle reactions, but the remaining four must be returned from the cytosol after their release by enzymes of the triose-P → sucrose sequence. Stitt (1986) has obtained evidence that the rate of return of these four P_i is so slow that it becomes the ultimate determinant of the rate of photosynthesis

Certainly the dependence of chloroplast CO_2 fixation on external P_i has been well demonstrated (Furbank et al. 1987; Robinson and Giersch 1987). There are indications that this dependence is manifested more strongly at lower temperatures (Leegood and Furbank 1986; Schnyder et al. 1986; Kobza and Edwards 1987b; Sage and Sharkey 1987), perhaps because low temperatures impede the transport properties of the chloroplast envelope (Fu and Gibbs 1987).

The utilization of triose-P exported from chloroplasts to the cytosol may be the rate-limiting step that determines the photosynthetic *capacity* of leaves, ie at light and CO_2 saturation (Stitt 1986). Capacities of 300 to 400 μmol CO_2 fixed/mg chlorophyll/h are usually measured (Dietz 1986; Stitt 1986; Chow and Anderson 1987; Kobza and Edwards 1987a; Heber et al. 1988; Terashima and Evans 1988), although values exceeding 700 are obtained from the figures presented by Begonia et al. (1987) for sunflowers, and assuming a leaf chlorophyll content of 490 mg/m^2 (Björkman 1981). However, rather high temperatures may have contributed to the attainment of these high rates. On the other hand, values exceeding 700 are not uncommon for marine microalgae; the reason why these microalgae possess photosynthetic capacities apparently greater than those of land plant leaves is not yet known (Kelly 1989), although rate limitation by the sucrose-synthesis pathway need not be considered if these microalgae, like *Chlamydomonas reinhardtii* (Klein 1987), do not form significant amounts of this sugar.

6. The Other Calvin Cycle Enzymes

Calvin cycle enzymes other than RuBP carboxylase have received intermittent attention, or none at all. Wolosiuk et al. (1986) found that the chloroplast's glyceraldehyde-3-P dehydrogenase activity is improved by the presence of organic solvents. This enzyme appears to play an important part during flowering (Bonzon et al. 1987). One of its counterparts in the cytosol, the nonphosphorylating glyceraldehyde-3-P dehydrogenase, has now been obtained in pure form (Iglesias et al. 1987; Iglesias and Losada 1988). In maize leaves this enzyme is restricted to the mesophyll (Valenti et al. 1986).

Separation of chloroplast aldolase from its slightly different cytosolic counterpart has been achieved with four plant species (Schnarrenberger and Krüger 1986; Jacobshagen and Schnarrenberger 1988). Once again, maize is unusual in that the cytosolic aldolase is missing in the bundle sheath, except in dark-grown leaves (Valenti et al. 1987). Ribulose-5-P kinase has been examined further, and the critical sulfydryl group at the active site identified as belonging to cysteine-16 (Porter et al. 1986; Krieger et al. 1987).

Aspects relating to the light-mediated activation of chloroplast FBPase (a process that requires both enzyme reduction and increases in stromal pH and Mg^{2+} concentration) have been addressed in several laboratories. Increased Mg^{2+} concentrations enhance association of the enzyme with thylakoid membranes (Rodríguez Andrés et al. 1987). This may facilitate its reduction via ferredoxin and thioredoxin; the latter protein binds to the enzyme vary rapidly (Soulié et al. 1987). A partially synthetic mini-electron-transport chain for light-mediated activation (in the absence of chloroplasts!) has been developed by Maheshwari et al. (1988). Returning to the normal chloroplast situation: the reduced FBPase gains activity following light-induced alkalisation of the stroma (Oja et al. 1986) because the increased pH induces a conformational change that favors activity (this seems to be the slowest step of the activation – Pradel and Aragnol 1986), and because the pH determines the concentration of the true substrate $MgFBP^{2-}$, and possibly also of an inhibitor, thought to be FBP^{3-} (Gardemann et al. 1986). Finally, a word about specificity. Past research tended to establish that FBP hydrolysis is catalyzed by a specific FBPase, and that a separate enzyme catalyzes hydrolysis of sedoheptulose-1,7-P_2 (see our previous reviews; Cadet et al. 1987a), but this conclusion is less secure, since the identification of blue-green algal (Gerbling et al. 1986) and spinach (Cadet et al. 1987b) enzymes able to hydrolyze both of these sugar bisphosphates.

Enzymologists are continually concerned by the nonphysiological environment to which enzymes necessarily are exposed during studies. It is not surprising then, that thoughts concerning the environment of Calvin cycle enzymes in the chloroplast stroma have led to experiments indicating that enzymes catalyzing consecutive reactions may be associated in some way, and that the product of one enzyme may be passed on directly to the enzyme that catalyzes the next reaction, i.e., metabolite channeling (Anderson 1987; Macioszek and Anderson 1987; Marques et al. 1987). These ideas are supported by the detection of loose multienzyme complexes from *Scenedesmus obliquus* (Nicholson et al. 1987) and from spinach leaves (Sainis and Harris 1986; Gontero et al. 1988); in the latter case a functional complex of the five enzymes from P-ribose isomerase to glyceraldehyde-3-P dehydrogenase was obtained.

7. Light-Mediated Enzyme Regulation

There is little more to add to the most thoroughly investigated facet of light-mediated enzyme regulation, viz. the reductive activation (or inactivation) that involves transfer of reducing equivalents from the photosynthetic electron transport chain's ferredoxin, through thioredoxin, to the enzyme. A ferredoxin/thioredoxin reductase that catalyzes the first step in this transfer has now been characterized; it is synonymous with the earlier reported "ferralterin" (Droux et al. 1987). The nature of the earlier reported "protein modulase" has also been clarified: it is a complex of the three small proteins ferredoxin, thioredoxin, and ferredoxin/thioredoxin reductase (Ford et al. 1987). The diversity of thioredoxin itself has been further illustrated by the report of a relatively larger form of this molecule in the green alga *Scenedesmus obliquus* (Langlotz et al. 1986; Langlotz and Follmann 1987).

Besides Calvin cycle enzymes (mainly RuBP carboxylase and FBPase), three other enzymes have been investigated with respect to light-mediated effects. Thioredoxin-mediated reductive activation of the chloroplast NADP-linked malate dehydrogenase (Miginiac-Maslow et al. 1988) is permitted only when the NADPH to $NADP^+$ ratio is so high that the export of reducing equivalents is feasible – the enzyme is centrally involved in this process (Scheibe et al. 1986; Scheibe 1987). A fascinating observation with this enzyme is that mild proteolytic truncation leads to activity with the oxidized (usually inactive) enzyme, and to a sixfold increase in the activity of the reduced enzyme (Fickenscher and Scheibe 1988). A second enzyme with activity influenced by illumination is the chloroplast coupling factor; thioredoxin-mediated reduction appears to be involved with this enzyme, too (Quick and Mills 1986; Noctor and Mills 1987; Selman-Reimer and Selman 1988). Finally, a different type of light-mediated regulation controls the cytosolic PEP carboxylase in C_4 plants and CAM plants. Protein phosphorylation results in increased activity. An odd (yet appropriate) aspect of this regulation is that light leads to activation (phosphorylation) in C_4 plant leaves, but to inactivation (dephosphorylation) in CAM plants (Brulfert et al 1986; Nimmo et al. 1987; Jiao and Chollet 1988). It will be a challnge to discover how the mechanism behind this regulatory phenomenon operates in opposite directions between these two groups of plants.

Our understanding of the role of Ca^{2+} in light-mediated regulation, and in the regulation of metabolism in general, is increasing (see Kauss 1987). The earlier demonstration of a light-induced uptake of Ca^{2+} by chloroplasts has been supplemented by a similar demonstration with protoplasts (Heimann et al. 1987). Within the chloroplast, several proteins, including ferredoxin, bind Ca^{2+} (Kreimer et al. 1987; Surek et al. 1987). The remaining unbound (free) Ca^{2+} enhances the light-mediated activation of FBPase (Kreimer et al. 1988). This observation may have implications for metabolic regulation beyond the chloroplast, in that Ca^{2+}, together with inositol-1,4,5-P_3 and diacylglycerol, from an important messenger system capable of transducing external signals into intracellular metabolic events (Heim and Wagner 1986; Reddy et al. 1987; Schumaker and Sze 1987; McMurray and Irvine 1988).

8. Starch Metabolism

The essentiality of the enzyme ADP-glucose pyrophosphorylase for starch synthesis has been emphasized by the acquisition of a starchless mutant of *Arabidopsis thaliana* lacking the enzyme (Lin et al. 1988a). This report is reminiscent of the normal lack of starch in *Allium* leaves, which also lack ADP-glucose pyrophosphorylase activity (Sakai 1974). A mutant of *Clarkia xantiana* with a low activity of chloroplast P-hexose isomerase exhibited reduced starch synthesis, pointing to the significance of this enzyme in the starch biosynthetic pathway (Jones et al. 1986).

Chloroplast starch degradation is initiated by either phosphorylase or endoamylase; the relative importance of each of these enzymes appears to depend on the plant speccies (Kakefuda et al. 1986; Goyal et al. 1987; Lin et al. 1988b). In leaves the ultimate product (in the chloroplast) is triose-P and/or glycerate-3-P which is exported to the cytosol; in *Chlamydomonas reinhardtii* glycerol-3-P is a major product (Klöck and Kruzberg 1987). Work with this green alga has also provided additional evidence that starch synthesis and degradation may occur simultaneously in the light (Klein 1987).

A continuing enigma is that separate forms of phosphorylase and endoamylase are found outside the chloroplast (Kakefuda et al. 1986; Steup and Schächtele 1986; Steup et al. 1986; Ziegler 1988). Resolution of this enigma may be assisted by the finding that, in the leaves of the C_4 plant maize, chloroplast phosphorylase appears to be restricted to the bundle sheath, while the extra-chloroplast phosphorylase is confined to the mesophyll (Mateyka and Schnarrenberger 1988).

9. Sucrose Metabolism: Fructose-2,6-P_2, and a Role of PP_i?

The regulatory metabolite fructose-2,6-P_2 has received an almost excessive amount of attention since its discovery 8 years ago. Its primary action in stimulating the activity of cytosolic FBPase (involved in sucrose synthesis) and in lowering the activity of PP_i-linked PFK (somehow involved in cytosolic carbohydrate metabolism) has been the subject of several reviews (including Black et al. 1987; Stitt et al. 1987). Evidence for its regulatory role has been confirmed for several crop species (Sicher et al. 1987). Details of the regulatory interactions involved remain as shown in Fig.3 of our preceding review. In addition, diurnal changes in the ratio of the enzyme activities synthesizing and degrading fructose-2,6-P_2 (Stitt et al. 1986) should be incorporated. These enzyme activities (PFK-2 and FBPase-2) are carried by the same protein molecule in mammalian tissues, thus the report by MacDonald et al. 1987 that in plant leaves the activities reside on different proteins came as something of a surprise, especially since Larondelle et al. 1986 did not report any separation of the spinach leaf activities. However, these authors, and also Walker and Huber (1987a,b) report that spinach leaves contain two forms of the bifunctional PFK-2/FBPase-2, one of which is activated by MgATP. In addition, in support of MacDonald et al. (1987), Kretschmer et al. (1987) found that PFK-2 and FBPase-2 were separable in yeast.

Besides the detection of diurnal changes in the activities of these two enzymes, diurnal rhythms in the activity of sucrose-P synthase have also been observed and credited with a controlling influence on sucrose synthesis (Kerr and Huber 1987; Usuda et al. 1987; Stitt et al. 1988). The mechanism behind the rhythm may involve some form of light-mediated regulation and/or the occurrence of two forms of this enzyme in leaves (Kalt-Torres et al. 1987a; Kerr et al. 1987; Köhler et al. 1988).

Research relating to sucrose breakdown in plant tissues necessarily involves a consideration of glycolysis. It has therefore been influenced by the radical alteration to our perception of plant glycolysis, and to its regulation, caused by the discovery some 9 years ago of an enzyme that catalyzes a freely reversible interconversion between fructose-6-P and FBP, viz. the PP_i-linked PFK (Carnal and Black 1979). With this discovery, old concepts concerning control of the opposing pathways of glycolysis and gluconeogenesis have had to be greatly modified, if not totally abandoned. The problem has been to find a replacement control system, and also to explain why plant cells need this extra enzyme when the traditional couple (ATP-linked PFK and FBPase) seemed perfectly adequate.

These problems have yet to be solved. Certainly the PP_i-linked PFK activity is regulated by fructose-2,6-P_2, as noted above, but both the forward and reverse reactions are similarly affected (Stitt and Vasella 1988). In the leaf cytosol, an appreciable concentration of PP_i exists (alkaline pyrophosphatase is restricted to the chloroplast), thus the PP_i-linked PFK reaction could be at equilibrium in vivo (Weiner et al. 1987). Given that the suspected presence of an ATP-linked PFK in the cytosol has been confirmed (Häusler et al. 1987a), and that FBPase is also present, the control of glycolysis in this compartment of the leaf cell is something of a mystery.

However, in mature, actively photosynthesizing leaves the major flow of carbon is from CO_2 to sucrose which is exported to sink tissues, and it is these sink tissues that have provided two new insights into the role of PP_i-linked PFK. Firstly, the relative activities of ATP-linked PFK and PP_i-linked PFK have been examined in several plant species, and in different tissues of the same species, and a trend was identified: the acitvity of ATP-linked PFK was more or less the same in all plant tissues, but the PP_i-linked PFK activity varied greatly, being highest in "sink" tissues expected to be importing and utilizing sucrose for growth, e.g., very young leaves, and roots (apRees et al. 1985; Xu et al. 1986; Black et al. 1987). This observation supported the second insight (illustrated in Fig.2): that the role of PP_i-linked PFK in these tissues might be to generate PP_i for the UDP-glucose pyrophosphorylase reaction, thereby permitting the metabolism of UDP-glucose and, in turn, allowing imported sucrose to be cleaved by sucrose synthase, which generates UDP-glucose as one of the products (apRees et al. 1985; Huber and Akazawa 1986; Morrell and apRees 1986). For hexose-P production from sucrose, this pathway is energetically twice as efficient as that in which the initial sucrose cleavage is catalyzed by invertase (Black et al. 1987).

Despite the appeal of this role for PP_i-linked PFK in sucrose-utilizing tissues, it would lose all value if the report from Salerno (1986), that developing maize endosperm contains UDP-glucose phosphorylase (catalyzing UDP-glucose + $P_i \rightarrow$ glucose-1-P + UDP), could be confirmed and extended to other tissues. Such an enzyme would short-circuit the scheme in Fig.2, and delete the requirement for PP_i-linked PFK. Finally, the presence of PP_i-linked PFK acctivity in mature leaves remains difficult to justify. Could it be that it is unnecessary, and that the leaf is unable to completely suppress synthesis of the enzyme as it attains maturity? Once again, C_4 leaves may provide some clue: maize leaf bundle sheath cells do lack the enzyme (Kruger et al. 1986).

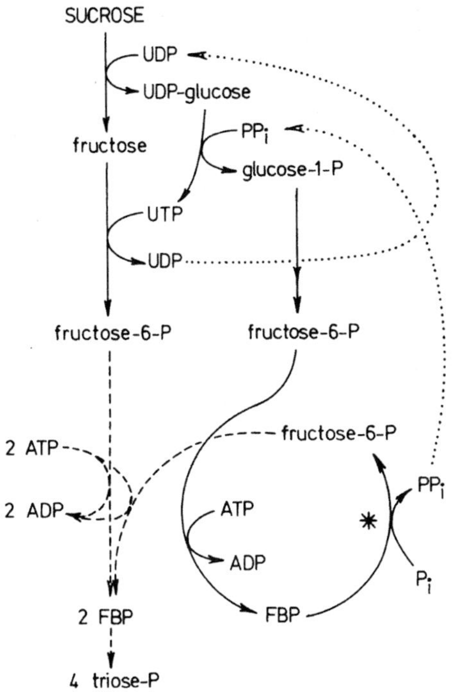

Fig. 2. Utilization of PP_i-linked PFK (reaction *) in an energy efficient pathway for the conversion of sucrose to triose-P. *Dotted arrows* show recycling of UDP and PP_i. The sum of the reactions is: sucrose + 3 ATP + P_i → 4 triose-P + 3 ADP. When hexose-P is the required end-product, then only one ATP is needed for the conversion of one sucrose to two hexose-P (omit reactions indicated by *dashed arrows*). Based upon apRees et al. (1985); Huber and Akazawa (1986), and Black et al. (1987)

10. Mitochondrial Respiration in Illuminated Photosynthetic Cells?

After years of procrastination upon the question of whether or not mitochondrial respiration occurs during photosynthesis, researchers have recently obtained several lines of evidence that it does indeed occur. Using the green microalga *Chlamydomonas reinhardtii*, Sueltemeyer et al. (1986) concluded that respiration was present, although of minor importance, in the light, but Husic and Tolbert (1987) came to regard respiration as being of greater importance in the light, since its inhibition led to the accumulation of intermediates of the tricarboxylic acid cycle. From studies with another green alga, Weger et al. (1988) concluded that both components of respiration, viz. the tricarboxylic acid cycle, and mitochondrial electron transport, occur in the light. Addition of NH_4^+ to cultures stimulated the former 2.5-fold, reflecting the role of this cycle in generating carbon skeletons for amino acid biosynthesis. A similar requirement could be envisaged in leaves, and indeed McCashin et al. (1988) recently reported that the rate of the tricarboxylic acid cycle in illuminated wheat leaves was 80% of that in darkened leaves.

In the past it has been argued that the other component of respiration, i.e., mitochondrial electron transport, would be unnecessary during photosynthesis because ATP generated from the coupled oxidative phosphorylation would be in excess of the leaf's requirements, given that ATP is already supplied by photosynthetic phosphorylation. This argument has been weakened by the experiments and calculations of Krömer et al. (1988). These authors found that oligomycin (an inhibitor of mitochondrial electron transport) inhibited the photosynthesis of barley leaf protoplasts, but not that of the chloroplasts isolated from these protoplasts. They concluded that, in the cell, chloroplast photosynthesis is dependent on simultaneous mitochondrial electron transport – this dependence might entail the export of photosynthetically generated reducing equivalents from chloroplasts, via an oxaloacetate/malate shuttle, to the mitochondria where their oxidation is coupled to the synthesis of ATP which is required in the cytosol for utilisation of the products of CO_2 fixation. This concept is illustrated in Fig.3. The value of the proposal is that production of this extra ATP is more energetically efficient: one ATP is produced in the mitochondrion per 1.05 photons of light harvested in the chloroplast, compared to one ATP per 3.08 photons when the ATP synthesis is directly coupled to photosynthetic electron transport (either cyclic or pseudocyclic) in the chloroplast. In addition, the system could be under the control of the enzyme reducing oxaloacetate in the chloroplast (NADP-linked malate dehydrogenase) in that this enzyme will be active only when the stromal NADPH/$NADP^+$ ratio is sufficiently high (Scheibe et al. 1986). Separate transporters for moving the dicarboxylates in and out of the mitochondria have recently been identified (Zoglowek et al. 1988).

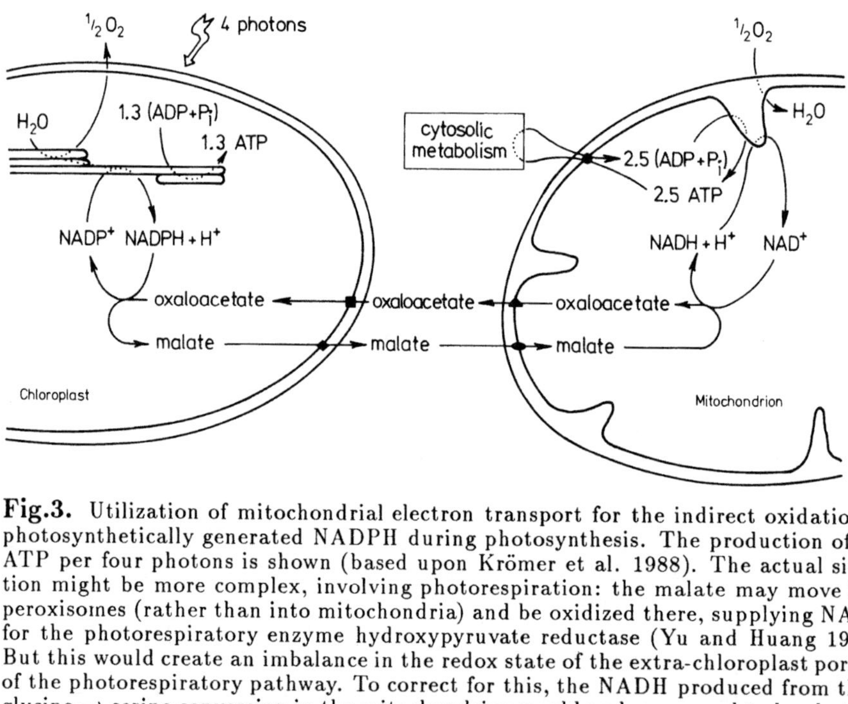

Fig.3. Utilization of mitochondrial electron transport for the indirect oxidation of photosynthetically generated NADPH during photosynthesis. The production of 3.8 ATP per four photons is shown (based upon Krömer et al. 1988). The actual situation might be more complex, involving photorespiration: the malate may move into peroxisomes (rather than into mitochondria) and be oxidized there, supplying NADH for the photorespiratory enzyme hydroxypyruvate reductase (Yu and Huang 1986). But this would create an imbalance in the redox state of the extra-chloroplast portion of the photorespiratory pathway. To correct for this, the NADH produced from the 2 glycine → serine conversion in the mitochondrion would no longer need to be shuttled to the peroxisome to supply this reductase; instead, it would presumably be oxidized by mitochondrial electron transport (Dry et al. 1987). The end result would be the same: 3.8 ATP per four photons.

Another somewhat related role for mitochondrial electron transport during photosynthesis is to accommodate the oxidation of at least some of the NADH produced during photorespiratory glycine oxidation in this organelle; the capacities of the dicarboxylate shuttles that can indirectly export this NADH appear to be too low to export all of it (Dry et al. 1987; see legend to Fig.3).

It remains to be determined whether the alternative, cyanide-resistant pathway of respiratory electron transport (Møller et al. 1988) is required by photosynthesizing cells. Musgrave et al. (1986) found that a pea hybrid lacking this pathway grew better when the CO_2 level in air was doubled, and thereby concluded that the pathway is energetically wasteful.

A third, least understood form of dark respiration, viz. chlororespiration, in which electrons are passed on to O_2 via chloroplast electron carriers, does seem to be inhibited in the light (Peltier et al. 1987); this is expected since the same electron carriers move electrons, presumably in the opposite direction, during photosynthesis. The light inhibition of chlororespiration is now thought to be the basis of the Kok effect (whereby the rate of O_2 uptake by photsynthetic cells is abruptly reduced when irradiance reaches about 15 μEinsteins/m^2/s) (Peltier and Sarrey 1988). Previously, this effect had been attributed to inhibition by light of the O_2 uptake associated with mitochondrial respiration.

11. Photorespiration

Measurements of photorespiration by a new procedure, in which any refixation of photorespired CO_2 is accounted for, show that underestimates result when this factor is ignored, and that photorespiratory CO_2 release is normally 30-50 % of the net photosynthesis in C_3 plants (Gerbaud and Andre 1987). The extent of refixation appears to be greater in C_3-C_4 intermediate species (Rawsthorne et al. 1988), thereby contributing to their C_4-like character. An interesting observation is that, in C_3 plants, a temperature decrease of 8°C or more causes a sudden evolution of CO_2, apparently from photorespiration (Tarłowski et al. 1986).

Hopes persist that plant productivity might be increased by decreasing or preventing photorespiration through biochemical means (Pierce 1988). Success in this direction is reported by VanAssche et al. (1987), who found that 0-p-nitrophenylhydroxylamine interacts with RuBP carboxylase/oxygenase, causing a conformational change that increases the ratio of carboxylase to oxygenase activities. When the chemical was applied to intact soybean leaves, CO_2 fixation was found to increase by about 50%.

Since CO_2 and O_2 are direct competitors for the enzyme's active site, increases in the intracellular CO_2 concentration (as occurs naturally in C_4 plants, and in aquatic plants that accummulate inorganic carbon – see Sects. 2 and 13) are expected to prevent photorespiration. However, this situation may not be so straightforward, at least for microalgae: Yokota and Kitaoka (1987) and Yokota et al. (1987) obtained evidence that when the concentration of CO_2 is raised, photosynthesis responds with an increase in rate which, of course, includes an increased rate of O_2 evolution. Consequently, the concentration of this gas in the vicinity of RuBP carboxylase/oxygenase may also be raised to a level that counteracts the rise in CO_2 concentration and once more permits photorespiration to occur.

While RuBP oxygenation is considered to be the principal source of glycolate, evidence for alternative sources is still being obtained. The latest comes from Zelitch (1988) and involves the formation of glycolate from pyruvate via isocitrate and glyoxylate. A combination of tricarboxylic acid cycle enzymes, isocitrate lyase, and glyoxylate reductase could achieve the overall conversion of pyruvate $\rightarrow$ CO_2

+ glycolate; one NADH and one $FADH_2$ would be generated in the mitochondrion. The pathway might involve the new, more active leaf glyoxylate reductase discovered by Kleczkowski et al. (1986), and might help to explain how the fraction of glycolate carbon photorespired as CO_2 can vary (it is greater at higher temperatures), while the fraction of RuBP oxygenated remains constant (Hanson and Peterson 1986). Incidentally, Kleczkowski and Randall (1988a) have also discovered a new form of the photorespiratory enzyme hydroxypyruvate reductase that prefers NADPH and, oddly enough, is not in the peroxisomes.

The conventional source of photorespiratory CO_2 is the reaction 2 glycine + $NAD^+ \rightarrow$ serine + NADH + H^+ + NH_4^+ + CO_2 in the mitochondrion. The multienzyme complex that contains the two activities that effect the overall reaction has now been isolated (Neuburger et al. 1986). The capacity of these two activities to accommodate observed rates of photorespiratory CO_2 release has been confirmed to be adequate (Walton and Woolhouse 1986; Lilley et al. 1987). However, as mentioned in the preceding section, regeneration of NAD^+ at the necessary rate would require that some NADH be oxidized through mitochondrial electron transport, since shuttle activities are inadequate to permit all NADH to be oxidized outside the mitochondrion at the required rate (Dry et al. 1987).

Cycling of the released NH_4^+ must also occur (see Fig.6 in our 1982 review). Briefly, it involves reassimilation by the GS/GOGAT (Avila et al. 1987) enzyme couple in the chloroplast, and export of the produced glutamate to peroxisomes for transamination of glyoxylate to glycine. The chloroplast-envelope system for exporting glutamate, and importing the α-ketoglutarate carbon skeleton, has now been shown to consist of two separate translocators, each using malate as the counter-metabolite (Woo et al. 1987). Further studies with mutant plants have confirmed the importance of this nitrogen cycling system (Murray et al. 1987; Wallsgrove et al. 1987), although thoughts that some NH_4^+ may be reassimilated directly in the mitochondrion by glutamate dehydrogenase still exist (Yamaya and Oaks 1987; but see Morris et al. 1988).

12. The Many Roles of O_2 in Photosynthetic Carbon Metabolism

The photosynthetic cleavage of water to O_2 (and reducing equivalents) is the only source of *net* O_2 production by plants, and it is to all intents and purposes the only means of O_2 production by the biosphere. This fundamental significance of water photolysis sometimes overshadows the many other reactions in photosynthetic cells that involve O_2 release or O_2 uptake. Release of O_2 also occurs when H_2O_2 is decomposed by catalase in peroxisomes – the H_2O_2 is generated by glycolate oxidase during photorespiration. The O_2 release is, of course, more than compensated for by the O_2 uptake of the oxidase. Other O_2-uptake processes of importance are mitochondrial respiration, chlororespiration, and the RuBP oxygenase reaction. These have been dealt with above (Sects. 10 and 11) and will not be further considered here.

A fourth major O_2 uptake reaction is that in which O_2 acts as an electron acceptor in photosynthesis, thus permitting formation of ATP coupled to photosynthetic electron transport, in the absence of $NADP^+$ reduction. Although this method of generating ATP is not as energetically efficient as the alternative involving mitochondrial respiration (see Sect. 10), it may in any event occur simply because of the inevitable presence of O_2 in photosynthesizing chloroplasts and the relatively reactive nature of this gas. Robinson (1988) states that rates of 10 to 40 μmol O_2 reduced/mg chlorophyll/h normally occur. ATP from the coupled photophosphorylation might be used by the Calvin cycle (which requires 3 ATP and 2 NADPH per CO_2 fixed, whereas linear electron transport to $NADP^+$ is understood to produce only 2.67 ATP along with each 2 NADPH), and also for other ATP-demanding processes, such as driving membrane transporters, e.g., inorganic carbon uptake by *Chlamydomonas reinhardtii* requires ATP (Sueltemeyer et al. 1986; Yokota et al. 1987). In the latter study it was calculated that this process would theoretically consume 43% of the photosynthetically generated ATP.

The product of O_2 photoreduction is the superoxide anion, which is dismutated to H_2O_2 by superoxide dismutase. Since H_2O_2 and the hydroxyl radical derived from it (Upham and Jahnke 1986; Morehouse and Mason 1988), are both sufficiently reactive to cause severe and indiscriminate damage to cell structures (Asada and Takahashi 1987; Husain et al. 1987; Ogawa et al. 1987), no positive value is attributed to them, and attention is restricted to metabolic processes that scavenge the H_2O_2. That is, with the exception of an unusual suggestion that H_2O_2 may be a naturally occurring electron donor for photosynthesis (Johansen 1988).

The favored system for H_2O_2 scavenging in leaves is that comprising ascorbate peroxidase, dehydroascorbate reductase, and glutathione reductase (see Fig.7 in our 1982 review). Evidence that this system operates in cyanobacteria has recently been obtained (Tel-Or et al. 1986). The activity of ascorbate peroxidase in leaves is greater when light intensities are higher (Gillham and Dodge 1987). In C_4 plant leaves the enzyme is present in both mesophyll and bundle-sheath cells (Nakano and Edwards 1987). While this support for a role of ascorbate peroxidase becomes greater, the roles for the other two enzymes are in question, since it appears that in vivo the peroxidase may release predominantly monodehydroascorbate which is converted back to ascorbate by an NAD(P)H-linked reductase (Hossain et al. 1984; Borraccino et al. 1986). In addition, the monodehydroascorbate is able to reactivate any ascorbate peroxidase inactivated in an ascorbate-depleted environment (Nakano and Asada 1987).

In mammalian tissues, H_2O_2 is scavenged by a different peroxidase, the selenoprotein glutathione peroxidase. Although it is generally thought that this enzyme is absent from plants, evidence that it is present in the diatom *Thalassiosira pseudonana* and in the green alga *Chlamydomonas reinhardtii* has been obtained (Price and Harrison 1988; Yokota et al. 1988). In the latter case, activity was induced by growing cells in a medium containing selenium at a level 1000 times that encountered naturally in water. Nevertheless, there is no doubt that certain microalgae benefit from the inclusion of much lower levels of selenium in growth media (Lindström and Rodhe 1978; Price et al. 1987). It will be intriguing to further examine the extent to which this lesser-known micronutrient is involved in the protection of photosynthetic cells against their toxic by-product, O_2.

13. C_4 Plants

Textbooks would have it that the pathways of carbon in C_4 plants are known. This is economy of the truth! The mechanisms of, and the capacities for, inter- and intracellular transport are, in general, poorly characterized. The simplified schemes usually drawn for PEP carboxykinase C_4 plants do not agree fully with the ^{14}C labeling patterns observed. Since our last review many of these uncertainties have been under concerted attack. Progress has included (1) the determination of the maximum size of compounds which can diffuse through the plasmodesmata that connect mesophyll and bundle-sheath cells (Burnell 1988), (2) the re-examination of the carbon pathway, enzyme complement and the inter- and intra-cellular transport of metabolites in PEP carboxykinaseC_4 plants, (3) characterization of the mechanism of pyruvate uptake by mesophyll chloroplasts, (4) determination of the mechanism by which inorganic carbon species are concentrated within the bundle-sheath cells, (5) characterization of a number of modulated enzymes, in particular PEP carboxylase, and (6) attempts to match various physiological models to the photosynthetic responses of C_3-C_4 intermediates.

a) Transport of Pyruvate

During photosynthesis in maize, the concentration of pyruvate in the mesophyll cells is higher than in the bundle-sheath cells (Leegood 1985; Stitt and Heldt 1985). Such a gradient is inconsistent with the widely held view that pyruvate diffuses along a concentration gradient from the bundle-sheath to the mesophyll cells. This apparent anomaly would be resolved if most of the pyruvate inside the mesophyll cells was compartmented inside the chloroplasts. Although the facilitated transport of pyruvate into chloroplasts from *Digitaria sanguinalis*, a NADP malic enzyme species, was demonstrated by Huber and Edwards in 1977, the concentration of pyruvate in the chloroplasts never exceeded that of the medium. Flügge et al. (1985) demonstrated that pyruvate can accumulate in maize chloroplasts if they are illuminated. Ohnishi and Kanai (1987b,c) have shown that in chloroplasts isolated from the NAD malic enzyme species *Panicum miliaceum*, the uptake of pyruvate is stimulated by light in the mesophyll but not the bundle-sheath chloroplasts.

Subsequent studies with *Panicum* provided the exciting observation that a Na^+ gradient could mimic the light-stimulated pyruvate uptake (Ohnishi and Kanai 1987a). Other mono- and divalent cations were ineffectual. It was proposed that (1) light induces alkalization in the stroma (2) a pH gradient or protonmotive force across the envelope forms a Na^+ gradient; possibly through the auspices of a Na^+/H^+ antiport, and (3) pyruvate is taken up by a Na^+/pyruvate$^-$ symport. If this mechanism is universal, and this has yet to be shown, then a reason for the mysterious Na micronutrient requirement of C_4 plants so long studied by Brownell and his colleagues may have been found.

The concept that all C_4 plants contain overwhelmingly a single decarboxylase is untrue. Burnell and Hatch (1988a,b) have shown in the bundle-sheath cells of the PEP carboxykinase plant *Urochloa panicoides* that both cytosolic PEP carboxykinase and mitochondrial NAD malic enzyme are involved in the decarboxylation of C_4 acids. The mitochondria provide the bulk of the ATP required by the PEP carboxykinase (Hatch et al. 1988). The presence of NAD malic enzyme catalyzed decarboxylation would not only explain the presence of pyruvate Pi dikinase in PEP carboxykinase C_4 species, but might also explain why these species exhibit a requirement for Na.

b) Transport of PEP

Mesophyll chloroplasts from C_4 plants transport PEP, Pi, glycerate-3-P and triose phosphates across the inner envelope membrane. Are these compounds transported by the same transporter? Rumpho et al. (1987, 1988) are of the opinion that they are. Nevertheless, the C_4 transporter is not identical to the transporter from C_3 plants. PEP and 2-phosphoglycolate inhibit glycerate-3-P dependent O_2 evolution by C_4 mesophyll chloroplasts but not C_3 chloroplasts. Phosphohydroxypyruvate inhibits O_2 evolution in both types. This suggests that the C_4 transporter recognizes certain organic phosphates with phosphate in the C-2 or C-3 positions but that the C_3 transporter requires the phosphate group to be in the C-3 position.

c) Transport of Malate

Surprisingly, the observation by Boag and Jenkins (1985) that the uptake of malate into bundle-sheath chloroplasts is stimulated by aspartate and glutamate has not been pursued.

d) Transport of Inorganic Carbon

It is widely accepted that the C_4 syndrome is a mechanism for pumping inorganic carbon into the bundle-sheath cells. There has been little consideration as to whether this carbon is delivered as CO_2 or HCO_3^- and what the ramifications regarding the intracellular ionic balance and the operation of the C_4 pathway might be. This uncertainty has been heightened by the allegation that HCO_3^-, not CO_2, is the inorganic carbon product of NADP malic enzyme in maize bundle-sheath chloroplasts (Asami et al. 1979). It is now clear that CO_2 is the product of the three C_4 acid decarboxylases, NADP malic enzyme, NAD malic enzyme, and PEP carboxykinase (Häusler et al. 1987b; Jenkins et al. 1987). Moreover, bundle-sheath cells from a variety of C_4 plants possess low levels of carbonic anhydrase, and thus a low capacity for converting CO_2 to HCO_3^- (Burnell and Hatch 1988c). At pH 8.0, 80-90% of the cytosolic inorganic carbon will be CO_2, rather than 2% if HCO_3^- and CO_2 are in equilibrium. This predominance of CO_2 is clearly vital for the absence of O_2 effects on the photosynthetic characteristics of C_4 plants (Furbank and Hatch 1987).

e) Enzymes

PEP carboxylase is modulated by light in a number of ways. De novo synthesis of the enzyme requires light. The enzyme from illuminated maize leaves is phosphorylated on one or more serine residues and is less sensitive to feedback inhibition by malate (Budde and Chollet 1986; Nimmo et al. 1987). PEP carboxylase is also modulated by metabolites (Huber and Sugiyama 1986; Doncaster and Leegood 1987). At low levels of PEP the enzyme is strongly stimulated by glucose-6-P, triose-P, alanine, serine, and glycine and is inhibited by organic acids. Activity of the enzyme is low in the presence of a simulated dark cytosol and high in the presence of a simulated illuminated cytosol.

A number of C_4 enzymes have been purified and characterized to varying degrees. These include adenylate kinase (Kleczkowski and Randall 1986), glycerate kinase (Kleczkowski and Randall 1988b,c), and PPi-dependent PFK (Kruger et al. 1983).

f) Control of Carbon Partitioning

Maize leaves export relatively more carbon during the day than do the C_3 species spinach and soybean. Moreover, carbon assimilation, enzyme activity, sucrose synthesis, and carbon export exhibit marked diurnal variations (Kalt-Torres et al. 1987b; Kalt-Torres and Huber 1987; Usuda et al. 1987).

Many C_4 species, when growing under natural conditions, synthesize sucrose in the mesophyll cells and starch in the bundle-sheath chloroplasts. The extent of this compartmentation is species-dependent (Ohsugi and Huber 1987). In maize this spatial separation is supported by the intracellular compartmentation of sucrose-P synthase, starch synthase, branching enzyme and ADP-glucose pyrophosphorylase (Furbank et al. 1985; Spilatro and Preiss 1987). The rates of sucrose formation and assimilate export are closely aligned to the rates of carbon fixation and the activation state of sucrose-P synthase (Huber et al. 1987).

It should be stressed that the patterns of carbon partitioning and regulation in maize may not be typical for all C_4 plants or even for all NADP malic enzyme C_4 plants.

14. C_3-C_4 Intermediates

Edwards and Ku (1987) and Hattersley (1987) have produced comprehensive and interpretive reviews on C_3-C_4 intermediates.

The rules for designating plants as C_3, C_4 (whether NADP-malic enzyme type, NAD-malic enzyme type or PEP carboxykinase type), or C_3-C_4 intermediate are, as is often the fate of dogma, under reassessment. There is also a new variant, the C_4-C_4 intermediate! The latter was suggested for *Bouteloua curtipendula*, which appears to be a PEP carboxykinase/NAD malic enzyme intermediate (Prendergast et al. 1987). The studies of Burnell and Hatch on *Urochloa* referred to earlier indicate that this is not an isolated phenomenon.

The C_4 acid cycle has been examined in detail in maybe only 15 species (Hattersley 1987), yet about 150 species have been typed on the basis of the predominant C_4 acid decarboxylase and many more have been determined on the basis of anatomy, $\delta^{13}C$ values, or compensation point (G). C_3-C_4 intermediates are known or suspected in the five dicotyledonous genera *Alternanthera* (Amaranthaceae), *Flaveria* (Asteraceae), *Mollugo* (Aizoaceae), *Moricandia* (Brassicaceae) and *Parthenium* (Asteraceae) and in the monocotyledonous genera *Neurachne* and *Panicum* (Rajendrudu et al. 1986; Edwards and Ku 1987; Hattersley 1987). It is now clear that it is no longer adequate to designate species as C_4 on the basis of G values alone or on anatomy, as C_3-C_4 intermediates may have G values or anatomical features which approach those of the C_4 condition (Hattersley et al. 1986; Prendergast et al. 1986). Similarly, a species should not be designated C_3 on the basis of $\delta^{13}C$ values alone, as most intermediates exhibit C_3-like $\delta^{13}C$ values.

The unique characteristics which identify C_3-C_4 intermediates are characteristic G responses to photon flux density and biphasic responses of G to O_2. The latter are in contrast to the linear positive responses of C_3 plants and the zero response of C_4 plants.

Although designated "intermediates", the expression of intermediate physiology is probably not due to the same mechanisms in all species. Certainly variety might be expected in plants which are likely to include incipient C_4 species, C_4 reversions and C_3-C_4 hybrids. Different species exhibit different activities and compartmentation of C_4 enzymes and exhibit characteristic O_2 and photon flux density responses of their compensation point values. The photosynthetic responses of intermediates may vary with developmental stage (Moore et al. 1987), growth conditions (Hattersley et al. 1986), or between individuals of a population growing under seemingly identical conditions (Monson et al. 1987).

Edwards and Ku (1987) and Peisker (1986) bravely speculated, principally on the basis of gradations within the genus *Flaveria*, on the evolutionary sequence from C_3 to C_4. The sequence begins with the development of organelles in the bundle-sheath cells followed by enhanced activities of C_4 enzymes and ends with the spatial separation of C_4 and Calvin Cycle processes. Current indications are that an efficient capacity for the reassimilation of photorespiratory CO_2 evolved prior to a CO_2-concentrating mechanism (Bauwe et al. 1987). The examination of hybrids of intermediate and C_4 or C_3 species may help clarify these uncertainties.

The physiology of intermediate species has been suggested to be due to (1) the possession of RuBP carboxylase-oxygenase with altered kinetic properties, (2) the efficient recycling of photorespiratory CO_2 via an unknown non-C_4 cycle, (3) the operation of a limited, cellularly compartmented C_4 acid cycle with the retention of the Calvin cycle in the mesophyll, and (4) the possession of a glycolate and/or glycine shuttle from mesophyll to bundle-sheath cells but no C_4 cycle. Of the four, only option (3) appears to explain the responses of G to O_2 and to photon flux density. However, the activities of C_4 enzymes and ^{14}C labeling patterns in *Panicum milioides*, *Moricandia* spp., and in *Parthenium hysterophorus* are difficult to explain in terms of a limited C_4 cycle (Hunt et al. 1987; Moore et al. 1987).

Hattersley (1987) considers that a plant does not have to be C_4-like biochemically for it to exhibit C_4-like G values, and it is possible that there may be species currently classified as C_4 on the basis of zero G values that may, in reality, be intermediate species. For example, *Flaveria brownii* is a C_3-C_4 species with a C_4-like $\delta^{13}C$ value and which exhibits only a slight O_2 inhibition of net photosynthesis (Cheng et al. 1986; Moore et al. 1987; Monson et al. 1987).

15. CAM Plants

a) Carbon Partitioning

Subsequent to the carboxylation of PEP at night, the transfer of carbon from malate to storage carbohydrate during the light involves gluconeogenesis. Energetic considerations require that in NADP malic enzyme CAM plants 1 in 19 of the malic acid molecules be oxidized in the mitochondria (Edwards et al. 1982). In PEP carboxykinase CAM plants the ratio is 1 in 9. Winter et al. (1986) have shown that mitochondria from CAM *Mesembryanthemum crystallinum*, a NADP malic enzyme CAM plant, have a high capacity for NAD malic enzyme catalyzed decarboxylation of malate, provided the pH of the matrix is less than 7.0.

The gluconeogenic transfer of carbon from pyruvate to carbohydrate requires the conversion of pyruvate to PEP in the chloroplasts. Chloroplasts from CAM *Mesembryanthemum crystallinum* are able to take up pyruvate from the surrounding medium, synthesize PEP and export the PEP (Neuhaus et al. 1988). The conversion of pyruvate to PEP is light-dependent, as one might expect if pyruvate Pi dikinase is involved. The export of PEP requires the presence of Pi, glycerate-3-P or triose-P in the external medium, suggesting that a Pi translocator with a high affinity for PEP is involved.

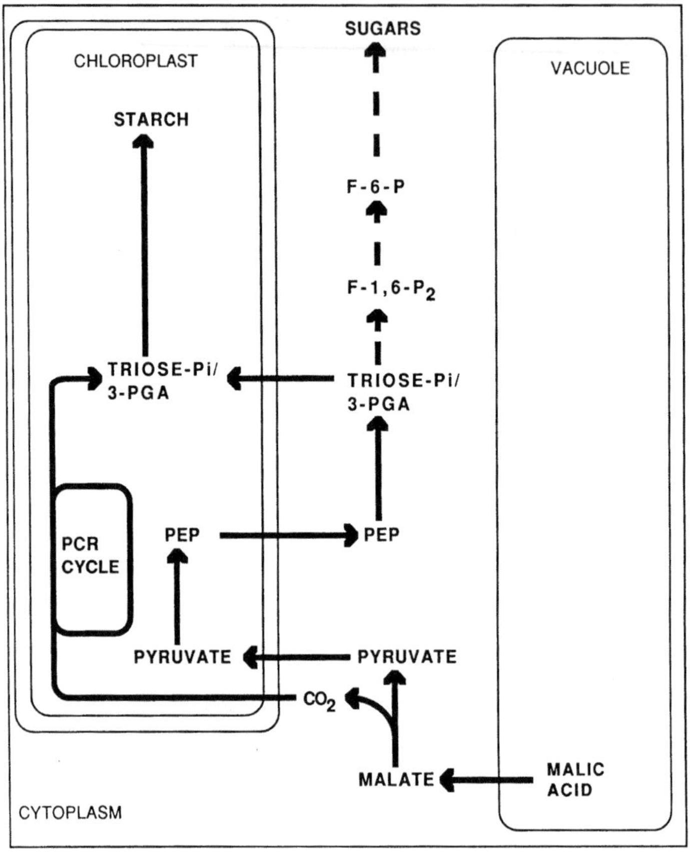

Fig.4A. Pathway postulated for the transfer of carbon from malic acid to storage carbohydrate during the light in CAM plants which use starch as a fluctuating storage carbohydrate. In these plants malic enzyme is the principal decarboxylase, pyruvate Pi dikinase is present, PEP and pyruvate transport across the chloroplast envelope is necessary, the levels of fructose-2,6-P_2 are low during deacidification and the capacity for the metabolism of fructose-1,6-P_2 in the cytoplasm is low

Fahrendorf et al. (1987) and Holtum et al.(1987) have postulated that CAM plants which store most of their fluctuating carbohydrate as starch in the chloroplasts tend to be malic enzyme plants, whilst plants, such as pineapple, which store large amounts of carbon in fluctuating extrachloroplastic carbohydrates tend to be PEP carboxykinase species. Studies of the levels of fructose-2,6-P_2, the transport of PEP and pyruvate across chloroplast envelopes, and the activities of PPi-dependent PFK and cytoplasmic fructose bisphosphatase are consistent with the pathways shown in Fig.4A and B. More species need to be examined to confirm this postulate.

Although all cells in leaves of *Kalanchoe daigremontiana* are chlorophyllous, the accumulation of malic acid and the distribution of starch are not uniform. The rates of CO_2 uptake are two to three fold greater through the lower than the upper surface. The lower half of the leaves accumulate almost three times the titratable acidity than does the upper half yet, when illuminated with a photon flux density of 670 μmol photons/m^2/s, only 5 μmol passed through the leaves (Winter 1987). The chloroplasts in the upper halves of the leaves contained less chlorophyll than those in the lower halves, much as do chloroplasts from sun and shade plants.

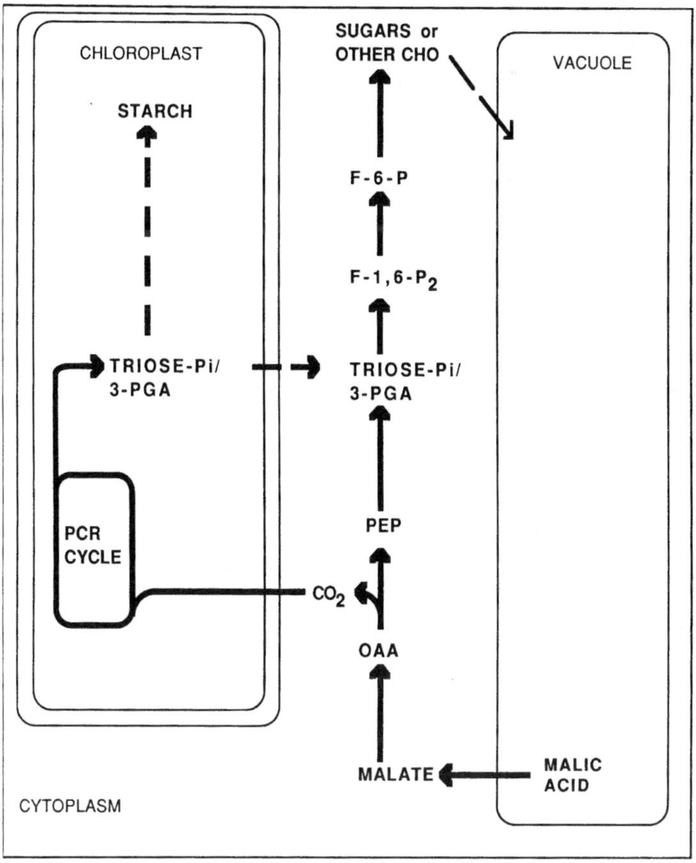

Fig.4B. Pathway postulated for the transfer of carbon from malic acid to storage carbohydrate during the light in CAM plants which possess a fluctuating extrachloroplastic storage carbohydrate pool. In these plants, about which there is a limited amount of information, PEP carboxykinase is the predominant decarboxylase, pyruvate Pi dikinase activity is low, there is little requirement for the transport of pyruvate or PEP across the chloroplast envelope, the levels of fructose-2,6-P_2 are high and the activities of PPi-dependent PFK are high (Holtum et al. 1987)

The development by Kluge et al. (1987) and Brulfert et al. (1987) of a green callus culture from *Kalanchoe blossfeldiana* capable of ^{14}C uptake patterns and malic acid fluctuations typical of CAM leaves, is potentially of great importance in the study of the physiology and biochemical control of CAM. Labeling and metabolite studies have long been hampered by either tightly shut stomates or the inability to kill the succulent tissues quickly enough to prevent the metabolism of small, rapidly turning-over metabolite pools. For regulatory studies the calli need to be completely autotrophic.

b) Malic Acid Accumulation

Evidence that the entry of malate^{2-} into CAM vacuoles is driven by a Mg^{2-}-dependent, electrogenic proton pump continues to accumulate. ATP-dependent uptake of malate has been demonstrated in vacuoles isolated from *Kalanchoe daigremontiana* (Nishida and Tominaga 1987) and in tonoplast vesicles from *Kalanchoe* and from CAM *Mesembryanthemum crystallinum* (Jochem and Lüttge 1987; Struve and Lüttge 1987; Balsamo and Uribe 1988). However, no substrate-specific ATPases were detected in vacuoles and tonoplast vesicles from *Sedum telphium* or *Ananas comosus* (Kenyon and Black 1986). Marquardt and Lüttge (1987) have also demonstrated the presence of a PPi-driven proton pump in tonoplasts from *Kalanchoe daigremontiana*.

References

Abe T, Tsuzuki M, Miyachi S (1987) Plant Cell Physiol 28:867-874 **Adam V, Rustin P** (1986) Physiol Veg 24:663-671 **Anderson LE** (1987) FEBS Lett 212:45-48 **Andrews TJ, Lorimer GH, Pierce J** (1986) J Biol Chem 261:12184-12188 **apRees T, Green JH, Wilson PM** (1985) Biochem J 227:299-304 **Asada K, Takahashi M** (1987) In: Kyle DJ, Osmond CB, Arntzen CJ (eds) Protoinhibition. Elsevier, Amsterdam, pp 227-287 **Asami S, Inoue K, Matsumoto K, Murachi A, Akazawa T** (1979) Arch Biochem Biophys 194:503-510 **Avila C, Botella JR, Cánovas FM, Núñez de Castro I, Valpuesta V** (1987) Plant Physiol 85:1036-1039.

Badger MR, Gallagher A (1987) Aust J Plant Physiol 14:189-201 **Bagge P, Larsson C** (1986) Physiol Plant 68:641-647 **Balsamo RA, Uribe EG** (1988) Planta 173:190-196 **Bauwe H, Keerberg O, Bassüner R, Pärnik T, Bassüner B** (1987) 172:214-218 **Begonia GB, Hesketh JD, Woolley JT, Peters DB** (1987) Photosynthetica 21:584-587 **Berry JA, Lorimer GH, Pierce J, Seemann JR, Meek J, Freas S** (1987) Proc Natl Acad Sci USA 84:734-738 **Björkman O** (1981) Encycl Plant Physiol 12A:57-107 **Black CC, Mustardy L, Sung SS, Kormanik PP, Xu DP, Paz N** (1987) Physiol Plant 69:387-394 **Boag S, Jenkins CLD** (1985) Plant Physiol 79:165-170 **Bonzon M, Simon P, Agosti RD, Greppin H, Wagner E** (1987) Physiol Plant 70:577-582 **Borraccino G, Dipierro S, Arrigoni O** (1986) Planta 167:521-526 **Brechignac F, Lucas WJ** (1987) Plant Physiol 83:163-169 **Brooks A** (1986) Aust J Plant Physiol 13:221-237 **Brulfert J, Vidal J, LeMarechal P, Gadal P, Queiroz O, Kluge M, Kruger I** (1986) Biochem Biophys Res Comm 136:151-159 **Brulfert J, Mricha A, Sossountzov L, Queiroz O** (1987) Plant Cell Environ 10:443-449 **Budde RJA, Chollet R** (1986) Plant Physiol 82:1107-1114 **Burnell JN** (1988) J Exp Bot (in press) **Burnell JN, Hatch MD** (1988a) Arch Biochem Biophys 260:177-186 **Burnell JN, Hatch MD** (1988b) Arch Biochem Biophys 260:187-199 **Burnell JN, Hatch MD** (1988c) Plant Physiol 87: (in press).

Cadet F, Meunier JC, Ferté N (1987a) Biochem J 241:71-74 **Cadet F, Meunier JC, Ferté N** (1987b) Eur J Biochem 162:393-398 **Camp PJ, Miernyk JA, Randall DD** (1988) Biochim Biophys Acta 933:269-275 **Carnal NW, Black CC** (1979) Biochem Biophys Res Comm 86:20-26 **Cerovic ZG, Cheesbrough JK, Walker DA** (1987) Plant Physiol 84:1249-1251 **Chapman MS, Se Won Suh, Cascio D, Smith WW, Eisenberg D** (1987) Nature (London) 329:354-356 **Chapman MS, Se Won Suh, Curmi PMG, Cascio D, Smith WW, Eisenberg DS** (1988) Science 241:71-74 **Cheng SH, Moore BD, Ku MSB** (1986) Plant Physiol 81s:55 **Chow WS, Anderson JM** (1987) Aust J Plant Physiol 14:1-8.

Denyer K, Smith AM (1988) Planta 173:172-182 **Dickerson RE** (1972) Sci Am 226:58-72 **Dietz KJ** (1986) Planta 167:260-263 **Dixon GK, Merrett MJ** (1988) New Phytol 109:47-51 **Dixon GK, Patel BN, Merrett MJ** (1987) Planta 172:508-513 **Doncaster HD, Leegood RC** (1987) Plant Physiol 84:82-87 **Droux M, Jacquot JP, Miginac-Maslow M, Gadal P, Huet JC, Crawford NA, Yee BC, Buchanan BB** (1987) Arch Biochem Biophys 252:426-439 **Dry IB, Dimitriadis E, Ward AD, Wiskich JT** (1987) Biochem J 245:669-675.

Echeverria E, Boyer CD, Thomas PA, Kang-Chien Liu, Shannon JC (1988) Plant Physiol 86:786-792 **Edwards GE, Ku MSB** (1987) In: Hatch MD, Boardman NK (eds) The Biochemstry of Plants. Vol 10. Academic Press, New York, pp 275-325 **Edwards GE, Foster JG, Winter K** (1982) In: Ting IP, Gibbs M (eds) Crassulacean Acid Metabolism. Am Soc Plant Physiol, Maryland, pp 92-111 **Entwistle G, Tyson RH, apRees T** (1988) Phytochemistry 27:993-996 **Espie GS, Colman B** (1987) Plant Physiol 83:113-117 **Evans JR** (1986) Planta 167:351-358.

Fahrendorf T, Holtum JAM, Mukherjee U, Latzko E (1987) Plant Physiol 84:182-187 **Farineau J, Suzuki A, Morot-Gaudry JF** (1988) Plant Sci 55:191-198 **Fickenscher K, Scheibe R** (1988) Arch Biochem Biophys 260:771-779 **Flügge UI, Stitt M, Heldt HW** (1985) FEBS Lett 183:335-339 **Fong FK, Butcher KA** (1988) Biochem Biophys Res Comm 150:399-404 **Ford DM, Jablonski PP, Mohamed AH, Anderson LE** (1987) Plant Physiol 83:628-632 **Foyer C, Spencer C** (1986) Planta 167:369-375 **Foyer CH, Furbank RT, Walker DA** (1987) Biochim Biophys Acta 894:157-164 **Fu CF, Gibbs M** (1987) Plant Physiol 83:849-855 **Furbank RT, Hatch MD** (1987) Plant Physiol 85:958-964 **Furbank RT, Stitt M, Foyer CH** (1985) Planta 164:172-178 **Furbank RT, Foyer CH, Walker DA** (1987) Biochim Biophys Acta 894:552-561.

Gardemann A, Schimkat D, Heldt HW (1986) Planta 168:536-545 **Gerbaud A, Andre M** (1987) Plant Physiol 83:933-937 **Gerbling KP, Steup M, Latzko E** (1986) Plant Physiol 80:716-720 **Gerbling KP, Steup M, Latzko E** (1986) Plant Physiol 80:716-720 **Gerhardt R, Stitt M, Heldt HW** (1987) Plant Physiol 83:399-407 **Gillham DJ, Dodge AD** (1987) Plant Sci 50:105-109 **Gontero B, Cárdenas ML, Ricard J** (1988) Eur J Biochem 173:437-443 **Goyal A, Brown AD, Gimmler H** (1987) J Plant Physiol 127:77-96 **Guitton C, Mache R** (1987) Eur J Biochem 166:249-254 **Gustafson SW, Raynes DA, Jensen RG** (1987) Plant Physiol 83:170-176 **Gutteridge S, Parry MAJ, Burton S, Keys AJ, Mudd A, Feeney J, Servaites JC, Pierce J** (1986a) Nature (London) 324:274-276 **Gutteridge S, Millard BN, Parry MAJ** (1986b) FEBS Lett 196:263-268.

Hanson KR, Peterson RB (1986) Arch Biochem Biophys 246:332-346 **Hanson KR, Peterson RB** (1987) Arch Biochem Biophys 252:591-605 **Hatch MD, Agostino A, Burnell JN** (1988) Arch Biochem Biophys 261:357-367 **Hattersley PW** (1987) In: Soderstron TR, Hilu KW, Campbell CS, Barkworth ME (eds) Grass Systematics and Evolution. Smithsonian Press, Washington, pp 49-64 **Hattersley PW, Wong SC, Perry S, Roksandic Z** (1986) Plant Cell Environ 9:217-233 **Häusler RE, Holtum JAM, Latzko E** (1987a) Plant Physiol 84:205-207 **Häusler RE, Holtum JAM, Latzko E** (1987b) Eur J Biochem 163:619-626 **Heber U, Neimanis S, Dietz KJ** (1988) Planta 173:267-274 **Heim S, Wagner KG** (1986) Biochem Biophys Res Comm 134:1175-1181 **Heimann K, Kreimer G, Melkonian M, Latzko E** (1987) Z Naturforsch 42c:283-287 **Holo H, Sirevåg R** (1986) Arch Microbiol 145:173-180 **Holtum JAM, Fahrendorf T, Neuhaus HE, Murkherjee U, Latzko E** (1987) In: Biggins J (ed) Progress in Photosynthesis Research. Vol 3. Nijhoff, Amsterdam, pp 735-738 **Hossain MA, Nakano Y, Asada K** (1984) Plant Cell Physiol 25:385-395 **Huber SC, Akazawa T** (1986) Plant Physiol 81:1008-1013 **Huber SC, Edwards GE** (1977) Biochim Biophys Acta 462:583-602 **Huber SC, Sugiyama T** (1986) Plant Physiol 81:674-677 **Huber SC, Kalt-Torres W, Usuda H, Bickett M** (1987) In: Biggins J (ed) Progress in Photosynthesis Research. Vol 3. Nijhoff, Amsterdam, pp 714-724 **Hunt S, Smith AM, Woolhouse HW** (1987) Planta 171:227-234 **Husain SR, Cillard J, Cillard P** (1987) Phytochemistry 26:2489-2491 **Husic DW, Tolbert NE** (1987) Proc Nat Acad Sci USA 84:1555-1559.

Iglesias AA, Losada M (1988) Arch Biochem Biophys 260:830-840 **Iglesias AA, Serrano A, Guerrero MG, Losada M** (1987) Biochim Biophys Acta 925:1-10 **Incharoensakdi A, Takabe T, Akazawa T** (1986) Arch Biochem Biophys 248:62-70.

Jacobshagen S, Schnarrenberger C (1988) Plant Physiol 87:78-82 **Jenkins CLD, Burnell JN, Hatch MD** (1987) Plant Physiol 85:952-957 **Jiao JA, Chollet R** (1988) Arch Biochem Biophys 261:409-417 **Jochem P, Lüttge U** (1987) J Plant Physiol 129:251-268 **Johansen J** (1988) Biochim Biophys Acta 933:406-412 **Jones TWA, Gottlieb LD, Pichersky E** (1986) Plant Physiol 81:367-371.

Kakefuda G, Duke SH, Hostak MS (1986) Planta 168:175-182 **Kalt-Torres W, Huber SC** (1987) Plant Physiol 83:294-298 **Kalt-Torres W, Kerr PS, Huber SC** (1987a) Physiol Plant 70:653-658 **Kalt-Torres W, Kerr PS, Usuda H, Huber SC** (1987b) Plant Physiol 83:283-288 **Kaplan A, Zenvirth D, Marcus Y, Omata T, Ogawa T** (1987) Plant Physiol 84:210-213 **Katz A, Kaback HR, Avron M** (1986) FEBS Lett 202:141-144 **Kaul R, Saluja D, Sachar RC** (1986) FEBS Lett 209:63-70 **Kauss H** (1987) Annu Rev Plant Physiol 38:47-72 **Kelly GJ** (1989) Oceanogr Mar Biol Annu Rev 26: (in press) **Kenyon WH, Black CC** (1986) Plant Physiol 82:916-924 **Kerr PS, Huber SC** (1987) Planta 170:197-204 **Kerr PS, Kalt-Torres W, Huber SC** (1987) Planta 170:515-519 **Kleczkowski LA, Randall DD** (1986) Plant Physiol 81:1110-1114 **Kleczkowski LA, Randall DD** (1988a) Biochem J 250:145-152 **Kleczkowski LA, Randall DD** (1988b) Planta 173:221-229 **Kleczkowski LA, Randall DD** (1988c) Phytochemistry 27:1269-1273 **Kleczkowski LA, Randall DD, Blevins DG** (1986) Biochem J 239:653-659 **Klein U** (1987) Plant Physiol 85:892-897 **Klöck G, Kreuzberg K** (1987) Z Naturforsch 42c:567-569 **Kluge M, Hell R, Pfeffer A, Kramer D** (1987) Plant Cell Environ 10:451-462 **Kobza J, Edwards GE** (1987a) Plant

Cell Physiol 28:1141-1152 **Kobza J, Edwards GE** (1987b) Plant Physiol 83:69-74 **Köhler J, Komor E, Thom M, Maretzki A** (1988) Phytochemistry 27:1605-1608 **Kreimer G, Surek B, Woodrow IE, Latzko E** (1987) Planta 171:259-265 **Kreimer G, Melkonian M, Holtum JAM, Latzko E** (1988) Plant Physiol 86:423-428 **Kretschmer M, Schellenberger W, Otto A, Kessler R, Hofmann E** (1987) Biochem J 246:755-759 **Krieger TJ, Mende-Mueller L, Miziorko HM** (1987) Biochim Biophys Acta 915:112-119 **Krömer S, Stitt M, Heldt HW** (1988) FEBS Lett 226:352-356 **Kruger NJ, Hemmingsen SM, Dennis DT** (1986) Plant Physiol 81:12-16 **Kruger NJ, Kombrink E, Beevers H** (1983) FEBS Lett 153:409-412.

Langlotz P, Follmann H (1987) Z Naturforsch 42c:1364-1366 **Langlotz P, Wagner W, Follmann H** (1986) Z Naturforsch 41c:275-283 **Larondelle Y, Mertens E, VanSchaftingen E, Hers HG** (1986) Eur J Biochem 161:351-357 **Leegood RCL** (1985) Planta 164:163-171 **Leegood RC, Furbank RT** (1986) Planta 168:84-93 **Liedvogel B, Bäuerle R** (1986) Planta 169:481-489 **Lilley RM, Ebbighausen H, Heldt HW** (1987) Plant Physiol 83:349-353 **Lin TP, Caspar T, Somerville C, Preiss J** (1988a) Plant Physiol 86:1131-1135 **Lin TP, Spilatro SR, Preiss J** (1988b) Plant Physiol 86:251-259 **Lindström K, Rodhe W** (1978) Mitt Int Ver Limnol 21:168-173 **Loza-Tavera H, Serrano B, Molina JD, Ortega-Delgado ML, Sanchez-DeJiménez E** (1987) Can J Bot 65:607-611.

MacDonald FD, Cséke C, Chou Q, Buchanan BB (1987) Proc Natl Acad Sci USA 84:2742-2746 **Mächler F, Lehnherr B, Nösberger J** (1986) Plant Physiol 82:1167-1168 **Mächler F, Oberson A, Grub A, Nösberger J** (1988) Plant Physiol 87:46-49 **Macioszek J, Anderson LE** (1987) Biochim Biophys Acta 892:185-190 **Maheshwari V, Bhardwaj R, Sharma D, Nagar S** (1988) Biochem Biophys Res Commun 152:668-673 **Marquardt G, Lüttge U** (1987) J Plant Physiol 129:269-286 **Marques IA, Ford DM, Muschinek G, Anderson LE** (1987) Arch Biochem Biophys 252:458-466 **Mateyka C, Schnarrenberger C** (1988) Plant Physiol 86:417-422 **McCashin BG, Cossins EA, Canvin DT** (1988) Plant Physiol 87:155-161 **McMurray WC, Irvine RF** (1988) Biochem J 249:877-881 **Miginiac-Maslow M, Cornic G, Jacquot JP** (1988) Planta 173:468-473 **Miller AG, Canvin DT** (1987) Plant Physiol 84:118-124 **Miller AG, Espie GS, Canvin DT** (1988) Plant Physiol 86:677-683 **Møller IM, Bérczi A, VanDerPlas LHW, Lambers H** (1988) Physiol Plant 72:642-649 **Monson RK, Schuster WS, Ku MSB** (1987) Plant Physiol 85:1063-1067 **Moore BD, Franceschi VR, Cheng SH, Wu J, Ku MSB** (1987) Plant Physiol 85:978-983 **Morehouse KM, Mason RP** (1988) J Biol Chem 263:1204-1211 **Moroney JV, Wilson BJ, Tolbert NE** (1986) Plant Physiol 82:821-826 **Moroney JV, Kitayama M, Togasaki RK, Tolbert NE** (1987) Plant Physiol 83:460-463 **Morrell S, apRees T** (1986) Phytochemistry 25:1579-1585 **Morris PF, Layzell DB, Canvin DT** (1988) Plant Physiol 87:148-154 **Murray AJS, Blackwell RD, Joy KW, Lea PJ** (1987) Planta 172:106-113 **Musgrave ME, Strain BR, Siedow JN** (1986) Proc Natl Acad Sci USA 83:8157-8161.

Nakano Y, Asada K (1987) Plant Cell Physiol 28:131-140 **Nakano Y, Edwards GE** (1987) Plant Physiol 85:294-298 **Neuburger M, Bourguignon J, Douce R** (1986) FEBS Lett 207:18-22 **Neuhaus HE, Holtum JAM, Latzko E** (1988) Plant Physiol 87:64-68 **Newman SM, Cattolico RA** (1987) Plant

Physiol 84:483-490 **Nicholson S, Easterby JS, Powls R** (1987) Eur J Biochem 162:423-431 **Nimmo GA, McNaughton GAL, Fewson CA, Wilkins MB, Nimmo HG** (1987) FEBS Lett 213:18-22 **Nishida K, Tominaga O** (1987) J Plant Physiol 127:385-393 **Noctor G, Mills JD** (1987) Biochim Biophys Acta 894:295-303.

Ogawa T, Kaplan A (1987) Plant Physiol 83:888-891 **Ogawa T, Omata T, Marcus Y, Kaplan A** (1987) Biochim Biophys Acta 893:219-224 **Ohnishi JI, Kanai R** (1987a) FEBS Lett 219:347-350 **Ohnishi JI, Kanai R** (1987b) Plant Cell Physiol 28:1-10 **Ohnishi JI, Kanai R** (1987c) Plant Cell Physiol 28:243-251 **Ohsugi R, Huber SC** (1987) Plant Physiol 84:1096-1101 **Oja V, Laisk A, Heber U** (1986) Biochim Biophys Acta 849:355-365 **Omata T, Ogawa T** (1986) Plant Physiol 80:525-530.

Paech C, Dybing CD (1986) Plant Physiol 81:97-102 **Peisker M** (1986) Plant Cell Environ 9:627-635 **Peltier G, Sarrey F** (1988) FEBS Lett 228:259-262 **Peltier G, Ravenel J, Verméglio A** (1987) Biochim Biophys Acta 893:83-90 **Peñarrubia L, Moreno J** (1987) Biochim Biophys Acta 916:227-235 **Pierce J** (1988) Physiol Plant 72:690-698 **Pilz I, Schwarz E, Pal GP, Saenger W** (1987) Z Naturforsch 42c:1089-1091 **Porter MA, Milanez S, Stringer CD, Hartman FC** (1986) Arch Biochem Biophys 245:14-23 **Portis AR, Salvucci ME, Ogren WL** (1986) Plant Physiol 82:967-971 **Pradel J, Aragnol D** (1986) Biochim Biophys Acta 871:293-301 **Prendergast HDV, Hattersley PW, Stone NE, Lazarides M** (1986) Plant Cell Environ 9:333-344 **Prendergast HDV, Hattersley PW, Stone NE** (1987) Aust J Plant Physiol 14:403-420 **Price NM, Harrison PJ** (1988) Plant Physiol 86:192-199 **Price NM, Thompson PA, Harrison PJ** (1987) J Phycol 23:1-9.

Quick WP, Mills JD (1986) Biochim Biophys Acta 851:166-172.

Rajendrudu G, Prasad JSR, Das VSR (1986) Plant Physiol 80:409-414 **Ramachandra Reddy A, Das VSR** (1987) Photosynthetica 21:341-348 **Rawsthorne S, Hylton CM, Smith AM, Woolhouse HW** (1988) Planta 173:298-308 **Reddy ASN, McFadden JJ, Friedmann M, Poovaiah BW** (1987) Biochem Biophys Res Commun 149:334-339 **Robinson JM** (1988) Physiol Plant 72:666-680 **Robinson SP** (1986) Photosynth Res 10:93-100 **Robinson SP, Giersch C** (1987) Aust J Plant Physiol 14:451-462 **Robinson SP, Portis AR** (1988a) Plant Physiol 86:293-298 **Robinson SP, Portis AR** (1988b) FEBS Lett 233:413-416 **Rodríguez Andrés A, Lázaro JJ, Chueca A, Hermoso R, López Gorgé J** (1987) Plant Sci 52:41-48 **Rumpho ME, Wessinger ME, Edwards GE** (1987) Plant Cell Physiol 28:805-813 **Rumpho ME, Edwards GE, Yousif AE, Keegstra K** (1988) Plant Physiol 86:1193-1198.

Sage RF, Sharkey TD (1987) Plant Physiol 84:658-664 **Sainis JK, Harris GC** (1986) Biochem Biophys Res Commun 139:947-954 **Sakai T** (1974) Shokubutsugaku Zasshi (Bot Mag Tokyo) 87:69-71 **Salerno GL** (1986) Plant Sci 44:111-117 **Salvucci ME, Anderson JC** (1987) Plant Physiol 85:66-71 **Salvucci ME, Portis AR, Ogren WL** (1985) Photosynth Res 7:193-201 **Salvucci ME, Portis AR, Ogren WL** (1986) Plant Physiol 80:655-659 **Salvucci ME, Werneke JM, Ogren WL, Portis AR** (1987a) Plant Physiol 84:930-936 **Salvucci ME, Portis AR, Heber U, Ogren WL** (1987b) FEBS Lett 221.215-220 **Salvucci ME, Holbrook GP, Anderson JC, Bowes G** (1988) FEBS Lett 231:197-201

Scheibe R (1987) Physiol Plant 71:393-400 **Scheibe R, Wagenpfeil D, Fischer J** (1986) J Plant Physiol 124:103-110 **Scherer S, Riege H, Böger P** (1988) Plant Physiol 86:769-772 **Schnarrenberger C, Krüger I** (1986) Plant Physiol 80:301-304 **Schnyder H, Mächler F, Nösberger J** (1986) J Exp Bot 37:1170-1179 **Schulze-Siebert D, Heintze A, Schultz G** (1987) Z Naturforsch 42c:570-580 **Schumaker KS, Sze H** (1987) J Biol Chem 262:3944-3946 **Seemann JR, Sharkey TD, Wang J, Osmond CB** (1987) Plant Physiol 84:796-802 **Selman-Reimer S, Selman BR** (1988) FEBS Lett 230:17-20 **Servaites JC, Parry MAJ, Gutteridge S, Keys AJ** (1986) Plant Physiol 82:1161-1163 **Sharkey TD, Seemann JR, Berry JA** (1986) Plant Physiol 81:788-791 **Sicher RC, Baysdorfer C, Kremer DF** (1987) Plant Physiol 83:768-771 **Sivak MN, Walker DA** (1986) New Phytol 102:499-512 **Somerville CR, Portis AR, Ogren WL** (1982) Plant Physiol 70:381-387 **Soulié JM, Riviere M, Buc J, Gontero B, Richard J** (1987) Eur J Biochem 162:271-274 **Spilatro SR, Preiss J** (1987) In: Biggins J (ed) Progress in Photosynthesis Research. Vol 3. Nijhoff, Amsterdam, pp 701-704 **Steup M, Schachtele C** (1986) Planta 168:222-231 **Steup M, Schächtele C, Melkonian M** (1986) Physiol Plant 66:234-244 **Stitt M** (1986) Plant Physiol 81:1115-1122 **Stitt M, Heldt HW** (1985) Biochim Biophys Acta 808:400-414 **Stitt M, Vasella A** (1988) FEBS Lett 228:60-64 **Stitt M, Mieskes G, Söling HD, Große H, Heldt HW** (1986) Z Naturforsch 41c:291-296 **Stitt M, Gerhardt R, Wilke I, Heldt HW** (1987) Physiol Plant 69:377-386 **Stitt M, Wilke I, Feil R, Heldt HW** (1988) Planta 174:217-230 **Streusand VJ, Portis AR** (1987) Plant Physiol 85:152-154 **Struve I, Lüttge U** (1987) Planta 170:111-120 **Sueltemeyer DF, Klug K, Fock HP** (1986) Plant Physiol 81:372-375 **Sukenik A, Bennett J, Falkowski P** (1987) Biochim Biophys Acta 891:205-215 **Sundblad LG, Palmqvist K, Samuelsson G** (1986) FEBS Lett 199:75-79 **Surek B, Kreimer G, Melkonian M, Latzko E** (1987) Planta 171:565-568 **Suzuki E, Tsuzuki M, Miyachi S** (1987a) Plant Cell Physiol 28:1377-1388 **Suzuki S, Nakamoto H, Ku MSB, Edwards GE** (1987b) Plant Physiol 84:1244-1248.

Tarłowski J Bzdęga T, Łoboda T, Nalborczyk E (1986) Physiol Plant 68:383-386 **Taylor SE, Terry N** (1986) Photosynth Res 8:249-256 **Tel-Or E, Huflejt ME, Packer L** (1986) Arch Biochem Biophys 246:396-402 **Terashima I, Evans JR** (1988) Plant Cell Physiol 29:143-155 **Thompson AG, Brailsford MA, Beechey RB** (1987) Biochem Biophys Res Commun 143:164-169 **Tomany MJ, Kent SS** (1986) Plant Physiol 80:1055-1058 **Treede HJ, Heise KP** (1986) Z Naturforsch 41c:1011-1017 **Treede HJ, Riens B, Heise KP** (1986) Z Naturforsch 41c:733-740.

Upham BL, Jahnke LS (1986) Photosynth Res 8:235-247 **Usuda H, Kalt-Torres W, Kerr PS, Huber SC** (1987) Plant Physiol 83:289-293.

Valenti V, Scagliarini S, Pupillo P (1986) J Exp Bot 37:606-614 **Valenti V, Puppilo P, Scagliarini S** (1987) J Exp Bot 38:1228-1237 **VanAssche CJ, Reinier N, Courtiade B, Chancel A, Huber S** (1987) Z Naturforsch 42c:837-844 **VonCaemmerer S, Edmondson DL** (1986) Aust J Plant Physiol 13:669-688 **Voordouw G, DeVries PA, VanDenBerg WAM, DeClerck EPJ** (1987) Eur J Biochem 163:591-598 **Vu JCV, Bowes G, Allen LH** (1986) Plant Sci 44:119-123 **Vu JCV, Allen LH, Bowes G** (1987) Plant Physiol 83:573-578.

Walker GH, Huber SC (1987a) FEBS Lett 213:375-380 **Walker GH, Huber SC** (1987b) Arch Biochem Biophys 258:58-64 **Wallsgrove RM, Turner JC, Hall NP, Kendall AC, Bright SWJ** (1987) Plant Physiol 83:155-158 **Walton NJ, Woolhouse HW** (1986) Planta 167:119-128 **Weger HG, Birch DG, Elrifi IR, Turpin DH** (1988) Plant Physiol 86:688-692 **Weiner H, Stitt M, Heldt HW** (1987) Biochim Biophys Acta 893:13-21 **Williams TG, Turpin DH** (1987) Plant Physiol 83:92-96 **Winter K** (1987) Planta 172:88-90 **Winter K, Arron GP, Edwards GE** (1986) Plant Cell Physiol 27:1533-1539 **Wolosiuk RA, Hertig CM, Busconi L** (1986) Arch Biochem Biophys 246:1-8 **Woo KC, Flügge UI, Heldt HW** (1987) Plant Physiol 84:624-632.

Xu DP, Sung SJS, Alvarez CA, Black CC (1986) Biochem Biophys Res Commun 141:440-445.

Yagawa Y, Muto S, Miyachi S (1987) Plant Cell Physiol 28:1253-1262 **Yamaya T, Oaks A** (1987) Physiol Plant 70:749-756 **Yokota A, Canvin DT** (1986) Plant Physiol 80:341-345 **Yokota A, Kitaoka S** (1987) Planta 170:181-189 **Yokota A, Iwaki T, Miura K, Wadano A, Kitaoka S** (1987) Plant Cell Physiol 28:1363-1376 **Yokota A, Shigeoka S, Onishi T, Kitaoka S** (1988) Plant Physiol 86:649-651 **Yu C, Huang AHC** (1986) Arch Biochem Biophys 245:125-133.

Zelitch I (1988) Plant Physiol 86:463-468 **Ziegler P** (1988) Plant Physiol 86:659-666 **Zima J, Šesták Z** (1985) In: Šesták Z (ed) Photosynthesis During Leaf Development. Academia, Praha, pp 145-156 **Zoglowek C, Krömer S, Heldt HW** (1988) Plant Physiol 87:109-115.

Dr Grahame J Kelly
CSIRO – Marine Laboratories
7000 Hobart, Tasmania
Australia
Present address:
Queensland University
of Technology
4000 Brisbane, Queensland
Australia

Dr. Joseph AM Holtum
Waite Agricultural
Research Institute
5064 Glen Osmond, SA
Australia

Professor Dr Erwin Latzko
Botanisches Institut der
Universität Münster
Schloßgarten 3
D-4400 Münster, FRG

Note added in proof (2005):
The second sentence of Sect. 9 should read: "Its primary action in inhibiting the activity of cytosolic FBPase (involved in sucrose synthesis) and in stimulating the activity of PPi-linked PFK …."

Photosynthesis.
Carbon Metabolism: On Regulation at the Cellular Level and at the Whole Plant Level, and Some Considerations Concerning the Interactions of These Regulatory Events with the Increasing Level of Atmospheric CO_2

By Grahame J. Kelly and Erwin Latzko

"Die anthropogenen CO_2-Emissionen verbleiben derzeit zu etwa 50 Prozent in der Atmosphäre. Es ist möglich, daß ein Teil des verbleibenden Rests aufgrund der zunehmenden Photosyntheseleistung der Pflanzen, die durch den steigenden CO_2-Gehalt hervorgerufen werden kann (CO_2-Düngungseffekt) in der terrestrischen Biosphäre gespeichert wird". (Deutscher Bundestag, 11. Wahlperiode, Drucksache 11/3246, 1988, S. 195)

1. Introduction

Since writing our last review in Volume 50 of Progress in Botany, a rather remarkable event has taken place. Photosynthetic carbon metabolism has become a public issue. People have become aware of the fact that the atmospheric CO_2 level is increasing "rapidly" (ca. 0.4% per year, and that this additional CO_2 will, theoretically, absorb additional outgoing infrared radiation, thereby warming the earth – the hypothesized "greenhouse effect". Given that additional atmospheric CO_2 will also, theoretically, stimulate photosynthetic CO_2 fixation, the reason for renewed public interest in photosynthesis becomes clear. It is quite another matter, however, to gather together what is known of photosynthesis and to predict, with a good degree of certainty, the extent to which the theoretical CO_2-stimulation of photosynthesis will actually take place. With good luck or with good counsel (we know not which), current emphasis in research on photosynthetic carbon metabolism is presently on the uptake of CO_2 by photosynthetic cells, and on the fixation of this CO_2 by ribulose-1,5-bisphosphate (RuBP) carboxylase and, in the case of C_4 plants and crassulacean acid metabolism (CAM) plants, by phosphoenolpyruvate (PEP) carboxylase. The results of this research are definitely assisting our efforts at prediction, as described below. It will also be important to have available the result of studies on the capacities of plant carbon sinks, both temporary (transitory starch, sucrose for translocation), and permanent (roots, fruits, seeds, and especially wood). Finally, the ability of plants to continue photosynthesis under adverse conditions, as may eventuate from future climate changes, may be ascertained from investigations of photosynthesis under stress conditions, and the parts played by photorespiration, CAM and other mechanisms in delaying the onset of the photoinhibition that ensues when there is nowhere safe for the light-generated reductant to go.

This review covers the research on these topics reported over the past 2 years, and concludes with some brief thoughts on the capacity of global photosynthetic CO_2 fixation to counter the predicted CO_2-induced global "greenhouse" warming.

G.J. Kelly and E. Latzko:
Thirty Years of Photosynthesis

Originally published in Progress in Botany 52

2. Uptake of CO_2

Aquatic microalgae continue to dominate research on CO_2 uptake by photosynthetic cells, largely because these algae possess membrane-located proteins capable of actively pumping inorganic carbon (CO_2 and/or HCO_3^-) into the cell, or from one cellular compartment to another (eg. into chloroplasts), while the uptake of CO_2 by terrestrial plant leaves is believed to occur by straightforward diffusion of CO_2 from the atmosphere to the cytoplasmic side of the cell membrane.

The confusion concerning whether CO_2 or HCO_3^- is pumped into cells has been reduced, at least for the cyanobacterium *Synechococcus*, by evidence that the cells possess separate, distinct transport systems for each form of inorganic carbon (Espie et al. 1988a; Price and Badger 1989a). The HCO_3^- transporter is Na^+-dependent, but the CO_2 transporter is more or less independent in that only micromolar concentrations of Na^+ are required (Espie et al. 1988b). Selective inhibition of the CO_2 transporter by H_2S or COS largely blocks CO_2 uptake, while HCO_3^- uptake remains unaffected (Espie et al. 1989; Miller et al. 1989). The CO_2 transporter appears to have a carbonic anhydrase function associated with it (Price and Badger 1989a) so that it is HCO_3^- that is delivered into the cell. CO_2 is regenerated from this HCO_3^- by a carboxysome-located carbonic anhydrase (Price and Badger 1989b).

In eukaryotic microalgae, the situation is clearly more complex. At least four proteins are devoted to accumulating inorganic carbon (Manuel and Moroney 1988; Spalding and Jeffrey 1989). Probably three different forms of carbonic anhydrase are involved (Husic et al. 1989), and three membrane transporters may be utilized: cell membrane CO_2 and HCO_3^- transporters, and a chloroplast envelope HCO_3^- transporter (Colman and Rotatore 1988; Sültemeyer et al. 1988; Goyal and Tolbert 1989; Thielmann et al. 1990).

On a global scale, inorganic carbon transporters are quantitatively more important in the marine environment. Most marine photosynthesis is performed by phytoplanktonic microalgae, and most experiments have shown that these cells actively transport CO_2 or HCO_3^- (Burns and Beardall 1987; Dixon and Merrett 1988; Muñoz and Merrett 1988, 1989). Where HCO_3^- is transported, its conversion to CO_2 in the cell may be facilitated by the internal pH of the cell being lower than the pH of seawater (Dixon et al. 1989; Kelly 1989; Muñoz and Merrett 1989).

3. Environmental Regulation of RuBP Carboxylase Activity

Great interest is being shown in the regulation of RuBP carboxylase (the Calvin cycle's CO_2-fixing enzyme) following the discovery of two "new" regulatory molecules. As reported in our last review, the carboxylase activity was found to be controlled not only by the CO_2-plus-Mg^{2+} activation process described some 15 years ago, but also by the tight binding of an inhibitory sugar-P identified as carboxyarabinitol-1-P (CA1P), and by a protein molecule named rubisco activase. The interest created by this extension of the list of principal regulatory molecules from one to three has been overwhelming, as evidenced by the appearance of six reviews devoted largely or entirely to the regulation of the carboxylase (Salvucci 1989; Sharkey 1989; Gutteridge 1990; Portis 1990; Seemann et al.

1990; Servaites 1990) and a seventh that gives it considerable attention (Woodrow and Berry 1988).

Regulation by CA1P occurs in only some plant species. For example, it is prominent in *Phaseolus vulgaris*, but absent in *Spinacea oleracea* (Kobza and Seemann 1988). In simple terms, it appears to be a nocturnal inhibitor, binding tightly to the catalytic site at night, but being released (by the activase? – see below) and enzymatically destroyed upon illumination. Holbrook et al. (1989) and Gutteridge and Julien (1989) have found that the CA1P-metabolizing enzyme is not an NADPH-dependent activity, as initially reported, but is rather a specific chloroplast phosphatase that produces carboxyarabinitol and Pi. However, the phosphatase is activated by NADPH.

Until very recently (Seemann et al. 1990), the origin of CA1P was unknown, but a promising possibility has now been raised by Beck et al. (1989) who realized that the inhibitor is identical to hamamelonate-2'-P, and therefore conceivably could be generated (via phosphatase and dehydrogenase reactions) from hamamelose-2', 5-P_2, which Gilck and Beck (1974) reported more than 15 years ago to be formed in leaves from fructose-1, 6-P_2. It will be interesting to determine whether or not this is indeed the CA1P biosynthetic pathway, and if so, how its operation is reduced during illumination.

Regulation by rubisco activase has been found in all plant species examined. The primary role of the activase appears to be the coordination of the carboxylase activity with irradiance and other factors (Salvucci 1989; Portis 1990), ie., it participates in "light-mediated activation" of the carboxylase. Purified activase has now been obtained by Robinson et al. (1988) and Robinson and Portis (1989a) and shown to (1) activate the carboxylase in an ATP-dependent process, and (2) hydrolyze ATP. It is an abundant protein, constituting about 2% of the soluble protein in spinach leaves. Exactly how light promotes the activase is not clear. Probably the supply of photosynthetically generated ATP is relevant, but additional unknown factors appear to be involved (Brooks et al. 1988; Campbell and Ogren 1990).

The mechanism by which activase activates the carboxylase must still be fully clarified, but it probably involves ATP hydrolysis (see above), facilitation of the CO_2-plus-Mg^{2+} activation (Portis 1990), and the displacement of phosphorylated inhibitors that bind to the enzyme (Parry et al. 1988; Robinson and Portis 1989b). Such an inhibitor has been suspected to cause the carboxylase activity to decrease over time during measurements in vitro in the absence of activase termed "fallover"), an event more noticeable with spinach-leaf carboxylase than with the *Euglena* enzyme (Yokota and Kitaoka 1989). The inhibitor has been considered to be (1) CA1P bound to the catalytic site, or (2) RuBP bound to a regulatory site of the CO_2-plus-Mg^{2+} activated enzyme (Yokota and Kitaoka 1989), or (3) RuBP bound to the catalytic site of an enzyme not yet activated by CO_2-plus-Mg^{2+} (Woodrow and Berry 1988; Cardon and Mott 1989; see Fig. 1). However, a recent report by Edmondson et al. (1990) has introduced a fourth regulatory molecule, viz. xylulose-1,5-P_2, into the carboxylase control scenario: they attribute "fallover" of activated spinach carboxylase to one in every 400 RuBPs being epimerized to xylulose-1,5-P_2, which then remains bound to the catalytic site and inhibits activity. It may be that this is another phosphorylated inhibitor loosened by the activase.

In recent years a better understanding of the interplay between the above regulatory mechanisms and environmental parameters has developed. With respect to light intensity, it is now more firmly established that regulation of RuBP carboxylase forms part of the

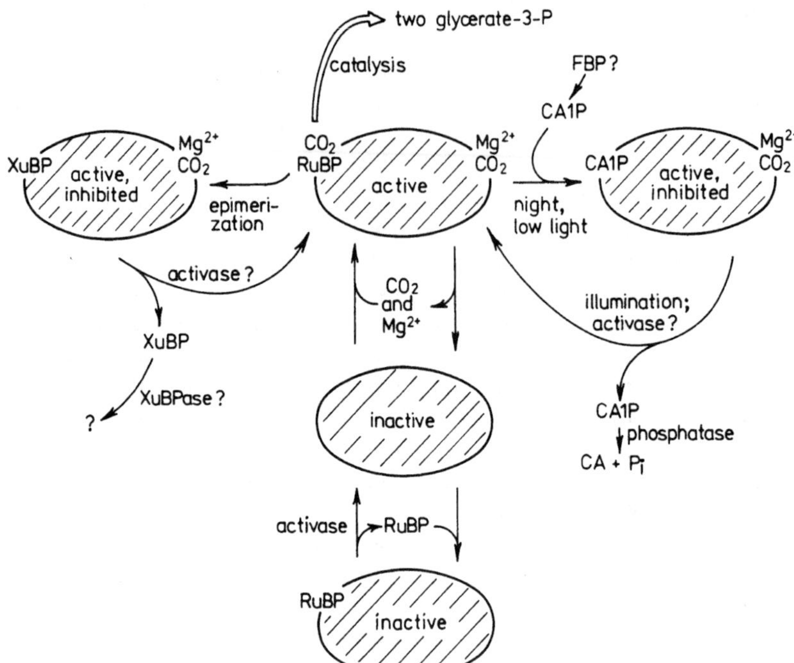

Fig. 1. Current concepts concerning the regulation of RuBP carboxylase activity. Enzyme activation/inactivation involves the gain or loss of a CO_2 that is covalently bound to the enzyme. Inhibition of the activity of the activated enzyme occurs when either CA1P or xylulose-1,5-P_2 (XuBP) attaches noncovalently to active sites. Stimulation of activity by rubisco activase may involve loosening of these inhibitory sugar-P's, and/or loosening of any RuBP attached to inactive enzyme (a prerequisite for CO_2-plus-Mg^{2+} activation)

mechanism whereby the rate of CO_2 fixation, and utilization of the products of CO_2 fixation, are matched to the rate of RuBP regeneration, which requires ATP and NADPH and is therefore ultimately dependent on the prevailing irradiance (Kobza and Seemann 1989a). Surprising is the fact that the carboxylase is regulated differently in different plant species: in *Phaseolus vulgaris*, a pseudo-steady-state concentration of the inhibitor CA1P, determined by the light intensity, is predominant (Kobza and Seemann 1989b) in *Beta vulgaris*, this mechanism operates together with changes in the extent of CO_2-plus-Mg^{2+} activation; in *Spinacia oleracea*, CO_2-plus-Mg^{2+} activation operates with some other tight-binding inhibitor (eg. RuBP) since CA1P is absent (Kobza and Seemann 1988, 1989a). In each case, release of the inhibitory sugar-P may be augmented by rubisco activase, but evidence for this is still not strong. Finally, a comparison of the sun species *Glycine max* with the shade species *Alocasia macrorrhiza* supported the roles of CA1P and CO_2-plus-Mg^{2+} activation in balancing the carboxylase activity to the light intensity, and demonstrated that full activation of the *A. macrorrhiza* carboxylase occurred at a lower light intensity compared to that of the *G. max* enzyme (Seemann et al. 1988; Seemann 1989).

Several studies on the effects of CO_2 and O_2 concentrations on the activity and activation state of the RuBP carboxylase in leaves have been made. Either raising CO_2 or lowering O_2 resulted in substantial deactivation of the carboxylase (Sage et al. 1988; Butz and Sharkey 1989), at least partly because starch and sucrose synthesis was reduced as the demand for Pi (to form ATP for the increase CO_2 fixation) increased (Sharkey and Vanderveer 1989; Sharkey and Vassey 1989). More interesting, but somewhat conflicting, results have been obtained when plants were first acclimated to high CO_2, at levels about threefold greater than that of the atmospheric level. The leaves of acclimated tomato species (*Lycopersicon esculentum* and *L. chmielewskii*) contained about half the RuBP carboxylase activity compared to those grown with atmospheric CO_2, although the percent of enzyme in the activated form was higher (Yelle et al. 1989a). However, in a survey of five other C_3 species, Sage et al. (1989) found that only two species developed a lower carboxylase content upon acclimation, but that in all five species the percent activation was in fact substantially reduced. Thus, from present indications, growth in high CO_2 appears to reduce the in vivo RuBP carboxylase activity, but as with light intensity (see above), different species may use different mechanisms to achieve this reduction.

Two other aspects of RuBP carboxylase deserve noting. The first is the calculation by Osafune et al. (1990) indicating that the carboxylase protein located in the pyrenoid of *Euglena* would need to be functioning in order to support the observed rate of CO_2 fixation. The second is the renewed attention given to the enzyme in stomatal guard cells. The consensus is that the amount of enzyme in these cells is relatively small (Vaughn and Vaughan 1988; Outlaw 1989; Shimazaki 1989, Tarczynski et al. 1989), and could not support syntheses of more than 2% of the solute required to effect stomatal opening (Reckmann et al. 1990).

4. Chloroplast Enzymes and Their Light-Mediated Regulation

Very little attention has been given to enzymes of chloroplast carbon metabolism (except RuBP carboxylase) over the past 2 years. In addition, except for one study of aldolase (Marsh et al. 1989) and another of ribulose-5-P kinase (Serra et al. 1989), most of the studies that have appeared have dealt with light-mediated regulation. Using barley leaves, Sibley and Anderson (1989) showed that light-mediated activation of ribulose-5-P kinase and NADP-linked glyceraldehyde-3-P dehydrogenase (NADP-GAPDH) could be detected at the earliest stage of leaf development, but similar activation of chloroplast fructose-1,6-bisphosphatase (FBPase) became apparent more gradually as leaf development ensued. In another study with barley, it was shown that when leaves grown in low light were suddenly transferred to high light (280 $\rightarrow$ 1400 μmol photons m^{-2} s^{-1}), a temporary reversal of light-mediated activation occurred due to the chloroplast stroma becoming relatively oxidized following photoinhibition (Dujardyn and Foyer 1989). This is no doubt because light-mediated regulation of chloroplast enzymes usually involves reduction of the enzyme by photosynthetically generated reducing equivalents passed through the small proteins ferredoxin and thioredoxin (see our preceding reviews). Exceptions include the C_4 enzyme pyruvate Pi dikinase (Roeske and Chollet 1989; Nakamoto and Young 1990) and possibly P-glucomutase (Sicher 1989), regulated by light-dependent enzyme phosphorylation or dephosphorylation. This mechanism is otherwise common where light-mediated regulation of extra-chloroplast enzymes occurs, e.g. sucrose-P synthase (Huber et al. 1989a), C_4 and CAM PEP carboxylases (Kluge et al. 1988; Jiao and Chollet 1989), and mitochondrial pyruvate dehydrogenase complex (Budde and Randall 1990).

Returning to the ferredoxin/thioredoxin system for reductive activation: the components thioredoxin and ferredoxin-thioredoxin reductase now have been demonstrated in *Chlamydomonas reinhardtii* and are thus presumably in other green algae (Huppe et al. 1990). Operation of the system has been "visualized" by Crawford et al. (1989) using a labeling reagent that, upon reacting with the enzyme sulfhydryl groups reductively generated, renders the enzyme fluorescent. Further studies of the reductive activation of sedoheptulose-1,7-bisphosphatase (Cadet and Meunier 1988) and the chloroplast coupling factor (CF_1; Noctor and Mills 1988) have been reported. In the latter case, CF_1 activation was stabilized by the concomitant proton gradient across the thylakoid membrane (the $\Delta\tilde{\mu}_{H+}$). Another thylakoid-membrane located enzyme, viz. ferredoxin-NADP oxidoreductase, undergoes light-mediated activation in a process involving the proton gradient (Pschorn et al. 1988), thus a similar $\Delta\tilde{\mu}_{H+}$-stabilized protein reduction might be involved.

The chloroplast FBPase has received detailed attention. Light-mediated activation, which may be assisted by the association of the enzyme with thylakoid membranes (Rodriguez Andrés et al. 1990), leads to modification of intramolecular hydrophobic interactions (Stein et al. 1989). These authors also propose a similar situation for NADP-GAPDH (Wolosiuk and Stein 1990). Such interactions may change relatively slowly in FBPase (Soulie et al. 1988). Elucidation of the complete amino acid sequence of the chloroplast FBPase subunit, and comparison with the pig-kidney enzyme, revealed a unique 16-residue sequence in the chloroplast enzyme, containing two cysteines separated by only four amino acid residues (Marcus et al. 1988; Marcus and Harrsch 1990). It is almost certain that this unique sequence is the site of light-mediated activation involving the reduction of a disulfide bridge previously formed between the two cysteines in darkness.

Computer-based modeling has been introduces as a means of evaluating the respective roles of potential regulatory mechanisms during photosynthetic carbon metabolism. Pettersson and Ryde-Pettersson (1988, 1989a, b) have considered photosynthesis under certain optimum conditions, with both light and CO_2 at saturating levels, and concluded that regulation of RuBP carboxylase and other Calvin cycle enzymes were of no significance under these conditions. They concluded that the catalytic capacities of ATP synthase (i.e., CF_1) and sedoheptulose-1,7-bisphosphatase determined the rate of photosynthesis. Other models place more emphasis on RuBP carboxylase (Farazdaghi and Edwards 1988; Collatz et al. 1990). Nevertheless, a role for ATP synthase needs to be kept in mind, given that Neuhaus and Stitt (1989) found intricate interactions between the activity of this enzyme, thylakoid energization, and most aspects of photosynthetic carbon metabolism.

5. Starch and Sucrose in Source and Sink Tissues

The photosynthetic assimilation of CO_2 produces a product that is, ultimately, new plant tissue. Since the development of this new plant tissue is not always perfectly synchronized with photosynthesis, interim assimilate sinks are necessary, such as chloroplasts (transitory starch), vacuoles (sucrose), and over long time periods, storage tissues such as seeds and tubers (starch and other carbohydrates). In view of the growing interest in the capability of plants to grow faster should photosynthesis be stimulated by increased atmospheric CO_2, it will be important to recall that photosynthetic carbon metabolism is influenced by metabolic events occurring in nonphotosynthetic sink tissues as well as in photosynthetic cells. As discussed below (Sect. 13), this aspect may be a key factor in determining the extent to which any long-term CO_2-stimulation of photosynthesis is realized.

Starch metabolism, in both source and sink tissues, has been reviewed in detail by Beck and Ziegler (1989). Most recent research has focused on starch degradation, which in the chloroplast occurs via a cooperative action of phosphorylase and hydrolytic enzymes such as amylases. The amylases also occur outside the chloroplast where there is no starch, thus their role is unknown (Beck and Ziegler 1989; Caspar et al. 1989).

Over the past 10 years it has been accepted that the hexose-Ps, generated when the chloroplast's transitory starch is degraded, are further metabolized in the chloroplast to triose-P which then moves, via the phosphate translocator, into the cytosol and becomes the substrate for sucrose synthesis; a minor amount of the triose-P may remain in the chloroplast and be further glycolytically metabolized to pyruvate (Schulze-Siebert and Schultz 1989). The identification of two forms of ATP-linked phosphofructokinase (ATP-PFK) in *Chlorella pyrenoidosa* (Kiss et al. 1989) suggests that this overall scheme is also applicable to green algae.

However, some recent experiments strengthen the likelihood that, in some plant species, the direct movement of hexose-Ps, or hexoses, from the chloroplast to the cytosol may occur, presumably mediated by some form of translocator (see Beck and Ziegler 1989). In leaves of *Beta vulgaris* and *Phaseolus vulgaris* under a normal sinusoidal light regime, starch was degraded during low-light periods (first and last 2 h of the photoperiod) in the presence of metabolite levels expected in the event that hexoses, rather than trioses, were exported (Servaites et al. 1989a). In sink tissues, where carbon flow is in the opposite direction (i.e., sugar units are imported by amyloplasts and used as the substrate for the synthesis of reserve starch), it is almost certain that hexoses, probably as hexose-Ps, are imported (Entwistle and ap Rees 1988; Tyson and ap Rees 1988; Hatzfeld and Stitt 1990) via a phosphate translocator that accepts glucose-6-P (Borchert et al. 1989). The hexose-Ps are generated from sucrose translocated into the cytosol surrounding the amyloplast, probably via the pathway shown in Fig. 2. An alternative pathway utilizing a reaction between UDP-glucose and Pi (see our preceding review) is no longer considered possible (Tyson and ap Rees 1989). Note that the scheme in Fig. 2 utilizes the pyrophosphate-linked PFK (PPi-PFK), and although this role for PPi-PFK in sink tissues is most important, a more central role, viz. maintaining the cytosolic PPi concentration, is being entertained (Dancer and ap Rees 1989; Hatzfeld et al. 1989), and may help to explain its co-existence with cytosolic ATP-PFK, e.g. in spinach leaves (Häusler et al. 1989)

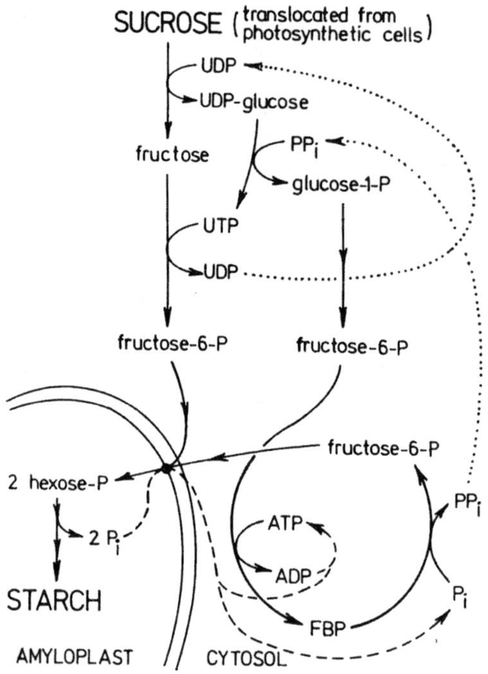

Fig. 2. Generation of amyloplast starch from sucrose in sink tissues. *Dotted arrows* show recycling of UDP and PPi; the *dashed arrows* indicate that Pi release is balanced by its uptake during (1) ATP regeneration, and (2) the reverse reaction of PPi-PFK. ●, plastid envelope translocator that accepts hexose-P (After Tyson and ap Rees 1988; Borchert et al. 1989)

The biosynthesis of sucrose in photosynthetic cells has received extensive attention. Further supporting evidence for the accepted biosynthetic pathway via UDP-glucose in the cytosol has been obtained (Dance et al. 1990). Regulation of pathway activity by the level of cytosolic fructose-2, 6-P_2 (Steingraber et al. 1988), an inhibitor of the cytosolic FBPase, has been more firmly established (Stitt and Grosse 1988; Neuhaus et al. 1989). However, a previous report that fructose-2, 6-P_2 also inhibits P-glucomutase has been found to be in error (Galloway et al. 1988). Discussion is continuing concerning the nature of the enzyme(s) responsible for synthesizing and degrading fructose-2, 6-P_2 (Larondelle et al. 1989; MacDonald et al. 1989).

Sucrose synthesis is also regulated at the reaction catalyzed by sucrose-P synthase. This enzyme undergoes light-mediated activation (see Sect. 4) in some species, e.g. spinach and barley, but not in others, e.g. soybean, pea, and cucumber (Huber et al. 1989a, b; Nielsen and Huber 1989; Walker and Huber 1989). Water stress also leads to increased enzyme activity (Quick et al. 1989). There are also indications that different forms of the enzyme with varying susceptibilities to allosteric regulation exist (Nielsen and Huber 1989; Siegl and Stitt 1990), but this needs further investigation in response to questions concerning the validity of the sucrose-P synthase assay (Crafts-Brandner and Salvucci 1989).

Once formed, sucrose is sometimes stored in leaf vacuoles before being translocated to sink tissues. The amount accumulated in leaves appears to be inversely proportional to the leaf's content of acid invertase (Huber 1989). Most sink tissues either metabolize sucrose to amyloplast starch (see above) or utilize it for the growth of new tissues, but some, e.g. sugar-cane and sugar beet, store the sucrose in vacuoles. A tonoplast-located mechanism considered earlier for moving the sucrose into these vacuoles (see Fig. 4 in our 1986 review) has now been disproven, and the mechanism whereby these vacuoles accumulate sucrose is still unknown (Preisser and Komor 1988). Of course, vacuoles have a finite capacity for storing sucrose. In photosynthetic cells, this capacity can be exceeded by CO_2 fixation under high light, forcing further photoassimilated carbon to travel to chloroplast starch (Servaites et al. 1989b). This process also has its limits, as often has been demonstrated when starch granules large enough to distort chloroplast structure have been observed after a period of sustained photosynthesis. Even a moderate stimulation of photosynthesis consequent upon the acclimation of plants to an increased level of atmospheric CO_2 can result in a marked accumulation of starch in the chloroplast (Dons 1988; Yelle et al. 1989b), thus the global diurnal turnover of chloroplast transitory starch may be expected to increase in the coming decades as atmospheric CO_2 increases.

6. Mitochondrial Respiration During Photosynthesis

The burgeoning enthusiasm for the concept that mitochondrial respiration continues to operate in photosynthetic cells when they become illuminated (see our preceding review) has been dampened by reports that photosynthetic activity modulates mitochondrial activity in *Lemna minor* (Nespoulous et al. 1989), and that the enzyme that delivers substrate into the Krebs cycle, the pyruvate dehydrogenase complex, is inactivated in a light-dependent manner in pea leaves (Budde and Randall 1990). Furthermore, from studies with the green microalgae *Dunaliella tertiolecta* (Bate et al. 1988) and *Chlorella vulgaris* (Shiraiwa et al. 1988), it was concluded that mitochondrial oxygen consumption in the light was of minor importance.

However, other studies with microalgae highlight the possibility that different facets of mitochondrial respiration can be switched on as needed during photosynthesis. Weger and Turpin (1989) were able to show that *Selenastrum minutum* developed increased Krebs cycle activity when assimilating NO_3^- or NH_4^+ (the Krebs cycle supplies carbon skeletons for amino acid biosynthesis). Lemaire et al. (1988) have developed a rather fascinating proposal of "mitochondrial-dependent photosynthesis" for a mutant of *Chlamydomonas reinhardtii* that lacks an active chloroplast ATP synthase (CF_1). In this proposal, reducing power photosynthetically generated in the chloroplast is shuttled into mitochondria, oxidized there using respiratory electron transport coupled to ATP production, and this ATP is shuttled back into the chloroplast to supply the Calvin cycle with the ATP that it requires. Finally, recent investigations with marine phytoplanktonic microalgae indicate that mitochondrial respiration is not only present in the light, but in fact is more active in the light than in darkness (Grande et al. 1989; Weger et al. 1989). This apparently odd situation might be related to the fact that marine phytoplankton are seldom CO_2-limited, but are often light-limited, and in addition they have no nonphotosynthetic cells to support. Whatever the explanation, observations such as these offer the hypothesis that the extend of mitochondrial respiration in illuminated photosynthetic cells may be highly variable, and depend on plant species, growth conditions, and physiological state.

Some further evidence for the somewhat enigmatic process termed chlororespiration has become available from Garab et al. (1989), although Klöck et al. (1989) question its significance. In chloroplasts of *Chlamydomonas reinhardtii* an appreciable activity of 21.6 µmol mg^{-1} chl h^{-1} has been measured for succinate dehydrogenase (Willeford et al. 1989), an enzyme classically considered to be restricted to the mitochondrion.

7. Photorespiration

Photorespiration is one aspect of photosynthetic carbon metabolism that will receive a more than average amount of attention as atmospheric CO_2 increases. This is because it is a process that is primarily determined by the simple competition between O_2 and CO_2 for the active site of RuBP carboxylase/oxygenase. Clearly, if CO_2 increases in the presence of an unaltered level of O_2, photorespiration will decrease and net carbon fixation will correspondingly increase. In fact, Sharkey (1988) has calculated that this has indeed happened, such that photorespiration has declined about 10% since its discovery in the mid-1950s. A further point is that, since the competition between O_2 and CO_2 can occur under all conditions, then increased CO_2 will lead to increased net carbon fixation by cells whose photosynthesis is primarily limited by some factor other than CO_2, e.g. light (Sharkey 1989). In other words, under light-limited conditions increased CO_2 stimulated CO_2 fixation because the set amount of NADPH and ATP made available under the prevailing limiting irradiance is used to support a greater amount of RuBP carboxylation and a correspondingly lesser amount of RuBP oxygenation. This assumes, of course, that other factors influencing intracellular O_2 and CO_2 levels, e.g. stomatal conductance, remain unaltered.

This assumption certainly does not apply to those groups of plants with biochemical CO_2-concentrating mechanisms e.g. C_4 plants, CAM plants, and microalgae. C_4 plants raise to CO_2 concentration in the vicinity of RuBP carboxylase/oxygenase to a value where photorespiratory CO_2 loss is only 2–3% of net photosynthesis, well below the value of ca. 30% for C_3 plants (de Veau and Burris 1989a; Jenkins et al. 1989). A similar situation applies for CAM plants, although near the end of the light period the CO_2 supply can approach exhaustion and photorespiration may then appear (Cote et al. 1989). Microalgae pump CO_2 and/or HCO_3^- across the plasma membrane (see Sect. 2) so that they also have low or nonexistent photorespiratory rates (de Veau and Burris 1989b).

For C_3 plants, several possibilities for "artificially" reducing photorespiration have been entertained: specific inhibition of the oxygenase activity (see our preceding review); manipulation of the CO_2/O_2 specificity of the bifunctional RuBP carboxylase/oxygenase by examining interspecies variation (Parry et al. 1989) or by genetic manipulation (Chen and Spreitzer 1989); and selection of mutants with altered metabolism of the products of RuBP oxygenation – such a mutant with high catalase activity has been identified by Zelitch (1990). However, no especially promising results have appeared, and at least with respect to RuBP carboxylase/oxygenase they may not be expected because the mechanisms of oxygenation and carboxylation are very similar (Kreckl et al. 1989).

Oxygenation of RuBP produces the glycolate that is metabolized by the glycolate pathway; two molecules of glycolate are metabolized (via glyoxylate, glycine, serine, and hydroxypyruvate) to one each of glycerate-3-P and CO_2. Reactions involving glyoxylate and hydroxypyruvate are located in the peroxisome; in addition, there is now evidence for "back-up" enzymes in the cytosol to metabolize these compounds in the event that leakage from the peroxisome occurs (Givan et al. 1988; Kleczkowski et al. 1988; Kleczkowski and Edwards 1989).

The two glycine → serine conversion occurs in mitochondria. This complex reaction releases NH_4^+, the photorespiratory CO_2, and produces NADH. Earlier research indicating that the NADH is oxidized in the mitochondrion to generate ATP has been supplemented with an investigation showing that this ATP is destined for the cytosol (Gardeström and Wigge 1988). The NH_4^+ is reassimilated into glutamine in chloroplasts, in a somewhat complex process involving three translocators that move glutamate, glutamine, and carboxylic acids in and out of the chloroplast (Flügge et al. 1988; Yu and Woo 1988). Nevertheless, sufficient NH_4^+ might accumulate in mitochondria to promote the inactivation of the pyruvate dehydrogenase complex, as observed by Schuller and Randall (1989), and thereby reduce the mitochondrial Krebs cycle activity during photorespiration.

Further indications that photorespiration can have a physiological role have been provided. Stuhlfauth et al. (1990) found that *Digitalis lanata*, a species adapted to a dry habitat, utilizes photorespiration to consume photosynthetically generated ATP and NADPH when CO_2 availability drops due to water stress-induced reduction of stomatal conductance. Heber et al. (1989) observed that chloroplasts are less stressed by a shortage of Pi when photorespiration can proceed and, consequently, some of the product of the Calvin cycle (i.e., glycolate-P) is hydrolyzed within the chloroplast, thus keeping the Pi in this organelle.

8. Metabolism in Stressed Photosynthetic Cells

The photosynthetic carbon metabolism of plants under stress may need more attention than it is currently receiving if increased atmospheric CO_2 leads to climate changes that make plant stress more commonplace. In general, stress may result from extremes in light, temperature, water supply, or nutrient supply. With respect to photosynthesis, most consideration has been given to excess light and the phenomenon of photoinhibition, whereas predicted climate changes relate more to temperature and water supply.

Photoinhibition is due to loss or damage to photosystem II protein and/or electron acceptors after more light energy is harvested than can be used for photosynthesis or can be otherwise dissipated (Critchley 1988; Kuhn et al. 1988; Schuster et al. 1988; Virgin et al. 1988; Neale and Melis 1990; but see Le Gouallec and Cornic 1988). Several energy dissipation processes that relieve photoinhibition are recognized, either predominantly biophysical – e.g. release of heat (Krause et al. 1988; Havaux 1989) – or predominantly biochemical, e.g. utilization of a smaller light-harvesting antenna (Neale and Melis 1990), or provision of metabolic pathways capable of reoxidizing NADPH when the generation of the reductant exceeds the capacity of photosynthesis to use it. Such pathways include photorespiration (Heber et al. 1989; see Sect. 7), recycling of respiratory CO_2 (Griffiths et al. 1989), and the ascorbate/glutathione system that concomitantly reduces O_2 (Mehler

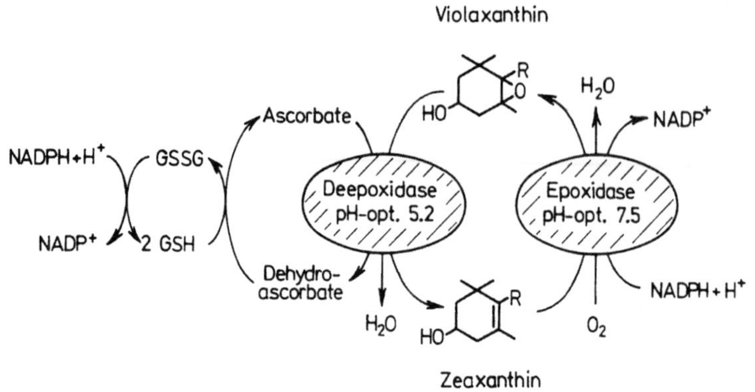

Fig. 3. Scheme for the xanthophyll cycle. The forward reaction (catalyzed by the de-epoxidase) and the reverse reaction (epoxidase) have very different pH optima and utilize different reductants (Hager 1975)

reaction; Cornic et al. 1989) to H_2O_2 and then, in effect, uses this H_2O_2 to oxidize NADPH (see Fig. 7 B_1 in our 1982 review). This latter system seems to be especially important when low temperature accompanies excess irradiance (Schöner and Krause 1990). Interestingly, the low temperature inactivates catalase (Volk and Feierabend 1989), probably leaving ascorbate peroxidase (one of the enzymes of the ascorbate/glutathione system) with a greater role in preventing H_2O_2 accumulation. The presence of both chloroplast and nonchloroplast forms of the enzyme in *Camellia sinensis* leaves would suit this role (Chen and Asada 1989), given that catalase is a peroxisomal enzyme. Finally, there is evidence indicating that the ascorbate/glutathione system also has a protective role during water stress (Smirnoff and Colombé 1988; Cornic et al. 1989), perhaps supplementing the protection afforded by photorespiration (Stuhlfauth et al. 1990) and a stimulation of sucrose synthesis (Quick et al. 1989).

An alternative ascorbate/glutathione system that could, in theory, dissipate excess harvested light energy is the xanthophyll cycle outlined by Hager (1975) over 15 years ago. This scheme (reproduced in Fig. 3) does not involve the potentially toxic H_2O_2 and, in addition, is of renewed interest following the discovery that leaves exposed to excess light energy respond with a dramatic increase in the level of zeaxanthin and a corresponding decrease in the level of violaxanthin; in contrast, the level of β-carotene is not altered (Demmig et al. 1987). Further experiments with the reductant dithiothreitol have added weight to the concept that zeaxanthin is involved in a process that affords protection against photoinhibition (Winter and Königer 1989; Demmig-Adams et al. 1990a). Other reports indicate that the process is widespread, being in the shade plant *Monstera deliciosa* (Demmig-Adams et al. 1989), lichens with green phycobionts (but not with blue-green phycobionts that lack zeaxanthin; (Demmig-Adams et al. 1990b), marine microalgae (Kana et al. 1988; Cunningham et al. 1989), and at an increased level in a mutant wheat deficient in superoxide dismutase (Antonielli et al. 1990). Although details of the photoprotective mechanism are unclear, it is intriguing that ascorbate is the reductant required to form the zeaxanthin (Hager 1975; Demmig-Adams et al. 1990a). Clearly, ascorbate is a multifunctional protectant in photosynthetic cells.

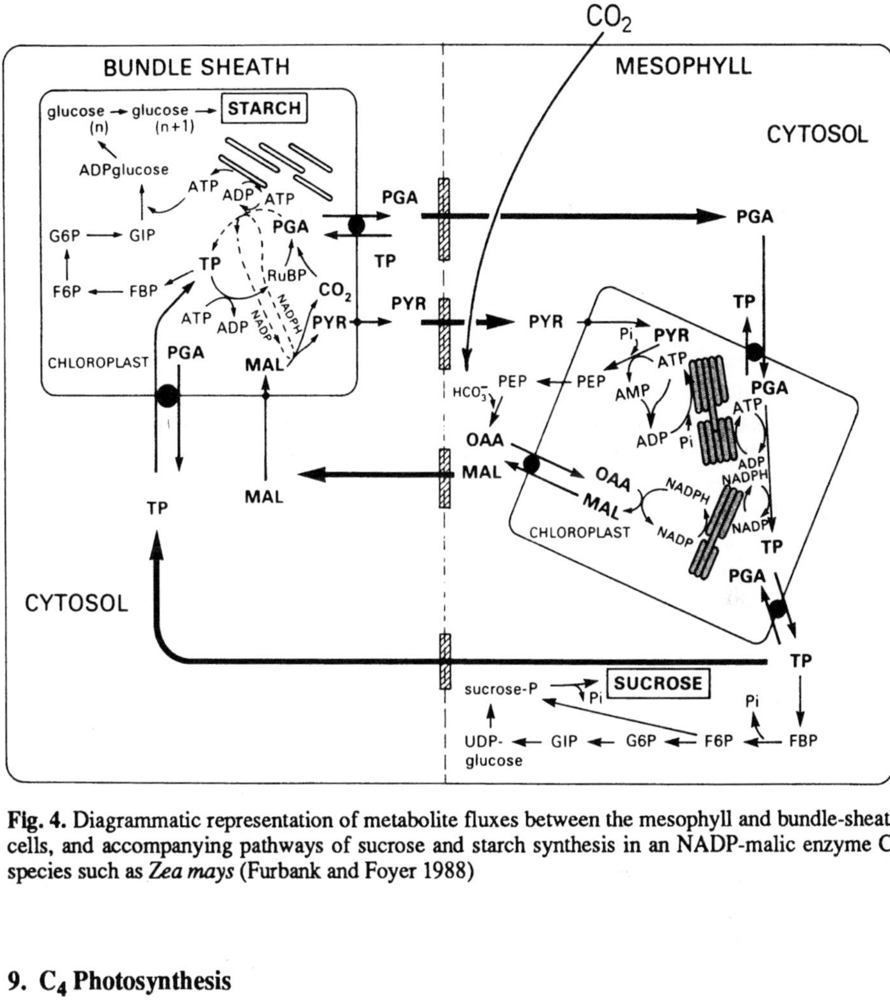

Fig. 4. Diagrammatic representation of metabolite fluxes between the mesophyll and bundle-sheath cells, and accompanying pathways of sucrose and starch synthesis in an NADP-malic enzyme C_4 species such as *Zea mays* (Furbank and Foyer 1988)

9. C_4 Photosynthesis

Details of the metabolic pathway that constitutes C_4 photosynthesis in C_4 plants have changed little since their elucidation during the 1970s. The process by which CO_2 is captured in mesophyll cells, and transported as part of a molecule of malate (or aspartate) into bundle sheath cells where it is released again (by one of three alternative decarboxylation enzymes) and fed into the Calvin cycle, still stands (Fig. 4). It is now clear that the malate or aspartate diffuses through the plasmodesmata that interconnect the mesophyll and bundle-sheath cells (Weiner et al. 1988). Otherwise, the permeability barrier formed by the location and packing of the bundle-sheath cells is sufficiently substantial that direct diffusion of CO_2 into or out of these cells is one to two orders of magnitude less than for photosynthetic cells of C_3 plants (Furbank et al. 1989). The predicted CO_2 concentration in bundle-sheath cells becomes 70 μM, compared to only 4 μM in mesophyll cells (Jenkins et al. 1989). This difference, however, depends on the relative impermeability of the bundle-sheath cells, a factor that seems to vary with leaf age (Moore and Edwards 1988) and may also vary genetically, as evidenced by genotype differences in carbon-isotope discrimination (Hubick et al. 1990).

One view of C_4 photosynthesis is that of a Calvin cycle connected to an ATP- dependent CO_2 "pump" (Furbank and Foyer 1988). The pump in effect requires that two molecules of ATP, in addition to the three required by the Calvin cycle, must be invested for each molecule of CO_2 fixed – the energy from these two ATP is used to synthesized phosphoenolpyruvate (PEP) from pyruvate (Fig. 4). Extra light energy would seem to be needed to generate this extra ATP, but calculations and considerations by Furbank et al. (1990) show that it is difficult to demonstrate such an increased light requirement. These authors therefore propose that, during photosynthetic electron transport, plastoquinone "cycles" and moves four, rather than two, protons into the thylakoid lumen as it transports each pair of electrons. This constitutes a Q-cycle, for which there is other evidence from experiments with chloroplasts and mitochondria. The greater proton gradient thus achieved leads to extra ATP sythesis by CF_1. This mechanism, together with some cyclic and/or pseudocyclic electron transport, could explain the observed efficiency of light utilization by C_4 plants.

Returning to carbon metabolism: regulation of this process in C_4 plants has been examined through studies of enzyme properties (see below; Sect. 12), membrane transport, and metabolite levels (Usuda 1988; Leegood and von Caemmerer 1989). An odd set of observations is that, like C_4 mesophyll chloroplasts (Gross et al. 1990), bundle-sheath chloroplasts possess phosphate translocators capable of transporting PEP, and glycolate/glycerate translocator (Ohnishi and Kanai 1988; Ohnishi et al. 1989), even though these chloroplasts are not required to transport either PEP or glycerate. Another aspect of transport is the loading of assimilates of C_4 photosynthesis into sieve tubes for movement to sink tissue. The sucrose produced in mesophyll cells (Fig. 4) is somehow directed to, and loaded into small bundles, and then moved via transverse veins into large bundles for longitudinal transport (Fritz et al. 1989). When the process of sucrose export was prevented in *Amaranthus edulis*, the level of sucrose in the leaves increased severalfold and C_4 photosynthesis was simultaneously reduced, indicating the presence of a feedback control mechanism (Blechschmidt-Schneider et al. 1989).

Besides PEP carboxylase (see Sect. 12), a number of other enzymes of central importance to C_4 photosynthesis have received recent attention. Kinetic, molecular, and immunological studies now permit more detailed descriptions to be made of the decarboxylation enzymes NAD-malic enzyme and NADP-malic enzyme, and have revealed some species-dependent variation in certain enzyme properties (Murata et al. 1989a, b; Fathi and Schnarrenberger 1990; Iglesias and Andreo 1990). The light/dark-mediated control of pyruvate Pi dikinase activity in mesophyll chloroplasts, which primarily involves enzyme phosphorylation/dephosphorylation, is thought by Nakamoto and Young (1990), but not by Usuda (1988), to be influenced by adenylate energy charge. Roeske and Chollet (1989) have an intermediate view. Incidentally, the adenylate pool in these chloroplasts is maintained in equilibrium by adenylate kinase (Usuda 1988), the activity of which increased fivefold upon greening of C_4 plant leaves, but remains constant in greening C_3 plant leaves (Kleczkowski and Randall 1988).

10. C_3–C_4 Intermediate Photosynthesis

The concept of a relatively strict categorization of plants into C_3, C_4, C_3–C_4, and CAM is being eroded as further studies have begun to reveal a spectrum of types between true C_3, on the one hand, and true C_4 or true CAM, on the other. A C_3–C_4 plant or tissue can now be thought of as any C_3 plant (or tissue) with a greater or lesser amount of C_4 metabolism. The genus *Flaveria* is providing some interesting examples. *F. ramosissima*, *F. floridana*, and *F. brownii* are C_3–C_4 species that possess so much C_4 metabolism that they are almost true C_4's – however, the C_4 character is reduced in *F. floridana* under mild water stress, and in *F. brownii* under low light (Apel and Peisker 1988; Chastain and Chollet 1989; Cheng et al. 1989). More unusual cases of C_3–C_4 include the glumes and lemmas of cereal grasses (Ziegler-Jöns 1989), the marine macroalga *Udotea flabellum* (Reiskind et al. 1988), and the amphibious plant *Eleocharis vivipara* that is C_4 under terrestrial conditions but C_3 when immersed (Ueno et al. 1988).

One characteristic of C_3–C_4 photosynthesis, viz. reduced photorespiration, has become a popular topic of investigation. In the C_3–C_4 *Flaveria* species that have strong C_4 character (see above), reduced photorespiration may be due, at least in part, to efficient refixation of any photorespired CO_2 by PEP carboxylase (Moore et al. 1988; Chastain and Chollet 1989; von Caemmerer and Hubick 1989). However, these and other studies (Moore and Edwards 1989; Reiskind et al. 1989) indicate that a compartmentation of photorespiratory metabolism may be equally important, not only in the strongly-C_4 C_3–C_4 plants, but also in typical C_3–C_4 species such as *F. linearis*, *Moricandia arvensis*, and *Panicum milioides*. The probable biochemical basis of this compartmentation has been elucidated by Rawsthorne et al. (1988) and Hylton et al. (1988) who demonstrated that glycine decarboxylase, the enzyme that catalyzes the release of photorespiratory CO_2, was restricted to bundle-sheath mitochondria in all of nine C_3–C_4 species examined. A similar enzyme localization occurred in C_4 plants, but in C_3 plants the enzyme was found in all cells. Combined with the observation that mitochondria are more partitioned into bundle-sheath cells in C_3–C_4 species compared to C_3 species (Brown and Hattersley 1989), this glycine decarboxylase localization not only provides a mechanism for explaining the reduced photorespiration of C_3–C_4 plants, but also may offer a qualitative biochemical definition of C_3–C_4 plants: those with mesophyll cells containing RuBP carboxylase but lacking glycine decarboxylase.

11. Photosynthesis and CAM

Like the C_4 pathway of photosynthesis, the metabolic pathway that constitutes CAM is now well established, and while some recent studies have concentrated on finer details of the pathway, most have related the pathway to environmental variables.

Briefly, CAM involves nocturnal carboxylation of PEP to form malate which accumulates in vacuoles until the following day, when decarboxylation of the malate supplies CO_2 for the Calvin cycle (which may thus operate behind closed stomates), and a 3C-compound that is usually converted to starch. The starch is mobilized for the following night's quota of PEP. However, in plants such as pineapple, soluble sugars rather than starch form this carbohydrate reserve, and for their nocturnal mobilization a high activity of pyrophosphate-linked PFK is utilized (Carnal and Black 1989). The malate formed at night is pumped into vacuoles, which must therefore possess tonoplast proteins for this task. Both an ATPase and a pyrophosphatase are being considered as the proteins supplying the energy for this accumulation process (Bremberger et al. 1988; Haschke et al. 1988). By translocating protons into the vacuole, they facilitate the import of malate by a malate transporter that is reasonably specific for the malate anion, although it will also accept fumarate (White and Smith 1989), a fact that may have bearing on the early observation that the malate produced during CAM is in equilibrium with fumarate through the fumarase reaction in mitochondria – this observation has been recently confirmed using $^{13}CO_2$ (Osmond et al. 1988). Another organic acid of probable significance to CAM is citrate which, like malate, exhibits day-night fluctuations in leaf content. Lüttge (1988) proposes some physiological functions for this citrate, although nocturnal storage of CO_2 is not one of them.

During the day, malate is returned from the vacuole to the cytoplasm. A protein "gate" that permits malate diffusion is considered to facilitate this return (Wilkins 1989). A cycle of nocturnal CO_2 uptake/malate synthesis and diurnal malate degradation/CO_2 emission is therefore established and continues even during prolonged darkness or illumination, indicating the presence of a circadian rhythm. A series of studies on *Bryophyllum fedtschenkoi* leaves by Anderson and Wilkins (1989) point to interactions between the synthesis and breakdown of malate, on the one hand, and malate inhibition of PEP carboxylase, on the other, as an integral part of the circadian oscillator.

Superimposed on this basic CAM rhythm are other metabolic events involving CO_2, to extents that are probably species-dependent. One of these is photorespiration, which is usually thought of as being suppressed in CAM plants by a high intracellular CO_2 concentration developed following malate decarboxylation. Cote et al. (1989) have shown that this is certainly true in pineapple leaves during the first part of the light period, but in fact is not true during the latter part when the CO_2 concentration becomes so low that photorespiration occurs. In other studies with the CAM bromeliads *Bromelia humilis* and *Aechmea nudicaulis* from South America, and the CAM fern *Pyrrosia piloselloides* from southeast Asia, the problem of keeping up the intracellular CO_2 concentration appears to be dealt with by recycling respiratory CO_2; without this, the plants would be threatened with photoinhibition (Griffiths 1988; Griffiths et al. 1989; Fetene et al. 1990). Such photoinhibition would, of course, be more likely under high irradiance. To help avoid this, CAM plants, like other plants, acclimate to high irradiance by reducing the leaf content of light-harvesting pigments (Winter and Awender 1989; Fetene et al. 1990). A notable feature of this response observed with the thick leaves of *Kalanchoë daigremontiana* was that CAM photosynthesis occurred predominantly in the lower (abaxial) layers of the leaf under high irradiance; adaxial photosynthesis predominated only when irradiance was quite low, e.g. 5% full sunlight (Winter and Awender 1989).

Further studies on the induction or enhancement of CAM by water and/or nutrient stress have been reported, but details of the mechanisms involved are still unclear (Brulfert et al. 1988; Guralnick and Ting 1988; Huerta and Ting 1988; Ota 1988; Paul and Cockburn 1990). However, the consensus of opinion seems to be that the stress-induced events involved are biochemical in nature.

12. The Other CO_2-Fixing Enzyme: PEP Carboxylase

All atmospheric CO_2 initially meets either RuBP carboxylase or PEP carboxylase on its way to being assimilated by plants. The predominance of PEP carboxylase in C_4 and CAM plants, and to some extent in C_3–C_4 plants, is well known, and the enzyme from these groups of plants has been the subject of numerous investigations. However, only minor attention has been given to the effect of CO_2 concentration on the enzyme, no doubt partly because the actual form of inorganic carbon used as substrate is HCO_3^-. Now three recent reports attend to the effects of CO_2. In the first two, Besford and Hand (1989) and Yelle et al. (1989a) show that the amounts of PEP carboxylase in the leaves of tomato and lettuce are unaffected when the plants are grown in a CO_2-enriched (ca. 900 µl/l) atmosphere. This contrasted with the response of RuBP carboxylase (see Sect. 3). It would be valuable to have a similar investigation made with a C_4 plant. In the third report, Samaras and Manetas (1988) found that the concentration of CO_2 influenced the extent of light-mediated activation of PEP carboxylase: there was a twofold activation with physiological CO_2, but no activation with CO_2 increased to 700 µl/l.

Further details of the light-mediated regulatory process have become available. In C_4-plant leaves, illumination leads to a protein kinase-catalyzed phosphorylation of a seryl residue on the PEP carboxylase, which then is less sensitive to feedback inhibition by malate (Jiao and Chollet 1989). Processing of the light signal to switch on the protein kinase may involve Ca^{2+} and calmodulin (Echevarria et al. 1988).

In CAM plants, a similar phosphorylation that produces enzyme less sensitive to malate occurs, but in darkness. The phosphorylation has now been reproduced in vitro (Kluge et al. 1988). In addition, the enzyme system for dephosphorylating the carboxylase upon illumination has been demonstrated (Carter et al. 1990). It is remarkable that the process of PEP carboxylase phosphorylation is enhanced by light in C_4 plants and by darkness in CAM plants.

Most other studies of the regulation of PEP carboxylase activity have involved effects of phosphorylated intermediates (including the substrate PEP) and malate. Generally, the form of PEP utilized as substrate is MgPEP, although the enzyme from CAM plants can use free PEP at pH 7 (Wedding et al. 1988; Outlaw 1990; Meyer et al. 1990). In addition, there is a separate binding site on the enzyme for activator molecules; this site will also accept PEP, resulting in stimulation of activity (Rustin et al. 1988a). It is free PEP, rather than MgPEP, that binds to the activator site of the maize-leaf enzyme (Rodríguez-Sotres and Muñoz-Clares 1990). Normally, the activator site is considered to bind glucose-6-P which, along with malate, is a principal regulator of the enzyme's activity (Walker et al. 1988; Wedding et al. 1989). In C_4 and CAM plants, but not C_3 plants, regulation by glucose-6-P may be modified by "energy charge" (i.e., AMP and ATP) regulation, especially during the day (Rustin et al. 1988b; Walker et al. 1988; Meyer et al. 1990).

Studies of malate inhibition have usually coincided with studies of the state of enzyme polymerization. This is because both the C_4 and CAM PEP carboxylases are tetrameric proteins that easily dissociate into two relatively inactive dimers, and this dissociation is promoted by malate; reassociation of the tetramer is promoted by the substrate MgPEP (McNaughton et al. 1989; Jawali 1990; Wedding et al. 1990; Willeford et al. 1990). The association/dissociation phenomenon is as might be expected, dependent on the protein concentration (Ngam-ek et al. 1989; Podestá and Andreo 1989), thus it will be important to establish its significance at high, physiological protein concentrations representative of in vivo conditions (Manetas 1990).

A reevaluation of some of the properties of PEP carboxylase may be necessary following several reports that question the procedure used to assay the enzyme activity. Possible misinterpretations may be caused by (1) changes in pH and/or effector concentrations causing changes in the concentration of the true substrate MgPEP although the total (free plus complexed) PEP concentration is constant (Outlaw 1990); (2) differences in enzyme activity dependent upon wether reactions are initiated with PEP or with enzyme (Angelopoulos et al. 1988); (3) possible interference by nonenzymatic decarboxylation of the product oxaloacetate and/or phosphatase-catalyzed degradation of PEP when the enzyme activity in crude or partially purified extracts is determined (Smith et al. 1989).

13. Photosynthetic Carbon Metabolism in a Future Atmosphere Containing a High Level of CO_2

For the 100 millenia prior to 1850, the level of atmospheric CO_2 fluctuated between about 200 and 280 µl/l (Pearman 1988). Anthropogenic activities have, since 1850, increased the level to 350 µl/l, and threaten to raise the level to over 600 µl/l sometime in the latter half of the next century. Will stimulated photosynthetic CO_2 fixation reduce the rate at which atmospheric CO_2 increases? Will stimulated plant growth lead to a greater global plant biomass, thereby reducing the final value reached for the concentration of atmospheric CO_2? These questions are addressed briefly in the context of the foregoing review of photosynthetic carbon metabolism.

Marine photosynthesis is unlikely to be stimulated by the increase in atmospheric CO_2 (i.e., as a result of the consequent increase in CO_2 dissolved in the photic zone of the ocean) if marine phytoplankton are already saturated with CO_2. The evidence for CO_2 and/or HCO_3^- pumps on marine microalgal cells (see Sect. 2) points to this being the case. Consequently, marine photosynthesis would not be expected to counteract any of the increase in atmospheric CO_2. However, this conclusion would not apply if the biomass of phytoplankton increased. This biomass is largely determined by the availability of limiting nutrients, principally N, P, and Fe (see Kelly 1989). The levels of these may indeed change as a result of altered ocean circulation patterns associated with any future climate changes.

Terrestrial photosynthesis almost certainly is being stimulated by the increased level of atmospheric CO_2, but the extent of this stimulation, and the consequences for the global carbon budget are still being addressed. An analysis of available data by Kimball (1983) provides overwhelming evidence that the photosynthesis of plants growing under ideal conditions is stimulated by increased CO_2, but whether the same applies to plants growing under more "natural" conditions where nutrient, water, or temperature stress are common is less clear. For example, Reddy et al. (1989) put some blame on nitrogen stress for the failure of cotton plants to utilize the increase in atmospheric CO_2 that has occurred over the past 15 years. In addition, many experiments have involved taking plants grown in a normal atmosphere and briefly exposing them to a high-CO_2 atmosphere, whereas realistic experiments (in the context of a future high-CO_2 environment)

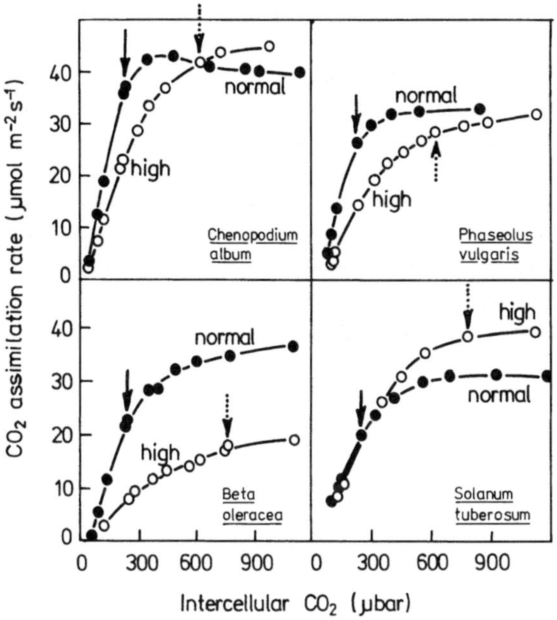

Fig. 5. Photosynthesis by four plant species grown in a normal CO_2 (300 µl/l; ●—●) or high CO_2 (900 to 1000 µl/l; ○—○) atmosphere. Under these atmospheres, the corresponding intercellular CO_2 levels were 200 to 250 µl/l and 620 to 850 µl/l, respectively. The effect of increasing CO_2 concentration on the rate of photosynthesis is shown. *Arrows* indicate the photosynthetic rates of plants grown and measured in normal CO_2 (→), and grown and measured in high CO_2 (···>) (Sage et al. 1989)

need to involve plants actually grown in a high-CO_2 atmosphere (Kramer 1981; Dahlman et al. 1985).

Of course, elevated CO_2 would be expected to increase terrestrial photosynthesis if the kinetic properties of RuBP carboxylase/oxygenase were solely responsible for determining the response. The increase would be attributed to (1) the direct effect of increasing the concentration of this subsaturating substrate, and (2) the indirect effect of this CO_2 displacing its competitor O_2, thereby decreasing photorespiration (Sharkey 1988, 1989; du Cloux et al. 1989). However, the enzyme's regulatory properties are also relevant, as indicated by a number of recent studies involving plants grown in a high-CO_2 atmosphere. These have indicated that the capacity of these plants to utilize the additional assimilated carbon limits photosynthesis, and the activity of RuBP carboxylase is correspondingly reduced by lessening the proportion of enzyme molecules in the CO_2-plus-Mg^{2+} activated form (Campbell et al. 1988; Sharkey et al. 1988; Besford and Hand 1989; Sage et al. 1989; Yelle et al. 1989a). Thus, the photosynthesis of plants grown and measured in high CO_2 may be little different to, or even less than, that of plants grown and measured in normal CO_2-examples taken from Sage et al. (1989) are shown in Fig. 5. While these responses of high CO_2-grown plants are being explained in terms of assimilate utilization (see below) it may be worth recalling the rather dramatic observation of Porter and Grodzinski (1984) that bean leaves grown in a high-CO_2 atmosphere contained only one-tenth of the carbonic anhydrase activity found in normal CO_2-grown leaves.

Many studies of the "CO_2-Düngungseffekt" have emphasized the importance of sink tissues to accommodate the increased quantity of assimilated carbon. In the first instance, such sinks are the sucrose and/or starch pools in the photosynthetic cells, and the rates at which these are filled can determine the photosynthetic rate (Sage and Sharkey 1987;

Stitt and Schreiber 1988) and cause the activities of RuBP carboxylase (see above) and photosynthetic electron transport (Sharkey et al. 1988) to be accordingly reduced. In the longer term, the carbon in these sucrose and starch pools needs to be translocated to support the growth of storage organs; where such organs are readily available, CO_2-stimulated photosynthesis is more easily accommodated (Bhattacharya et al. 1989; Diethelm and Shibles 1989; Idso and Kimball 1989). Where sink organs are less available, the additional assimilated carbon may instead be used to support the initiation and early growth of new shoots (Nijs et al. 1988, 1989; Poorter et al. 1988; du Cloux et al. 1989), although where competition and/or stress factors exist, this stimulation may be reduced or absent (Williams et al. 1988). Where additional plant growth does not occur, excess carbohydrate may be simply respired (Musgrave et al. 1986), or it may accumulate in excess, especially in chloroplasts as starch grains that can become so large that the organelle is physically distorted, an event that probably contributes to the failure of high CO_2 to increase photosynthesis in some high CO_2-acclimated plants (DeLucia et al. 1985; Dons 1988; Yelle et al. 1989b; Wong 1990).

In summary, the above information points to the availability of carbon sinks as a prime determinant of whether or not the increasing level of atmospheric CO_2 stimulates terrestrial photosynthesis. In quantitative terms, the predominant carbon sink is wood, thus it is the photosynthetic carbon metabolism of trees that needs to be addressed when attempting to ascertain whether stimulated photosynthesis will counteract anthropogenic CO_2 release. In a recent review of the effects increased of CO_2 on tree growth, Eamus and Jarvis (1989) concluded that the maintenance of active sinks for assimilates is a necessary precondition for any CO_2 stimulation of tree photosynthesis. They also pointed out that most experiments have involved very young trees for a relatively short (e.g. 12 month) period of their life cycle, and since the response to increased CO_2 decreases or ceases as plants grow older (Kramer 1981), it may be that the actual stimulation of global photosynthesis is less than that indicated by these experiments with young trees. Another factor is whether or not stress might hamper positive responses of trees to CO_2. Experiments with *Quercus alba* seedlings showed that CO_2 enrichment increased both water-use efficiency and nutrient-use efficiency (Norby et al. 1986; Norby and O'Neill 1989), thereby counteracting some of the stress effects. However, Conroy et al. (1988) found that phosphate deficiency eliminate the growth response of *Pinus radiata* to high CO_2, and Mousseau and Enoch (1989) observed that *Castanea sativa* seedlings suffered an unusual nutrient stress brought about by nutrient dilution following the rapid growth induced by CO_2 enrichment; this caused premature stem elongation and leaf senescence.

Undoubtedly, the response of trees to CO_2 enrichment will be found to vary, just as the responses of nontree plants do (Fig. 5). Understanding the variation will require studies equal to those that have revealed the unexpected variability of the mechanisms for regulating photosynthetic carbon metabolism, e.g. in some species RuBP carboxylase is regulated by carboxyarabinitol-1-P, in others it is not (see Sect. 3). It is not inconceivable, to put our current knowledge of photosynthetic carbon metabolism into the context of tree biomass accumulation in a high CO_2 atmosphere, that oak leaves and eucalyptus leaves will replace spinach leaves as the preferred material for laboratory studies in the years to come.

Finally, it is salutatory to note that, no matter what the outcome of such future investigations may be, the very existence of a carbon cycle in nature precludes the possibility that all anthropogenic CO_2 release could be counteracted by tree photosynthesis. There could be no contribution whatsoever from removing mature forests and then replanting them (Harmon et al. 1990), while growing trees in areas not now forested is equivalent to removing from the atmosphere (and sequestering as wood) that carbon originally injected into the atmosphere by the decay or combustion of the wood obtained by human deforestation of the same areas sometime in the past. Thus, during the coming centuries there is no prospect that photosynthesis will remove from the atmosphere any of the CO_2 that has originated from the combustion of fossil fuels – an amount equal to about half of the anthropogenic CO_2 release since 1750 (Gifford 1988), plus practically all of the release predicted for the future.

Acknowledgments. We thank Dr. Simon P. Robinson for helpful discussion, and Ms. Wendy Schipper for excellent assistance.

References

Anderson CM, Wilkins MB (1989) Planta 180:61–73. **Angelopoulos K**, Stamatakis K, Manetas Y, Gavalas NA (1988) Photosyn Res 18:317–325. **Antonielli M**, Pocceschi N, Lupattelli M (1990) Physiol Plant 78:187–192. **Apel P**, Peisker M (1988) Biochem Physiol Pflanzen 183:439–442.

Bate GC, Sültemeyer DF, Fock HP (1988) Photosyn Res 16:219–231. **Beck E**, Ziegler P (1989) Annu Rev Plant Physiol Plant Mol Biol 40:95–117. **Beck E**, Scheibe R, Reiner J (1989) Plant Physiol 90:13–16. **Besford RT**, Hand DW (1989) J Exp Bot 40:329–336. **Bhattacharya NC**, Bhattacharya S, Strain BR, Biswas PK (1989) J Plant Physiol 135:261–266. **Blechschmidt-Schneider S**, Ferrar P, Osmond CB (1989) Planta 177:515–525. **Borchert S**, Große H, Heldt HW (1989) FEBS Lett 253:183–186. **Bremberger C**, Haschke H-P, Lüttge U (1988) Planta 175:465–470. **Brooks A**, Portis AR, Sharkey TD (1988) Plant Physiol 88:850–853. **Brown RH**, Hattersley PW (1989) Plant Physiol 91:1543–1550. **Brulfert J**, Kluge M, Güçlü S, Queiroz O (1988) J Plant Physiol 133:222–227. **Budde RJA**, Randall DD (1990) Proc Natl Acad Sci USA 87:673–676. **Burns BD**, Beardall J (1987) J Exp Mar Bio Ecol 107:75–86. **Butz ND**, Sharkey TD (1989) Plant Physiol 89:735–739.

Cadet F, Meunier J-C (1988) Biochem J 253:243–248. **Caemmerer S von**, Hubick KT (1989) Planta 178:475–481. **Campbell WJ**, Ogren WL (1990) Plant Physiol 92:110–115. **Campbell WJ**, Allen LH, Bowes G (1988) Plant Physiol 88:1310–1316. **Cardon ZG**, Mott KA (1989) Plant Physiol 89:1253–1257. **Carnal NW**, Black CC (1989) Plant Physiol 90:91–100. **Carter PJ**, Nimmo HG, Fewson CA, Wilkins MB (1990) FEBS Lett 263:233–236. **Caspar T**, Lin T-P, Monroe J, Bernhard W, Spilatro S, Preiss J, Somerville C (1989) Proc Natl Acad Sci USA 86:5830–5833. **Chastain CJ**, Chollet R (1989) Planta 179:81–88. **Chen G-X**, Asada K (1989) Plant Cell Physiol 30:987–998. **Chen Z**, Spreitzer RJ (1989) J Biol Chem 264:3051–3053. **Cheng S-H**, Moore BD, Wu J, Edwards GE, Ku MSB (1989) Plant Physiol 89:1129–1135. **Cloux H du**, André M, Gerbaud A, Daguenet A (1989) Photosynthetica 23:145–153. **Collatz GJ**, Berry JA, Farquhar GD, Pierce J (1990) Plant Cell Environ 13:219–225. **Colman B**, Rotatore C (1988) J Exp Bot 39:1025–1032. **Conroy JP**, Kuppers M, Kuppers B, Virgona J, Barlow EWR (1988) Plant Cell Environ 11:91–98. **Cornic G**, Le Gouallec J-L, Briantais JM, Hodges M (1989) Planta 177:84–90. **Cote FX**, Andre M, Folliot M, Massimino D, Daguenet A (1989) Plant Physiol 89:61–68. **Crafts-Brandner SJ**, Salvucci ME (1989) Plant Physiol 91:469–472. **Crawford NA**, Droux M, Kosower NS, Buchanan BB (1989) Arch Biochem Biophys 271:223–229. **Critchley C** (1988) Aust J Plant Physiol 15:27–41. **Cunningham FX**, Dennenberg RJ, Mustardy L, Jursinic PA, Gantt E (1989) Plant Physiol 91:1179–1187.

Dahlman RC, Strain BR, Rogers HH (1985) J Environ Qual 14:1–8. **Dancer JE**, ap Rees T (1989) J Plant Physiol 135:197–206. **Dancer J**, Neuhaus HE, Stitt M (1990) Plant Physiol 92:637–641. **DeLucia EH**, Sasek TW, Strain BR (1985) Photosyn Res 7:175–184. **Demmig B**, Winter K, Krüger A, Czygan F-C (1987) Plant Physiol 84:218–224. **Demmig-Adams B**, Winter K, Krüger A, Czygan F-C (1989) Plant Physiol 90:881–886. **Demmig-Adams B**, Adams WW, Heber U, Neimanis S, Winter K, Krüger A, Czygan F-C, Bilger W, Björkman O (1990a) Plant Physiol 92:293–301. **Demmig-Adams B**, Adams WW, Czygan F-C, Schreiber U, Lange OL (1990b) Planta 180:582–589. **Diethelm R**, Shibles R (1989) J Plant Physiol 134:70–74. **Dixon GK**, Merrett MJ (1988) New Phytol 109:47–51. **Dixon GK**, Brownlee C, Merrett MJ (1989) Planta 178:443– 449. **Dons C** (1988) Photosynthetica 22:328–334. **Dujardyn M**, Foyer CH (1989) Plant Physiol 91:1562–1568.

Eamus D, Jarvis PG (1989) Adv Ecol Res 19:1–55. **Echevarria C**, Vidal J. LeMaréchal P, Brulfert J, Ranjeva R, Gadal P (1988) Biochem Biophys Res Commun 155:835–840. **Edmondson DL**, Kane HJ, Andrews TJ (1990) FEBS Lett 260:62–66. **Entwistle G**, ap Rees T (1988) Biochem J 255:391–396. **Espie GS**, Miller AG, Birch DG, Canvin DT (1988a) Plant Physiol 87:551–554. **Espie GS**, Miller AG, Canvin DT (1988b) Plant Physiol 88:757–763. **Espie GS**, Miller AG, Canvin (1989) Plant Physiol 91:387–394.

Farazdaghi H, Edward GE (1988) Plant Cell Environ 11:789–798. **Fathi M**, Schnarrenberger C (1990) Plant Physiol 92:710–717. **Fetene M**, Lee HSJ, Lüttge U (1990) New Phytol 114:399–406. **Flügge IU**, Woo KC, Heldt HW (1988) Planta 174:534–541. **Fritz E**, Evert RG, Nasse H (1989) Planta 178:1–9. **Furbank RT**, Foyer CH (1988) New Phytol 109:265–277. **Furbank RT**, Jenkins CLD, Hatch MD (1989) Plant Physiol 91:1364–1371. **Furbank RT**, Jenkins CLD, Hatch MD (1990) Aust J Plant Physiol 17:1–7.

Galloway CM, Dugger WM, Black CC (1988) Plant Physiol 88:980–982. **Garab G**, Lajkó F, Mustárdy L. Márton L (1989) Planta 179:349–358. **Gardeström P**, Wigge B (1988) Plant Physiol 88:69–76. **Gifford RM** (1988) Interactions with vegetation In: Pearman GI (ed) Greenhouse: planning for climate change. CSIRO, Melbourne, pp 83–89. **Gilck H**, Beck E (1974) Z Pflanzenphysiol 72:395–409. **Givan CV**, Tsutakawa S, Hodgson JM, David N, Randall DD (1988) J Plant Physiol 132:593–599. **Goyal A**, Tolbert NE (1989) Plant Physiol 89:1264–1269. **Grande KD**, Marra J, Langdon C, Heinemann K, Bender ML (1989) J Expt Mar Biol Ecol 129:95–120. **Griffiths H** (1988) Plant Cell Environ 11:603–611. **Griffiths H**, Ong BL, Avadhani PN, Goh CJ (1989) Planta 179:115–122. **Gross A**, Brückner G, Heldt HW, Flügge U-I (1990) Planta 180:262–271. **Guralnick LJ**, Ting IP (1988) Plant Cell Environ 11:811–818. **Gutteridge S** (1990) Biochem Biophys Acta 1015:1–14. **Gutteridge S**, Julien B (1989) FEBS Lett 254:225–230.

Hager A (1975) Ber Dtsch Bot Gesell 88:27–44. **Harmon ME**, Ferrell WK, Franklin JF (1990) Science 247:699–702. **Haschke H-P**, Grötsch S, Lüttge U (1988) J Plant Physiol 132:604–607. **Hatzfeld W-D**, Stitt M (1990) Planta 180:198–204. **Hatzfeld W-D**, Dancer J. Stitt M (1989) FEBS Lett 254:215–218. **Häusler RE**, Holtum JAM, Latzko E (1989) Plant Physiol 90:1498–1505. **Havaux M** (1989) Plant Physiol 89:286–292. **Heber U**, Viil J, Neimanis S, Mimura T, Dietz K-J (1989) Z Naturforsch 44c:524–536. **Holbrook GP**, Bowes G, Salvucci ME (1989) Plant Physiol 90:673–678. **Huber SC** (1989) Plant Physiol 91:656–662. **Huber JLA**, Huber SC, Neilsen TH (1989a) Arch Biochem Biophys 270:681–690. **Huber SC**, Neilsen TH, Huber JLA, Pharr DM (1989b) Plant Cell Physiol 30:277–285. **Hubick KT**, Hammer GL, Farquhar GD, Wade LJ, von Caemmerer S, Henderson SA (1990) Plant Physiol 92:534–537. **Huerta AJ**, Ting IP (1988) Plant Physiol 88:183–188. **Huppe HC**, de Lamotte-Guéry F, Jacquot JP, Buchanan BB (1990) Planta 180:341–351. **Husic HD**, Kitayama M, Togasaki RK, Moroney JV, Morris KL, Tolbert NE (1989) Plant Physiol 89:904–909. **Hylton CM**, Rawsthorne S, Smith AM, Jones DA, Woolhouse HW (1988) Planta 175:452–459.

Idso SB, Kimball BA (1989) Environ Exp Bot 29:135–139. **Iglesias AA**, Andreo CS (1990) Plant Physiol 92:66–72.

Jawali N (1990) Arch Biochem Biophys 277:61–68. **Jenkins CLD**, Furbank RT, Hatch MD (1989) Plant Physiol 91:1372–1381. **Jiao J-A**, Chollet R (1989) Arch Biochem Biophys 269:526–535.

Kana TM, Glibert PM, Goericke R, Welschmeyer NA (1988) Limnol Oceanogr 33:1623–1627. **Kelly GJ** (1989) Oceanogr Mar Biol Annu Rev 27:11–44. **Kimball B** (1983) Agron J 75:779–788. **Kiss F**, Johnson TC, Klecan AL, Vincze G, Buchanan BB, Balogh A (1989) Photosyn Res 21:123–128. **Kleczkowski LA**, Edwards GE (1989) Plant Physiol 91:278–286. **Kleczkowski LA**, Randall DD (1988) Photosynthetica 22:112–115. **Kleczkowski LA**, Givan CV, Randall DD, Loboda T (1988) Physiol Plant 74:763–769. **Klöck G**, Sültemeyer DF, Fock HP, Kreuzberg K (1989) Physiol Plant 75:109–113. **Kluge M**, Maier P, Brulfert J, Faist K, Wollny E (1988) J Plant Physiol 133:252–256. **Kobza J**, Seemann JR (1988) Proc Natl Acad Sci USA 85:3815–3819. **Kobza J**, Seemann JR (1989a) Plant Physiol 89:918–924. **Kobza J**, Seemann JR (1989b) Plant Physiol 89:174–179. **Kramer PJ** (1981) Bioscience 31:29–33. **Krause GH**, Laasch H, Weis E (1988) Plant Physiol Biochem 26:445–452. **Kreckl W**, Kexel H, Melzer E, Schmidt H-L (1989) J Biol Chem 264:10982–10986. **Kuhn M**, Thiel A, Böger P (1988) Z Naturforsch 43c:413–417.

Larondelle Y, Mertens E, Van Schaftingen E, Hers H-G (1989) Plant Physiol 90:827–834. **Leegood RC**, von Caemmerer S (1989) Planta 178:258–266. **Le Gouallec J-L**, Cornic G (1988) Plant Physiol Biochem 26:705–712. **Lemaire C**, Wollman F-A, Bennoun P (1988) Proc Natl Acad Sci USA 85:1344–1348. **Lüttge U** (1988) Plant Cell Environ 11:445–451.

MacDonald FD, Chou Q, Buchanan BB, Stitt M (1989) J Biol Chem 264:5540–5544. **Manetas Y** (1990) Physiol Plant 78:225–229. **Manuel LJ**, Moroney JV (1988) Plant Physiol 88:491–496. **Marcus F**, Harrsch PB (1990) Arch Biochem Biophys 279:151–157. **Marcus F**, Moberly L, Latshaw SP (1988) Proc Natl Acad Sci USA 85:5379–5383. **Marsh JJ**, Wilson KJ, Lebherz (1989) Plant Physiol 91:1393–1401. **McNaughton GAL**, Fewson CA, Wilkins MB, Nimmo HG (1989) Biochem J 261:349–355. **Meyer CR**, Rustin P, Black MK, Wedding RT (1990) Arch Biochem Biophys 278:365–372. **Miller AG**, Espie GS, Canvin DT (1989) Plant Physiol 90:1221–1231. **Moore Bd**, Edwards GE (1988) Plant Physiol 88:125–130. **Moore Bd**, Edwards GE (1989) Plant Sci 60:155–161. **Moore Bd**, Monson RK, Ku MSB, Edwards GE (1988) Plant Cell Physiol 29:999–1006. **Mousseau M**, Enoch HZ (1989) Plant Cell Environ 12:927–934. **Muñoz J**, Merrett MJ (1988) Planta 175:460–464. **Muñoz J**, Merrett MJ (1989) Planta 178:450–455. **Murata T**, Ikeda J-I, Takano M, Ohsugi R (1989a) Plant Cell Physiol 30:429–437. **Murata T**, Ohsugi R, Matsuoka M, Nakamoto H (1989b) Plant Physiol 89:316–324. **Musgrave ME**, Strain BR, Siedow JN (1986) Proc Natl Acad Sci USA 83:8157–8161.

Nakamoto H, Young PS (1990) Plant Cell Physiol 31:1–6. **Neale PJ**, Melis A (1990) Plant Physiol 92:1196–1204. **Nespoulous C**, Peltier G, Gans P (1989) Plant Physiol Biochem 27:863–871. **Neuhaus HE**, Stitt M (1989) Planta 179:51–60. **Neuhaus HE**, Kruckeberg AL, Feil R, Stitt M (1989) Planta 178:110–122. **Ngam-ek A**, Seery TAP, Amis EJ, Grover SD (1989) Plant Physiol 91:954–960. **Nielsen TH**, Huber SC (1989) Physiol Plant 76:309–314. **Nijs I**, Impens I, Behaeghe T (1988) Photosynthetica 22:44–50. **Nijs I**, Impens I, Behaeghe T (1989) J Exp Bot 40:353–359. **Noctor G**, Mill JD (1988) Biochem Biophys Acta 935:53–60. **Norby RJ**, O'Neill EG (1989) New Phytol 111:491–500. **Norby RJ**, O'Neill EG, Luxmoore RJ (1986) Plant Physiol 82:83–89.

Ohnishi J-i, Kanai R (1988) J Plant Physiol 133:119–121. **Ohnishi J-i**, Flugge UI, Heldt HW (1989) Plant Physiol 91:1507–1511. **Osafune T**, Yokota A, Sumida S, Hase E (1990) Plant Physiol 92:802–808. **Osmond CB**, Holtum JAM, O'Leary MH, Roeske C, Wong OC, Summons RE, Avadhani PN (1988) Planta 175:184–192. **Ota K** (1988) Plant Cell Physiol 29:801–806. **Outlaw WH** (1989) Physiol Plant 77:275–281. **Outlaw WH** (1990) Plant Physiol 92:528–530.

Parry MAJ, Keys AJ, Foyer CH, Furbank RT, Walker DA (1988) Plant Physiol 87:558–561. **Parry MAJ**, Keys AJ, Gutteridge S (1989) J Exp Bot 40:317–320. **Paul MJ**, Cockburn W (1990) New Phytol 114:391–398. **Pearman GI** (1988) Greenhouse gases: evidence for atmospheric changes and anthropogenic causes In: Pearman GI (ed) Greenhouse: planning for climate changes. CSIRO, Melbourne, pp 3–21. **Pettersson G**, Ryde-Pettersson U (1988) Eur J Biochem 177:351–355. **Pettersson G**, Ryde-Pettersson U (1988a) Eur J Biochem 182:373–377. **Pettersson**

G, Ryde-Pettersson U (1989b) Eur J Biochem 186:683–687. **Podestá FE**, Andreo CS (1989) Plant Physiol 90:427–433. **Poorter H**, Pot S, Lambers H (1988) Physiol Plant 73:553–559. **Porter MA**, Grodzinski B (1984) Plant Physiol 74:413–416. **Portis AR** (1990) Biochem Biophys Acta 1015:15–28. **Preisser J**, Komor E (1988) Plant Physiol 88:259–265. **Price GD**, Badger MR (1989a) Plant Physiol 89:37–43. **Price GD**, Badger MR (1989b) Plant Physiol 91:505–513. **Pschorn R**, Rühle W, Wild A (1988) Z Naturforsch 43c:207–212.

Quick P, Siegl G, Neuhaus E, Feil R, Stitt M (1989) Planta 177:535–546.
Rawsthorne S, Hylton CM, Smith AM, Woolhouse HW (1988) Planta 173:298–308. **Reckmann U**, Scheibe R, Raschke K (1990) Plant Physiol 92:246–253. **Reddy VR**, Baker DN, McKinion JM (1989) J Environ Qual 18:427–432. **Reiskind JB**, Seamon PT, Bowes G (1988) Plant Physiol 87:686–692. **Reiskind JB**, Berg RH, Salvucci ME, Bowes G (1989) Plant Sci 61:43–52. **Robinson SP**, Portis AR (1989a) Arch Biochem Biophys 268:93–99. **Robinson SP**, Portis AR (1989b) Plant Physiol 90:968–971. **Robinson SP**, Streusand VJ, Chatfield JM, Portis AR (1988) Plant Physiol 88:1008–1014. **Rodríguez Andrés A**, Lázaro JJ, Chueca A, Hermoso R, López Gorgé J (1990) Physiol Plant 78:409–413. **Rodríguez-Sotres R**, Muñoz-Clares RA (1990) Arch Biochem Biophys 276:180–190. **Roeske CA**, Chollet R (1989) Plant Physiol 90:330–337. **Rustin P**, Meyer CR, Wedding RT (1988a) J Biol Chem 263:17611–17614. **Rustin P**, Meyer C, Wedding R (1988b) Plant Physiol 88:153–157.

Sage RF, Sharkey TD (1987) Plant Physiol 84:658–664. **Sage RF**, Sharkey TD, Seemann JR (1988) Planta 174:407–416. **Sage RF**, Sharkey TD, Seemann JR (1989) Plant Physiol 89:590–596. **Salvucci ME** (1989) Physiol Plant 77:164–171. **Samaras Y**, Manetas Y (1988) Photosyn Res 18:299–305. **Schöner S**, Krause GH (1990) Planta 180:383–389. **Schuller KA**, Randall DD (1989) Plant Physiol 89:1207–1212. **Schulze-Siebert D**, Schultz G (1989) Plant Sci 59:167–174. **Schuster G**, Timberg R, Ohad I (1988) Eur J Biochem 177:403–410. **Seemann JR** (1989) Plant Physiol 91:379–386. **Seemann JR**, Kirschbaum MUF, Sharkey TD, Pearcy RW (1988) Plant Physiol 88:148–152. **Seemann JR**, Kobza J, Moore Bd (1990) Photosyn Res 23:119–130. **Serra JL**, Llama MJ, Rowell P, Stewart WDP (1989) Plant Sci 59:1–9. **Servaites JC** (1990) Plant Physiol 92:867–870. **Servaites JC**, Geiger DR, Tucci MA, Fondy BR (1989a) Plant Physiol 89:403–408. **Servaites JC**, Fondy BR, Li B, Geiger DR (1989b) Plant Physiol 90:1168–1174. **Sharkey TD** (1988) Physiol Plant 73:147–152. **Sharkey TD** (1989) Philos Trans R Soc Lond B 323:435–448. **Sharkey TD**, Vanderveer PJ (1989) Plant Physiol 91:679–684. **Sharkey TD**, Vassey TL (1989) Plant Physiol 90:385–387. **Sharkey TD**, Berry JA, Sage RF (1988) Planta 176:415–424. **Shimazaki K** (1989) Plant Physiol 91:459–463. **Shiraiwa Y**, Bader KP, Schmidt GH (1988) Z Naturforsch 43c:709–716. **Sibley MH**, Anderson LE (1989) Plant Physiol 91:1620–1624. **Sicher RC** (1989) Plant Physiol 89:557–563. **Siegl G**, Stitt M (1990) Plant Sci 66:205–210. **Smirnoff N**, Colombé SV (1988) J Exp Bot 39:1097–1108. **Smith AM**, Hylton CM, Rawsthorne S (1989) Plant Physiol 89:982–985. **Soulie J-M**, Riviere M, Ricard J (1988) Eur J Biochem 176:111–117. **Spalding MH**, Jeffrey M (1989) Plant Physiol 89:133–137. **Stein M**, Lazaro JJ, Wolosiuk RA (1989) Eur J Biochem 185:425–431. **Steingraber M**, Outlaw WH, Hampp R (1988) Planta 175:204–208. **Stitt M**, Grosse H (1988) Plant Physiol 133:129–137. **Stitt M**, Schreiber U (1988) J Plant Physiol 133:263–271. **Stuhlfauth T**, Scheuermann R, Fock HP (1990) Plant Physiol 92:1053–1061. **Sültemeyer DF**, Klöck G, Kreuzberg K, Fock HP (1988) Planta 176:256–260.

Tarczynski MC, Outlaw WH, Arold N, Neuhoff V, Hampp R (1989) Plant Physiol 89:1088–1093. **Thielmann J**, Tolbert NE, Goyal A, Senger H (1990) Plant Physiol 92:622–629. **Tyson RH**, ap Rees T (1988) Planta 175:33–38. **Tyson RH**, ap Rees T (1989) Plant Sci 59:71–76.

Ueno O, Samejima M, Muto S, Miyachi S (1988) Proc Natl Acad Sci USA 85:6733–6737. **Usuda H** (1988) Plant Physiol 88:1461–1468.

Vaughn KC, Vaughan MA (1988) Physiol Plant 74:409–413. **Veau EJ de**, Burris JE (1989a) Plant Physiol 90:500–511. **Veau EJ de**, Burris JE (1989b) Plant Physiol 91:1085–1093. **Virgin I**, Styring S, Andersson B (1988) FEBS Lett 233:408–412. **Volk S**, Feierabend J (1989) Plant Cell Environ 12:701–712.

Walker JL, Huber SC (1989) Planta 177:116–120. **Walker GH**, Ku MSB, Edwards GE (1988) J Plant Physiol 133:144–151. **Wedding RT**, Rustin P, Meyer CR, Black MK (1988) Plant Physiol 88:976–979. **Wedding RT**, Black MK, Meyer CR (1989) Plant Physiol 90:648–652. **Wedding RT**, Black MK, Meyer CR (1990) Plant Physiol 92:456–461. **Weger HG**, Turpin DH (1989) Plant Physiol 89:409–415. **Weger HG**, Herzig R, Falkowski PG, Turpin DH (1989) Limnol Oceanogr 34:1153–1161. **Weiner H**, Burnell JN, Woodrow IE, Heldt HW, Hatch MD (1988) Plant Physiol 88:815–822. **White PJ**, Smith JAC (1989) Planta 179:265–274. **Wilkins MB** (1989) J Exp Bot 40:1315–1321. **Willeford KO**, Gombos Z, Gibbs M (1989) Plant Physiol 90:1084–1087. **Willeford KO**, Wu M-X, Meyer CR, Wedding RT (1990) Biochem Biophys Res Commun 168:778–785. **Williams WE**, Garbutt K, Bazzaz FA (1988) J Exp Bot 28:123–130. **Winter K**, Awender G (1989) Plant Physiol 90:948–954. **Winter K**, Königer M (1989) Planta 180:24–31. **Wolosiuk RA**, Stein M (1990) Arch Biochem Biophys 279:70–77. **Wong S-C** (1990) Photosyn Res 23:171–180. **Woodrow IE**, Berry JA (1988) Annu Rev Plant Physiol 39:533–594.

Yelle S, Beeson RC, Trudel MJ, Gosselin A (1989a) Plant Physiol 90:1473–1477. **Yelle S**, Beeson RC, Trudel MJ, Gosselin A (1989b) Plant Physiol 90:1465–1472. **Yokota A**, Kitaoka S (1989) Plant Cell Physiol 30:183–191. **Yu J**, Woo KC (1988) Plant Physiol 88:1048–1054.

Zelitch I (1990) Plant Physiol 92:352–357. **Ziegler-Jöns A** (1989) Planta 178:164–175.

Dr. Grahame J. Kelly
Centre for Biological Population Management
Queensland University of Technology
4000 Brisbane, Queensland
Australia

Prof. Dr. Erwin Latzko
Botanisches Institut der Universität Münster
Schloßgarten 3
W-4400 Münster, FRG

Note added in proof (2005):
Mechanisms for the accumulation of sucrose in sugar-beet and sugar-cane vacuoles (Sect. 5) are reviewed in the 1996 review (Sect. 2.f) and 1998 review (Sect. 4.b).

Photosynthesis: Carbon Metabolism
Twenty Years of Following Carbon Cycles in Photosynthetic Cells[1]

By Grahame J. Kelly and Erwin Latzko

"The first man I saw was of a meagre aspect, with sooty hands and face, his hair and beard long, ragged and singed in several places. His clothes, shirt and skin were all of the same colour. He had been eight years upon a project for extracting sunbeams out of cucumbers, which were to be put into vials hermetically sealed, and let out to warm the air in raw inclement summers. He told me, he did not doubt in eight years more, that he should be able to supply the Governor's gardens with sunshine at a reasonable rate..."

Jonathan Swift in *Gulliver's Travels*, ca. 1727

1. Introduction

We do not know whether or not the gentleman described above completed his planned 16 years of research on cucumber photosynthesis. If he did, he still would not have matched our efforts for *Progress in Botany*, in which our reviews of photosynthetic carbon metabolism have now reached their 20th year. We began at a time when the Calvin cycle was well established (although sufficient activities of all of its enzymes were not), and the processes of C_4 photosynthesis and photorespiration had been only recently discovered. During the ensuing 20 years, many details concerning the operation and regulation of these carbon cycles have been revealed. Seven stand out: (1) the demonstration of adequate activities for all Calvin cycle enzymes; (2) the discovery of the chloroplast envelope's phosphate translocator; (3) the CO_2-plus-Mg^{2+} activation of ribulose-1,5-bisphosphate (RuBP) carboxylase and, more recently, this enzyme's regulation by the protein "rubisco activase" and a seemingly endless array of novel sugar phosphates; (4) the demonstration of the light-mediated activation of Calvin cycle enzymes, mediated by the small proteins ferredoxin and thioredoxin; (5) the discoveries that the enzymes phosphoenolpyruvate (PEP) carboxylase, pyruvate Pi dikinase, and sucrose-P synthase are regulated by light-mediated phosphorylation/dephosphorylation; (6) the discoveries of the enzyme PPi-phosphofructokinase (PPi–PFK) and the regulatory compound fructose-2,6-P_2 in the cytosol of photosynthetic cells – their roles in regulating sucrose metabolism are

[1] This review is dedicated to M. Gibbs on the occasion of his retirement as Editor-in-Chief of *Plant Physiology*.

G.J. Kelly and E. Latzko:
Thirty Years of Photosynthesis

Originally published in Progress in Botany 54

still being elucidated; (7) the discovery (reported in this review) that the chloroplast envelope's adenylate translocator accommodates the uptake of ADP-glucose.

Clearly, the 1970s and 1980s were an exciting 20 years, but it is worth noting that no major new metabolic cycles were discovered, as happened in the 1950s and 1960s. It would seem that there are no new cycles to discover. Then what of the future? Will research be devoted to the acquisition of evermore detailed descriptions of the Calvin cycle, C_4 photosynthesis, crassulacean acid metabolism (CAM), sucrose biosynthesis, photorespiration, and the enzymes that catalyze their component reactions? Perhaps it will, but we suspect that the consequences will be anything but boring, if for no other reason than the introduction of the new tools of biotechnology. This is our first review in which we mention transgenic plants, for example: tobacco plants containing yeast invertase (Stitt et al. 1990), tomato plants containing maize sucrose-P synthase (Worrell et al. 1991) and *Synechococcus* containing human carbonic anhydrase (Price and Badger 1991). The construction of such transgenic plants, together with the techniques of gene sequencing and site-directed mutagenesis, will without doubt play a major (perhaps *the* major) role in the next 20 years of research into photosynthetic carbon metabolism.

2. Uptake of CO_2

It is generally believed that CO_2 enters the leaf cells of terrestrial plants by simple diffusion across the cell membrane. However, it is tempting to speculate that one or more transporter proteins may be involved, given that aquatic plants possess these (see below). A similarity between the conductivities to CO_2 of algal and higher-plant protoplast membranes has led Gimmler et al. (1990) to suggest that such transporters do indeed exist in higher plants.

A fundamental parameter is the concentration of CO_2 in the photosynthesizing chloroplast. This has been calculated by von Caemmerer and Evans (1991) to be about one-half of the ambient partial pressure, i.e., close to 200 ppm for leaves in air.

a) Cyanobacteria

The cell membranes of cyanobacteria possess mechanisms for the active uptake of both CO_2 and HCO_3^-. The CO_2-uptake system is present irrespective of the level of inorganic carbon (CO_2 plus HCO_3^-) in the growth medium, but the HCO_3^--uptake system is only present when the level of inorganic carbon is relatively low, e.g., air levels (Espie et al. 1991a, b). In addition, HCO_3^- transport appears in two forms: a Na^+-dependent form in cultures bubbled with air, and a Na^+-independent form in standing cultures (Espie et al. 1991a; Espie and Kandasamy 1992).

There is some confusion concerning how distinct these mechanisms are. Certainly more than one protein is involved (Ogawa 1990), but whether or not the CO_2- and HCO_3^--uptake systems are all part of one multi-protein complex (Badger and Price 1990), and whether or not carbonic anhydrase has a role (Bédu et al. 1990; Price and

Badger 1991), remain to be determined. Several reviews of cyanobacterial inorganic carbon uptake have appeared, including Miller et al. (1990) and Kaplan et al. (1991).

b) Eukaryotes

Studies with unicellular freshwater algae have shown that, like cyanobacteria, they possess plasmalemma-located mechanisms for the active uptake of both CO_2 and HCO_3^-; the inorganic carbon accumulates internally to a concentration that can suppress photorespiration (Gehl et al. 1990; Palmqvist et al. 1990; Spalding 1990; Rotatore and Colman 1991a, b). Energy to drive the CO_2 uptake is derived from photosynthetic, but not respiratory, ATP (Sültemeyer et al. 1991). It seems unlikely that additional uptake mechanisms are located on the chloroplast envelope (Rotatore and Colman 1991c), although a special chloroplast protein does appear when *Chlamydomonas reinhardtii* is grown on a low level of CO_2 (Mason et al. 1990).

Marine microalgae and freshwater and marine macrophytes also possess HCO_3^- and CO_2-uptake systems able to suppress photorespiration (Beer et al. 1990, 1991; Booth and Beardall 1991; Drechsler and Beer 1991). Extensive investigations with the freshwater macrophyte *Elodea nuttallii* have resulted in the proposal that it utilizes two separate transport systems for CO_2 and HCO_3^- (Jahnke et al. 1991). However, other freshwater macrophytes, the marine macrophyte *Ulva rotundata*, and the marine microalga *Emiliania huxleyi* appear to have a limited or even nonexistent ability to transport inorganic carbon (Beer et al. 1991; Levavasseur et al. 1991; Nimer et al. 1992).

c) Carbonic Anhydrase

Investigations of CO_2/HCO_3^- uptake have always been complicated by the enzyme carbonic anhydrase which catalyzes the reversible reaction $H_2O + CO_2 \leftrightarrow H_2CO_3$; the H_2CO_3 immediately equilibrates with H^+ and HCO_3^- according to the prevailing pH. Unicellular algae contain one or more intracellular carbonic anhydrases, and some species produce an additional, ionically-bound *external* carbonic anhydrase, the amount of which usually increases when cells are transferred from high-CO_2 to air-level CO_2 media (Husic and Quigley 1990; Sültemeyer et al. 1990; Coleman et al. 1991; Goyal et al. 1992). The induction of this external enzyme seems to be influenced by the intracellular metabolic status (Martínez and Orús 1991; Umino and Shiraiwa 1991). Utilization of HCO_3^- via an extracellular carbonic anhydrase and subsequent entry of the generated CO_2 into cells also has been demonstrated for a macroalga, *Ulva rigida* (Björk et al. 1992).

These carbonic anhydrases not only confuse the issue of whether CO_2 or HCO_3^- is the species transported across the plasmalemma (indeed, interconversion might be part of the transport process! – see Badger and Price 1990), but also raise the question of what role, if any, they have in the events that occur between the active uptake process at the plasmalemma and the assimilation of CO_2 inside the chloroplast. Indications are that the enzyme's importance depends on the species of alga (Gehl et al. 1990; Nara et al. 1990; Palmqvist et al. 1990). This interpretation also applies, in a sense, to terrestrial plants, in that all of the carbonic anhydrase activity present in C_4 plant leaves is needed for C_4 photosynthesis to proceed at the rates observed (Hatch and Burnell 1990; see Sect. 10). Finally, some studies may need re-evaluation following the report by Hatch (1991) of a problem with the conventional carbonic anhydrase assay system.

3. RuBP Carboxylase

The plethora of research on RuBP carboxylase has continued unabated. The carboxylase is the first enzyme of the Calvin cycle to which the tools of so-called molecular biology have been applied. The genes encoding the large and small subunits of the enzyme were the first plant genes to be sequenced; this occurred about 12 years ago (see Raines et al. 1991). Four years ago Rodermel et al. (1988) transformed tobacco plants with a nuclear antisense gene for the small subunit, thereby reducing the level of RuBP carboxylase in the plants. The photosynthesis by similarly transformed tobacco plants now has been characterized by two groups (Quick et al. 1991; Hudson et al. 1992) and the results provide qualified support to the concept that the activity of RuBP carboxylase is one of the prime determinants of photosynthetic capacity. The qualification is that, with these tobacco plants, the amount of carboxylase protein had to be reduced by at least 50% before a drop in photosynthesis, measured at a relatively low irradiance, could be detected (Quick et al. 1991), although with high irradiance the contribution of the carboxylase to control was greater (Stitt et al. 1991; Fig. 1). When the carboxylase content was reduced to only 18% of normal, and irradiance was high, it limited photosynthesis even at high CO_2 levels (Hudson et al. 1992). These results infer that natural variation of RuBP carboxylase, such as the range of 15 to 50% of total soluble protein in barley varieties (Metodiev and Demirevska-Kepova 1992), would certainly alter the relative contribution of the enzyme to photosynthetic control. In addition, the results are not unexpected, given that leaves contain sufficient potential RuBP carboxylase activity such that, in situations where it is unlikely to be limiting photosynthesis, it is actually downregulated so that production and consumption of RuBP are balanced (Sage et al. 1990a). Possible situations include high CO_2 and/or lower growth temperature (Pi regeneration during starch and/or sucrose synthesis become limiting) and low light (electron transport becomes limiting); the diversion of photorespiratory glycolate away from 3-P-glycerate formation (see Fig. 6 in our 1982 review) modifies these responses (Sage et al. 1990b; Harley and Sharkey 1991).

RuBP carboxylase has been the subject of a range of other studies. One of the more fascinating has been the isolation of a previously undescribed chloroplast protein that catalyzes the addition of methyl groups to the lysyl residue at position 14 of the enzyme's large subunit (Houtz et al. 1991). This residue is normally methylated in some species, e.g., tomato, tobacco, and pea, but not in others, e.g., wheat and spinach; the reason for this difference is not yet known (Houtz et al. 1992). A separate study with a mutant *Chlamydomonas reinhardtii* RuBP carboxylase also provided evidence that some separate protein could influence the carboxylase's structure (Chen et al. 1990).

The enzyme's affinity for RuBP has been halved by either cleaving off the large subunit's C-terminal amino acid (valine; Portis 1990), or by replacing this subunit's lysine 11 with leucine, us-

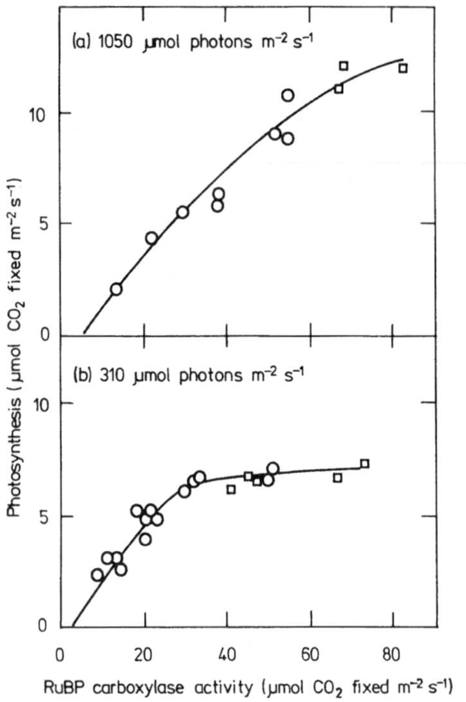

Fig. 1. Photosynthesis by tobacco leaves containing varying quantities of RuBP carboxylase. □, Wild-type plants; 0, plants transformed to contain reduced levels of the carboxylase. Measurement conditions were: 350 ppm CO_2, 22 °C, 65% relative humidity, and irradiances as shown – approximately two-thirds of full sunshine in **(a)**, and one-fifth in **(b)**. (Redrawn from Stitt et al. 1991)

ing site-directed mutagenesis (Kettleborough et al. 1991). An odd observation is that the enzyme blunders 0.7% of the time, and produces pyruvate instead of glycerate-3-P (Andrews and Kane 1991). In storage, the enzyme suffers inactivation in the cold (Heuer and Portis 1990), a phenomenon that also occurs in vivo, such that long-term chilling may lower photosynthetic capacity (Brüggemann et al. 1992).

Evidence for a strong association of a proportion of the chloroplast's RuBP carboxylase with membranes has been reported for red clover, pea, and soybean (Grub and Mächler 1990; Crafts-Brandner et al. 1991; Makino and Osmond 1991). In the latter case, fruit removal acted as a stimulus for the enzyme to move from a soluble to an insoluble form. An obvious candidate membrane for binding the enzyme is the thylakoid membrane, thus the observation by Süss (1990) that the thylakoid-coupling factor forms a specific complex with the carboxylase is of particular interest.

Contemporary environmental issues have persuaded researchers to examine the level of RuBP carboxylase in plants exposed to high levels of CO_2, and to increased UV-B radiation. In general, acclimation of plants to a CO_2 level more or less double the current atmospheric level tends to involve a decline in the leaf content of RuBP carboxylase, more so if the availability of sinks for increased photosynthate is low (Bowes 1991). Such a response has been documented for tomato (Besford et al. 1990) and rice (Rowland-Bamford et al. 1991). However, this decline, as seen in cultivated species, may be less common in other plants for which consistently higher photosynthetic rates have been measured during prolonged exposure to high CO_2 (Arp and Drake 1991; Idso and Kimball 1991; Ziska et al. 1991).

One disturbing report is that the content of RuBP carboxylase in pea leaves dropped dramatically upon exposure of the leaves to supplementary UV-B radiation (Jordon et al. 1992). The dosage was rather high (2.5 times normal at 297 nm), but the response was dramatic: enzyme activity declined by 71% after 3 days, while the mRNAs for the enzyme subunits declined by more than 80% after only 4 h. Similar studies involving lower UV-B irradiances, comparable to those expected next century as a consequence of stratospheric ozone depletion, need to be made available. Another study involving light quality has provided an intriguing insight into fern RuBP carboxylase (Eilenberg et al. 1991): the enzyme from gametophytes is two to three times more active when the gametophytes are grown under blue light than when they are grown under red light, and the small subunits of the enzyme from each of these light qualities differ immunologically. This latter aspect is especially interesting, given that the small subunit is intimately involved in ensuring that the six-carbon intermediate involved in the reaction catalyzed by the carboxylase is not lost (Smrcka et al. 1991).

a) Regulation of RuBP Carboxylase Activity

No new additions have been made to the exciting discoveries of rubisco activase, carboxyarabinitol-1-P (CA1P), and xylulose-1,5-P_2 (XuBP), reported in our last two reviews. However, further details concerning the operation of these regulators of RuBP carboxylase have come to hand. Rubisco activase, which occurs in two isoforms (Shen et al. 1991), is more effective at activating the carboxylase when photosynthetic electron transport through photosystem I is operating, but the reasons for this are not yet clear (Campbell and Ogren 1990a). The effectiveness is decreased by certain sugar-Ps, e.g, RuBP and fructose-1,6-P_2 (FBP; Lilley and Portis 1990). It also seems to be decreased when leaves are exposed to prolonged high light intensity (Suzuki et al. 1990).

A further study has confirmed that CA1P inhibition of the carboxylase is confined to certain plants such as bean, but absent from others such as wheat (Moore et al. 1991). The CA1P is degraded by a phosphatase that is, in turn, stimulated by sugar-Ps such as RuBP and FBP (Holbrook et al. 1991), a result in contrast to the effect of these sugar-Ps on the activase (see above). The origin of CA1P is still unknown, but it seems clear that the carboxylase is not involved; an end-product of photosynthesis is the more likely precursor (Moore and Seemann 1990), thus FBP still has a chance (see our preceding review).

With respect to regulation by XuBP, and by the substrate RuBP, a relatively more complex and somewhat unclear picture is emerging. Originally, the phenomenon of "fallover" (initial gradual decrease in carboxylase activity until a final linear rate becomes established) was attributed to XuBP, produced by a side-reaction of the carboxylase and then bound to the catalytic site (Edmondson et al. 1990a). Binding of *RuBP* to a separate *regulatory* site was not believed to be involved (Edmondson et al. 1990b), but Yokota (1991a) has demonstrated the existence of just such a regulatory site, and attributes fallover to a slow conformational change in the carboxylase caused when RuBP occupies this site. More recently, Yokota and Tsujimoto (1992) proposed, from studies of the spinach enzyme, that RuBP needs to be bound to the regulatory site before rubiso activase can be effective. Activase was envisaged to bind to one of the eight large subunits of the carboxylase carrying regulatory RuBP, thereby stimulating the well-known CO_2-plus-Mg^{2+} activation. Furthermore, XuBP on the catalytic site can be used as a substrate for CO_2 fixation and, consequently, removed – but not before causing some fallover by removing the activator-CO_2 (Yokota 1991b) that is involved in the CO_2-plus-Mg^{2+} activation. Zhu and Jensen (1991) have also related the binding of XuBP to the absence of this activator CO_2. Clearly, more studies are needed in this area. They may be assisted by the novel estimate of Purohit et al. (1991) that the pH of the microenvironment of the catalytic site jumps from <2 to >8 upon activation by Mg^{2+}.

Finally, two simple metabolites have been implicated in RuBP carboxylase regulation. These are Pi (Sawada et al. 1990) and glyoxylate (Campbell and Ogren 1990b). In the latter case, it was concluded that glyoxylate inhibited an unidentified component of the RuBP carboxylase activation system. Apparently, the list of regulators is not yet complete.

4. Other Calvin Cycle Enzymes

Calvin cycle enzymes other than RuBP carboxylase have attracted a moderate increase in researchers' attention over the past 2 years. The first study of a marine algal FBPase has been made (Yu 1989), and two microalgal ribulose-5-P (Ru5P) kinases, from *Chlamydomonas reinhardtii* (Roesler and Ogren 1990) and *Selenastrum minutum* (Lin and Turpin 1992), have been purified; the latter is somewhat unique in being a heterodimer. The classes of aldolases in algae have been evaluated by Jacobshagen and Schnarrenberger (1990) who thereby obtained molecular evidence that the charophytes form an evolutionary link between chlorophytes and higher plants.

Two unusual effects of metal ions on the higher plant chloroplast bisphosphatases have been detected. Oxidized FBPase (i.e., the "dark-inactivated" form) was found to possess activity in the presence of the substrate FBP and a relatively unphysiological concentration of Ca^{2+} (Chardot and Meunier 1990). On the other hand, oxidized sedoheptulose bisphosphatase (SBPase) in the presence of Fe^{2+} was not only inactive, but could not be activated by dithiothreitol while the Fe^{2+} remained (Queiroz-Claret and Meunier 1992). Other studies, somewhat more physiologically relevant, have shown that the bisphosphatases are sites for the fine control of the Calvin cycle during illumination (being regulated by prevailing levels of glycerate-3-P and sedoheptulose-7-P; Schimkat et al. 1990), and that they are relatively unique in that they were the only two of nine Calvin cycle enzymes to increase in activity in the leaves of soybean plants exposed to low phosphate nutrition – the activities of the other seven enzymes decreased (Fredeen et al. 1990).

A new outlook on chloroplast aldolase is required following the discovery by Iwaki et al. (1991) that the enzyme is strongly inhibited by RuBP when operating in the photosynthetic direction (FBP formation) and with physiological concentrations of its triose-P substrates.

Two groups have simultaneously separated and purified the leaf chloroplast and cytosolic 3-P-glycerate kinases: from spinach (Köpke-Secundo et al. 1990) and from barley

(McMorrow and Bradbeer 1990). In both species at least 90% of the total activity was in the chloroplasts. There is evidence that, within the chloroplast, this kinase interacts with the next enzyme in the Calvin cycle sequence, viz. glyceraldehyde-3-P dehydrogenase (GAPDH) (Macioszek et al. 1990), and that the other kinase, Ru5P kinase, interacts with its nearest neighbor, viz. P-ribose isomerase (Skrukrud et al. 1991). In fact, the latter interaction can occur even when the isomerase is from yeast (Skrukrud and Anderson 1991). Finally, interaction between this kinase and chloroplast GAPDH occurs in a nonreducing environment such that the two enzymes copurify as a complex (Clasper et al. 1991).

a) Light-Mediated Regulation

The chloroplast enzymes FBPase, SBPase, Ru5P kinase, and GAPDH are capable of undergoing light-mediated reductive activation (Fig. 2) because their peptide sequences include "regulatory sections" that have evolved for this purpose; these sections are absent

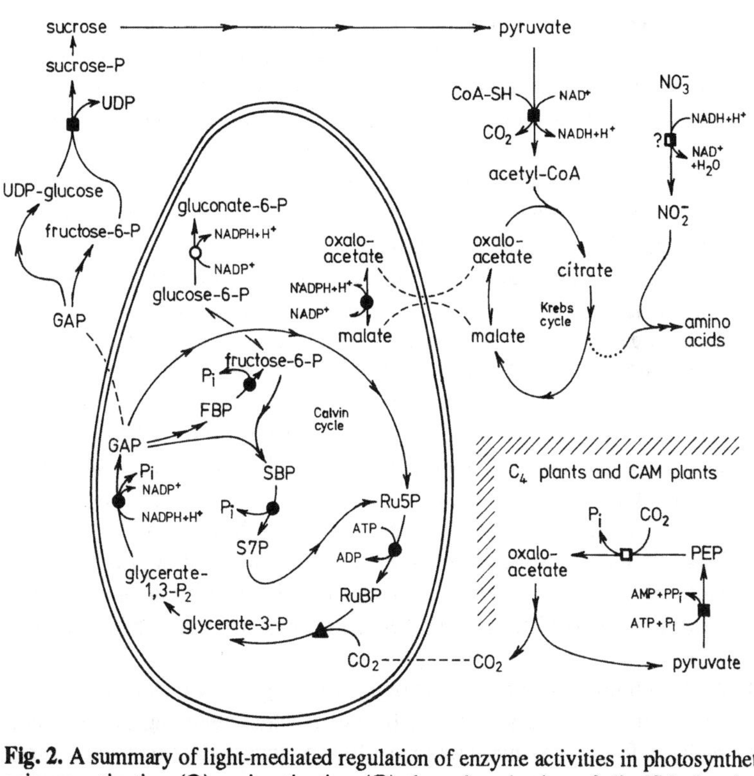

Fig. 2. A summary of light-mediated regulation of enzyme activities in photosynthetic cells. Mechanisms: activation (●) or inactivation (○) through reduction of disulfide bonds to sulfhydryls, mediated by thioredoxin; ▲, rubisco activase-promoted carbamylation and/or dissociation of sugar-P inhibitor; ■, phosphorylation resulting in inactivation; □, phosphorylation resulting in activation. *S7P* Sedoheptulose-7-P. (After Budde and Randall 1990; Scheibe 1991)

from nonchloroplast FBPases and GAPDHs (Scheibe 1990, 1991). The in vivo activation of all four enzymes in the cyanobacterium *Nostoc* sp. has recently been demonstrated (Austin et al. 1992). A novel observation is that the activity of three of these enzymes (SBPase was not tested) can be enhanced by high hydrostatic pressure, whereas pressure did not influence nonchloroplast FBPase and GAPDH, or the chloroplast's RuBP carboxylase (Prat-Gay et al. 1991). The authors argue that noncovalent interactions are an important component of light-mediated activation. However, Porter and Hartmann (1990) did not consider that conformational changes were necessary in the case of Ru5P kinase regulation.

Minor attention has been given to the bisphosphatases. Their light-mediated activation in tomato leaves is impaired by chilling, and this seems to be the primary cause of the reduced photosynthetic capacity of these leaves (Sassenrath et al. 1990). An odd observation is that FBPase reduced by dithiothreitol behaves differently to that reduced by the physiological reductant, thioredoxin (Chardot and Meunier 1991).

A collection of studies surrounds the sequence between Ru5P kinase and GAPDH. Light-mediated activation of these two enzymes is more rapid than that of RuBP carboxylase; the latter is not a reductive process, but involves rubisco activase (Servaites et al. 1991; see Sect. 3.a). The most rapid activation seems to be that of GAPDH, which is completely inactive in the dark, but fully activated even in low light (Siebke et al. 1991; Laisk et al. 1992). Ru5P kinase activation may be more dependent on light intensity, as evidenced by its poor activation in vivo when photochemical capacity was reduced by iron deficiency (Arulanantham et al. 1990). This was despite the probability that the kinase was integrated in a multi-enzyme complex, an event that facilitates its activation (Rault et al. 1991).

Five enzymes *not* associated with the Calvin cycle have activities that undergo light-mediated regulation in photosynthetic cells. One of these, viz. the chloroplast $NADP^+$-linked malate dehydrogenase, is activated by reduction of one disulfide bridge per subunit (Scheibe et al. 1991). But the other four are regulated through phosphorylation/dephosphorylation mechanisms (Fig. 2). These are chloroplast pyruvate Pi dikinase (in C_4 plants), the mitochondrial pyruvate dehydrogenase complex, and the cytosolic enzymes PEP carboxylase and sucrose-P synthase (Budde and Randall 1990). A fifth contender is the cytosolic nitrate reductase (Kaiser and Spill 1991) that is rapidly inactivated upon darkening, thereby preventing accumulation of toxic nitrite (Riens and Heldt 1992). This mechanism complements another in the light, in which the presence of nitrate indirectly stimulates the quick utilization of nitrite through nitrate enhancement of the light-mediated phosphorylative activation of PEP carboxylase, thereby directing carbon flow towards amino acid synthesis (Van Quy et al. 1991).

Returning to light-mediated reductive activation: electrons for this process are delivered to target enzymes in the chloroplast via ferredoxin, ferredoxin/thioredoxin reductase, and thioredoxin (reviewed by Buchanan 1991). Besides this well-established system, which uses either of two types of thioredoxin, viz. thioredoxin$_f$, (which is present in appreciable amounts; Hermoso et al. 1992) or thioredoxin$_m$, recent evidence points to the existence of a comparable *cytosolic* system of NADPH, NADP/thioredoxin reductase, and a third type of thioredoxin, viz. thioredoxin$_h$, in leaves, algae, and nonphotosynthetic plant cells (Florencio et al. 1988; Huppe et al. 1991; Marcus et al. 1991; Van Langendonckt and Driessche 1992). Its function is unknown.

5. Chloroplast Starch

For many years the regulation of starch synthesis in plastids, including chloroplasts, has been considered to take place predominantly at the reaction catalyzed by ADP-glucose (ADPG) pyrophosphorylase, principally through 3-P-glycerate activation and Pi inhibition of this enzyme (reviewed by Kleczkowski et al. 1991). A mutant of *Chlamydomonas reinhardtii* that accumulates very little starch and contains an enzyme with a dramatically lowered sensitivity to 3-P-glycerate supports this view (Ball et al. 1991a). However, by using the wrinkle-seeded and round-seeded genotypes of pea (that differ in leaf content of starch-branching enzyme), and mutants of *Arabidopsis thaliana* constructed to contain very low activities of either ADPG pyrophosphorylase or P-glucomutase in chloroplasts, Smith et al. (1990) and Neuhaus and Stitt (1990) found that the contemporary view of the importance of the pyrophosphorylase was only applicable to plants grown in low light; under high light, the other enzymes became important in controlling the rate of starch synthesis.

Accumulation of starch benefits plant growth mostly when other factors, e.g., nitrogen supply, are not limiting (Schulze et al. 1991). The rate of nocturnal mobilization of starch is somehow determined by the amount finally accumulated at day's end (Hendrix and Grange 1991). The initial steps in mobilization may include phosphorylase, amylases (see our earlier reviews), α-glucosidase (Beers et al. 1990), and an "unidentified regulatory factor" (Casper et al. 1991). Incidentally, a first clue to the role of the *extra*-chloroplastic phosphorylase has been obtained by Yang and Steup (1990) who found that it strongly interacts with a large complex polysaccharide composed of at least four different monosaccharides.

Studies with nonphotosynthetic plastids have had a major repercussion on our view of photosynthetic carbon metabolism. Recent efforts to elucidate the pathway between sucrose (imported into nonphotosynthetic cells) and starch synthesized in the plastids of these cells have added support to the concept that carbon is transported into the plastids in the form of a hexose unit, and not a triose unit (Frehner et al. 1990; Hatzfeld and Stitt 1990; Hill and Smith 1991). The fact that such plastids lack FBPase (Entwistle and ap Rees 1990) gives strength to this concept. Until recently, the import of hexose units in the form of glucose-6-P, and their mediation by a phosphate translocator (Flügge and Heldt 1991), was entertained. Then came the communication from Pozueta-Romero et al. (1991a) that the adenylate translocator of amyloplasts could transport ADPG, thereby supplying to the plastid the immediate substrate of the starch synthase enzyme. Coupled with a demonstration that sucrose synthase might be capable of generating ADPG from sucrose in the cytosol (Pozueta-Romero et al. 1991b), a relatively concise picture of the sucrose-to-starch conversion can be envisaged (Fig. 3). Note, however, that the fructose liberated by the action of sucrose synthase would need to be processed by a hexokinase (Schnarrenberger 1990), and a straightforward pathway for the incorporation of this half of the sucrose molecule into starch would still require utilization of the phosphate translocator for glucose-6-P import (Fig. 3).

The repercussion on photosynthesis came when Pozueta-Romero et al. (1991c) reported that chloroplasts also are capable of importing ADPG. It can be questioned whether or not any significant quantity of ADPG is imported by photosynthesizing chloroplasts already capable of generating hexose-P from CO_2 internally. If ADPG, gen-

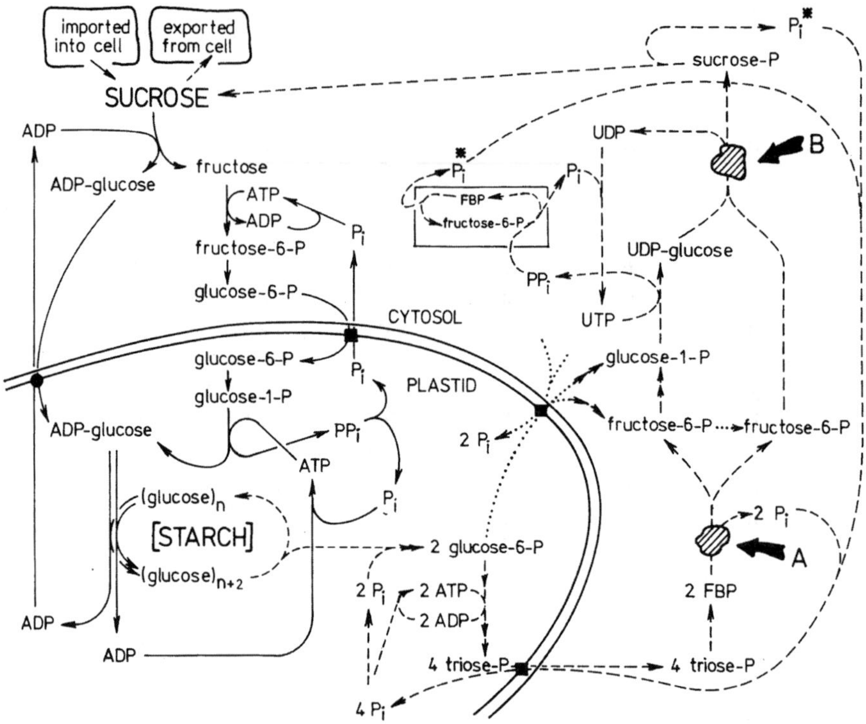

Fig. 3. A composite picture of starch and sucrose metabolism in plastid-containing plant cells. —, Plastid starch synthesis from cytosolic sucrose, utilizing ADPG transport via the adenylate translocator (●) (Pozueta-Romero et al. 1991a, c) and glucose-6-P transport via the phosphate translocator (■). The sum of the reactions is: sucrose + $(glucose)_n \rightarrow (glucose)_{n+2}$. This pathway is probably important for starch synthesis by amyloplasts; its activity with chloroplasts might be limited by the lower capacity of the chloroplast's phosphate translocator to transport glucose-6-P (Flügge and Heldt 1991). -----, Cytosolic sucrose synthesis from plastid starch, utilizing triose-P export by the phosphate translocator (■). The sum of the reactions is: $(glucose)_{n+2} \rightarrow (glucose)_n$ + sucrose. Not all of the plastid reactions are shown. This pathway is probably important for the nocturnal mobilization of chloroplast starch. Note that the cytosol lacks pyrophosphatase, but the combination of FBPase and PPi-linked P-fructokinase catalyzes, overall, the same reaction *(boxed small print)*. *A* and *B* point to the cytosolic enzymes FBPase and sucrose-P synthase (respectively), which are the principal regulatory enzymes of sucrose synthesis in photosynthetic cells (Neuhaus et al. 1990a). Mobilization of *amyloplast* starch may involve a modification (and simplification) of this pathway, whereby two glucose-6-P are exported (....) via the phosphate translocator in exchange for the net two Pi liberated (shown *Pi**) when the molecule of sucrose is synthesized

erated from sucrose, was imported by photosynthesizing chloroplasts that were simultaneously exporting triose-P to manufacture the sucrose, a new cycle for photosynthetic carbon metabolism would be set up (Fig. 3), but it might prove difficult to discover whether it achieves anything more significant than the consumption of 4 ATP and 1 UTP per sucrose cycled (Fig. 3). Regulation might prevent the cycle from being quantitatively significant; such regulation could be uncomplicated since the enzymes and membrane proteins involved in the sucrose-to-starch conversion are mostly separate from those that convert starch to sucrose; only chloroplast P-glucomutase, cytosolic P-hexose isomerase, and the chloroplast's phosphate translocator are common to both sectors (Fig. 3).

6. Cytosolic Sucrose Biosynthesis: Control by Cytosolic FBPase and Sucrose-P-Synthase

Practically all recent research into sucrose metabolism has centered on two enzymes: cytosolic FBPase and sucrose-P synthase (SPS). The respective locations of these enzymes in the pathway of sucrose synthesis are designated by A and B in Fig. 3.

FBPase activity is controlled via inhibition by the specially manufactured regulator molecule fructose-2,6-P_2. Its concentration is greater in tissues not expected to be actively synthesizing sucrose, e.g., the "sink" portion of leaves (Nielsen and Veierskov 1990), and leaves diverting carbohydrate to form the carbon skeletons for amino acid synthesis (Champigny et al. 1992). Blocking the enzyme that normally removes this inhibitor leads, as expected, to a reduction in sucrose synthesis (Brauer and Stitt 1990). However, fructose-2,6-P_2 may not be the only regulator of FBPase: inhibition by micromolar concentrations of Ca^{2+} has been reported from two laboratories (Brauer et al. 1990; Prado et al. 1991). Either or both of these inhibitors, acting on the FBPase, may form part of the mechanism whereby sucrose accumulation in leaves inhibits further sucrose synthesis – a phenomenon recently demonstrated with transgenic tobacco expressing yeast invertase in cell walls (Stitt et al. 1990) – but the overall mechanism is still something of a mystery (Krause and Stitt 1992). Perhaps studies of FBPase from plant species evolutionarily distant from spinach and tobacco may provide clues; in this context, the recent discovery of cytosolic FBPase in the green alga *Selenastrum minutum* (Botha and Turpin 1990) may prove valuable.

Another mystery associated with FBPase is the role of the "competitor" enzyme PPi-linked P-fructokinase (which catalyzes the reversible reaction FBP + Pi $\rightleftharpoons$ fructose-6-P + PPi). In the direction of FBP formation it could, in a roundabout way, convert the PPi generated during UDP-glucose synthesis to two Pi, but the significance of such a role seems uncertain (Neuhaus et al. 1990b; Neuhaus and Stitt 1991). On the other hand, it could boost carbon flow from FBP to sucrose (via fructose-6-P) if FBPase activity was low; evidence for such a role in the grass *Lolium temulentum* has been obtained by Collis and Pollock (1991). In studies with nonphotosynthetic tissues, the enzyme was judged to be of value in the direction of FBP formation in potato tubers (and also *Euglena gracilis* under anaerobiosis), but not in banana fruit or carrot root; in the latter, a role comparable to that in leaves, viz. dealing with PPi, was envisaged (Enomoto et al. 1990; Ball et al. 1991b; Hajirezaei and Stitt 1991). Evidence from Wong et al. (1990) that more than one form of the enzyme exists, and that they have different susceptibilities to fructose-2,6-P_2 [which activates this enzyme (Kiss et al. 1991), in contrast to its inhibition of FBPase], helps to complicate the picture. Especially enigmatic is the apparent contradiction that fructose-2,6-P_2 can retard sucrose synthesis by *inhibiting* FBPase, yet it *stimulates* the PPi-linked P-fructokinase which is capable of achieving the same FBP $\rightarrow$ fructose-6-P conversion that is supposed to be prevented through FBPase inhibition.

The importance of SPS, the other regulatory enzyme of sucrose synthesis, has been highlighted by the report that a transgenic tomato plant, with total SPS activity elevated up to sixfold with maize SPS, synthesized 2.5 times the usual amount of sucrose in its leaves (Worrell et al. 1991). The enzyme has been more highly purified (Salvucci et al. 1990; Bruneau et al. 1991), and its light-mediated activation by dephosphorylation (and corresponding inactivation by phosphorylation in darkness) has been further studied (Budde and Randall 1990). The phosphorylation is probably mediated by a protein kinase, the identity of which is just beginning to appear (Huber and Huber 1991a). Interestingly, it seems that dephosphorylation can go too far, in that highly active (and therefore "dephosphorylated") enzyme is inactivated if further phosphate groups, not normally involved in regulation in vivo, are cleaved from it (Huber et al. 1991). Another unusual observation was that the normal inactive (phosphorylated) enzyme was, in fact, quite active when assayed in the presence of high ionic strength, but only if the enzyme was obtained from the leaves of a C_4 plant (Huber and Huber 1991b).

Triose-P for sucrose synthesis is delivered to the cytosol from the chloroplast by the phosphate translocator (Fig. 3). This translocator is the major protein of the chloroplast envelope; it constitutes 12% of the total envelope protein (Flügge et al. 1991). It is possibly universal in species containing chloroplasts: there is now evidence for its presence in the green macroalga *Bryopsis corticulans* (Yu and Pedersén 1991). A review of this and other metabolite translocators has recently appeared (Flügge and Heldt 1991). Operation of the translocators is indirectly regulated by reactions on both sides of the membrane (Heineke et al. 1991).

7. Mitochondrial Respiration

The problem of whether or not the mitochondria in photosynthesizing cells respire continues to be intractable. Earlier impressive experiments in which barley protoplast photosynthesis was shown to be inhibited by oligomycin (an inhibitor of mitochondrial electron transport) have been extended by those workers (Krömer and Heldt 1991a, b) to provide additional evidence that ATP generated by mitochondrial respiration and delivered to the cytosol is required for photosynthesis to continue, perhaps by assisting in the disposal of excess photosynthetic energy. It might even be possible for some of the mitochondrial ATP to be delivered to the chloroplast stroma when growth temperatures are high (Tarchevskii and Konyukhova 1989). Supportive evidence from studies with pea protoplasts has been reported by Vani et al. (1990) and Reddy et al. (1991), who described how a preceding period of respiration (dark) stimulated a subsequent short period of photosynthesis (light) and vice versa. However, the complexities of oxygen metabolism in photosynthetic cells have led others to question what is being observed when oxygen uptake in the light is measured: mitochondrial respiration, photosynthetic oxygen reduction (Mehler reaction), or even chlororespiration (Avelange and Rébeillé 1991; Avelange et al. 1991), although the latter is predominantly a dark process (Peltier and Schmidt 1991).

A matter related to mitochondrial respiration is the theoretical possibility that a reversal of the Krebs cycle could substitute for the Calvin cycle in photosynthesis. Such a situation is now more or less established for the primitive green anaerobic photosynthetic sulfur bacterium *Chlorobium limicola* forma *thiosulphatophilum* (Buchanan and Arnon 1990). A first indication that it might also be possible in eukaryotic cells has been provided by Chen and Gibbs (1992) who found that a *Chlamydomonas reihardtii* mutant lacking Ru5P kinase displayed acetate-enhanced CO_2 fixation, and contained three of the requisite enzymes.

8. Photorespiration

Much of the recent research on photorespiration has been directed at the individual steps between RuBP oxygenation and glycerate entry into the chloroplast. The oxygenase activity of RuBP carboxylase will no doubt receive special attention as biotechnologists dream of using site-directed mutagenesis to reduce or prevent it, thereby increasing plant productivity (Chen and Spreitzer 1989). To date, it appears that the small subunits have little to do with determining the enzyme's CO_2/O_2 specificity (Gutteridge 1991), but there may be some other nuclear-encoded protein that does (Chen et al. 1990).

RuBP oxygenation produces glycolate that is transported out of the chloroplast (Howitz and McCarty 1991) and further metabolized in peroxisomes. Acquisition of high quality preparations of peroxisomes is now possible (Turley and Trelease 1990). Although devoid of any visible internal structure, some structural arrangement of enzymes does appear to be present in peroxisomes, so that "metabolite channeling" is possible (Heupel et al. 1991). Nevertheless, variation in enzyme content is possible, as evidenced by the odd glyoxysomal-type peroxisomes in the xanthophycean algae *Bumilleriopsis* and *Vaucheria* (Backeshoff and Stabenau 1990).

The first half of the peroxisome's contribution to photorespiration is to effect the glycolate → glyoxylate → glycine conversion. Glycine is then transported to mitochondria. A barley mutant apparently deficient in the transport system accumulates glycine (Blackwell et al. 1990). Once in the mitochondria, two molecules of glycine are converted to one each of serine, CO_2 and NH_4^+ by a multi-enzyme complex that constitutes one-third of the total soluble protein of the mitochondrial matrix (Oliver et al. 1990), and possibly exists in a distinct "photorespiratory metabolic domain" of the mitochondrion (Wiskich et al. 1990). Reassimilation of released NH_4^+ occurs in the chloroplast (Yu and Woo 1991). If prevented, photosynthesis is inhibited, not so much by accumulated NH_4^+, but by accumulated glyoxylate that inhibits RuBP carboxylase (Wendler et al. 1990; see Sect. 3.a).

The second half of the peroxisome's contribution to photorespiration occurs after the serine (generated above) is transported to this organelle. It is deaminated to hydroxypyruvate which is then reduced to glycerate either in the peroxisome, or perhaps in the cytosol (Kleczkowski et al. 1992). The glycerate returns to the chloroplast in exchange for another glycolate (Howitz and McCarty 1991).

The only known, quantitatively important fate of glycolate in plant leaves is its metabolism to glycerate-plus-CO_2 via the glycolate pathway. However, in algae, two alternative fates appear to be possible: (1) a short-circuit from glyoxylate to glycerate via tartronic semialdehyde (see our 1978 review), and (2) excretion from the cell. Evidence for a transporter protein that facilitates excretion from *Chlamydomonas reinhardtii* cells has been obtained by Wilson and Tolbert (1991). Algae also differ from plant leaves in that glycolate, when metabolized , is not always oxidized to glyoxylate by a peroxisomal oxidase. Current indications are that brown macroalgae, the unusual *Cyanophora paradoxa*, and chromophyte microalgae other than diatoms utilize the oxidase; but green, prasinophyte, and cryptophyte microalgae, and diatoms, utilize a dehydrogenase located in the mitochondrion (Gross 1990; Suzuki et al. 1991; Betsche et al. 1992). Categories are not clear-cut: the green microalga *Dunaliella primolecta* contains both the oxidase and the dehydrogenase (Nimer et al. 1990). Finally, questions are raised by the apparent total lack of glycolate-oxidizing enzymes in dinoflagellates and prymnesiophytes (Suzuki et al. 1991).

The function of photorespiration in plants has received little recent attention, apart from a commanding study by Wu et al. (1991) in which photorespiration was evaluated to be superior to the photoreduction of oxygen (Mehler reaction and subsequent H_2O_2 metabolism) for protection against photoinhibition. The strength of this conclusion will have considerable bearing on the survivability of any future crop plants engineered to contain low (or zero) RuBP oxygenase activity.

9. Stress and Photosynthetic Carbon Metabolism

Photosynthetic organisms seldom grow in ideal environments. Temperature, light level, water supply, nutrient availability, and soil pH are five prominent external environmental parameters that plants must deal with. Most commonly, the problem is that one (or more) of these is too low. When the light level is low, plants respond by mimicking shade plants: RuBP carboxylase is down-regulated, and the quantity of light-harvesting machinery is increased. But when temperature or water supply is too low, a major requirement is to deal with an excess of harvested light energy that follows the consequent retardation of photosynthetic carbon metabolism. Failure to deal with this problem is common, and leads to "overreduction"-induced photoinhibition.

Recent studies with the C_3 plant *Helianthus annuus* (sunflower) and the CAM plant *Kalanchoe pinnata* indicate that, under normal growth conditions, photosynthetic electron transport and enzyme activities, coupled with temporary engagements of oxygen reduction (Mehler reaction) are sufficient to fully utilize the harvested light energy and thus avoid photoinhibition (Siebke et al. 1990, 1992; Luttge et al. 1991). But a fascinating study of the C_3 plant *Gossypium hirsutum* (cotton) by Winter and Königer (1991) modifies this view, and may fundamentally alter our perception of the photosynthetic carbon metabolism of plant leaves. In essence, these authors found that the potential photosynthesis of cotton (a "sun" plant) is underutilized in full sunlight when the temperature is "slightly suboptimal", e.g., 27 °C. When increased to 33–34 °C, extremely high growth rates of up to 0.26 $g \cdot g^{-1} \cdot day^{-1}$ were measured – we calculate that this represents a *field* rate of around 100 $\mu mol\ CO_2\ fixed \cdot mg^{-1}\ chlorophyll \cdot h^{-1}$. Additionally, these authors found that, with irradiance reduced to about one-half of full sunshine, the dramatic difference between 33 and 27 °C plants was not observed. They concluded that high light limits the growth of sun plants because even relatively warm temperatures of 25–27 °C are suboptimal. This proposed "subtle photoinhibition" somehow must have been prevented from increasing to a catastrophic level, otherwise plants would have died. Königer and Winter (1991) obtained evidence that the reversible de-epoxidation of violaxanthin to zeaxanthin was central to the process of prevention, but the exact mechanism by which zeaxanthin assists in the dissipation of excess harvested light energy still has not been experimentally elucidated (Demmig-Adams 1990; Young 1991).

Of course, stress situations can increase to a point where they ultimately overwhelm protective mechanisms, leading to severe photoinhibition. Two recently reported examples are a mutant of *Synechocystis* that contains greatly reduced Ru5P kinase activity (Su and Bogorad 1991), and *Valerianella locusta* protoplasts exposed to a low temperature of 4 °C (van Wijk and Krause 1991). In the latter case, support for the conclusions of Wu et al. (1991), (see Sect. 8) were obtained: energy dissipation by photoreduction of O_2 was found to accentuate, rather than relieve, photoinhibition because metabolism of the products of oxygen reduction was too slow, and these reactive products then damaged the photosynthetic apparatus of photosystem II.

Little attention has been given to the induction of photoinhibition by low water supply. Rather, most studies have concentrated on whether the reduction in photosynthesis caused by water stress is due to stomatal closure, or altered metabolism, or a combination of both. Stomatal closure was concluded to cause most of the reduction of photosynthesis with leaves of bean, lupin, sunflower, and *Eucalyptus globulus*, but a metabolic component appeared to be important in grapevine leaves (Cornic and Briantais 1991; Quick et al. 1992), and even with bean leaves a metabolic response to water stress was detected, viz. a reduction in SPS activity (Vassey et al. 1991). Of course, it can be argued that metabolic responses may be a consequence of the reduced availability of CO_2 for photosynthesis following stomatal closure. However, four studies with sunflower, and one with soybean (Frederick et al. 1989; Graan and Boyer 1990; Wise et al. 1990, 1991; Gimez et al. 1992) have not supported this interpretation; rather, water stress-induced perturbations of metabolism appeared to be, at least in part, primarily responsible for the reductions in photosynthesis. These perturbations were looked upon as being normal adaptations to accommodate the unpredictable nature of water availability, and may center on a reduction in photophosphorylation and consequent shortfall in the capacity to regenerate RuBP.

The shortfall is less severe with plants acclimatized to water stress (Santakumari and Berkowitz 1991), and the observation that gradual development of water stress had little effect on the photosynthetic *capacity* of soybean leaves (Frederick et al. 1990) again raises questions concerning the relative contributions of metabolism and stomata. Interactions between these may be possible (Wise et al. 1991), and the question of whether or not the partial closure of stomata (due to water stress) results in lowered levels of CO_2 in the leaf needs to be critically addressed (Cornic and Briantais 1991).

10. C_4 Photosynthesis

Two notable events influenced the biosphere about 60 million years ago. One was the catastrophic impact of a large asteroid with the earth, probably upon present-day Mexico, resulting in a temporary but drastic climate change that halted photosynthesis, and caused many species, including dinosaurs, to become extinct (McKinnon 1992). Less well known, and less dramatic, was the adaptation of plants to a gradually declining level of CO_2 in the atmosphere, from something like 3000 ppm (120 million years ago) to a value similar to today's, about 350 ppm (60 million years ago). Ehleringer et al. (1991) proposed that the evolutionary appearance of C_4 plants probably took place at this time, when the benefits of the C_4 pathway's CO_2 concentrating mechanism first would have been of benefit to photosynthesis.

Atmospheric CO_2 has remained near this low level for the past 60 million years, thus today's C_4 plants continue to benefit from possession of the C_4 "CO_2-pump". This pump is so efficient that, even at very limiting CO_2 concentrations, photosynthesis is streamlined in such a way that both photorespiration and photoreduction of O_2 are kept low (Krall and Edwards 1990). With such efficiency, one would not expect C_4 plants to benefit from a future increase in atmospheric CO_2 (as is occurring due to human activities), and indeed this expectation has often been demonstrated (most recently by Ziska et al. 1990, 1991). An exception was *Echinochloea frumentacea*, which did show some response to increased CO_2 when grown with a low supply of nitrogen (Wong and Osmond 1991).

When CO_2 enters the C_4 plant leaf it is initially fixed by PEP carboxylase which, however, uses HCO_3^- as substrate. It is therefore dependent on rapid generation of HCO_3^- from entering CO_2 by the enzyme carbonic anhydrase. Measurements have shown that the activity of this enzyme is just sufficient to accommodate the observed carbon flux (Burnell et al. 1990; Hatch and Burnell 1990). One thought was that the unique Na^+ requirement of C_4 plants might be related to this enzyme, but it was not (Brownell et al. 1991). More likely, Na^+ is required to be cotransported with pyruvate during the latter's uptake by mesophyll chloroplasts, but even this explanation is incomplete because it does not seem to apply to one of the three groups of C_4 plants (NADP-malic enzyme group) which possibly uses H^+/pyruvate cotransport instead (Ohnishi et al. 1990).

A surprise addition to the list of C_4 plants is imminent. Reiskind and Bowes (1991) report that the marine green macroalga *Udotea flabellum* displays some of the physiological and biochemical characteristics of a C_4 plant, including minimal photorespiration, although it lacks a Kranz anatomy. Also, capture of cytosolic CO_2 is not by PEP carboxylase, but by PEP carboxykinase. Nevertheless, it seems that this enzyme participates in the development of a pool of malate that moves to the chloroplast where decarboxylation elevates the CO_2 concentration in the vicinity of RuBP carboxylase. Furthermore, it is not inconceivable that a comparable system operates in the brown alga *Ascophyllum nodosum* (Johnston and Raven 1989).

Returning to the standard C_4 plant: Doncaster and Leegood (1990) have clarified why cytosolic malate dehydrogenase does not interfere with the substantial flow of malate from mesophyll cells to bundle-sheath cells. Pyruvate produced following malate decarboxylation in the bundle sheath mostly returns to the mesophyll, but a few percent can be diverted for nitrogen assimilation; it is aminated to alanine, which constitutes one-quarter of all amino acids in the phloem of maize (Valle and Heldt 1991).

Enzymes associated with C_4 photosynthesis have continued to attract attention. At least six of them undergo reversible polymerization in vitro – the possible significance of this to in vivo regulation has been reviewed by Podestá et al. (1990). The polymerization state of NADP-linked malic enzyme changes noticeably in the physiological pH range of 7 to 8 (Iglesias et al. 1991). PEP carboxylase has received most attention, and is dealt with below (Sect. 10 a). The mesophyll chloroplast's NADP-linked malate dehydrogenase and the bundle sheath's NADP-linked malic enzyme both have been sequenced from clones of their complementary DNAs (Metzler et al. 1989; Börsch and Westhoff 1990). The dehydrogenase, which is reported to exist as isoenzymes in *Echinochloa crusgalli* (Vairinhos and Simon 1990), must undergo at least 50% of its potential reductive activation before any activity appears (Miginiac-Maslow et al. 1990). C_4 plants that require NAD-linked malic enzyme apparently require contrasting regulation: in "NAD malic enzyme-type" C_4 plants, the enzyme is inhibited by physiological levels of AMP, ADP and especially ATP, but in "PEP carboxykinase-type" C_4 plants, ATP activates the enzyme (Furbank et al. 1991).

A characteristic, and possibly limiting (Reddy and Das 1990) enzyme of C_4 photosynthesis is pyruvate Pi dikinase which generates the CO_2-acceptor PEP (and AMP and PPi) from pyruvate, Pi, and ATP. The form of Pi used has been identified by Nakamoto and Edwards (1990) as HPO_4^{2-}. Assay of the enzyme activity is difficult; improvements recommended by Salahas et al. (1990a, b) are inclusion of ample, glucose-6-P-activated coupling PEP carboxylase, and glycerol to stabilize the enzyme. Regulation of the dikinase in vivo is centered on light-mediated activation by dephosphorylation of a threonyl-P residue (reviewed by Budde and Randall 1990; Bennett 1991), a process that is relatively independent of the normal temperatures experienced by C_4-plant leaves (Grammatikopoulos and Manetas 1990).

a) PEP Carboxylase

Almost all investigations of PEP carboxylase over the past 2 years have involved the C_4 leaf enzyme. Genetic analysis indicates that its synthesis is directed by a subgroup of genes rather than by one gene, at least in the C_4 dicotyledonous *Flaveria trinervia* (Poetsch et al. 1991). Past studies with the enzyme in vitro may need re-evaluation: (1) the *Cynodon dactylon* enzyme was found to be relatively insensitive to glucose-6-P activation but more sensitive to malate inhibition when assayed with glycerol at a concentration that stimulated in vivo concentrations of solutes (Stamatakis et al. 1990), and (2) its specific activity after ion-exchange chromotography that involved sulfate or phosphate anions was more than twice that measured when chloride ions were involved (Angelopoulos and Gavalas 1991). More realistic properties are considered to be seen when the enzyme is assayed immediately after release from protoplasts or single cells. Such tests have confirmed the high sensitivity to feedback inhibition by malate for the maize leaf enzyme (Petropoulou et al. 1990), and for the enzyme from the guard cells of C_3-plant stomata (Tarczynski and Outlaw 1990; Willmer et al. 1990).

Maize leaf PEP carboxylase has been the focus of five studies dealing with effects of pH and temperature on hysteretic (i.e., slowly changing) kinetic properties and related reversible enzyme polymerization (Kleczkowski and Edwards 1990, 1991; López-Pozos et al. 1990; Wu et al. 1990). Overall, hysteresis was attributed to slow reversible changes in the degree of polymerization. Cold storage or alkaline pH favored inactive polymers, while neutral pH, warmer temperatures, and the substrate PEP favored the active tetramer. However, the effects of high pH may be complicated by enzyme sensitivity to contaminating heavy metal ions at this pH (Wedding and Black 1990).

Real advances have been made in unraveling the complex mechanism of light-mediated activation of PEP carboxylase, which in turn is a major determinant of the functional properties of the enzyme (Arrio-Dupont et al. 1992). Jiao and Chollet (1991) argue that it is far more important than tetramer/dimer interconversions, although the latter may have some value in CAM plants. As reported in our last review, activation of the C_4-plant enzyme involves phosphorylation of a seryl residue, catalyzed by a protein kinase (Fig. 4). The residue involved is near the N-terminus of the polypeptide chain: serine-15 in the maize enzyme, serine-8 in the sorghum enzyme (Jiao et al. 1991a), quite distant from the enzyme's catalytic site (Jiao and Chollet 1990). It seems that the phosphorylation is light-mediated because the protein kinase that catalyzes it is "light-activated" (Echevarría et al. 1990) by a process that, at its simplest, involves light-dependent, short-term degradation and resynthesis of the kinase in the cytosol (Jiao et al. 1991b; Jiao and Chollet 1992; see Fig. 4). The messenger(s) that connects the light signal to kinase resynthesis remains to be found. Other intriguing facets must also be kept in mind: (1) phosphorylative activa-

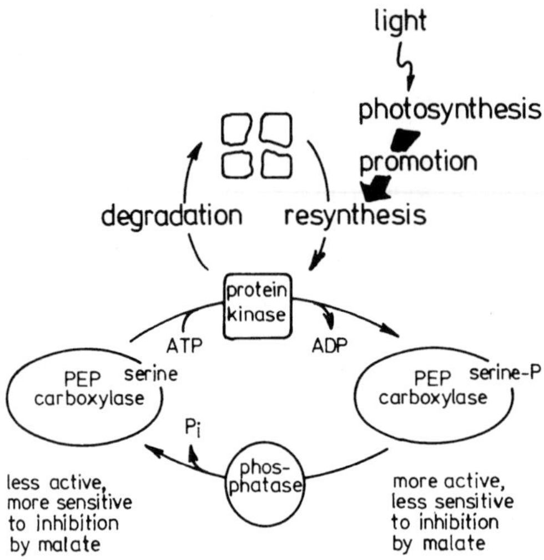

Fig. 4. Known aspects of the mechanism of light-mediated activation of PEP carboxylase in C_4-plant leaves. (After Echevarría et al. 1990; Jiao et al. 1991b; and Jiao and Chollet 1991, 1992)

tion of PEP carboxylase is *dark*-mediated in CAM plants (see review by Jiao and Chollet 1991), and (2) it does not occur at all in C_3-plant guard cells (Willmer et al. 1990).

11. C_3–C_4 Intermediate Photosynthesis

Not a lot has been added to our understanding of those plants that display both C_3 and C_4 characteristics of photosynthesis. An earlier suggestion that *Manihot esculenta* (cassava) belongs to this group has been contested by Edwards et al. (1990), but the concept that certain photosynthetic tissues in the inflorescence of cereal grasses are C_3–C_4 has been extended to the rice panicle; it will be difficult to contest this conclusion since substantial activity of the C_4-enzyme pyruvate Pi dikinase was found in the panicle (Imaizumi et al. 1990).

The photosynthetic responses of those C_3–C_4 species more extensively studied during the past decade were monitored by Brown et al. (1991) after leaves were fed a PEP-analog that inhibits PEP carboxylase. Surprisingly, none of *Panicum milioides, Moricandia arvensis,* and *Neurachne minor* were affected; these species responded to the inhibitor as if they were C_3! However, inhibition of photosynthesis was detected for *Flaveria linearis;* this response was qualitatively similar to those of C_4-like and true-C_4 species of *Flaveria.* Indeed, the *Flaveria* genus is remarkable in that a continuous gradation from C_3 to C_4 exists, e.g., *F. pringlei* (C_3), *F. sonorensis* (C_3–C_4, but lacking a C_4 cycle), *F. floridana* (C_3–C_4, with C_4 cycle), *F. brownii* (C_4-like), *F. bidentis* (C_4) (Krall et al. 1991; Ku et al. 1991). In addition, a C_3–C_4/C_4-like hybrid has been reported (Araus et al. 1991).

The gradation of photosynthetic characteristics in *Flaveria* was matched by a gradual decrease in the level of photorespiratory enzymes, leading Ku et al. (1991) to propose that, evolutionarily, photorespiration was first combated by recapture of photorespiratory CO_2 (as occurs in C_3–C_4 plants), and was then repressed when the true C_4 pathway's

CO_2-pump generated CO_2 levels that overwhelmed the oxygenase activity of RuBP carboxylase. Thus, despite the apparent uselessness of PEP carboxylase evidenced from the experiments of Brown et al. (1991; see above), situations where it could efficiently recapture photorespiratory CO_2 (e.g., warm leaf temperatures) have been identified (Schuster and Monson 1990), and the high pools of glycine and serine in C_3–C_4 plant leaves (Rawsthorne and Hylton 1991) attest to a special photorespiratory metabolism in these plants. Indeed, the pivotal discovery of Hylton et al. (1988) that the CO_2-releasing enzyme of photorespiration (glycine decarboxylase) is confined to the bundle-sheath cells of C_3–C_4 plants is now more significant in that it can explain why PEP carboxylase is not necessary for reducing photorespiration in these plants: quite simply, photorespired CO_2, being released only in bundle-sheath cells, is directly recaptured by the RuBP carboxylase in these cells (Brown et al. 1991).

12. Crassulacean Acid Metabolism (CAM)

Just as more plant species have been discovered with photosynthesis intermediate between C_3 and C_4, so also more species with photosynthesis between C_3 and CAM have been found. Haag-Kerwer et al. (1992) have described a C_3-CAM intermediate, *Clusia minor*, and Herrera et al. (1991) report that *Talinum triangular* displays an inducible weak CAM that assists survival during drought. The better-known case of inducible CAM is that shown by *Mesembryanthemum crystallinum*. Environmental factors, such as high salinity, water stress, and photoperiod, are more important than developmental factors for inducing CAM in this species (Cheng and Edwards 1991; Piepenbrock and Schmitt 1991). Interestingly, the induction by water stress could be mimicked by applying abscisic acid to leaves (Chu et al. 1990). After CAM has been induced, *M. crystallinum* retains the capacity for photorespiration (Whitehouse et al. 1991).

In the standard interpretation of CAM, nocturnal CO_2 fixation by PEP carboxylase generates a pool of malate that is stored in the vacuole until the next day, when the malate is returned to the cytoplasm and decarboxylated, thereby making the "stored CO_2" available for photosynthesis behind closed stomates. Water loss is consequently minimized. The PEP carboxylase activity reflects the degree of CAM: it fluctuates seasonally, and with the degree of water stress (Pilon-Smits et al. 1990, 1991). The fluctuation results form enzyme synthesis (Pilon-Smits et al. 1990). In *Sedum telephium*, which displays inducible CAM, a new molecular species of PEP carboxylase is synthesized (Groenhof et al. 1990). A method for the improved recovery of CAM-plant PEP carboxylase for future studies has become available (Silva et al. 1991).

A variation of CAM that differs from the standard, and may be considered as very C_3-like, is "CAM cycling". Species prominently engaged in CAM cycling, such as those in the Portulacaceae, take up atmospheric CO_2 during the day, and have stomata open by day and closed by night – these are C_3 plant features, and contrast to CAM characteristics. Their CAM attribute derives from their recapture of respiratory CO_2 to form malic acid that accumulates nocturnally (Harris and Martin 1991), and can be used to supply CO_2 for photosynthesis the next day behind *open* stomates.

However, as may be envisaged, there is no reason why respiratory CO_2 may not be recaptured by standard CAM plants, and indeed this does occur, especially if higher nocturnal temperatures enhance respiration, or water stress becomes so severe that stomatal opening even in the dark compromises water economy (Kalt et al. 1990; Fetene and Lüttge 1991). In extreme circumstances, total stomatal closure may occur, but daytime photoinhibition is avoided because sufficient CO_2 for photosynthesis is made available from respiratory CO_2. Storage of citrate, rather than malate, has been detected, and is considered to be energetically advantageous for this process (Lüttge 1990; Haag-Kerwer et al. 1992).

Two details of CAM biochemistry have been further investigated. The first relates to malate transport in and out of the vacuole. Uptake by the vacuole is active, and is linked to H^+ transport which is complex in that three activities are involved: ATP-dependent, PPi-dependent, and pyrophosphatase-activated ATP-dependent (Marquardt-Jarczyk and Lüttge 1990). Efflux of malate is passive, and may be slowed if low temperature early in the morning reduces tonoplast fluidity; this can cause a lag in photosynthesis (Kluge et al. 1991). The second relates to the necessity to cycle one 3C residue through the PEP $\rightarrow$ malate $\rightarrow$ pyruvate $\rightarrow$ reserve polysaccharide $\rightarrow$ PEP circuit for each CO_2 captured nocturnally and subsequently fixed diurnally. If the products of net CO_2 fixation are not mobilized to sink tissues, excessive storage of the reserve polysaccharide (starch) can distort chloroplasts, ultimately reducing CAM activity (Mayoral et al. 1991). In *Fourcroya humboldtiana* the reserve polysaccharide is not starch, rather it is a low molecular weight fructan (Olivares and Medina 1990).

The requirement of this voluminous traffic of a 3C residue would intuitively suggest that the potential productivity of CAM plants might be less than that of C_3 and C_4 plants. Indeed, CAM plants have the highest requirement for ATP: about 6 ATP per CO_2 fixed, compared to 5 ATP (C_4 plants) and 3 ATP (C_3 plants). However, a series of calculations and predictions by Nobel (1991), nevertheless, indicates that the potential productivity of CAM plants can equal, or even exceed that of C_3 and C_4 plants, partly because CAM plants avoid photorespiratory CO_2 loss and are better adapted to continue photosynthesis when water stress is substantial. Quite recently, Nobel et al. (1992) have demonstrated this potential by recording productivities averaging 46 t dry $wt.ha^{-1} \cdot year^{-1}$ (which are among the highest ever reported for any plant species) for *Opuntia amyclea* and *Opuntia ficus-indica* growing in Mexico.

13. Concluding Comment

In our reviewing of 20 years of literature, we have given much emphasis to enzymes. We make no apologies for maintaining this emphasis during a time when DNA-based biotechnology became "the flavor of the month". After all, the basic aim of this technology is to introduce new enzymes (and other proteins) into cells. Of all the biomolecules, the enzymes are most alive, as Kornberg (1980) pointed out about 12 years ago: "You know a cell's enzymes can sustain it without DNA. The cell may have no future without DNA but it still can exist. By contrast, a cell with DNA but without enzymes cannot exist at all".

There is no doubt that the study of enzymes will continue to hold a prominent place in research on photosynthetic carbon metabolism during the next 20 years. But proteins with other functions, which have received less attention in the past, will also be worth watching. These include: (1) membrane proteins that transport metabolites between cellular compartments – the role of these in photosynthesis has already received detailed attention; (2) proteins designed to directly influence the function of other proteins, e.g., rubisco activase; (3) membrane receptor proteins, about which little is known in relation to photosynthesis; and (4) hormones – the first report of a plant hormone that is a polypeptide was recently published (Pearce et al. 1991).

The currently fascinating picture of how photosynthetic carbon metabolism is run by the molecules of life – proteins – will almost certainly be more intricate, and more wonderful, in 2012.

References

Andrews TJ, Kane HJ (1991) J Biol Chem 266:9447–9452. **Angelopoulos K,** Gavalas NA (1991) J Plant Physiol 137:363–367. **Araus JL,** Brown HR, Byrd GT, Serret MD (1991) Planta 183:497–504. **Arp WJ,** Drake BG (1991) Plant Cell Environ 14:1003–1006. **Arrio-Dupont M,** Bakrim N, Echevarría C, Gadal P, Le Maréchal P, Vidal J (1992) Plant Sci 81:37–46. **Arulanantham AR,** Rao IM, Terry N (1990) Plant Physiol 93:1466–1475. **Austin PA,** Ross IS, Mills JD (1992) Biochim Biophys Acta 1099:226–232. **Avelange M-H,** Rébeillé F (1991) Planta 183:158–163. **Avelange M-H,** Thiéry JM, Sarrey F, Gans P, Rébeillé F (1991) Planta 183:150–157.

Backeshoff K, Stabenau H (1990) Bot Acta 103:190–196. **Badger MR,** Price GD (1990) Plant Physiol 94:35–39. **Ball S,** Marianne T, Dirick L, Fresnoy M, Delrue B, Decq A (1991a) Planta 185:17–26. **Ball KL,** Green JH, ap Rees T (1991b) Europ J Biochem 197:265–269. **Bédu S,** Peltier G, Sarrey F, Joset F (1990) Plant Physiol 93:1312–1315. **Beer S,** Israel A, Drechsler Z, Cohen Y (1990) Plant Physiol 94:1542–1546. **Beer S,** Sand-Jensen K, Vindbaek Madsen T, Nielsen SL (1991) Oecologia 87:429–434. **Beers EP,** Duke SH, Henson CA (1990) Plant Physiol 94:738–744. **Bennett J** (1991) Annu Rev Plant Physiol Plant Mol Biol 42:281–311. **Besford RT,** Ludwig LJ, Withers AC (1990) J Exp Bot 41:925–931. **Betsche T,** Schaller D, Melkonian M (1992) Plant Physiol 98:887–893. **Björk M,** Haglund K, Ramazanov Z, Garcia-Reina G, Pedersén M (1992) Planta 187:152–156. **Blackwell RD,** Murray AJS, Lea PJ (1990) Plant Physiol 94:1316–1322. **Booth WA,** Beardall J (1991) Phycologia 30:220–225. **Börsch D,** Westhoff P (1990) FEBS Lett 273:111–115. **Botha FC,** Turpin DH (1990) Plant Physiol 93:1460–1465. **Bowes G** (1991) Plant Cell Environ 14:795–806. **Brauer M,** Stitt M (1990) Physiol Plant 78:568–573. **Brauer M,** Sanders D, Stitt M (1990) Planta 182:236–243. **Brown RH,** Byrd GT, Black CC (1991) Plant Physiol 97:985–989. **Brownell PF,** Bielig LM, Grof CPL (1991) Aust J Plant Physiol 18:589–592. **Brüggemann W,** van der Kooij TAW, van Hasselt PR (1992) Planta 186:179–187. **Bruneau J-M,** Worrell AC, Cambou B, Lando D, Voelker TA (1991) Plant Physiol 96:473–478. **Buchanan BB** (1991) Arch Biochem Biophys 288:1–9. **Buchanan BB,** Arnon DI (1990) Photosyn Res 24:47–53. **Budde RJA,** Randall DD (1990) Plant Physiol 94:1501–1504. **Burnell JN,** Suzuki I, Sugiyama T (1990) Plant Physiol 94:384–387.

Campbell WJ, Ogren WL (1990a) Plant Physiol 94:479–484. **Campbell WJ,** Ogren WL (1990b) Photosyn Res 23:257–268. **Casper T,** Lin T-P, Kakefuda G, Benbow L, Preiss J, Somerville C (1991) Plant Physiol 95:1181–1188. **Champigny ML,** Brauer M, Bismuth E, Manh CT, Siegl G, Quy LV, Stitt M (1992) J Plant Physiol 139:361–368. **Chardot T,** Meunier J-C (1990) Plant Sci 70:1–9. **Chardot T,** Meunier J-C (1991) Biochem J 278:787–791. **Chen C,** Gibbs M (1992) Plant

Physiol 98:535–539. **Chen Z,** Spreitzer RJ (1989) J Biol Chem 264:3051–3053. **Chen Z,** Green D, Westhoff C, Spreitzer RJ (1990) Arch Biochem Biophys 283:60–67. **Cheng S-H,** Edwards GE (1991) Plant Cell Environ 14:271–278. **Chu C,** Dai Z, Ku MSB, Edwards GE (1990) Plant Physiol 93:1253–1260. **Clasper S,** Easterby JS, Powls R (1991) Eur J Biochem 202:1239–1246. **Coleman JR,** Rotatore C, Williams TG, Colman B (1991) Plant Physiol 95:331–334. **Collis BE,** Pollock CJ (1991) New Phytol 119:483–489. **Cornic G,** Briantais J-M (1991) Planta 183:178–184. **Crafts-Brandner SJ,** Salvucci ME, Egli DB (1991) Planta 183:300–306.

Demmig-Adams B (1990) Biochim Biophys Acta 1020:1–24. **Doncaster HD,** Leegood RC (1990) J Plant Physiol 136:330–335. **Drechsler Z,** Beer S (1991) Plant Physiol 97:1439–1444.

Echevarría C, Vidal J, Jiao J, Chollet R (1990) FEBS Lett 275:25–28. **Edmondson DL,** Badger MR, Andrews TJ (1990a) Plant Physiol 93:1390–1397. **Edmondson DL,** Badger MR, Andrews TJ (1990b) Plant Physiol 93:1376–1382. **Edwards GE,** Sheta E, Moore BD, Dai Z, Franceschi VR, Cheng S-H, Lin C-H, Ku MSB (1990) Plant Cell Physiol 31:1199–1206. **Ehleringer JR,** Sage RF, Flanagan LB, Pearcy RW (1991) Trends Ecol Evol 6:95–99. **Eilenberg H,** Beer S, Gepstein S, Geva N, Tadmor O, Zilberstein A (1991) Plant Physiol 95:298–304. **Enomoto T,** Ohyama H, Inui H, Miyatake K, Nakano Y, Kitaoka S (1990) Plant Sci 67:161–167. **Entwistle G,** ap Rees T (1990) Biochem J 271:467–472. **Espie GS,** Kandasamy RA (1992) Plant Physiol 98:560–568. **Espie GS,** Miller AG, Kandasamy RA, Canvin DT (1991a) Can J Bot 69:936–944. **Espie GS,** Miller AG, Canvin DT (1991b) Plant Physiol 97:943–953.

Fetene M, Lüttge U (1991) J Exp Bot 42:25–31. **Florencio FJ,** Yee BC, Johnson TC, Buchanan BB (1988) Arch Biochem Biophys 266:496–507. **Flügge U-I,** Heldt HW (1991) Annu Rev Plant Physiol Plant Mol Biol 42:129–144. **Flügge U-I,** Weber A, Fischer K, Lottspeich F, Eckerskorn C, Waegemann K, Soll J (1991) Nature 353:364–367. **Fredeen AL,** Raab TK, Rao IM, Terry N (1990) Planta 181:399–405. **Frederick JR,** Alm DM, Hesketh JD (1989) Photosynthetica 23:575–584. **Frederick JR,** Alm DM, Hesketh JD, Below FE (1990) Photosyn Res 25:49–57. **Frehner M,** Pozueta-Romero J, Akazawa T (1990) Plant Physiol 94:538–544. **Furbank RT,** Agostino A, Hatch MD (1991) Arch Biochem Biophys 289:376–381.

Gehl KA, Colman B, Sposato LM (1990) J Exp Bot 41:1385–1391. **Gimenez C,** Mitchell VJ, Lawlor DW (1992) Plant Physiol 98:516–524. **Gimmler H,** Weiss C, Baier M, Hartung W (1990) J Exp Bot 41:785–795. **Goyal A,** Shiraiwa Y, Tolbert NE (1992) Plant Physiol 98:578–583. **Graan T,** Boyer JS (1990) Planta 181:378–384. **Grammatikopoulos G,** Manetas Y (1990) Physiol Plant 80:593–597. **Groenhof AC,** Smirnoff N, Bryant JA (1990) Plant Cell Environ 13:437–445. **Gross W** (1990) J Phycol 26:381–383. **Grub A,** Mächler F (1990) J Exp Bot 41:1293–1301. **Gutteridge S** (1991) J Biol Chem 266:7359–7362.

Haag-Kerwer A, Franco AC, Lüttge U (1992) J Exp Bot 43:345–352. **Hajirezaei M,** Stitt M (1991) Plant Sci 77:177–183. **Harley PC,** Sharkey TD (1991) Photosyn Res 27:169–178. **Harris FS,** Martin CE (1991) Plant Physiol 96:1118–1124. **Hatch MD** (1991) Anal Biochem 192:85–89. **Hatch MD,** Burnell JN (1990) Plant Physiol 93:825–828. **Hatzfeld W-D,** Stitt M (1990) Planta 180:198–204. **Heineke D,** Riens B, Grosse H, Hoferichter P, Peter U, Flügge U-I, Heldt HW (1991) Plant Physiol 95:1131–1137. **Hendrix DL,** Grange RI (1991) Plant Physiol 95:228–233. **Hermoso R,** Fonollá J, de Felipe MR, Vivó A, Chueca A, Lázaro JJ, López-Gorgé J (1992) Plant Physiol Biochem 30:39–46. **Herrera A,** Delgado J, Paraguatey I (1991) J Exp Bot 42:493–499. **Heuer B,** Portis AR (1990) Plant Physiol 93:1511–1513. **Heupel R,** Markgraf T, Robinson DG, Heldt HW (1991) Plant Physiol 96:971–979. **Hill LM,** Smith AM (1991) Planta 185:91–96. **Holbrook GP,** Galasinski SC, Salvucci ME (1991) Plant Physiol 97:894–899. **Houtz RL,** Royer M, Salvucci ME (1991) Plant Physiol 97:913–920. **Houtz RL,** Poneleit L, Jones SB, Royer M, Stults JT (1992) Plant Physiol 98:1170–1174. **Howitz KT,** McCarty RE (1991) Plant Physiol 96:1060–1069. **Huber SC,** Huber JL (1990) Arch Biochem Biophys 282:421–426. **Huber SC,** Huber JL (1991a) Plant Cell Physiol 32:319–326. **Huber SC,** Huber JL (1991b) Plant Cell Physiol

32:327–333. **Huber JL**, Hite DRC, Outlaw WH, Huber SC (1991) Plant Physiol 95:291–297. **Hudson GS**, Evans JR, von Caemmerer S, Arvidsson YBC, Andrews TJ (1992) Plant Physiol 98:294–302. **Huppe HC**, Picaud A, Buchanan BB, Miginiac-Maslow M (1991) Planta 186:115–121. **Husic HD**, Quigley EA (1990) Plant Physiol 94:380–383. **Hylton CM**, Rawsthorne S, Smith AM, Jones DA, Woolhouse HW (1988) Planta 175:452–459.

Idso SB, Kimball BA (1991) Plant Physiol 96:990–992. **Iglesias AA**, Spampinato CP, Andreo CS (1991) Arch Biochem Biophys 290:272–276. **Imaizumi N**, Usuda H, Nakamoto H, Ishihara K (1990) Plant Cell Physiol 31:835–843. **Iwaki T**, Wadano A, Yokota A, Himeno M (1991) Plant Cell Physiol 32:1083–1091.

Jacobshagen S, Schnarrenberger C (1990) J Phycol 26:312–317. **Jahnke LS**, Eighmy TT, Fagerberg WR (1991) Plant Cell Environ 14:147–156. **Jiao J**, Chollet R (1990) Arch Biochem Biophys 283:300–305. **Jiao J**, Chollet R (1991) Plant Physiol 95:981–985. **Jiao J**, Chollet R (1992) Plant Physiol 98:152–156. **Jiao J**, Vidal J, Echevarría C, Chollet R (1991a) Plant Physiol 96:297–301. **Jiao J**, Echevarría C, Vidal J, Chollet R (1991b) Proc Nat Acad Sci USA 88:2712–2715. **Johnston AM**, Raven JA (1989) J Phycol 25:568–576. **Jordan BR**, He J, Chow WS, Anderson JM (1992) Plant Cell Environ 15:91–98.

Kaiser WM, Spill D (1991) Plant Physiol 96:368–375. **Kalt W**, Osmond CB, Siedow JN (1990) Plant Physiol 94:826–832. **Kaplan A**, Schwarz R, Lieman-Hurwitz J, Reinhold L (1991) Plant Physiol 97:851–855. **Kettleborough CA**, Phillips AL, Keys AJ, Parry MA (1991) Planta 184:35–39. **Kiss F**, Wu M-X, Wong JH, Balogh Á, Buchanan BB (1991) Arch Biochem Biophys 287:337–340. **Kleczkowski LA**, Edwards GE (1990) Z Naturforsch 45c:42–46. **Kleczkowski LA**, Edwards GE (1991) Plant Physiol Biochem 29:9–17. **Kleczkowski LA**, Villand P, Lönneborg A, Olsen O-A, Lüthi E (1991) Z Naturforsch 46c:605–612. **Kleczkowski LA**, Edwards GE, Randall DD (1992) Phytochemistry 31:51–54. **Kluge M**, Wolf H, Fischer A (1991) Plant Physiol Biochem 29:83–90. **Königer M**, Winter K (1991) Oecologia 87:349–356. **Köpke-Secundo E**, Molnar I, Schnarrenberger C (1990) Plant Physiol 93:40–47. **Kornberg A** (1980) Can J Biochem 58:93–96. **Krall JP**, Edwards GE (1990) Aust J Plant Physiol 17:579–588. **Krall JP**, Edwards GE, Ku MSB (1991) Aust J Plant Physiol 18:369–383. **Krause K-P**, Stitt M (1992) Phytochemistry 31:1143–1146. **Krömer S**, Heldt HW (1991a) Plant Physiol 95:1270–1276. **Krömer S**, Heldt HW (1991b) Biochim Biophys Acta 1057:42–50. **Ku MSB**, Wu J, Dai Z, Scott RA, Chu C, Edwards GE (1991) Plant Physiol 96:518–528.

Laisk A, Kiirats O, Oja V, Gerst U, Weis E, Heber U (1992) Planta 186:434–441. **Levavasseur G**, Edwards GE, Osmond CB, Ramus J (1991) J Phycol 27:667–672. **Lilley RMcC**, Portis AR (1990) Plant Physiol 94:245–250. **Lin M**, Turpin DH (1992) Plant Physiol 98:82–88. **López-Pozos R**, Rodríguez-Sotres R, Muñoz-Clares RA (1990) J Plant Physiol 136:451–457. **Lüttge U** (1990) Plant Cell Environ 13:977–982. **Lüttge U**, Ball E, Fetene M (1991) J Plant Physiol 137:268–272.

Macioszek J, Anderson JB, Anderson LE (1990) Plant Physiol 94:291–296. **Makino A**, Osmond B (1991) Photosyn Res 29:79–85. **Marcus F**, Chamberlain SH, Chu C, Masiarz FR, Shin S, Yee BC, Buchanan BB (1991) Arch Biochem Biophys 287:195–198. **Marquardt-Jarczyk G**, Lüttge U (1990) Bot Acta 103:203–213. **Martínez F**, Orús MI (1991) Plant Physiol 95:1150–1155. **Mason CB**, Manuel LJ, Moroney JV (1990) Plant Physiol 93:833–836. **Mayoral ML**, Medina E, Garcia V (1991) J Exp Bot 42:1123–1129. **McKinnon WB** (1992) Nature 357:15–16. **McMorrow EM**, Bradbeer JW (1990) Plant Physiol 93:374–383. **Metodiev M**, Demirevska-Kepova K (1992) J Exp Bot 43:155–158. **Metzler MC**, Rothermel BA, Nelson T (1989) Plant Mol Biol 12:713–722. **Miginiac-Maslow M**, Decottignies P, Jacquot J-P, Gadal P (1990) Biochim Biophys Acta 1017:273–279. **Miller AG**, Espie GS, Canvin DT (1990) Can J Bot 68:1291–1302. **Moore Bd**, Seemann JR (1990) Plant Cell Physiol 31:969–974. **Moore Bd**, Kobza J, Seemann JR (1991) Plant Physiol 96:208–213.

Nakamoto H, Edwards GE (1990) Plant Physiol Biochem 28:553–559. **Nara M,** Shiraiwa Y, Hirokawa T (1990) Plant Cell Physiol 31:961–967. **Neuhaus HE,** Stitt M (1990) Planta 182:445–454. **Neuhaus HE,** Stitt M (1991) Plant Sci 76:49–55. **Neuhaus HE,** Quick WP, Siegl G, Stitt M (1990a) Planta 181:583–592. **Neuhaus HE,** Krause K–P, Stitt M (1990b) Phytochemistry 29:3411–3415. **Nielsen TH,** Veierskov B (1990) Plant Physiol 93:637–641. **Nimer NA,** Sallal A-KJ, Al-Hasan RH, Merrett MJ (1990) Planta 181:374–377. **Nimer NA,** Dixon GK, Merrett MJ (1992) New Phytol 120:153–158. **Nobel PS** (1991) New Phytol 119:183–205. **Nobel PS,** García-Moya E, Quero E (1992) Plant Cell Environ 15:329–335.

Ogawa T (1990) Plant Physiol 94:760–765. **Ohnishi J,** Flügge U-I, Heldt HW, Kanai R (1990) Plant Physiol 94:950–959. **Olivares E,** Medina E (1990) J Exp Bot 41:579–585. **Oliver DJ,** Neuburger M, Bourguignon J, Douce R (1990) Plant Physiol 94:833–839.

Palmqvist K, Ramazanov Z, Samuelsson G (1990) Physiol Plant 80:267–276. **Pearce G,** Strydom D, Johnson S, Ryan CA (1991) Science 253:895–898. **Peltier G,** Schmidt GW (1991) Proc Natl Acad Sci USA 88:4791–4795. **Petropoulou Y,** Manetas Y, Gavalas NA (1990) Physiol Plant 80:605–611. **Piepenbrock M,** Schmitt JM (1991) Plant Physiol 97:998–1003. **Pilon-Smits EAH,** 't Hart H, van Brederode J (1990) J Plant Physiol 136:155–160. **Pilon-Smits EAH,** 't Hart H, van Brederode J (1991) J Plant Physiol 138:581–586. **Podestá FE,** Iglesias AA, Andreo CS (1990) Photosyn Res 26:161–170. **Poetsch W,** Hermans J, Westhoff P (1991) FEBS Lett 292:133–136. **Porter MA,** Hartmann FC (1990) Arch Biochem Biophys 281:330–334. **Portis AR** (1990) Arch Biochem Biophys 283:397–400. **Pozueta-Romero J,** Frehner M, Viale A, Akazawa T (1991a) Proc Natl Acad Sci USA 88:5769–5773. **Pozueta-Romero J,** Yamaguchi J, Akazawa T (1991b) FEBS Lett 291:233–237. **Pozueto-Romero J,** Ardila F, Akazawa T (1991c) Plant Physiol 97:1565–1572. **Prado FE,** Lázaro JJ, Gorgé JL (1991) Plant Physiol 96:1026–1033. **Prat-Gay G,** Paladini A, Stein M, Wolosiuk RA (1991) J Biol Chem 266:20913–20921. **Price GD,** Badger MR (1991) Can J Bot 69:963–973. **Purohit US,** Verma NC, Bhagwat AS (1991) FEBS Lett 284:23–26.

Queiroz-Claret C, Meunier JC (1992) Phytochemistry 31:751–754. **Quick WP,** Schurr U, Fichtner K, Schulze E-D, Rodermel SR, Bogorad L, Stitt M (1991) Plant J 1:51–58. **Quick WP,** Chaves MM, Wendler R, David M, Rodrigues ML, Passaharinho JA, Pereira JS, Adcock MD, Leegood RC, Stitt M (1992) Plant Cell Environ 15:25–35.

Raines CA, Lloyd JC, Dyer TA (1991) Photosyn Res 27:1–14. **Rault M,** Gontero B, Ricard J (1991) Eur J Biochem 197:791–797. **Rawsthorne S,** Hylton CM (1991) Planta 186:122–126. **Reddy AR,** Das VSR (1990) Biochem Intern 20:25–35. **Reddy MM,** Vani T, Raghavendra AS (1991) Plant Physiol 96:1368–1371. **Reiskind JB,** Bowes G (1991) Proc Natl Acad Sci USA 88:2883–2887. **Riens B,** Heldt HW (1992) Plant Physiol 98:573–577. **Rodermel SR,** Abbott MS, Bogorad L (1988) Cell 55:673–681. **Roesler KR,** Ogren WL (1990) Plant Physiol 93:188–193. **Rotatore C,** Colman B (1991a) Plant Cell Environ 14:371–375. **Rotatore C,** Colman B (1991b) Plant Cell Environ 14:377–382. **Rotatore C,** Colman B (1991c) Can J Bot 69:1025–1031. **Rowland-Bamford AJ,** Baker JT, Allen LH, Bowes G (1991) Plant Cell Environ 14:577–583.

Sage RF, Sharkey TD, Seemann JR (1990a) Plant Physiol 94:1735–1742. **Sage RF,** Sharkey TD, Pearcy RW (1990b) Aust J Plant Physiol 17:135–148. **Salahas G,** Manetas Y, Gavalas NA (1990a) Photosyn Res 24:183–188. **Salahas G,** Manetas Y, Gavalas NA (1990b) Photosyn Res 26:9–17. **Salvucci ME,** Drake RR, Haley BE (1990) Arch Biochem Biophys 281:212–218. **Santakumari M,** Berkowitz GA (1991) Photosyn Res 28:9–20. **Sassenrath GF,** Ort DR, Portis AR (1990) Arch Biochem Biophys 282:302–308. **Sawada S,** Usuda H, Hasegawa Y, Tsukui T (1990) Plant Cell Physiol 31:697–704. **Scheibe R** (1990) Bot Acta 103:327–334. **Scheibe R** (1991) Plant Physiol 96:1–3. **Scheibe R,** Kampfenkel K, Wessels R, Tripier D (1991) Biochim Biophys Acta 1076:1–8. **Schimkat D,** Heineke D, Heldt HW (1990) Planta 181:97–103. **Schnarrenberger C** (1990) Planta 181:249–255. **Schulze W,** Stitt M, Schulze E-D, Neuhaus HE, Fichtner K (1991) Plant Physiol 95:890–895. **Schuster WS,** Monson RK (1990) Plant Cell Environ 13:903–912. **Servaites JC,**

Shieh W-J, Geiger DR (1991) Plant Physiol 97:1115–1121. **Shen JB,** Orozco EM, Ogren WL (1991) J Biol Chem 266:8963–8968. **Siebke K,** Laisk A, Oja V, Kiirats K, Heber U (1990) Planta 182:513–522. **Siebke K,** Laisk A, Neimanis S, Heber U (1991) Planta 185:337–343. **Siebke K,** Yin Z-H, Raghavendra AS, Heber U (1992) Planta 186:526–531. **Siegl G,** MacKintosh C, Stitt M (1990) FEBS Lett 270:198–202. **Silva AB,** Arrabaca MC, Cabral JMS (1991) Plant Sci 76:29–34. **Skrukrud CL,** Anderson LE (1991) FEBS Lett 280:259–261. **Skrukrud CL,** Gordon IM, Dorwin S, Yuan X-H, Johansson G, Anderson LE (1991) Plant Physiol 97:730–735. **Smith AM,** Neuhaus HE, Stitt M (1990) Planta 181:310–315. **Smrcka AV,** Bohnert HJ, Jensen RG (1991) Arch Biochem Biophys 286:14–19. **Spalding MH** (1990) Photosyn Res 24:245–252. **Stamatakis K,** Skaliora I, Gavalas NA, Manetas Y (1990) Aust J Plant Physiol 17:407–411. **Stitt M,** von Schaewen A, Willmitzer L (1990) Planta 183:40–50. **Stitt M,** Quick WP, Schurr U, Schulze E-D, Rodermel SR, Bogorad L (1991) Planta 183:555–566. **Su X,** Bogorad L (1991) J Biol Chem 266:23698–23705. **Sültemeyer DF,** Fock HP, Canvin DT (1990) Plant Physiol 94:1250–1257. **Sültemeyer DF,** Fock HP, Canvin DT (1991) Can J Bot 69:995–1002. **Süss K-H** (1990) Z Naturforsch 45c:633–637. **Suzuki A,** Lelandais M, Morot-Gaudry J-F (1990) Phytochemistry 29:3067–3072. **Suzuki K,** Iwamoto K, Yokoyama S, Ikawa T (1991) J Phycol 27:492–498.

Tarchevskii IA, Konyukhova TM (1989) Photosynthetica 23:472–478. **Tarczynski MC,** Outlaw WH (1990) Arch Biochem Biophys 280:153–158. **Turley RB,** Trelease RN (1990) Physiol Plant 79:570–578.

Umino Y, Shiraiwa Y (1991) J Plant Physiol 139:41–44.

Vairinhos F, Simon J-P (1990) Plant Sci 71:173–177. **Valle EM,** Heldt HW (1991) Plant Physiol 95:839–845. **Vani T,** Reddy MM, Raghavendra AS (1990) Physiol Plant 80:467–471. **Van Langendonckt A,** Driessche TV (1992) Arch Biochem Biophys 292:156–164. **Van Quy L,** Foyer C, Champigny M-L (1991) Plant Physiol 97:1476–1482. **van Wijk KJ,** Krause GH (1991) Planta 186:135–142. **Vassey TL,** Quick WP, Sharkey TD, Stitt M (1991) Physiol Plant 81:37–44. **von Caemmerer S,** Evans JR (1991) Aust J Plant Physiol 18:287–305.

Wedding RT, Black MK (1990) Arch Biochem Biophys 282:284–289. **Wendler C,** Barniske M, Wild A (1990) Photosyn Res 24:55–61. **Whitehouse DG,** Rogers WJ, Tobin AK (1991) J Exp Bot 42:485–492. **Willmer CM,** Petropoulou Y, Manetas Y (1990) J Exp Bot 41:1103–1107. **Wilson BJ,** Tolbert NE (1991) FEBS Lett 279:313–315. **Winter K,** Königer M (1991) Oecologia 87:190–197. **Wise RR,** Frederick JR, Alm DM, Kramer DM, Hesketh JD, Crofts AR, Ort DR (1990) Plant Cell Environ 13:923–931. **Wise RR,** Sparrow DH, Ortiz-Lopez A, Ort DR (1991) Plant Sci 74:45–52. **Wiskich JT,** Bryce JH, Day DA, Dry IB (1990) Plant Physiol 93:611–616. **Wong SC,** Osmond CB (1991) Aust J Plant Physiol 18:137–152. **Wong JH,** Kiss F, Wu M-X, Buchanan BB (1990) Plant Physiol 94:499–506. **Worrell AC,** Bruneau J-M, Summerfelt K, Boersig M, Volker TA (1991) Plant Cell 3:1121–1130. **Wu J,** Neimanis S, Heber U (1991) Bot Acta 104:283–291. **Wu M-X,** Meyer CR, Willeford KO, Wedding RT (1990) Arch Biochem Biophys 281:324–329.

Yang Y, Steup M (1990) Plant Physiol 94:960–969. **Yokota A** (1991a) J Biochem 110:246–252. **Yokota A** (1991b) Plant Cell Physiol 32:755–762. **Yokota A,** Tsujimoto N (1992) Eur J Biochem 204:901–909. **Young AJ** (1991) Physiol Plant 83:702–708. **Yu S** (1989) J Phycol 25:608–612. **Yu S,** Pedersén M (1991) Bot Mar 34:323–325. **Yu J,** Woo KC (1991) Aust J Plant Physiol 18:583–588.

Zhu G, Jensen RG (1991) Plant Physiol 97:1354–1358. **Ziska LH,** Drake BG, Chamberlain S (1990) Oecologia 83:469–472. **Ziska LH,** Hogan KP, Smith AP, Drake BG (1991) Oecologia 86:383–389.

Dr. Grahame J. Kelly
Centre for Biological Population Management
School of Life Science
Queensland University of Technology
4000 Brisbane, Queensland Australia

Professor Dr. Erwin Latzko
Institut für Allgemeine Chemie und Biochemie
Technische Universität München
W-8050 Freising, Germany.

Photosynthesis

Carbon Metabolism: The Chloroplast's Sesquicentenary, and Some Thoughts on the Limits to Plant Productivity

By Grahame J. Kelly and Erwin Latzko

"At ten p.m. or thereabout,
the little bean began to sprout.
By morning it had grown so tall
you couldn't see the top at all".
(Anonymous 1734)

1. Introduction

The anonymous author of the above quotation, which is taken from her (or his) legendary children's story "Jack and the Beanstalk", commented that it is the kind of thing that "proceeds from the maggots in a madman's brain". It does indeed require little worldly experience to recognise that such a prodigious rate of plant growth is unheard of, and the addition of a primary-school level of education would (hopefully!) impart a sufficient knowledge of photosynthesis as to help one realise that, assuming the bean seed was of normal size, such growth would be impossible during the darkness between 22.00 h and morning. Nevertheless, if an innocent child (just exposed to the Jack-and-the-beanstalk tale) was to reflect upon reality and ask why plants do not grow faster than they do, then even a good photosynthesis researcher would find it difficult to frame a concise answer.

Indeed, *we* are not ashamed to confess that we are unsure of the precise answer. What we do know is that photosynthesis research, since its inception in 1648 by Van Helmont (Walker 1992a; Huzisige and Ke 1993), has revealed hundreds of components of the green cell's photosynthetic machinery, and that the characteristics of these are now sufficiently well understood that a certain few can be identified as being intimately involved in determining "photosynthetic capacity". Moreover, current research on photosynthesis (including photosynthetic carbon metabolism) is directed towards obtaining more refined explanations, apparently for two reasons: (1) to discover whether some form of biotechnology, e.g. genetic engineering, can be used to increase photosynthetic capacity and, consequently, agricultural productivity, and (2) to ascertain the responses of plants to predicted changes in the global environment – especially to the increased level

G.J. Kelly and E. Latzko:
Thirty Years of Photosynthesis

Originally published in Progress in Botany 56

of atmospheric CO_2. Ironically, it may be argued that these investigations are at the expense of those aimed at discovering more components of the photosynthetic machinery: the literature (for the past 2 years) on chloroplasts and the Calvin cycle is meagre. Alternatively, perhaps the remaining components to be discovered are few.

This review covers research published since our last chapter in Volume 54 of Progress in Botany, and concludes by attempting to relate these studies to a consideration of the biophysical and biochemical factors that limit plant productivity. Meantime, within the review, we pay particular attention to events in and on the chloroplast, not only because this assists our interpretation of the limits to productivity, but also because this year (according to Huzisige and Ke 1993) we celebrate the 150th anniversary of the discovery of chloroplasts by von Mohl (1844).

2. New Locations and Pathways

Without doubt, essentially all photosynthetic CO_2 fixation is accomplished by the Calvin cycle in the green cells of plants, algae and cyanobacteria. The leaf is the major location of plants' green cells, but other localities are not uncommon, e.g. potentially photosynthetic chloroplasts have been found under the bark and wood of poplar twigs, in the dimly illuminated pith cells (van Cleve et al. 1993), and they have been induced to develop in cultured *Acmella oppositifolia* roots which, consequently, became green and autotrophic (Flores et al. 1993). More exotic are reports of genetically engineered *Escherichia coli* producing ribulose-1,5-bisphosphate (RuBP) and sucrose-P (Hudson et al. 1992; Sonnewald et al. 1993), metabolites that in nature are the proprietary of photosynthetic cells.

Obscure anaerobic photosynthetic bacteria are the source of three alternatives to the Calvin cycle: the reductive citric acid cycle, the reductive acetyl CoA pathway, and the reductive 3-hydroxypropionate cycle; the latter was recently proposed to accommodate photosynthesis by *Chloroflexus aurantiacus* (Strauss and Fuchs 1993). For higher plants, Janave et al. (1993) have further evidence that potato tuber chloroplasts can directly reduce CO_2 to formate and thence glyoxylate, but the H-donor for the reduction is still unclear. Interestingly, a major soluble protein in mitochondria from *non*photosynthetic plant cells appears to have the opposite function, i.e. formate oxidation (Colas des Francs-Small et al. 1993).

3. Uptake of Inorganic Carbon

The uptake of CO_2 and/or HCO_3^- by photosynthetic cells has always focussed on aquatic species, but within this group the tide has turned: algae, rather than cyanobacteria, have received most attention in the past 2 years. For cyanobacteria, an NADH dehydrogenase seems to be part of the mechanism supplying ATP to pump inorganic carbon (Ci) into cells (Ogawa 1992). Either CO_2 or HCO_3^- may be pumped in, although the activity of the HCO_3^- pump seems to be highest (Yu et al. 1994). Once inside the cell, the HCO_3^- becomes the source of CO_2 for RuBP carboxylase. Recent work has shown that the efficient generation of a high concentration of CO_2 from this HCO_3^- occurs in intracellular structures, called carboxysomes, that contain RuBP carboxylase and a distinctive carbonic anhydrase (Badger and Price 1992; Marcus et al. 1992; Price et al. 1992).

More complex Ci-concentrating mechanisms are possessed by some eukaryotic microalgal cells. External and internal carbonic anhydrases (Shiraiwa et al. 1993; Williams and Colman 1993; Palmqvist et al. 1994a), three unidentified chloroplast-envelope proteins (Thielmann et al. 1992; Ramazanov et al. 1993), and a plasma-membrane ATPase that appears to be part of an active CO_2-uptake mechanism (Rotatore et al. 1992; Karlsson et al. 1994), all seem to be important when Ci accumulation is carried out by these cells. A new twist is that links to photorespiration and stress responses have appeared: Ramazanov and Cárdenas (1992a) found that a functional photorespiratory glycolate pathway was required before the induction of a Ci-transport mechanism occurred in *Chlamydomonas reinhardtii*, while Sültemeyer et al. (1993) concluded that the ATP to power active Ci uptake is generated by a "pseudocyclic" electron transport in which O_2 is the electron acceptor and, consequently, superoxide radical ions are produced. This reactive form of oxygen can be removed by enzymes of the ascorbate/glutathione system (Sect. 4g), and indeed the activities of these enzymes increased 4- to 20-fold when *C. reinhardtii* cells were induced to begin active Ci uptake.

Experiments with marine algae indicate that a variety of approaches to Ci uptake exist. The green microalga *Dunaliella* has almost no external carbonic anhydrase (Goyal et al. 1992), and probably transports CO_2 across the plasma membrane (Ramazanov and Cárdenas 1992b). However, the green macroalga *Ulva* appears to use an anion transporter to permit entry of HCO_3^- (which, at 2 mM, is abundant in seawater) (Drechsler et al. 1993). Studies with seven species of red macroalgae indicate that only some species possess active Ci-uptake mechanisms, and that only some can use HCO_3^-, although all can use CO_2 (Haglund et al. 1992; Israel and Beer 1992; Johnston et al. 1992). Finally, chromophyte microalgae, which make up a good portion of the ocean's phytoplankton, may be limited by Ci supply despite the abundance of total Ci in seawater: both diatoms and coccolithophorids depended on simple diffusive entry of CO_2 for photosynthesis, and since the concentration of CO_2 in seawater is the same as in air, viz. $10 \mu M$, but its rate of diffusion in seawater is less than 0.1% of that in air, slow growth ensues (Nimer et al. 1992; Riebesell et al. 1993). These conclusions are supported by the generally low levels of carbonic anhydrase in these microalgae – the enzyme was restricted to the chloroplast in one examined coccolithophorid (Quiroga and González 1993) – and contrast with the reported active uptake of CO_2 by the freshwater diatom *Navicula pelliculosa* (Rotatore and Colman 1992). However, this contrast may not reflect any particular inability of the marine microalgae to synthesise carbonic anhydrase. Rather, it may be a consequence of the fact that carbonic anhydrase is a zinc-containing protein, but there is not enough zinc in ocean waters to support the synthesis of adequate amounts of the enzyme. In a very recent report, Morel et al. (1994) provide strong evidence that this is the case. Their report is of special significance because it positions carbonic anhydrase as a central component in the major issues of (1) the global carbon cycle, and (2) the determinants of the standing biomass of oceanic phytoplankton.

On land, plants obtain Ci via CO_2 diffusion through air. There is no recent evidence for any active Ci-uptake mechanism on the plasma membrane of photosynthetic cells of higher plants (Loreto et al. 1992; Palmqvist et al. 1994b). Nevertheless, the terrestrial plant community is not totally devoid of the Ci-uptake phenomenon, in that the cyanobacterial photobionts of lichens have been shown to operate Ci-concentrating mechanisms (Badger et al. 1993; Palmqvist 1993). In addition, carbonic anhydrase is in the leaves of all land plants, and this represents the presence of one component of the algal systems that assist the delivery of CO_2 to RuBP carboxylase. We are still impressed by the observation of Porter and Grodzinski (1984) that bean leaves grown in a high-CO_2 atmosphere contained only one-tenth of the carbonic anhydrase activity found in leaves from plants grown in a normal atmosphere; now we are intrigued to learn that wheat leaves *normally* contain a low activity of carbonic anhydrase, equivalent to about one-tenth of that in the leaves of rice, spinach, pea and bean (Makino et al. 1992).

4. Events In and On the Chloroplast

As mentioned in the Introduction, this year marks the 150th anniversary of the publication of one of the reports from the discoverer of chloroplasts, von Mohl (1844). Given this anniversary, it is fitting that we emphasise the obvious central role of chloroplasts to photosynthesis. A good beginning is the reminder from Winter et al. (1993) that these photosynthetic organelles are not only qualitatively, but also quantitatively important: in barley leaf mesophyll cells, the combined volume of the chloroplasts is easily twice the volume of the cytosol.

a) The Calvin Cycle

Essentially all global CO_2 assimilation occurs via the Calvin cycle. Further evidence has appeared that the enzymes catalysing the 13 reactions of this cycle are not randomly distributed through the chloroplast stroma, but rather organised in some fashion. This concept is attractive, given the high protein concentration and consequent limited water availability in the stroma (Sainis and Srinivasan 1993). A multienzyme complex of the five enzymes that use ATP and NADPH to convert ribose-5-P and CO_2 to glyceraldehyde-3-P (GAP) was detected by Gontero et al. (1993), while Süss et al. (1993) found a similar complex closely associated with the surface of thylakoid membranes, especially in the vicinity of thylakoid membrane proteins involved in the generation of the NADPH and ATP used to drive the Calvin cycle. On the other hand, Scagliarini et al. (1993) have reason to doubt that GAP dehydrogenase could be part of any membrane-bound multienzyme complex.

b) RuBP Carboxylase

The most abundant protein of the chloroplast, the CO_2-fixing enzyme RuBP carboxylase, has continued to attract researchers' attention, principally with respect to details of previously identified regulatory mechanisms (see Fig. 1 in our 1990 review). These mechanisms have been reviewed more recently by Portis (1992). One mechanism, viz. activation by interaction with a regulatory protein called rubisco activase, clearly can be important in vivo: tobacco plants genetically engineered to contain a low content of activase were found to photosynthesise poorly, presumably because the activase-mediated promotion of RuBP carboxylase carbamylation (i.e. covalent attachment of an activator CO_2) was incomplete (Mate et al. 1993). However, the activase does more than promote carbamylation. It also promotes dissociation of the inhibitors RuBP and carboxyarabinitol-1-P (CA1P) from the carboxylase (Wang and Portis 1992; Mate et al. 1993) and somehow is itself activated in a light-dependent process (Woodrow and Mott 1992) that may be related to its ability to hydrolyse ATP (Campbell and Ogren 1992). Furthermore, self-association of activase enhances these functions, and may well occur in the chloroplast stroma (Salvucci 1992).

A second carboxylase regulator of contemporary interest is the sugar-P CA1P that inhibits enzyme activity. As noted above, activase causes it to dissociate from the enzyme, and it is then destroyed by a phosphatase (Kingston-Smith et al. 1992) that releases carboxyarabinitol. This carboxy-sugar now has been identified in leaves, and appears to be a precursor of CA1P resynthesis in darkened leaves (Moore and Seemann 1992) although the responsible enzymes(s) has not yet been detected. Overall, the contribution of CA1P to carboxylase regulation is species-variable. Its role is minor in C_4 plants such as *Chenopodium album* (Sage et al. 1993). Even in C_3 plants where it has a major role, e.g. in *Phaseolus vulgaris*, this role is predominately one of maintaining the carboxylase in an inactive state in darkened leaves; adjustment of the carboxylase activity to variable light intensities during the day still involves modulation of the extent of carbamylation (Sage 1993; Sage et al. 1993).

Other studies of RuBP carboxylase regulation refer to activation by Pi that may be more significant in sink-limited plants (Sawada et al. 1992); activation when 6-P-gluconate binds to regulatory sites which otherwise can bind RuBP (Yokota et al. 1992, 1994); and a novel phosphorylation of *Cicer arietinum* RuBP carboxylase that stabilised the enzyme's quaternary structure – comparable phosphorylation of the enzyme from tobacco was neither detected, nor required for stability (Aggarwal et al. 1993).

In normal air, RuBP carboxylase sometimes utilises O_2 instead of CO_2. This so-called oxygenase activity initiates photorespiration (see Sect. 4g) and reduces plant productivity. An enticing objective of research is, therefore, to improve the enzyme's specificity for CO_2. However, initial efforts in this direction have met with only limited success. Two approaches have been tried. In the first, Read and Tabita (1992) constructed "hybrid" enzymes composed of large subunits from a cyanobacterial enzyme and small subunits from either of two marine microalgae, *Cylindrotheca* sp. or *Olisthodiscus luteus* (normal RuBP carboxylase is composed of eight large subunits and eight small subunits). Indeed, the cyanobacteria/*Cylindrotheca* hybrid enzyme showed a 60% better specificity than the cyanobacterial enzyme. However, its specificity was still lower than that of the native *Cylindrotheca* enzyme, and the overall catalytic activity of the hybrid enzyme was greatly reduced. In the second approach, site-directed mutagenesis was used. The first report of an improvement in specificity resulting from this technique is that from Harpel and Hartman (1992) who replaced serine-368 with alanine in the *Rhodospirillum rubrum* RuBP carboxylase. However, only 2% of the enzyme's activity remained. Consequently, the outlook for increasing the specificity of crop-plant RuBP carboxylases, while maintaining their catalytic capacity, is not especially promising (Spreitzer 1993).

c) Other Calvin Cycle Enzymes

Minor attention has been paid to the remaining enzymes of the Calvin cycle. Chloroplast glycerate-3-P kinase (and its cytosolic counterpart) have now been purified from a eukaryotic alga, viz. *Selenastrum minutum* (Lin and Turpin 1993). The in vivo state of the next Calvin cycle enzyme, GAP dehydrogenase, has been considered by Scagliarini et al. (1993), who concluded that dark/light-mediated association/dissociation of four enzyme molecules possibly occurs, the changes being assisted by regulator molecules that include the product glycerate-1,3-P_2 (Trost et al. 1993).

Incidentally, some interesting concepts concerning GAP dehydrogenases have arisen from studies with nonchloroplast forms of the enzyme. The cytosolic NADP-linked *non*reversible GAP dehydrogenase, present in all photosynthetic organisms except cyanobacteria (Mateos and Serrano 1992), seems to be more important to photosynthetic carbon metabolism than previously envisaged. Specifically, it is thought to have a central role in the cytosolic biosynthesis of sugar alcohols, such as mannitol (Fig. 1) and sorbitol (Everard et al. 1993), and since these sugar alcohols constitute some 30% of the global translocated carbon, its role is quite significant.

The nonreversible GAP dehydrogenase does not seem to be closely related to the more conventional, reversible GAP dehydrogenases that use NADP (photosynthetic cells) or NAD (cells of all organisms) (Michels et al. 1994). It is thus fascinating that Corbier et al. (1992), using site-directed mutagenesis, were able to transform the conventional

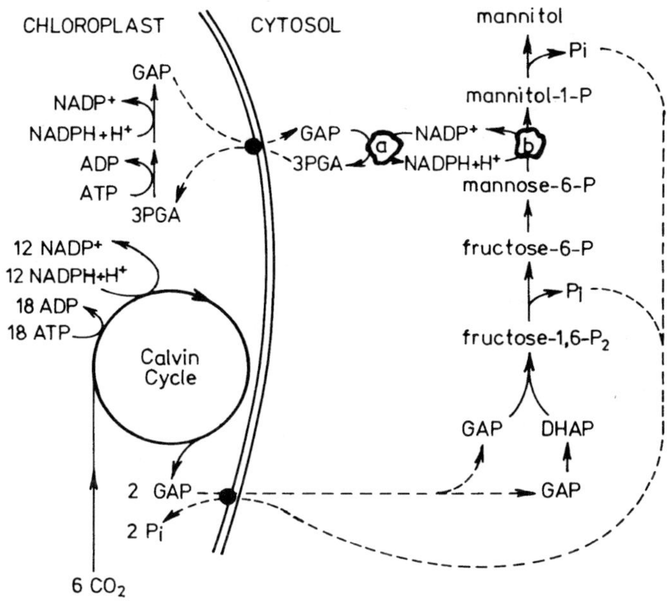

Fig. 1. Central role of nonreversible GAP dehydrogenase *a* in the synthesis of sugar alcohols for translocation in the phloem sap. The enzyme participates in a shuttle system for the indirect transport of photosynthetically generated NADPH from the chloroplast to the cytosol. The cytosolic NADPH is, in turn, used by mannose-6-P reductase. *b* ● Phosphate translocator of the inner membrane of the chloroplast envelope (see Sect. 4e). *DHAP* Dihydroxyacetone-P; *3PGA* glycerate-3-P. (After Everard et al. 1993)

NAD-linked enzyme of *Escherichia coli* into an NADP-linked nonreversible form simply by altering cysteine-149 to alanine. In other studies with the NAD-linked enzyme from animal cells, additional possible functions for the enzyme have been revealed: it may act as a "binding protein" for the export of tRNA from the nucleus (Singh and Green 1993), and it may be involved in cellular signalling processes after an unusual modification in which the greater portion of the substrate NAD (specifically, ADP-ribose; i.e. NAD after removal of the nicotinamide base) is covalently attached to it (Kots et al. 1993). It remains to be seen whether or not plant GAP dehydrogenases possess these functions.

Pea chloroplasts have been found to contain not one, but two, aldolase isoenzymes, and although they are very similar (Razdan et al. 1992), it is intriguing to reflect that the Calvin cycle includes two different aldolase-catalysed reactions, for the synthesis of (respectively) fructose-1,6-P_2 and sedoheptulose-1,7-P_2. These sugar-P_2s are then hydrolysed by specific bisphosphatases. The chloroplast fructose-1,6-bisphosphatase undergoes light-mediated activation in a process that involves the small protein thioredoxin; the properties of this protein seem to vary between plant species (Prado et al. 1992). Separation of chloroplast fructose-1, 6-bisphosphatase from its cytosolic counterpart has been achieved for *Chlorella kessleri* (Grotjohann et al. 1993) and for spinach and pea (Fonollá et al. 1994);

in the latter case, the two forms of the enzyme were shown to be immunologically distinct. Finally, a notable observation is that the deactivation of stromal enzymes such as the bisphosphatases during the brief but often frequent exposures of leaves to low light (as is typical in a field plant canopy) results in a definite limitation on photosynthetic carbon uptake (Sassenrath-Cole and Pearcy 1992).

d) Starch Metabolism

When chloroplasts in plant leaves are well illuminated, photosynthetic CO_2 fixation can exceed the immediate capacity of sucrose synthesis in the cytosol and/or the capacity of sink tissues to utilise sucrose. If this occurs, the excess photosynthate may be diverted into the synthesis of a transitory reserve of starch that can be remobilised, after the sun goes down, to extend net sucrose synthesis into the night. An endogenous circadian shift ensures that the "sunset shift" away from starch synthesis occurs effectively (Li et al. 1992a). The metabolic pathway for the daytime starch synthesis is GAP →→ hexose-P →→ ADP-glucose → starch. Regulation of this pathway may involve several enzymes, including starch synthase which can undergo a Ca^{2+}/calmodulin-dependent phosphorylation and consequent alteration in activity (Dreier et al. 1992). However, the best-known regulatory enzyme of starch synthesis is ADP-glucose pyrophosphorylase (Stark et al. 1992). The pivotal role of this enzyme has been used by Okita (1992) to argue against any alternative starch synthesis pathway not involving it, including the proposed pathway in which ADP-glucose is generated in the cytosol by sucrose synthase, and then transported into the chloroplast by a chloroplast-envelope located adenylate translocator (see Fig. 3 in our preceding review). In any event, this pathway no longer can be seriously considered because, although all types of plastids appear to be capable of importing ADP-glucose (Ardila et al. 1993), sucrose synthase in fact cannot directly generate ADP-glucose (Perata et al. 1992). In addition, the *affinity* of the envelope's adenylate translocator for ADP-glucose (as determined with pea-root plastids) appears to be too low to permit a reasonable rate of ADP-glucose import (Schunemann et al. 1993).

Initial steps in the degradation of chloroplast starch probably use a debranching enzyme and at least four amylases (Li et al. 1992b; Ghiena et al. 1993), as well as phosphorylase (see our last review).

Starch is also synthesised and degraded in plastids other than chloroplasts, and these nongreen plastids have received more attention than chloroplasts during the past 2 years. Etioplasts, the immediate precursors of chloroplasts, synthesise starch from triose-P imported by a typical chloroplast-type phosphate translocator of the inner envelope membrane (Batz et al. 1992). However, other nongreen plastids, including maize-seed, potato-tuber and cauliflower-bud amyloplasts lack fructose-1,6-bisphosphatase and therefore could not generate starch from triose-P; instead, glucose-P is imported as the direct substrate for starch synthesis (Kossmann et al. 1992; Tobias et al. 1992; Neuhaus et al. 1993) via a recently identified translocator with a high affinity for hexose-P (Batz et al. 1993). Furthermore, a similar mechanism of hexose-P import operates on other plastids that form little or no starch, such as leucoplasts and root plastids. These oxidise the hexose-P to CO_2: root plastids use the oxidative pentose-P pathway, thereby generating NADPH that, in turn, reduces NO_2^- to NH_4^+ (Borchert et al. 1993; Coates and ap Rees 1994).

e) Envelope Transporters

Certain proteins of the inner membrane of plastids act as transporters, controlling the entry and exit of metabolites to and from the plastids. The best known is the chloroplast phosphate translocator that, during photosynthesis, allows exit of triose-P in exchange for imported Pi. As mentioned above in Sect. 4g, the etioplast phosphate translocator operates similarly, but the equivalent amyloplast translocator also accepts hexose-P's. Interestingly, Overlach et al. (1993) have found that chloroplasts in guard cells are like amyloplasts in that their phosphate translocators also accept hexose-P; in addition, these chloroplasts (like amyloplasts) lack fructose-1,6-bisphosphatase. It would appear that the specificity of this translocator is easily adjusted – indeed, Fischer et al. (1994) have concluded that only minor exchanges of amino acids are required for the translocator to accept P-enolpyruvate, as is required in C_4 plants.

The phosphate translocator can also mediate triose-P/glycerate-3-P shuttles for the indirect export of ATP and NAD(P)H from chloroplasts to the cytosol. This shuttle ensures that, upon illumination, the cytosolic ATP/ADP ratio rises rapidly in a matter of seconds, reflecting the similar rise that occurs in the chloroplast (Gardeström 1993). The phosphate translocator, which in evolutionary terms is as old as the chloroplast (Schlichting and Bothe 1993), is not the only envelope translocator. Others include the adenylate translocator (see Sect. 4d above), a recently identified pyrophosphate (PPi) translocator (Lunn and Douce 1993) that may import cytosolic PPi generated during sucrose synthesis (Sect. 6), an ATP-driven H^+ pump that moves H^+ from the chloroplast stroma to the cytosol, thus facilitating a suitable stroma:cytosol ΔpH (Berkowitz and Peters 1993), and a transporter that exchanges outgoing glycolate for incoming glycerate during photorespiration (Yu and Woo 1992; Young and McCarty 1993).

f) Photoinhibition

It sometimes happens that the rate of CO_2 assimilation into carbohydrates during photosynthesis is markedly less than the capacity for photosynthetic electron transport from water to NADP. Examples include sun plants exposed to low-water or low-temperature stress, and shade plants exposed to full sunlight. A consequence is that the photosynthetic apparatus of the chloroplast thylakoids becomes over-reduced, which causes it to become altered in such a way that, upon return of the plant to normal conditions, CO_2 fixation is depressed, i.e. photoinhibited. Recent examples include protoplasts of the brown alga *Laminaria saccharina* exposed to high light (Benet et al. 1994) and *Sphagnum* mosses exposed to moderately high light intensities, similar to those that they often experience in the field (Murray et al. 1993).

The alteration that the thylakoid undergoes has several components. One is that some photosystem II reaction centres and/or light-harvesting antennae become inactive and aggregate somewhat in the membrane; another is that one of the proteins of the photosystem II reaction centre, termed the D1 protein, fails to undergo the relatively rapid rate of turnover (i.e. degradation and resynthesis) that is normal for it (Aro et al. 1993, 1994; Schnettger et al. 1994). Interference with turnover of the D1 protein leads to severe photoinhibition, but inactivation of photosystem II centres can be viewed as a normal adjustment of electron flow to match a lower demand for NADPH (Walker 1992b; Huner et al. 1993). The down regulation of photosystem II centres is in turn promoted by increased acidity in the thylakoid lumen (due to accumulation of H^+ which has difficulty exiting via the coupling factor because insufficient ADP + P_i is returned from a slowly turning Calvin cycle), and by the carotenoid pigment zeaxanthin which is formed from violaxanthin and ascorbate in this low pH environment (Bailey and Walker 1992; Gilmore and Yamamoto 1992; Choudhury et al. 1993; Demmig-Adams and Adams 1993; Havaux and Gruszecki 1993; Johnson et al. 1993; Jahns and Krause 1994; Neubauer and Yamamoto 1994). All of these changes favour the dissipation of excess harvested light energy as heat, and therefore help to avoid damage to the photochemical apparatus.

Incidentally, in a report just published, Quiñones and Zeiger (1994) provide evidence that zeaxanthin may be the long-sought blue light photoreceptor of plants.

g) The Mehler Reaction, and Photorespiration

It may be questioned why the chloroplast needs to resort to the above largely biophysical mechanisms for energy dissipation under stress situations when two alternative biochemical mechanisms exist, viz. O_2 reduction (Mehler reaction) and photorespiration. The short answer relates to capacities: neither of these processes can proceed as fast as CO_2 fixation (Heber and Walker 1992), and therefore neither can prevent photoinhibition if the light intensity is high but CO_2 is unavailable, as is possible if water stress leads to stomatal closure on a sunny day. Ultimately, both processes must be viewed primarily as responses to the "interference" by O_2 in metabolism, and the fact that those responses involve the (possibly beneficial) dissipation of harvested light energy is, presumably, a secondary consequence.

The product of the Mehler reaction is the superoxide anion. This anion is processed in the chloroplast by four enzymes of the so-called ascorbate/glutathione system (Fig. 2, enzymes e, b, c and d; see also Fig. 7 of our 1982 review). The flow of electrons from H_2O to O_2 ("pseudocyclic" electron transport) and subsequent processing of superoxide anions by the ascorbate/glutathione system not only directly dissipate harvested light energy, but also influence zeaxanthin-dependent energy dissipation (see Sect. 4f above) in two somewhat opposing ways: on the one hand, it supports electron transport that, at the plastoquinone step, moves protons from the stroma into the thylakoid lumen, thereby creating the low pH that favours zeaxanthin-dependent energy dissipation (Neubauer and

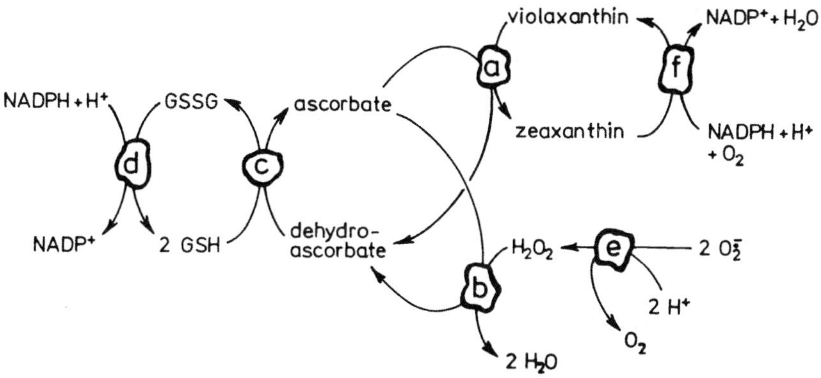

Fig. 2. Competition between *a* violaxanthin deepoxidase and *b* ascorbate peroxidase for chloroplast ascorbate. Other enzymes: *c* dehydroascorbate reductase; *d* glutathione reductase; *e* superoxide dismutase and *f* zeaxanthin epoxidase. *GSH* and *GSSG* reduced and oxidized glutathione respectively. (After Neubauer and Yamamoto 1994)

Yamamoto 1992); on the other hand, if "locks up" a good portion of the chloroplast's ascorbate, thereby restricting the deepoxidase that also needs ascorbate to generate the zeaxanthin in the first place (Neubauer and Yamamoto 1994; see Fig. 2).

One point that has become dramatically clear in the past 2 years is that plants respond to stress situations by increasing the activities of the ascorbate/glutathione-system enzymes, and/or increasing the pool sizes of ascorbate and glutathione. Recent examples include drought stress of pea and other species (Smirnoff 1993; Mittler and Zilinskas 1994), high light/low temperature stress of tomato, wheat and pea (Mishra et al. 1993; Walker and McKersie 1993; Edwards et al. 1994), low nutrient stress of bean and spruce (Cakmak and Marschner 1992; Polle et al. 1992), pollutant or herbicide stress of pumpkin and maize (Ranieri et al. 1993; Pastori and Trippi 1993), and O_2 stress of rice and wheat (Ushimaru et al. 1992; Albrecht and Wiedenroth 1994). Moreover, in two instances where the activity of an enzyme was boosted by genetic engineering, the resultant transgenic tobacco plants displayed increased tolerance to stress (Aono et al. 1993; Gupta et al. 1993). Clearly, there is scope for biotechnologists to increase the productivity of plants growing in stressful environments.

A final fascinating report is that from Sültemeyer et al. (1993). These authors report one of the largest jumps in the activities of ascorbate/glutathione system enzymes (two- to ten-fold) when *Chlamydomonas reinhardtii* cells were transferred from 5% CO_2 to air-level CO_2. It was proposed that the extra activities can deal with the extra superoxide anions formed when the rate of "pseudocyclic" electron transport is increased to supply (via the associated pseudocyclic photophosphorylation) ATP to drive this alga's CO_2 uptake (see Sect. 3).

The second process that reflects the "interference" of O_2 in the metabolism of photosynthetic cells is photorespiration. An intriguing proposal is that inhibition of photorespiration may be the biochemical explanation for the increased growth recorded following a brief application of methanol to the leaves of plants in an arid environment (Nonomura and Benson 1992).

The initial reaction of photorespiration is that in which O_2 substitutes for CO_2 on RuBP carboxylase (see Sect. 4b), leading to the formation of one glycerate-3-P and one glycolate-P. The latter is hydrolysed in the chloroplast to glycolate and Pi. The remainder of the process occurs outside the chloroplast, and achieves, overall, the conversion of two molecules of glycolate to one each of CO_2 and glycerate; the latter is moved into the chloroplast (Sect. 4e) where it is phosphorylated prior to reentering the Calvin cycle. However, this may not be the only major metabolic role of glycerate; considerable amounts are produced and simply stored in *Vicia faba* (El Bouhmadi and Larher 1993).

Minor attention has been given to the extra-chloroplastic reactions of photorespiration. Most occur in the peroxisome, mediated by enzymes that appear to be organised into multienzyme complexes (Heupel and Heldt 1994). However, the CO_2 evolving reaction, viz. 2 glycine + NAD $\rightarrow$ serine + CO_2 + NH_4^+ + NADH, takes place in the mitochondria. Reoxidation of the NADH is coupled to ATP formation (Wigge et al. 1993). This ATP, and also the NH_4^+, in turn contribute to inactivation of the Krebs cycle's pyruvate dehydrogenase complex (Gemel and Randall 1992), implying that active photorespiration retards the classical "dark" respiration in illuminated photosynthetic cells.

There has been some confusion concerning the nature and location of the first extra-chloroplast reaction of photorespiration (i.e. glycolate oxidation) in green algae, but now Stabenau et al. (1993) claim to have settled the matter by reporting that only the charophycean line contains, like higher plants, glycolate oxidase. Members of the other two lines (Chlorophyceae and Ulvophyceae), including *Dunaliella* spp. previously reported to contain this oxidase, in fact contain a mitochondrial glycolate dehydrogenase only.

5. Mitochondrial Respiration

The perceived importance of mitochondrial respiration to photosynthesis continues to fluctuate. As mentioned above, Gemel and Randall (1992) have evidence that entry of pyruvate into the Krebs cycle might be inhibited in the light, and indeed Kent et al. (1992) have calculated that illumination somehow reduces mitochondrial respiration by about 20%. Nevertheless, features of mitochondrial respiration that seem to be essential to photosynthesis continue to be reported. The most popular is that it might act as an "energy release valve", thereby lessening the likelihood of photoinhibition (Saradadevi and Raghavendra 1992; Krömer et al. 1993; Shyam et al. 1993). This role is easier to accept if the alternative, cyanide-resistant system for NADH oxidation is utilised (Weger and Dasgupta 1993). Another role is that vital metabolic links exist between chloroplasts, mitochondria and the cytosol (Hanning and Heldt 1993; Krömer et al. 1993), especially in stomatal guard cells (Mawson 1993). The most comprehensive picture is that from Hanning and Heldt (1993) who envisage an abbreviated Krebs cycle

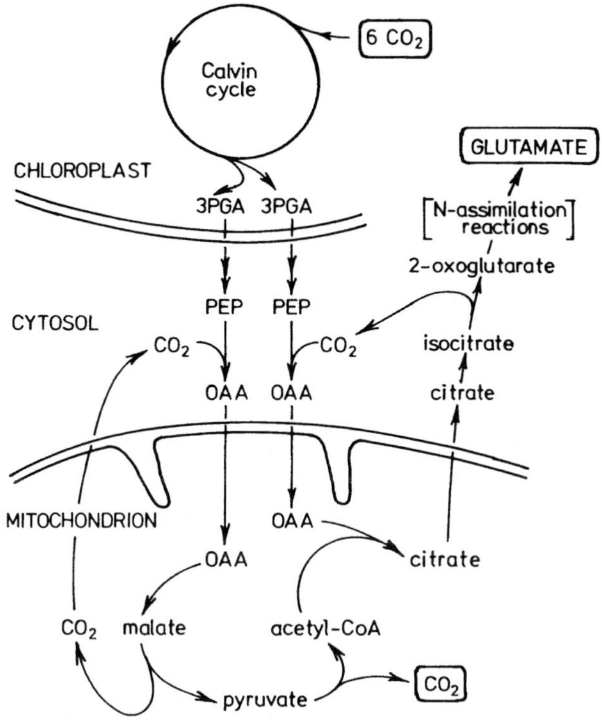

Fig. 3. Carbon balance for the biosynthesis of glutamate during the cooperation of chloroplasts mitochondria and cytosol in photosynthetic cells. The sum of the reactions is: 6 CO_2 → glutamate + CO_2. *OAA* Oxaloacetate; *PEP* phosphoenolpyruvate; *3PGA* glycerate-3-P. (Adapted from Hanning and Heldt 1993)

that assists in the manufacture of amino-acid carbon skeletons from the products of the chloroplast's CO_2 fixation (Fig. 3).

6. Sucrose Biosynthesis

Sucrose is the major form of carbohydrate exported out of photosynthetic tissues and translocated to sink tissues. It is manufactured in the cytosol from GAP exported from the chloroplast via the phosphate translocator. The biosynthesis pathway from GAP to fructose-6-P is as shown in Fig. 1 for mannitol synthesis; subsequent reactions are (in order): fructose-6-P → glucose-6-P → glucose-1-P, glucose-1-P + UTP → UDP-glucose + PPi, UDP-glucose + (a second) fructose-6-P → sucrose-P + UDP (this reaction catalysed by sucrose-P synthase [SPS]), sucrose-P → sucrose + Pi.

The final reaction is catalysed by a phosphatase that Echeverria and Salerno (1993) confirmed to be, as expected, in the cytosol. This report, and another indicating that the amount and activity of the cytosolic fructose-1,6-bisphosphatase required to generate fructose-6-P might be light modulated (Khayat et al. 1993), are two of the few studies that do not deal with SPS. However, before reviewing the research on SPS, a few words relating to the preceding reaction that generates PPi are in order. It had been thought that efficient conversion of this PPi to 2 Pi might be one role for the PPi-linked P-fructokinase in the cytosol of all higher plant photosynthetic cells so far investigated [including C_4-plant bundle sheath cells (Clayton et al. 1993), previously reported to lack the enzyme]. However, this role is now more dubious for three reasons. Firstly, there is a translocator for importing PPi into chloroplasts (Lunn and Douce 1993) where pyrophosphatase is located – indeed, the PPi contents of cytosol and chloroplast are similar (Eberl et al. 1992). Secondly, when *Escherichia coli* pyrophosphatase was genetically engineered into the cytosol of tobacco-leaf and potato-leaf cells, carbon flow increased in the direction of sucrose, at the expense of chloroplast starch (Jelitto et al. 1992). Thirdly, plants such as celery export much of the newly fixed carbon from their leaves in the form of sugar alcohols which are synthesised via a pathway that does not involve PPi formation (Everard et al. 1993; see Fig. 1). However, we are not aware of any reports concerning the PPi-linked P-fructokinase status of celery leaves. This matter clearly deserves attention, if only to clarify the role of the enzyme. The only recent reports on the enzyme relate to nonphotosynthetic tissues, viz. the potato tuber enzyme's net catalysis is in the glycolytic direction (Hajirezaei et al. 1994), and the tomato fruit enzyme needs a reducing environment for fructose-2,6-P_2-activation to be effective (Kiss et al. 1991).

The further investigations of SPS have confirmed that the enzyme is activated upon illumination by a dephosphorylation catalysed by a phosphatase (Weiner et al. 1992, 1993), and is inactivated upon darkness by a phosphorylation catalysed by a kinase; the critical phosphorylation/dephosphorylation site is serine-158 (Huber and Huber 1992; McMichael et al. 1993). The modifying enzymes are in turn influenced by light and metabolites, especially Pi (Weiner et al. 1992, 1993); Pi-supply also markedly influences the amount of SPS in leaves (Usuda and Shimogawara 1993).

Two other enzymes regulated through a phosphorylation/dephosphorylation cycle are nitrate reductase and phosphoenolpyruvate (PEP) carboxylase. The former is similar to SPS, undergoing activation via a phosphatase-catalysed dephosphorylation (Huber et al. 1992), but PEP carboxylase shows the opposite response, being activated via a kinase-catalysed phosphorylation (see Sect. 7d). Links have been established by Quy and Champigny (1992) who found that NO_3^- enhances the activities of the kinases that regulate both SPS and PEP carboxylase, but since the effects are opposite (inactivation and activation, respectively) carbon flow in the cytosol might be expected to occur in the direction of PEP and amino-acid C-skeletons (thus the NO_3^- could be put to good use) rather than in the direction of sucrose. The NO_3^- effect on PEP carboxylase seems to be mediated by glutamine activation of the kinase that catalyses its activation (Manh et al. 1993).

The fate of sucrose manufactured in the cytosol is export into vacuoles via a transporter protein (Thom et al. 1992) or export into the phloem via an apoplastic and/or symplastic route, depending on species (van Bel et al. 1994). In the former, a H^+/sucrose cotransporter found only on phloem cells loads sucrose into the phloem (Riesmeier et al. 1993a). When the amount of this cotransporter in potato leaves was reduced using cDNA-based antisense technology, the effect was dramatic: leaves displayed a 20-fold increase in soluble carbohydrate content, a 5-fold increase in chloroplast starch, but reduced photosynthesis. Clearly, translocation was mostly blocked; indeed, root and tuber growth was poor (Riesmeier et al. 1994).

Similar results were obtained when Heineke et al. (1992) expressed a foreign enzyme, viz. yeast invertase, in the cell walls of potato leaves. Once again, photosynthesis dropped after sucrose exported from cells was hydrolysed to glucose and fructose that, not being recognised by the phloem's H^+/sucrose contransporter, simply accumulated, raising the osmolality of the cell sap. The authors attributed the drop in photosynthesis to this osmolality increase, consistent with another study in which invertase secreted by a pathogenic fungus on barley leaves also caused an accumulation of sugars and decreased photosynthesis (Scholes et al. 1994).

Interference with the normal flow of carbon inside the photosynthetic cell also causes major (and somewhat predictable) changes. For example, when the amount of the potato chloroplast's phosphate translocator was reduced using antisense technology, photosynthesis was halved and, as expected, fixed carbon was trapped in the chloroplast, boosting the chloroplast starch level three-fold (Riesmeier et al. 1993b). Similarly a *Flaveria linearis* mutant that lacked most of the normal cytosolic fructose-1,6-bisphosphatase activity was also found to be incapable of moving all fixed carbon into sucrose, with a comparable result: more carbon moved to chloroplast starch (Sharkey et al. 1992).

These drastic changes to photosynthetic physiology usually resulted in reduced photosynthesis, which is not unexpected, but identifying the principal regulatory step(s) that mediate such a reduction is not easy. In fact, even with normal plants it is known that a period of rapid photosynthesis can lead to an accumulation of sucrose (and/or other major photosynthetic products) that somehow inhibits subsequent photosynthesis. Goldschmidt and Huber (1992) do not believe that sucrose, starch or hexoses are directly responsible (consistent with the observation that the leaves of some plants, e.g. cassava, are avid accumulators of sucrose; Angelov et al. 1993). They suggest that some intermediate metabolic step is critical, a point that might implicate SPS. However, earlier suggestions that this enzyme might be regulated by sucrose appear to be unfounded (Mitra and Srivastava 1992), and when the activity was boosted in tomato leaves by genetic engineering, photosynthesis in a normal environment was little affected (Galtier et al. 1993). Incidentally, the activity is also increased by low temperature (Guy et al. 1992) and by application of gibberellins (Cheikh et al. 1992). In the latter case, protein synthesis was involved. Perhaps also the sucrose-induced drop in photosynthesis is a consequence of altered protein synthesis/gene expression (Krapp et al. 1993). But, unless sucrose directly interacts with genes, unknown intermediate regulatory steps must still exist.

7. C_4 Metabolism

a) C_4 Plants

The first thing that happens to CO_2 upon entering the leaf of a C_4 plant is its conversion to HCO_3^- in the mesophyll cells. This is catalysed by carbonic anhydrase, a good portion of which is in the plasma membrane of these cells (Utsunomiya and Muto 1993). The consequent low concentration of CO_2 results in the cytostic pH being higher than usual (Yin et al. 1993). In the cytosol, HCO_3^- is combined with PEP (Sect. 7d) to yield oxaloacetate that is rapidly converted to a 4C-compound (malate or aspartate). This is moved into inner bundle sheath cells where it is decarboxylated (by one of three possible enzymes), releasing and concentrating the initial CO_2 at the site of the Calvin cycle (which is restricted to these cells). The remaining 3C-piece is moved back into mesophyll cells to regenerate PEP.

C_4 plants are subtyped according to the decarboxylation enzyme. Hattersley (1992) has shown interesting correlations between subtype and rainfall for Australian C_4 grasses: NADP-malic enzyme and PEP carboxykinase subtypes were slightly more plentiful in higher rainfall areas, but the proportional representation of the NAD-malic enzyme subtype was much greater in low rainfall areas.

Regeneration of PEP takes place in mesophyll cell chloroplasts. The returning 3C-piece must first be converted to pyruvate that, in turn, must be transported into the chloroplast. This is effected by a chloroplast-envelope pyruvate transporter that is stimulated by Na^+ (or, less usually, H^+; Aoki et al. 1992); direct Na^+ stimulation of the protein that transports pyruvate may occur, or alternatively, the protein is a Na^+/pyruvate symporter (Murata et al. 1992). Once inside the chloroplast, pyruvate is converted to PEP by the enzyme pyruvate Pi dikinase. Like SPS and other enzymes mentioned above (Sect. 6), this enzyme also undergoes light-mediated regulation via a phosphorylation/dephosphorylation system; the kinase/phosphatase that acts on the enzyme constitutes only a tiny fraction of the total chloroplast protein (Smith et al. 1994), not a sizeable fraction as reported earlier.

The classical picture of enzyme distribution between mesophyll and bundle-sheath cells in C_4 plant leaves has been challenged by Ueno (1992), who describes a C_4 grass, *Aristida latifolia*, that has two bundle sheaths. Cells of the outer sheath contain *both* RuBP carboxylase and pyruvate Pi dikinase, a rare and possibly unique event since these enzymes usually occur separately (in bundle-sheath cells and mesophyll cells, respectively).

The benefits of C_4 photosynthesis have been considered for today's environment, and for those of other geological periods. In the current climate, a major advantage is that the act of concentrating CO_2 at the site of RuBP carboxylase reduces the propensity for RuBP oxygenation and, hence, "wasteful" photorespiration. Further studies have confirmed that photorespiration is low or absent in C_4-plant leaves (Brown and Byrd 1993; Dai et al. 1993; Oberhuber and Edwards 1993), even though photorespiratory enzymes are present (e.g. P-glycolate phosphatase; Baldy 1993) and the potential for photorespiration exists, e.g. when the CO_2 supply is very low (Dai et al. 1993; Raghavendra et al. 1993).

It might be envisaged that the "CO_2 pump" of the C_4 plants could keep photorespiration at an insignificant level even if atmospheric CO_2 was considerably reduced below the current level of 357 ppm; of course C_3 plants, which already photorespire under 357 ppm CO_2, would photorespire even more and thus be at a competitive disadvantage. This line of thought has been used to explain (1) the unexpected abundance of C_4 plants in a temperate and moist Himalayan location (where CO_2 is low due to altitude; Rama Das and Vats 1993), (2) the rapid expansion of C_4 plants some 7 million years ago when the atmospheric CO_2 level became relatively low (Cerling et al. 1993; Morgan et al. 1994), and (3) the partial decline of C_4 plants 7000–9000 years ago when the retreat of the last Ice Age was accompanied by a rapid increase in atmospheric CO_2 from about 200 to 280 ppm (Cole and Monger 1994); quite probably, the decline in the number of C_4 plants was matched by an equivalent increase in the number of C_3 plants as they benefited from the increased CO_2 (Polley et al. 1993). Finally, readers may be intrigued to learn that, because of C_4 plants, the terrestrial biosphere will continue for at least another 0.9 billion years, even though C_3 plants will cease to grow when atmospheric CO_2 falls below 150 ppm about 0.1 billion years from now (Caldeira and Kasting 1992).

b) C_3–C_4 Intermediate Plants

Minor attention has been given to those plants with photosynthetic characteristics intermediate between true C_3 plants and true C_4 plants. Rawsthorne (1992), in a review of what is known, was unhappy with the term "C_3–C_4" but could not find an all-embracing alternative. The difficulty is apparent from the report by Brown et al. (1993) that *Flaveria brownii*, a C_3–C_4 species that is so close to being C_4 that it is termed "C_4-like", has been crossed with *Flaveria linearis*, a more conventional C_3–C_4.

A salient feature of C_3–C_4 plants is that photorespiration is reduced. The reduction is probably assisted by efficient recapture of photorespired CO_2 by PEP carboxylase (a process that increases the photon requirement of these plants; Osborne 1994), by overall reductions in the activities of photorespiratory enzymes (Devi and Raghavendra 1993), and by restriction of the CO_2-evolving enzyme, glycine decarboxylase, to bundle-sheath cells (Morgan et al. 1993).

c) Crassulacean Acid Metabolism (CAM)

The classical CAM plant captures CO_2 at night, then photosynthetically reduces it to carbohydrate the next day, behind closed stomates. Transpirational water loss is thereby minimised. Nocturnal CO_2 capture is catalysed by PEP carboxylase and malate dehydrogenase (Sect. 7d), and as the resultant malate accumulates it is stored in the large vacuole typical of these plants. Daytime decarboxylation of malate releases the CO_2 for the Calvin cycle, while the remaining 3C-piece is metabolised to starch which becomes the source of the next night's PEP.

This diurnal metabolic cycle persists even if the plant is disturbed by continuous light or darkness. Continuation of the disturbance results, however, in a damping of the rhythm due to excessive accumulation of malate in darkness (Wilkins and Thomas 1993), or to a drift toward chaotic metabolism in light (Lüttge and Beck 1992).

Normal CAM requires high activities of starch degrading enzymes to generate the required amount of PEP at night (Paul et al. 1993). It also seems to require respiratory activity: by night to permit malate accumulation (Watanabe et al. 1992), and by day respiration is increased by engagement of the alternative, cyanide-resistant pathway, perhaps to allow some mitochondrial metabolism of malate without upsetting the cell's ATP/ADP ratio (Robinson et al. 1992).

Tonoplast proteins that contribute to the movement of malate into and out of the vacuole have received more attention. Active malate uptake is driven by an H^+/ATPase, as evidenced from experiments with the orchid *Epidendrum ibaguense* (White and Smith 1992) and the facultative CAM plant *Mesembryanthemum crystallinum*; in this latter case, a previously identified H^+/pyrophosphatase appeared to have minor importance for malate uptake (Bremberger and Lüttge 1992). The H^+/ATPase has been solubilized and then reconstituted, in an active state, into proteoliposomes (Behre et al. 1992). It even has been visualised and quantified on freeze-fracture replicas of *M. crystallinum* tonoplasts; its density in the membrane increased after CAM was induced, further establishing its central role in CAM (Klink and Lüttge 1992).

While H^+/ATPase presumably establishes an H^+ gradient across the tonoplast and thereby drives malate movement, less is known of the proteins that actually translocate the malate. Two channel proteins for malate^{2-}, possibly one for uptake and the other for efflux, have been detected in *Graptopetalum paraguayense* by Iwasaki et al. (1992). One clear aspect is that control of malate movement and storage is dependent on a relatively low level of tonoplast fluidity, which in turn can be induced by environmental factors that promote CAM (Kliemchen et al. 1993).

Although CAM is designed to equip plants for survival under harsh conditions, especially those where water deficit is common, its capacity to protect the plant from stress is still limited, thus it is not unusual to find CAM plants with metabolism altered as a response to severe stress. For example, the Madagascan

CAM plant *Xerosicyos dunguyi* switches to CAM-idling when water stress becomes particularly severe: stomata shut permanently, but a low diurnal fluctuation in organic acids continues internally (Bastide et al. 1993). A similar response is seen when water stress is experienced by Central American species of *Clusia* that can quickly (within 24 h) switch from C_3 to CAM photosynthesis (Zotz and Winter 1993). A notable feature is a marked accumulation of citrate (Borland et al. 1993) that could play a pivotal role in preventing photoinhibition during prolonged drought (Franco et al. 1992) by participating in a CAM-idling metabolic cycle that consumes energy. One possibility is suggested in Fig. 4, in which 10 ATP are consumed for each molecule of citrate cycled. However, the system could additionally consume the equivalent of 4 NADH per citrate if the NADH and $FADH_2$ generated in the Krebs cycle during the citrate-to-pyruvate conversion was reoxidised by the alternative electron transport pathway that releases heat instead of ATP. As mentioned above, this pathway was found to be engaged by day in *Kalanchoë daigremontiana* (Robinson et al. 1992).

Experiments with the facultative CAM plant *M. crystallinum* have shown that the switch from C_3-photosynthesis to CAM photosynthesis that is induced by water stress cannot occur until young plants have passed from the seedling to the juvenile growth phase (Herppich et al. 1992). The switch is accompanied

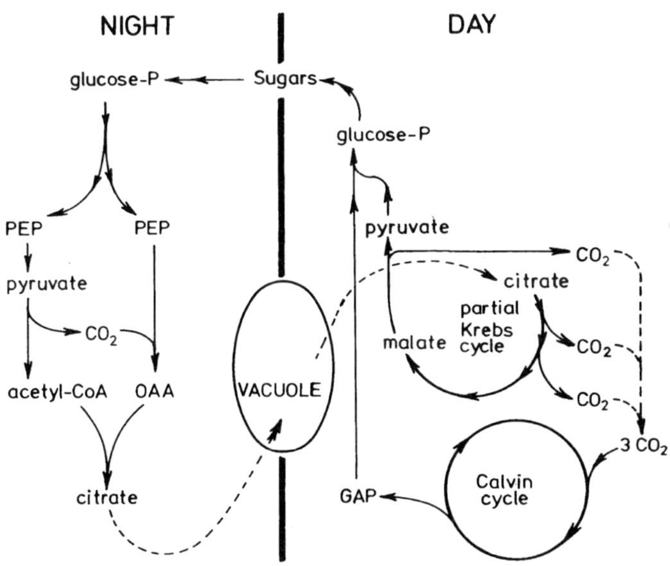

Fig. 4. Carbon flow in stressed CAM plants. By night, net conversion of reserve polysaccharide to citrate occurs. By day, each molecule of citrate is converted to three of CO_2 and one of pyruvate; the 3 CO_2 are then photosynthetically reduced to one of GAP which, with the pyruvate, replenishes the pool of reserve polysaccharide. There is no *net* fixation of CO_2 in these stressed plants, and therefore no growth. Overall, the entire diurnal cycle dissipates the energy of 10 ATP per molecule of citrate cycled. Scheme developed from observations of Franco et al. (1992)

by marked increases in the activities of starch-degrading enzymes (Paul et al. 1993), as is required to initiate the generation of PEP, and in the activity of PEP carboxylase. A new observation is that events between the environmental signal (e.g. water stress) and the synthesis of new protein (e.g. PEP carboxylase) involve well-known plant hormones such as abscisic acid and cytokinin (Schmitt and Piepenbrock 1992; Dai et al. 1994). Hormonal effects are possibly also involved in "reversing" stomatal behaviour, as observed when CAM was induced in the facultative CAM plant *Portulaca afra* (Lee and Assman 1992).

d) PEP Carboxylase and Malate Dehydrogenase

A central feature of C_4 metabolism is the capture of CO_2 by PEP carboxylase. The product is the 4C acid oxaloacetate which, in the majority of cases, is rapidly reduced to malate by malate dehydrogenase (although a significant alternative is transamination to asparatate). Both the carboxylase and dehydrogenase have recently received considerable attention.

The light-mediated activation of the cytosolic enzyme PEP carboxylase in C_4 plant leaves has been a primary point of interest. The activation is achieved by attachment of a phosphate group to a serine residue near the N-terminus of the protein; the phosphate is transferred from ATP in a reaction catalysed by a "PEP-carboxylase kinase" (PEPC-kinase). It is the negative charge on the phosphate that is critical for activation (Wang et al. 1992). Light-induced alkalisation of the cytosol, ATP formation, and Ca^{2+} all appear to be important for the activation to proceed (Devi and Raghavendra 1992; Pierre et al. 1992). The exact mechanism by which light is connected to PEPC-kinase activity is still unclear, although it is likely that a light-induced signal originates in the chloroplast, and cytosolic protein synthesis is involved (Bakrim et al. 1992; Li and Chollet 1993). The role of Ca^{2+} is also unclear: a Ca^{2+}-dependent PEPC-kinase has been identified (Ogawa et al. 1992), but Ca^{2+}-independent forms of the enzyme have also been obtained and are believed to be more important (Bakrim et al. 1992; Wang and Chollet 1993a).

A similar phosphorylative activation of PEP carboxylase occurs in CAM plants, but in darkness (see our last review); the activation has now been demonstrated for the facultative CAM plant *M. crystallinum* (Baur et al. 1992). Recent evidence indicates that the activation is widespread, in that phosphorylation of PEP carboxylase from tobacco leaves (Wang and Chollet 1993b), wheat leaves (Manh et al. 1993), bean stomatal cells (Schnabl et al. 1992), and even soybean root nodules (Schuller and Werner 1993) has been reported. However, the extent of the resultant activation may be less for the C_3-plant enzyme compared to the C_4-plant enzyme (Rajagopalan et al. 1993).

Other regulatory aspects of PEP carboxylase have been briefly considered. Reversible association/dissociation of enzyme subunits may be involved in inactivation of the enzyme upon binding to chloroplast membranes (Wu and Wedding 1992) and in kinetic hysteresis in vitro (Avilan 1993), but is unlikely to be important for light/dark regulation (Weigend and Hincha 1992). The CAM-plant enzyme easily associates with malate dehydrogenase, resulting in a change in activity (Queiroz-Claret and Queiroz 1992). Several similar, but not identical, forms of PEP carboxylase have been reported for *M. crystallinum* leaves (Slocombe et al. 1993) and for bean guard cells, but only one form was in bean-leaf mesophyll cells (Schulz et al. 1992). Interestingly, two C_4 species of *Panicum* contained PEP carboxylases that totally differed with respect to cold inactivation in vitro (Krall and Edwards 1993).

The chloroplast's NADP-linked malate dehydrogenase has been the object of site-directed mutagenesis aimed at cysteine residues implicated in the light-dependent reductive activation process. Results show that the reduction of two disulfides per enzyme monomer are involved for the sorghum enzyme (Issakidis et al. 1992) and probably also for the pea enzyme (Reng et al. 1993); other approaches led to a comparable conclusion for the maize enzyme (Hatch and Agostino 1992). Prevention of disulfide formation by replacing cysteines with serines results in a "permanently active" enzyme, unable to undergo light-mediated regulation (Issakidis et al. 1993). Incidentally, light-mediated activation of the native enzyme increases its hydrophobicity, an event that has practical and, perhaps, physiological consequences (Scheibe et al. 1993).

8. The Limits to Plant Productivity

We return to our question: why can't plants grow faster than they do? We have addressed this question previously in our 1988 review, where two levels were distinguished: (1) external factors that limit growth in the field – principally temperature and the availabilities of light, water, CO_2 and nutrients, and (2) plant physiological/biochemical factors that determine the photosynthetic capacity, i.e. the growth rate when temperature is optimum and the supplies of the other four external factors are adequate. In trying to identify the factor(s) that ultimately determine maximum plant productivity, we confess to finding the literature somewhat confusing. However, before explaining our predicament, it is worthwhile to briefly review some of the recent relevant research, since this provides clues as to what the ultimate determinant(s) of plant productivity might be.

a) CO_2

Much attention has been given to CO_2 supply because the level of CO_2 in the atmosphere is well below that required to saturate photosynthesis, but *less* so each year because fossil fuel combustion adds a further 10 billion tonnes to the atmosphere annually (Walker 1992a). The number of reports confirming that this increased CO_2 supply is stimulating photosynthesis and plant growth is itself increasing exponentially; recent studies revealed positive responses for crops and grasses (Nie et al. 1992; Baker and Allen 1993; Holbrook et al. 1993; Betsche 1994), the CAM plant *Opuntia ficus-indica* (Cui et al. 1993), and trees (Gunderson et al. 1993; Idso and Kimball 1992, 1993; Tissue et al. 1993). It is not inconceivable that the stimulation of tree growth is occurring on a global scale (Gifford 1994). Two comprehensive reviews of CO_2-stimulation have been published (Bowes 1993; Poorter 1993); in the latter, the growth stimulation of 156 plant species by doubled atmospheric CO_2 was, on average, 37%.

However, responses to increased CO_2 are not straightforward, as we emphasised earlier (see our 1990 review). Stimulation is less if there are inadequate sinks for photosynthate to travel to (Poorter 1993); in many of these cases, excess photosynthetic product simply accumulates as chloroplast starch (Körner and

Arnone 1992; Morin et al. 1992; Holbrook et al. 1993) or some other nonstructural carbohydrate (Smart et al. 1994). Another point is that growth stimulation and altered assimilate partitioning may not be entirely due to increased photosynthesis: increased CO_2 also reduces dark respiration (Idso and Kimball 1993; Bunce 1994) which, of course, translates into extra growth, and it reduces photorespiration which, in turn, appears to influence whether carbon flows to starch or sucrose (Morin et al. 1992). A third point concerns diffusion of CO_2 from the atmosphere to the chloroplast stroma. In the past, the rate of this diffusion was thought to be so high that it could not significantly limit photosynthesis. While this may be true for some plants, it may not apply to others, especially woody species for which rather low conductances have been measured (Lloyd et al. 1992; Parkhurst 1994). In particular, *Citrus aurantium* had a very low conductance, which may explain why it has a low growth rate (Baker and Allan 1993), and why its growth is so effectively stimulated by increased CO_2 (Idso and Kimball 1992). A final point concerns the CO_2-response of plants in an ecosystem context: in two recent studies, CO_2-stimulation of plant growth was followed by an increase in soil microflora activity that led, in one instance, to a doubling of CO_2 evolution from the soil (Körner and Arnone 1992), and in the other, to a sequestration of soil nutrients that retarded further plant growth (Diaz et al. 1993).

b) Water

Although it is well known that severe and prolonged water stress results in stomatal closure, lack of CO_2, inhibited photosynthesis and onset of photoinhibition, the consequences of milder water stress are less clear-cut. It seems that the availability of CO_2 for photosynthesis is little affected, but photosynthesis is nevertheless reduced via altered biochemistry; however, the alteration does not directly involve RuBP carboxylase (Lauer and Boyer 1992; Gunasekera and Berkowitz 1993).

c) Light

Concepts concerning the effect of light intensity on photosynthesis are not as clear as might be thought. One common misconception is that, at low intensities, the photosynthetic light-response curve (i.e. curve of photosynthesis versus light intensity) for "shade" plants is steeper than that for "sun" plants. In fact, Walker (1992a,b) has pointed out that, given the almost invariable quantum yield of 9 (i.e. 9 photons utilised per molecule of CO_2 fixed) for C_3 plants when not stressed and under good conditions, the slopes of these light-response curves at low light intensities are identical and simply reflect the fact that both types of plant are using all available light as best they can, and close to the maximum efficiency dictated by the Z-scheme of photochemistry, i.e. 8 photons required to generate the 2 NADPH needed to reduce 1 CO_2 to carbohydrate in the Calvin cycle.

At high light intensities, explanations are more difficult. A common observation is that the slope of the light-response curve becomes zero when light saturation is reached. This occurs at about one-half of full sunshine for sun plants, but at a much lower light intensity for shade plants (Walker 1992b). The lower value for shade plants reflects the lower capacity to fix CO_2, due at least in part to a reduced content of RuBP carboxylase; this reduction in the carboxylase-protein content spares available nitrogen, which is invested instead into additional pigment-protein complexes of light-harvesting antennae (Woodward 1990). An explanation, in biophysical terms, of why several to many hundreds of accessory light-harvesting pigment molecules and their associated proteins are required per reaction centre chlorophyll, is given by Glazer (1989).

Sun plants, on the other hand, are free to invest a greater proportion of their nitrogen into the provision of RuBP carboxylase (Makino et al. 1992), thus the question arises as to what limits their photosynthetic ability. In normal air, their light-response curves do show light saturation, but when CO_2 is increased, either artificially (Bugbee and Salisbury 1988; Fig. 5) or naturally (C_4 plants), light saturation is difficult to detect. Results from the impressive experiments of Bugbee and Salisbury (1988; see also Bugbee and Monje 1992) led these authors to conclude that, in essence, the dry matter production of wheat plants growing under ideal conditions was limited by an *external* factor, i.e. light supply, *even when this supply was well in excess of full sunshine.* Presumably, all of the plants

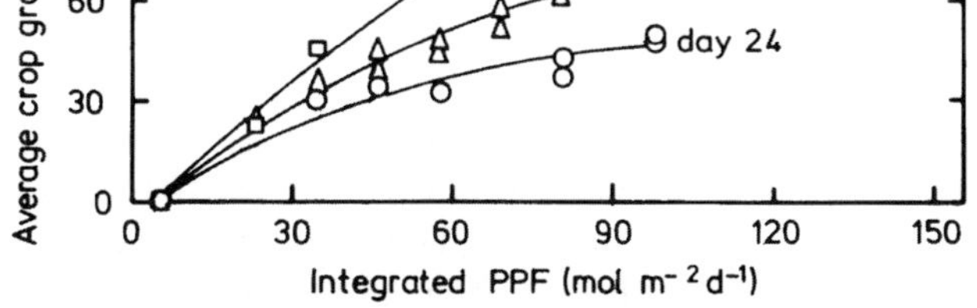

Fig. 5. Effect of daily PPF (photosynthetic photon flux 400–700 nm) on average wheat growth rate at three different harvest dates. An integrated PPF of 150mol m^{-2} d^{-1} equates to a light intensity of 2080 μmol photons m^{-2} s^{-1} for 20 h. This is more than twice the PPF from summer sunlight in the field. Redrawn from Bugbee and Salisbury (1988)

internal physiological/biochemical components could accomodate the prodigious growth rates achieved (see below).

d) Determinants of Photosynthetic Rate

How, then, should we view experiments in which the capacity of one or other *biochemical* component is considered in the context of the photosynthetic rate? Three components are often considered: (1) the rate of photosynthetic electron transport (as detemined by its limiting step, the rate of plastoquinol diffusion within the thylakoid membrane; Blackwell et al. 1994), (2) the activity of RuBP carboxylase, and (3) the utilisation of exported GAP in the cytosol and the return of the Pi, liberated from the GAP during this process, back to the chloroplast; proteins that are contenders for the rate-limiting step in this process are the cytosolic fructose-1,6-bisphosphatase (Daie 1993), SPS (Galtier et al. 1993), and the chloroplast envelope's phosphate translocator (Riesmeier et al. 1993b).

There are numerable instances where one or other of these components has been recognised as imposing a primary limitation on photosynthesis (Bowes 1993; Ögren and Evans 1993; Scholes et al. 1994). It is beyond the scope of this review to give further consideration to electron transport, but we shall briefly consider the other two components. Firstly, the evidence that RuBP carboxylase activity is relevant. Certainly, the carboxylation of RuBP in plants grown in *normal* air can contribute to the determination of photosynthetic rate (Jiang et al. 1993; Lauerer et al. 1993; Masle et al. 1993; Stitt and Schulze 1994), but this is as much a reflection of the low CO_2 supply as it is of the RuBP carboxylase level (Bowes 1993; Woodrow 1994). There is now ample evidence that, when plants are grown in a high-CO_2 atmosphere, photosynthesis increases (Poorter 1993; see Sect. 8a). However, while some species appear to make full use of the extra CO_2 by allowing their normal RuBP carboxylase (i.e. that present in air-grown plants) to operate correspondingly faster, at least an equal number reduce the activity by regulatory means (Sect. 4b), or by synthesising less of it (Bowes 1993; Socias et al. 1993; Tissue et al. 1993; Sicher et al. 1994). One is left with the impression that the RuBP carboxylase activity, while not excessive, is still adequate.

The case for the other option (cytosolic utilisation of GAP and return of Pi to the chloroplast) as being the determinant of photosynthetic rate under high light and saturating CO_2 is stronger (Bowes 1993), but less attention has been devoted to it. Scholes et al. (1994) concluded that it applied to barley leaves. The activity of SPS appeared to be important in bean leaves (Socias et al. 1993). Notably, the photosynthetic capacity of tomato leaves was increased by 20% after the SPS activity was elevated by expressing a maize SPS gene in the leaves; however, the activity of the maize enzyme could not be regulated by the tomato leaf cells (Galtier et al. 1993), and it is questionable whether a similar increase would be measured if the extra SPS was from spinach (Sonnewald et al. 1994).

Finally, we conclude by addressing the fundamental question of whether or not the capacities of the above three biochemical components exceed those required to fully utilise the light energy available in full sunshine. If they do [and the above-mentioned experiments of Bugbee and Salisbury (1988) suggest that they do in wheat], then the biochemical/biophysical component that determines photosynthetic *capacity* will not be found by growing plants under full sunshine, because the light level will still be below saturation.

In addition, if these plants convert light energy into biomass with an efficiency close to the theoretical maximum, then we may conclude that the component determining photosynthetic capacity is not yet limiting photosynthesis; rather, light supply is. Indeed, measurements consistent with appropriately high efficiencies have been made. Bugbee and Salisbury (1988) report that the photosynthetic machinery of wheat was able to convert light energy into biomass at close to the maximum expected efficiency, even when the daily supply of light energy was more than twice that in normal summer sunshine. They calculated an efficiency of 7 to 10% for wheat grown up to 79 d. This compares with the proposed theoretical maximum of 15.6% (Bugbee and Monje 1992). The associated growth rate under the equivalent of full sunshine was about 60 g dry matter m^{-2} d^{-1} (Fig. 5). An important point is that these are not isolated values; the best reported growth rates for plants in the field (and presumably at ambient CO_2) were at least one-half of the CO_2-saturated rate for wheat (Bugbee and Salisbury 1988), and more recent reports include growth rates for cacti in Mexico (Nobel et al. 1992) and for a C_4 plant on an Amazon floodplain (Piedade et al. 1994) that we calculate to be equivalent to 13 and 30 g dry matter m^{-2} d^{-1}. Furthermore, leaf photosynthesis rates of over 60 μmol CO_2 fixed m^{-2} s^{-1} have been observed for castor bean (Dai et al. 1992) and calculated feasible for other species (Wullschleger 1993); such rates are comparable to a growth rate of 60 g dry matter m^{-2} d^{-1}. Finally, it is worth recalling that rapidly growing microalgal cultures under high light and saturating CO_2 (the latter due to their CO_2-concentrating mechanism; Sect. 3) have been reported to have high photon-to-biomass conversion efficiencies, e.g. 13% for *Phaeodactylum tricornutum* (Thomas et al. 1984).

Collectively, these reports indicate that, for some sun plants at least, growth is ultimately limited by the availability of solar energy. The arrangement of the photochemical apparatus and the nature of light (Walker 1992b) dictate that the maximum growth rate possible (and demonstrated to be practicable) is 60 g dry matter m^{-2} d^{-1}.

Therefore, if we knew the area upon which Jack's proverbial beanstalk grew, we could now predict the maximum rate at which it could have grown upwards, and we could explain that it could not have grown faster because the sun is not bright enough to power faster growth!

References

Aggarwal KK, Saluja D, Sachar RC (1993) Phytochemistry 34:329–335. **Albrecht G**, Wiedenroth E-M (1994) J Exp Bot 45:449–455. **Angelov MN**, Sun J, Byrd GT, Brown RH, Black CC (1993) Photosyn Res 38:61–72. **Aoki N**, Ohnishi J, Kanai R (1992) Plant Cell Physiol 33:805–809. **Aono M**, Kubo A, Saji H, Tanaka K, Kondo N (1993) Plant Cell Physiol 34:129–135. **Ardila F**, Pozueta-Romero J, Akazawa T (1993) Plant Cell Physiol 34:237–242. **Aro E-M**, McCaffery S, Anderson JM (1993) Plant Physiol 103:835–843. **Aro E-M**, McCaffery S, Anderson JM (1994) Plant Physiol 104:1033–1041. **Avilan L** (1993) Photosynthetica 28:317–320.

Badger MR, Price GD (1992) Physiol Plant 84:606–615. **Badger MR**, Pfanz H, Büdel B, Heber U, Lange O (1993) Planta 191:57–70. **Bailey KJ**, Walker DA (1992) Plant Physiol 99:124–129. **Baker JT**, Allen LH (1993) Vegetatio 104/105:239–260. **Bakrim N**, Echevarria C, Cretin C, Arrio-Dupont M, Pierre JN, Vidal J, Chollet R, Gadal P (1992) Eur J Biochem 204:821–830. **Baldy P** (1993) J Plant Physiol 141:633–635. **Bastide B**, Sipes D, Hann J, Ting IP (1993) Plant Physiol 103:1089–1096. **Batz O**, Scheibe R, Neuhaus HE (1992) Plant Physiol 100:184–190. **Batz O**, Scheibe R, Neuhaus HE (1993) Biochem J 294:15–17. **Baur B**, Dietz K-J, Winter K (1992) Eur J Biochem 209:95–101. **Behre B**, Ratajczak R, Lüttge U (1992) Bot Acta 105:260–265. **Benet H**, Bruss U, Duval J-C, Kloareg B (1994) J Exp Bot 45:211–220. **Berkowitz GA**, Peters JS (1993) Plant Physiol 102:261–267. **Betsche T** (1994) Environ Exp Bot 34:75–86. **Blackwell M**, Gibas C, Gygax S, Roman D, Wagner B (1994) Biochim Biophys Acta 1183:533–543. **Borchert S**, Harborth J, Schünemann D, Hoferichter P, Heldt HW (1993) Plant Physiol 101:303–312. **Borland AM**, Griffiths H, Broadmeadow MSJ, Fordham MC, Maxwell C (1993) Oecologia 95:444–453. **Bowes G** (1993) Annu Rev Plant Physiol Plant Mol Biol 44:309–332. **Bremberger C**, Lüttge U (1992) Planta 188:575–580. **Brown RH**, Byrd GT (1993) Plant Physiol 103:1183–1188. **Brown RH**, Byrd GT, Bouton JH, Bassett CL (1993) Plant Physiol 101:825–831. **Bugbee B**, Monje O (1992) Bioscience 42:494–502. **Bugbee BG**, Salisbury FB (1988) Plant Physiol 88:869–878. **Bunce JA** (1994) Physiol Plant 90:427–430.

Cakmak I, Marschner H (1992) Plant Physiol 98:1222–1227. **Caldeira K**, Kasting JF (1992) Nature 360:721–723. **Campbell WJ**, Ogren WL (1992) Plant Cell Physiol 33:751–756. **Cerling TE**, Wang Y, Quade J (1993) Nature 361:344–345. **Cheikh N**, Brenner ML, Huber JL, Huber SC (1992) Plant Physiol 100:1238–1242. **Choudhury NK**, Choe HT, Huffaker RC (1993) J Plant Physiol 141:551–556. **Clayton H**, Ranson J, ap Rees T (1993) Arch Biochem Biophys 301:151–157. **Coates SA**, ap Rees T (1994) Phytochemistry 35:881–883. **Colas des Francs-Small C**, Ambard-Bretteville F, Small ID, Remy R (1993) Plant Physiol 102:1171–1177. **Cole DR**, Monger HC (1994) Nature 368:533–536. **Corbier C**, Della Seta F, Branlant G (1992) Biochemistry 31:12532–12535. **Cui M**, Miller PM, Nobel PS (1993) Plant Physiol 103:519–524.

Dai Z, Edwards GE, Ku MSB (1992) Plant Physiol 99:1426–1434. **Dai Z**, Ku MSB, Edwards GE (1993) Plant Physiol 103:83–90. **Dai Z**, Ku MSB, Zhang D, Edwards GE (1994) Planta 192:287–294. **Daie J** (1993) Photosyn Res 38:5–14. **Demmig-Adams B**, Adams WW (1993) Plant Physiol 103:1413–1420. **Devi MT**, Raghavendra AS (1992) J Plant Physiol 139:431–435. **Devi MT**, Raghavendra AS (1993) J Exp Bot 44:779–784. **Diaz S**, Grime JP, Harris J, McPherson E (1993) Nature 364:616–617. **Drechsler Z**, Sharkia R, Cabantchik ZI, Beer S (1993) Planta 191:34–40. **Dreier W**, Preusser E, Gründel M (1992) Biochem Physiol Pflanzen 188:81–96.

Eberl D, Preissler M, Steingraber M, Hampp R (1992) Physiol Plant 84:13–20. **Echeverria E**, Salerno G (1993) Physiol Plant 88:434–438. **Edwards EA**, Enard C, Creissen GP, Mullineaux PM (1994) Planta 192:137–143. **El Bouhmadi K**, Larher F (1993) Plant Sci 94:99–107. **Everard JD**, Franceschi VR, Loescher WH (1993) Plant Physiol 102:345–356.

Fischer K, Arbinger B, Kammerer B, Busch C, Brink S, Wallmeier H, Sauer N, Eckerskorn C, Flügge U-I (1994) Plant J 5:215–226. **Flores HE**, Dai Y, Cuello JL, Maldonado-Mendoza IE, Loyola-Vargas VM (1993) Plant Physiol 101:363–371. **Fonollá J**, Hermoso R, Carrasco JL, Chueca A, Lázaro JJ, Prado FE, López-Gorgé J (1994) Plant Physiol 104:381–386. **Franco AC**, Ball E, Lüttge U (1992) Plant Cell Environ 15:821–829.

Galtier N, Foyer CH, Huber J, Voelker TA, Huber SC (1993) Plant Physiol 101:535–543. **Gardeström P** (1993) Biochim Biophys Acta 1183:327–332. **Gemel J**, Randall DD (1992) Plant Physiol 100:908–914. **Ghiena C**, Schulz M, Schnabl H (1993) Plant Physiol 101:73–79. **Gifford RM** (1994) Aust J Plant Physiol 21:1–15. **Gilmore AM**, Yamamoto HY (1992) Proc Natl Acad Sci USA 89:1899–1903. **Glazer AN** (1989) J Biol Chem 264:1–4. **Goldschmidt EE**, Huber SC (1992) Plant Physiol 99:1443–1448. **Gontero B**, Mulliert G, Rault M, Giudici-Orticoni M-T, Ricard J (1993) Eur J Biochem 217:1075–1082. **Goyal A**, Shiraiwa Y, Husic HD, Tolbert NE (1992) Mar Biol 113:349–355. **Grotjohann N**, Schneider G, Kowallik W (1993) Z Naturforsch 48c:22–27. **Gunasekera D**, Berkowitz GA (1993) Plant Physiol 103:629–635. **Gunderson CA**, Norby RJ, Wullschleger SD (1993) Plant Cell Environ 16:797–807. **Gupta AS**, Heinen JL, Holaday AS, Burke JJ, Allen RD (1993) Proc Natl Acad Sci USA 90:1629–1633. **Guy CL**, Huber JLA, Huber SC (1992) Plant Physiol 100:502–508.

Haglund K, Björk M, Ramazanov Z, García-Reina G, Pedersén M (1992) Planta 187:275–281. **Hajirezaei M**, Sonnewald U, Viola R, Carlisle S, Dennis D, Stitt M (1994) Planta 192:16–30. **Hanning I**, Heldt HW (1993) Plant Physiol 103:1147–1154. **Harpel MR**, Hartman FC (1992) J Biol Chem 267:6475–6478. **Hatch MD**, Agostino A (1992) Plant Physiol 100:360–366. **Hattersley PW** (1992) In: Chapman GP (ed) Desertified grasslands: their biology and management. Linnean Soc Symp Ser No 13. Academic Press, London, pp 181–212. **Havaux M**, Gruszecki WI (1993) Photochem Photobiol 58:607–614. **Heber U**, Walker DA (1992) Plant Physiol 100:1621–1626. **Heineke D**, Sonnewald U, Büssis D, Günter G, Leidreiter K, Wilke I, Raschke K, Willmitzer L, Heldt HW (1992) Plant Physiol 100:301–308. **Herppich W**, Herppich M, von Willert DJ (1992) Bot Acta 105:34–40. **Heupel R**, Heldt HW (1994) Eur J Biochem 220:165–172. **Holbrook GP**, Hansen J, Wallick K, Zinnen TM (1993) Environ Exp Bot 33:313–321. **Huber JLA**, Huber SC (1992) Biochem J 283:877–882. **Huber SC**, Huber JL, Campbell WH, Redinbaugh MG (1992) Plant Cell Physiol 33:639–646. **Hudson GS**, Morell MK, Arvidsson YBC, Andrews TJ (1992) Aust J Plant Physiol 19:213–221. **Huner NPA**, Öquist G, Hurry VM, Krol M, Falk S, Griffith M (1993) Photosyn Res 37:19–39. **Huzisige H**, Ke B (1993) Photosyn Res 38:185–209.

Idso SB, Kimball BA (1992) Plant Physiol 99:341–343. **Idso SB**, Kimball BA (1993) J Plant Physiol 141:166–171. **Israel A**, Beer S (1992) Mar Biol 112:697–700. **Issakidis E**, Miginiac-Maslow M, Decottignies P, Jacquot J-P, Cretin C, Gadal P (1992) J Biol Chem 267:21577–21583. **Issakidis E**, Decottignies P, Miginiac-Maslow M (1993) FEBS Lett 321:55–58. **Iwasaki I**, Arata H, Kijima H, Nishimura M (1992) Plant Physiol 98:1494–1497.

Jahns P, Krause GH (1994) Planta 192:176–182. **Janave MT**, Ramaswamy NK, Nair PM (1993) Eur J Biochem 214:889–896. **Jelitto T**, Sonnewald U, Willmitzer L, Hajirezeai M, Stitt M (1992) Planta 188:238–244. **Jiang C-Z**, Rodermel SR, Shibles RM (1993) Plant Physiol 101:105–112. **Johnson GN**, Young AJ, Scholes JD, Horton P (1993) Plant Cell

Environ 16:673–679. **Johnston AM**, Maberly SC, Raven JA (1992) Oecologia 92:317–326.

Karlsson J, Ramazanov Z, Hiltonen T, Gardeström P, Samuelsson G (1994) Planta 192:46–51. **Kent SS**, André M, Cournac L, Farineau J (1992) Plant Physiol Biochem 30:625–637. **Khayat E**, Harn C, Daie J (1993) Plant Physiol 101:57–64. **Kingston-Smith AH**, Major I, Parry MAJ, Keys AJ (1992) Biochem J 287:821–825. **Kiss F**, Wu M-X, Wong JH, Balogh A, Buchanan BB (1991) Arch Biochem Biophys 287:337–340. **Kliemchen A**, Schomburg M, Galla H-J, Lüttge U, Kluge M (1993) Planta 189:403–409. **Klink R**, Lüttge U (1992) Bot Acta 105:414–420. **Körner C**, Arnone JA (1992) Science 257:1672–1675. **Kossmann J**, Müller-Röber B, Dyer TA, Raines CA, Sonnewald U, Willmitzer L (1992) Planta 188:7–12. **Kots AY**, Sergienko EA, Bulargina TV, Severin ES (1993) FEBS Lett 324:33–36. **Krall JP**, Edwards GE (1993) Plant Cell Physiol 34:1–11. **Krapp A**, Hofmann B, Schäfer C, Stitt M (1993) Plant J 3:817–828. **Krömer S**, Malmberg G, Gardeström P (1993) Plant Physiol 102:947–955.

Lauer MJ, Boyer JS (1992) Plant Physiol 98:1310–1316. **Lauerer M**, Saftic D, Quick WP, Labate C, Fichtner K, Schulze E-D, Rodermel SR, Bogorad L, Stitt M (1993) Planta 190:332–345. **Lee DM**, Assmann SM (1992) Physiol Plant 85:35–42. **Li B**, Chollet R (1993) Arch Biochem Biophys 307:416–419. **Li B**, Geiger DR, Shieh W-J (1992a) Plant Physiol 99:1393–1399. **Li B**, Servaites JC, Geiger DR (1992b) Plant Physiol 98:1277–1284. **Lin M**, Turpin DH (1993) J Phycol 29:777–786. **Lloyd J**, Syvertsen JP, Kriedemann PE, Farquhar GD (1992) Plant Cell Environ 15:873–899. **Loreto F**, Harley PC, Di Marco G, Sharkey TD (1992) Plant Physiol 98:1437–1443. **Lunn JE**, Douce R (1993) Biochem J 290:375–379. **Lüttge U**, Beck F (1992) Planta 188:28–38.

Makino A, Sakashita H, Hidema J, Mae T, Ojima K, Osmond B (1992) Plant Physiol 100:1737–1743. **Manh CT**, Bismuth E, Boutin J-P, Provot M, Champigny M-L (1993) Physiol Plant 89:460–466. **Marcus Y**, Berry JA, Pierce J (1992) Planta 187:511–516. **Masle J**, Hudson GS, Badger MR (1993) Plant Physiol 103:1075–1088. **Mate CJ**, Hudson GS, von Caemmerer S, Evans JR, Andrews TJ (1993) Plant Physiol 102:1119–1128. **Mateos MI**, Serrano A (1992) Plant Sci 84:163–170. **Mawson BT** (1993) Plant Cell Environ 16:207–214. **McMichael RW**, Klein RR, Salvucci ME, Huber SC (1993) Arch Biochem Biophys 307:248–252. **Michels S**, Scagliarini S, Della Seta F, Carles C, Riva M, Trost P, Branlant G (1994) FEBS Lett 339:97–100. **Mishra NP**, Mishra RK, Singhal GS (1993) Plant Physiol 102:903–910. **Mitra S**, Srivastava GC (1992) Photosynthetica 26:87–89. **Mittler R**, Zilinskas BA (1994) Plant J 5:397–405. **Moore BD**, Seemann JR (1992) Plant Physiol 99:1551–1555. **Morel FMM**, Reinfelder JR, Roberts SB, Chamberlain CP, Lee JG, Yee D (1994) Nature 369:740–742. **Morgan CL**, Turner SR, Rawsthorne S (1993) Planta 190:468–473. **Morgan ME**, Kingston JD, Marino BD (1994) Nature 367:162–165. **Morin F**, André M, Betsche T (1992) Plant Physiol 99:89–95. **Murata S**, Kobayashi M, Matoh T, Sekiya J (1992) Plant Cell Physiol 33:1247–1250. **Murray KJ**, Tenhunen JD, Nowak RS (1993) Oecologia 96:200–207.

Neubauer C, Yamamoto HY (1992) Plant Physiol 99:1354–1361. **Neubauer C**, Yamamoto HY (1994) Photosyn Res 39:137–147. **Neuhaus HE**, Henrichs G, Scheibe R (1993) Plant Physiol 101:573–578. **Nie D**, He H, Kirkham MB, Kanemasu ET (1992) Photosynthetica 26:189–198. **Nimer NA**, Dixon GK, Merrett MJ (1992) New Phytol 120:153–158. **Nobel PS**, García-Moya E, Quero E (1992) Plant Cell Environ 15:329–335. **Nonomura AM**, Benson AA (1992) Proc Natl Acad Sci USA 89:9794–9798.

Oberhuber W, Edwards GE (1993) Plant Physiol 101:507–512. **Ogawa T** (1992) Plant Physiol 99:1604–1608. **Ogawa N**, Okumura S, Izui K (1992) FEBS Lett 302:86–88. **Ögren E**, Evans JR (1993) Planta 189:182–190. **Okita TW** (1992) Plant Physiol 100:560–564.

Osborne BA (1994) Plant Cell Environ 17:143–152. **Overlach S**, Diekmann W, Raschke K (1993) Plant Physiol 101:1201–1207.

Palmqvist K (1993) Planta 191:48–56. **Palmqvist K**, Yu J-W, Badger MR (1994a) Physiol Plant 90:537–547. **Palmqvist K**, Ögren E, Lernmark U (1994b) Plant Cell Environ 17:65–72. **Parkhurst DF** (1994) New Phytol 126:449–479. **Pastori GM**, Trippi VS (1993) Physiol Plant 87:227–231. **Paul MJ**, Loos K, Stitt M, Ziegler P (1993) Plant Cell Environ 16:531–538. **Perata P**, Pozueta-Romero J, Yamaguchi J, Akazawa T (1992) FEBS Lett 309:283–287. **Piedade MTF**, Long SP, Junk WJ (1994) Oecologia 97:193–201. **Pierre JN**, Pacquit V, Vidal J, Gadal P (1992) Eur J Biochem 210:531–537. **Polle A**, Chakrabarti K, Chakrabarti S, Seifert F, Schramel P, Rennenberg H (1992) Plant Physiol 99:1084–1089. **Polley HW**, Johnson HB, Marino BD, Mayeux HS (1993) Nature 361:61–64. **Poorter H** (1993) Vegetatio 104/5:77–97. **Porter MA**, Grodzinski B (1984) Plant Physiol 74:413–416. **Portis AR** (1992) Annu Rev Plant Physiol Plant Mol Biol 43:415–437. **Prado FE**, Lázaro JJ, Hermoso R, Chueca A, López Gorgé J (1992) Planta 188:345–353. **Price GD**, Coleman JR, Badger MR (1992) Plant Physiol 100:784–793.

Queiroz-Claret C, Queiroz O (1992) J Plant Physiol 139:385–389. **Quiñones MA**, Zeiger E (1994) Science 264:558–561. **Quiroga O**, González EL (1993) J Phycol 29:321–324. **Quy LV**, Champigny M-L (1992) Plant Physiol 99:344–347.

Raghavendra AS, Yin Z-H, Heber U (1993) Planta 189:278–287. **Rajagopalan AV**, Devi MT, Raghavendra AS (1993) Photosyn Res 38:51–60. **Rama Das VS**, Vats SK (1993) Photosynthetica 28:91–97. **Ramazanov Z**, Cárdenas J (1992a) Physiol Plant 84:502–508. **Ramazanov Z**, Cárdenas J (1992b) Physiol Plant 85:121–128. **Ramazanov Z**, Mason CB, Geraghty AM, Spalding MH, Moroney JV (1993) Plant Physiol 101:1195–1199. **Ranieri A**, Lencioni L, Schenone G, Soldatini GF (1993) J Plant Physiol 142:286–290. **Rawsthorne S** (1992) Plant J 2:267–274. **Razdan K**, Heinrikson RL, Zurcher-Neely H, Morris PW, Anderson LE (1992) Arch Biochem Biophys 298:192–197. **Read BA**, Tabita FR (1992) Biochemistry 31:5553–5560. **Reng W**, Riessland R, Scheibe R, Jaenicke R (1993) Eur J Biochem 217:189–197. **Riebesell U**, Wolf-Gladrow DA, Smetacek V (1993) Nature 361:249–251. **Riesmeier JW**, Hirner B, Frommer WB (1993a) Plant Cell 5:1591–1598. **Riesmeier JW**, Flügge U-I, Schulz B, Heineke D, Heldt H-W, Willmitzer L, Frommer WB (1993b) Proc Natl Acad Sci USA 90:6160–6164. **Riesmeier JW**, Willmitzer L, Frommer WB (1994) EMBO J 13:1–7. **Robinson SA**, Yakir D, Ribas-Carbo M, Giles L, Osmond CB, Siedow JN, Berry JA (1992) Plant Physiol 100:1087–1091. **Rotatore C**, Colman B (1992) J Exp Bot 43:571–576. **Rotatore C**, Lew RR, Colman B (1992) Planta 188:539–545.

Sage RF (1993) Photosyn Res 35:219–226. **Sage RF**, Seemann JR (1993) Plant Physiol 102:21–28. **Sage RF**, Reid CD, Moore BD, Seemann JR (1993) Planta 191:222–230. **Sainis JK**, Srinivasan VT (1993) J Plant Physiol 142:564–568. **Salvucci ME** (1992) Arch Biochem Biophys 298:688–696. **Saradadevi K**, Raghavendra AS (1992) Plant Physiol 99:1232–1237. **Sassenrath-Cole GF**, Pearcy RW (1992) Plant Physiol 99:227–234. **Sawada S**, Usuda H, Tsukui T (1992) Plant Cell Physiol 33:943–949. **Scagliarini S**, Trost P, Pupillo P, Valenti V (1993) Planta 190:313–319. **Scheibe R**, Geissler A, Rother T (1993) Arch Biochem Biophys 300:635–640. **Schlichting R**, Bothe H (1993) Bot Acta 106:428–434. **Schmitt JM**, Piepenbrock M (1992) Plant Physiol 99:1664–1669. **Schnabl H**, Denecke M, Schulz M (1992) Bot Acta 105:367–369. **Schnettger B**, Critchley C, Santore UJ, Graf M, Krause GH (1994) Plant Cell Environ 17:55–64. **Scholes JD**, Lee PJ, Horton P, Lewis DH (1994) New Phytol 126:213–222. **Schuller KA**, Werner D (1993) Plant Physiol 101:1267–1273. **Schulz M**, Hunte C, Schnabl H (1992) Physiol Plant 86:315–321. **Schünemann D**, Borchert S, Flügge U-I, Heldt HW (1993) Plant Physiol 103:131–137. **Sharkey TD**, Savitch LV, Vanderveer PJ, Micallef BJ (1992) Plant

Physiol 100:210–215. **Shiraiwa Y**, Goyal A, Tolbert NE (1993) Plant Cell Physiol 34:649–657. **Shyam R**, Raghavendra AS, Sane PV (1993) Physiol Plant 88:446–452. **Sicher RC**, Kremer DF, Rodermel SR (1994) Plant Physiol 104:409–415. **Singh R**, Green MR (1993) Science 259:365–368. **Slocombe SP**, Whitelam GC, Cockburn W (1993) Plant Cell Environ 16:403–411. **Smart DR**, Chatterton NJ, Bugbee B (1994) Plant Cell Environ 17:435–442. **Smirnoff N** (1993) New Phytol 125:27–58. **Smith CM**, Duff SMG, Chollet R (1994) Arch Biochem Biophys 308:200–206. **Socias FX**, Medrano H, Sharkey TD (1993) Plant Cell Environ 16:81–86. **Sonnewald U**, Quick WP, MacRae E, Krause K-P, Stitt M (1993) Planta 189:174–181. **Sonnewald U**, Lerchl J, Zrenner R, Frommer W (1994) Plant Cell Environ 17:649–658. **Spreitzer RJ** (1993) Annu Rev Plant Physiol Plant Mol Biol 44:411–434. **Stabenau H**, Winkler U, Säftel W (1993) Planta 191:362–364. **Stark DM**, Timmerman KP, Barry GF, Preiss J, Kishore GM (1992) Science 258:287–292. **Stitt M**, Schulze D (1994) Plant Cell Environ 17:465–487. **Strauss G**, Fuchs G (1993) Eur J Biochem 215:633–643. **Sültemeyer D**, Biehler K, Fock HP (1993) Planta 189:235–242. **Süss K-H**, Arkona C, Manteuffel R, Adler K (1993) Proc Natl Acad Sci USA 90:5514–5518.

Thielmann J, Goyal A, Tolbert NE (1992) Plant Physiol 100:2113–2115. **Thom M**, Getz HP, Maretzki A (1992) Physiol Plant 86:104–114. **Thomas WH**, Seibert DLR, Alden M, Neori A, Eldridge P (1984) Biomass 5:181–209. **Tissue DT**, Thomas RB, Strain BR (1993) Plant Cell Environ 16:859–865. **Tobias RB**, Boyer CD, Shannon JC (1992) Plant Physiol 99:146–152. **Trost P**, Scagliarini S, Valenti V, Pupillo P (1993) Planta 190:320–326.

Ueno O (1992) Physiol Plant 85:189–196. **Ushimaru T**, Shibasaka M, Tsuji H (1992) Plant Cell Physiol 33:1065–1071. **Usuda H**, Shimogawara K (1993) Plant Cell Physiol 34:767–770. **Utsunomiya E**, Muto S (1993) Physiol Plant 88:413–419.

van Bel AJE, Ammerlaan A, van Dijk AA (1994) Planta 192:31–39. **van Cleve B**, Forreiter C, Sauter JJ, Apel K (1993) Planta 189:70–73. **Von Mohl H** (1844) Botan Z II:273–277.

Walker D (1992a) Energy, plants and man. Oxygraphics Ltd, Brighton, 277 pp. **Walker D** (1992b) New Phytol 121:325–345. **Walker MA**, McKersie BD (1993) J Plant Physiol 141:234–239. **Wang Y-H**, Chollet R (1993a) Arch Biochem Biophys 304:496–502. **Wang Y-H**, Chollet R (1993b) FEBS Lett 328:215–218. **Wang ZY**, Portis AR (1992) Plant Physiol 99:1348–1353. **Wang Y-H**, Duff SMG, Lepiniec L, Crétin C, Sarath G, Condon SA, Vidal J, Gadal P, Chollet R (1992) J Biol Chem 267:16759–16762. **Watanabe S**, Tamai N, Hoshina S, Sanada Y, Nishida K, Wada K (1992) J Plant Physiol 140:129–133. **Weger HG**, Dasgupta R (1993) J Phycol 29:300–308. **Weigend M**, Hincha DK (1992) J Plant Physiol 140:653–660. **Weiner H**, McMichael RW, Huber SC (1992) Plant Physiol 99:1435–1442. **Weiner H**, Weiner H, Stitt M (1993) FEBS Lett 333:159–164. **White PJ**, Smith JAC (1992) J Plant Physiol 139:533–538. **Wigge B**, Krömer S, Gardeström P (1993) Physiol Plant 88:10–18. **Wilkins MB**, Thomas SL (1993) J Exp Bot 44:901–906. **Williams TG**, Colman B (1993) Plant Physiol 103:943–948. **Winter H**, Robinson DG, Heldt HW (1993) Planta 191:180–190. **Woodrow IE** (1994) Plant Cell Environ 17:277–286. **Woodrow IE**, Mott KA (1992) Plant Physiol 99:298–303. **Woodward FI** (1990) Trends Ecol Evol 5:111–115. **Wu M-X**, Wedding RT (1992) Plant Physiol 100:382–387. **Wullschleger SD** (1993) J Exp Bot 44:907–920.

Yin Z-H, Heber U, Raghavendra AS (1993) Planta 189:267–277. **Yokota A**, Higashioka M, Wadano A (1992) Eur J Biochem 208:721–727. **Yokota A**, Higashioka M, Taira T, Usuda H, Wadano A, Murayama H (1994) Plant Cell Physiol 35:317–321. **Young XK**, McCarty RE (1993) Plant Physiol 101:793–799. **Yu J-W**, Woo KC (1992) Aust J Plant

Physiol 19:671–678. **Yu J-W**, Price GD, Badger MR (1994) Aust J Plant Physiol 21:185–195.

Zotz G, Winter K (1993) Plant Physiol 102:835–841.

Dr. Grahame J Kelly
Centre for Molecular Biotechnology
School of life Science
Queensland University of Technology
Brisbane, Queensland 4000, Australia

Prof. Dr. Dr. Erwin Latzko
Institut für Allgemeine Chemie und Biochemie
Technische Universität München
Freising Weihenstephan, Germany

Note added in proof (2005):
Increased CO_2 does not reduce dark respiration (Sect. 8.a) but rather has no effect on it (see Sect. 4, 2002 review).

Photosynthesis. Carbon Metabolism: The Carbon Metabolisms of Unstressed and Stressed Plants

By G. J. Kelly and E. Latzko

"My investigations lead me to show how water and air contribute more to the formation of the dry matter of plants growing in a fertile soil than does the humus matter that they absorb, in aqueous solution, through their roots ..."

(Nicolas de Saussure ca. 1804)

1. Introduction

Following on from the pioneer experiments of investigators such as Joseph Priestly and Jan Ingenhousz, Nicolas de Saussure must have felt close to a full explanation of the growth of plants when he concluded that plant dry matter originated *predominantly* from water and air. Indeed, if we ignore biochemical details, we must conclude that he was doing well for "plants growing in an [adequately watered and] fertile soil". Most subsequent studies have addressed the question of *how* an actively growing plant converts water and air (and sunshine) into its dry matter, i.e. how good plants perform photosynthesis. But what about poor plants? There is no doubt that other plant scientists have pondered upon, and studied, the responses of plants to adverse conditions, but it is doubtful whether the plant-growth scientists realised the extent to which adversity existed in even "normal" growing conditions, and the degree to which photosynthesis has had to adjust. de Saussure could not have imagined how insidiously stresses influence photosynthetic cells, although (ironically) he is credited with the discovery of crassulacean acid metabolism (see Walker 1992), probably the best-known plant adaptation to water deficit.

By contrast, in recent years there has been an exponential increase in plant stress research, spurred on not only by the discovery of biochemical mechanisms that presumably play protective roles, but also (we assume) by the global realisation that the biosphere as a whole is becoming increasingly stressed, and that understanding and solutions are needed. This review updates our conception of the details of photosynthetic carbon metabolism (a recent informative review of regulatory aspects is

G.J. Kelly and E. Latzko:
Thirty Years of Photosynthesis

Originally published in Progress in Botany 58

available, Stitt and Sonnewald 1995), and then examines plant stress in the limited context of how stress influences this photosynthetic metabolism.

2. Production of Photosynthate, and Some Comments on its Movement and Deposition

a) RuBP Carboxylase

Ribulose-1,5-bisphosphate (RuBP) carboxylase, the Calvin cycle's CO_2-fixing enzyme, is at the centre of one facet of plant stress, viz. the low availability of CO_2. Impressive as this enzyme's affinity for CO_2 might be (in vivo Km in air is 18 µM) this is still not sufficient to permit it to operate at capacity because the atmospheric CO_2 concentration is only sufficient to generate unaided, 13 µM CO_2 in solution. Moreover, even if atmospheric CO_2 was to increase (as is indeed happening; see Sect. 3.d), the enzyme's low turnover number (3.5 s^{-1}) when operating at capacity gives it the dubious distinction of being the predominant determinant of photosynthetic capacity in normally grown C_3 plants (Furbank and Taylor 1995).

The above in vivo Km and turnover-number values were obtained by an ingenious application of antisense technology (von Caemmerer et al. 1994). When the same technology was used to construct tobacco plants with a range of low RuBP carboxylase contents, and these tobacco plants were pushed to their growth limits in Portugal, the predominant influence of the enzyme on photosynthetic capacity was clearly seen (Krapp et al. 1994).

The attention given to the *regulation* of RuBP carboxylase activity has been more subdued during the past 2 years. An interesting parallel has been presented between the chemistry of the CO_2-plus-Mg^{2+} activation (the "primary" mechanism of activation), and a similar CO_2-plus-Ni^{2+} activation of urease (Lippard 1995). The carboxylase's activation is, in turn, influenced by "secondary" regulators (see Gutteridge and Gatenby 1995). One of these is inorganic phosphate (Pi), which promotes the binding of the activating CO_2 (Anwaruzzaman et al. 1995). A second is the enigmatic (Foyer and Parry 1995) regulatory protein "rubisco activase" that prevents, in an ATP-dependent manner, the binding of RuBP to enzyme that has not undergone the CO_2-plus-Mg^{2+} activation (Portis et al. 1995); such binding otherwise inhibits the enzyme's activity. Incidentally, Campbell and Ogren (1995) report that rubisco activase does not undergo light-mediated activation, as was reported earlier. A third regulator is the sugar-phosphate carboxyarabinitol-1-P (CA1P), whose main role seems to be to bind to the carboxylase and inhibit it at night (Moore and Seemann 1994). The extent to which CA1P does this varies remarkably between plant species, and even between the cultivars (Holbrook et al. 1994) and the stages of leaf development (Moore et al. 1995) within one species. The CA1P can represent more than 2% of the photosynthate in a darkened leaf (Andralojc et al. 1994), but it is degraded when the leaf is exposed to even relatively low light (ca. 30 µmol photons $m^{-2}s^{-1}$; Moore and Seemann 1994).

Besides this short-term regulation, RuBP carboxylase activity adjusts in the long term to stimuli such as sink strength and CO_2 concentration. Removal of sinks lowers activity, via a reduction in the activator Pi (Sawada et al. 1995) or some other mechanism: Crafts-Brandner and Salvucci (1994) are intrigued by a "rubisco complex protein" that accumulates in sink-deprived soybean plants and can form an insoluble complex with the carboxylase, but any regulatory function is obscured by the fact that the protein is cytosolic, not chloroplastic. High external CO_2 during plant growth sometimes results in a reduction of the RuBP carboxylase activity, presumably through a feedback down-regulation of the expression of the gene for the enzyme's small subunit, but this may not occur in all plant species (Van Oosten and Besford 1994; Xu et al. 1994; Habash et al. 1995; Jacob et al. 1995; see Sect. 3.d). Poor Pi availability for the growing plant also reduces the carboxylase activity indirectly by lowering the supply of sugar-Ps, one of which is the substrate RuBP (Rao and Terry 1995).

The specificity factor of RuBP carboxylase, i.e. its propensity to use CO_2 rather than O_2 [in the latter case photorespiration (Sect. 3.c.) is initiated], has still not been improved in experiments with *Chlamydomonas reinhardtii* mutants (Gotor et al. 1994; Spreitzer et al. 1995), although the possibility of locating carboxylases with better specificity factors by surveying a range of terrestrial plants has been raised (Kent and Tomany 1995; Delgado et al. 1995), and the naturally superior factors for the enzyme from non-green marine microalgae have been reemphasised (Read and Tabita 1994). Microalgae have provided what is probably the biggest surprise in recent RuBP carboxylase research: three marine dinoflagellates were found to contain RuBP carboxylase proteins that are quite distinct from other eukaryotic carboxylases, apparently lacking small subunits and resembling more closely the RuBP carboxylase from the prokaryote *Rhodospirillum rubrum* (Morse et al. 1995; Whitney and Yellowlees 1995).

b) Other Calvin Cycle Enzymes

Of the remaining ten enzymes of the Calvin cycle, only five have received significant attention. Glycerate-3-P kinase was found to occur as one form in cyanobacteria, but as two forms (chloroplastic and cytosolic) in a broad range of vascular plants (Shah and Bradbeer 1994). Aldolase, which similarly occurs as two forms in higher plants (Boldt et al. 1994), presents a complex picture in protists: a chloroplast aldolase for the Calvin cycle was always found, but cytosolic aldolase was not detected (e.g., *Chlamydomonas reinhardtii*) or was found as the "class II", M^{2+}-dependent fungal-type aldolase (e.g., *Euglena gracilis*; Pelzer-Reith et al. 1994; Schnarrenberger et al. 1994). Two further reports support the concept that these two enzymes, together with several others (including glyceraldehyde-3-P [GAP] dehydrogenase) may be associated into functional multienzyme complexes in vivo (Anderson et al. 1995; Süss et al. 1995).

Inverted cDNAs encoding GAP dehydrogenase, fructose-1,6-bisphosphatase (FBPase) and phosphoribulokinase (PRK) have been constructed and used to transform potato and tobacco plants (Koßmann et al. 1994; Paul et al. 1995a; Price et al. 1995; see Furbank and Taylor 1995). This "antisense" technology creates plants with lowered contents of the respective enzyme; a range of reductions can be selected and tested. For the above three enzymes, lowered photosynthesis was not noticed until the enzyme level had been reduced by at least 60% (FBPase) or 80% (GAP dehydrogenase and PRK). This contrasts with the greater sensitivity of photosynthesis to RuBP carboxylase reduction (Sect. 2.a). These results are not surprising, given that the in vivo activities of RuBP carboxylase and FBPase are generally lower than those of GAP dehydrogenase and PRK.

The light-mediated regulation of chloroplast enzymes has been recently reviewed (Scheibe 1994) and interpreted to a greater level of sophistication (Geiger and Servaites 1994; Ocheretina and Scheibe 1994; Faske et al. 1995). It may be a primary determinant of the rate of photosynthesis by the lower, partially shaded leaves in a crop canopy (Sassenrath-Cole and Pearcy 1994). Additional mechanistic details have been revealed for four enzymes that are reductively (disulfide → 2 sulfhydryls) activated:

1. for GAP dehydrogenase, a concomitant activation by glycerate-1,3-P_2 is involved (Baalmann et al. 1994);
2. for FBPase, reductive activation is easier after the enzyme's conformation has been altered by solvents other than water (Ballicora and Wolosiuk 1994);
3. for sedoheptulose-1,7-bisphosphatase, the natural chloroplast-thylakoid/ferredoxin/thioredoxin system is a more effective activator than the artificial reductant dithiothreitol (Queiroz-Claret and Meunier 1995), and
4. for PRK, activation involves two pairs of cysteine residues, one pair on each of two enzyme subunits.

Forms of PRK with none, one or both of the cysteine pairs oxidised (and thus connected by a disulfide bond) have been detected (Clasper et al. 1994). The distance between the two cysteines of a pair is 40 amino acid residues in algae and higher plants, but only 20 in cyanobacteria (Wadano et al. 1995). However, even between algae and higher plants the enzymes seem to differ: soybean PRK was fully light-activated by quite low levels of light (Sassenrath-Cole and Pearcy 1994), but *Chlamydomonas reinhardtii* PRK activation required light levels similar to those needed for photosynthesis (Farr et al. 1994). This difference is not an isolated one: certain marine red algae and cyanobacteria possess a chloroplast GAP dehydrogenase that lacks one of the cysteine residues that is central to the light-mediated activation process, and indeed in these cases the activation cannot be detected (Pacold et al. 1995a, b). Furthermore, photosynthesis by green algae and cyanobacteria was found by Takeda et al. (1995) to be more resistant than spinach-chloroplast photosynthesis to inhibition by H_2O_2, implying that sulfhydryl groups, and therefore light-mediated activation, are less important to the algal and cyanobacterial enzymes.

c) Plastids and Their Envelope Transporters

Most recent studies with plastids have dealt with their envelope-transporter proteins, with one fascinating exception: Schnarrenberger et al. (1995) claim, from enzyme localisation studies, that the complete oxidative pentose-P pathway of spinach leaves is restricted to the chloroplast. This is contrary to most textbook presentations showing it to be in the cytosol, but (in retrospect) it is not surprising given

1. past instances where suspected cytosolic activities were discovered to be instead plastidic,
2. the calculations by Winter et al. (1994) showing that the volume of chloroplast stroma in spinach and barley leaf mesophyll cells is easily twice the volume of the cytosol, and
3. evidence for a functional oxidative pentose-P pathway in green pepper-fruit chloroplasts (Thom and Neuhaus 1995).

Returning to transporters: it used to be that the majority of attention was devoted to the movement of 3-carbon sugar-Ps (triose-Ps such as GAP, and glycerate-3-P) and Pi in and out of chloroplasts and other plastids via the unidirectionally-oriented (Schwarz et al. 1994) phosphate translocator; after all, the exit of GAP in counter-exchange with Pi is the chloroplast/cytosol link whereby photosynthate is directed toward sucrose biosynthesis in the cytosol (Fig. 1A); but in recent years non-photosynthetic plastids such as amyloplasts have been obtained in good condition, and similar tests with these implied that glucose-Ps were accepted by the translocator (see our last review). The most recent studies point to a "spectrum of plasticity" in the degree to which different sugar-Ps can be transported. However, now we are not sure whether or not a separate translocator moves these glucose-Ps. Results consistent with the "classical" phosphate translocator transporting glucose-6-P across the envelopes of pea-embryo plastids (Hill and Smith 1995), potato-tuber and cauliflower-bud amyloplasts (Möhlmann et al. 1995; Schott et al. 1995), and tomato-fruit chromoplasts and chloroplasts (Schünemann and Borchert 1994) have appeared, but a novel translocator that exchanges glucose-6-P and Pi (Thom and Neuhaus 1995) has been proposed to exist in the envelopes of green pepper (*Capsicuum anuum*) fruit chloroplasts, and even chloroplasts from spinach leaves previously fed glucose for several days (Batz et al. 1995; Quick et al. 1995). Moreover, the other glucose-P, i.e. glucose-1-P, which is able to cross amyloplast envelopes and become the substrate for starch synthesis, is clearly transported by a protein that is different from any protein that transports either glucose-6-P or triose-Ps (Kosegarten and Mengel 1994; Tetlow et al. 1994; Möhlmann et al. 1995; Schott et al. 1995; see Fig. 1B). In wheat-seed amyloplasts, the starch synthesis from glucose-1-P is dependent on the concomitant import of ATP by yet another protein, the adenylate translocator (Tetlow et al. 1994). This plastid translocator turns out to be somewhat exotic in that DNA sequence data point to its closest relative being a similar ATP translocator in the membrane of the

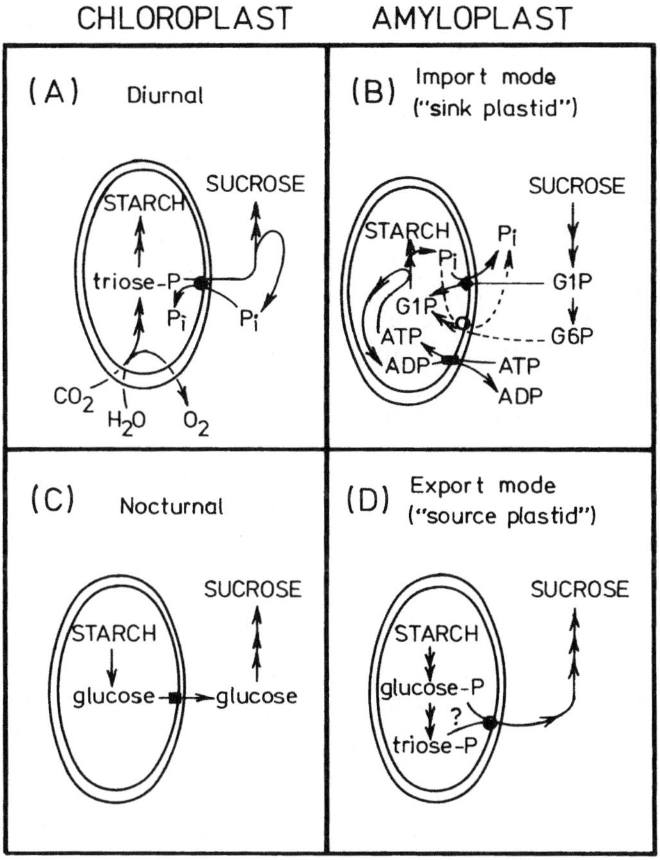

Fig. 1A-D. A summary of the major roles of known plastid-envelope transporter proteins: **A** Photosynthesizing chloroplast. **B** Starch-synthesizing amyloplast; fruit chloroplasts, and the chloroplasts from sugar-fed leaves, behave similarly. **C** Chloroplast mobilising transitory starch. **D** Amyloplast mobilising storage starch. Based on references in Sect. 2.c Abbreviations: G1P, glucose-1-P; G6P, glucose-6-P

pathogenic bacterium that causes typhus in humans (Kampfenkel et al. 1995).

Starch deposited in plastids is ultimately destined to be mobilised by metabolism to sugars that are returned to the cytosol. In leaf chloroplasts this occurs nocturnally. It used to be thought that starch was metabolised to triose-Ps that were exported via the phosphate translocator (i.e. a starch → triose-P → sucrose variation of Fig. 1A), but now it seems that the long-forgotten hexose transporter plays the predominant role so that, quite simply, starch is hydrolysed (probably with the help of α-glucosidase; Sun et al. 1995) and the resultant glucose is exported (Trethewey and ap Rees 1994a, b; Fig. 1C). Support for this concept comes from experiments in which the amount of the classical phosphate translocator in chloroplasts was reduced using antisense technology. The appearance and photosynthesis of these plants appeared normal, but by day the transport of photosynthate out of chloroplasts was (as expected) hindered, resulting in a substantial rise in the amount of starch stored temporarily inside the chloroplast (Barnes et al. 1994; Heineke et al. 1994a). This necessitated a greater-than-normal nocturnal flux of photosynthate out of the chloroplast and, despite the low phosphate translocator activity, this was possible because that translocator was not totally involved; instead, the hexose transporter (Fig. 1C) assumed a part (or all?) of the role (Heineke et al. 1994a). Presumably, cytosolic hexokinase facilitated entry of hexose into the sucrose biosynthetic pathway. A remaining question is whether starch mobilisation in amyloplasts occurs similarly. These plastids do possess a hexose transporter (Möhlmann et al. 1995), but one study indicates that it is a sugar-P that is exported (Neuhaus et al. 1995; Fig. 1D).

d) Starch and Fructans

The best-known form of stored photosynthate is starch: it is stored on a short-term basis (diurnal) in chloroplasts, and on a long-term basis (usually seasons) in amyloplasts. Short-term storage is greater in spinach than in barley leaves (Riens et al. 1994); the latter is a member of the temperate grasses which are noted for an alternative "intermediate-term" photosynthate storage as fructans (see below). Three reports suggest that starch synthesis is something of a luxury rather than a necessity, in that two tobacco mutants (Eichelmann and Laisk 1994; Geiger et al. 1995) and a transgenic potato plant with only ca. 20% of the usual level of ADP-glucose pyrophosphorylase (Leidreiter et al. 1995), all deficient in the ability to synthesise starch, could overcome this handicap by simply increasing the flux of carbon towards cytosolic sucrose. However, the capacity to do this was insufficient if the leaf's photosynthetic capacity was engaged (i.e. under high light and saturating CO_2, Eichelmann and Laisk 1994).

ADP-glucose pyrophosphorylase still receives the most attention of the starch-biosynthesis enzymes. Typical properties include inhibition by Pi and activation by glycerate-3-P (as recently shown for the *Chlamydomonas reinhardtii* enzyme; Iglesias et al. 1994). There is evidence that further unidentified regulatory mechanisms may exist (Jeannette and Prioul 1994). The higher-plant enzyme is a tetramer of two pairs of subunits of slightly different size, the larger having the regulatory role and the smaller the catalytic (Ball and Preiss 1994; Ballicora et al. 1995).

An often neglected but still quantitatively important storage form of photosynthate is fructan. Fructans are common in the leaves and stems of temperate grasses (but not tropical C_4 grasses; Lunn and Hatch 1995), and the bulbs and tubers of certain other plants such as onion. They are relatively short polymers of ca. 10 to 100 fructose units attached to a terminal glucose. Sucrose, transported from leaf photosynthetic cells, is the ultimate fructosyl donor for their synthesis, which occurs in the cell vacuole. At least three fructosyl transferase enzymes are involved: the first transferring a fructose unit from one sucrose to another to form the fructose$_2$-glucose trisaccharide named kestose (and liberating a glucose), the second mediating further polymerisation by transferring fructosyl residues between kestose and longer glucose-(fructose)$_n$ polymers (Koops and Jonker 1994; Penson and Cairns 1994; Lüscher and Nelson 1995), and a third recently characterized enzyme that transfers fructose units from sucrose to developing fructan polymers and, unlike the other two enzymes, forms β(2–6) linkages between the fructose groups (Duchateau et al. 1995); the other two enzymes form β(2–1) linkages. A feature of fructan-accumulating plants is that their chloroplasts store less transitory starch than those of other plants; indeed, onion leaves effectively lack starch, and only recently has the presence of starch in onions been reported (Ernst and Bufler 1994). Finally, there is a startling report from van der Meer et al. (1994) that potato plants, which normally never make fructans, accumulated up to 30% of the dry weight of their leaves as abnormally large fructan molecules after being genetically engineered to produce bacterial fructosyltransferases.

e) Sucrose Production

There are as yet no detailed proposed changes to the pathway by which photosynthate exported from the chloroplast is converted to sucrose in the cytosol. None may be needed for daytime sucrose synthesis, but if by night the export process is as shown in Fig. 1C, then the pathway will be shorter, cytosolic hexokinase will be required and the quagmire of regulatory effects that operate between fructose-6-P and fructose-1,6-P$_2$ (see below) would be irrelevant.

Let us begin near the end. An important regulatory role is still envisioned for the penultimate enzyme, sucrose-P synthase, which is susceptible to a protein kinase-catalysed phosphorylation (reducing activity) and a phosphatase-catalysed dephosphorylation (increasing activity). Weiner (1995) has obtained antibodies that distinguish between these two forms of the enzyme. The protein kinase involved may be a different protein to the protein kinase that similarly regulates another enzyme, nitrate reductase (McMichael et al. 1995). At first sight, this reversible phosphorylation of sucrose-P synthase might appear to be a "light/dark" regulation, and indeed in leaves the activation by dephosphorylation seems to be light-mediated, and vice versa; but this viewpoint needs to be modified for two reasons:

1. some activity must remain in the dark if the chloroplast starch → cytosolic sucrose conversion (Fig. 1C) is to proceed, and
2. the enzyme in stored potato tubers (that normally never see the light of day) is similarly regulated (Reimholz et al. 1994).

An intriguing observation is that the total activity in developing potato tubers exceeds the total activity in leaves (Misra et al. 1994). Finally, two more startling reports: tobacco leaves, that normally manufacture sucrose, were induced to behave like celery or apple leaves and synthesise mannitol or sorbitol (respectively) after being genetically engineered to produce one additional enzyme, which was all that was needed for glucose-6-P to be converted to mannitol-1-P or sorbitol-6-P (Tarczynski et al. 1992; Tao et al. 1995). In the former case, the mannitol conferred increased resistance to salinity stress (Tarczynski et al. 1993). Incidentally, in celery, which normally produces some mannitol, salinity stress increases its production and storage (Stoop and Pharr 1994).

It is probably premature to consider in detail the regulation of enzymes interconverting fructose-6-P and fructose-1,6-P_2 in the cytosol. Three enzymes are involved (the classical ATP-linked P-fructokinase, PPi-linked P-fructokinase [PPi–PFK], and cytosolic FBPase), each regulated by a range of metabolites; fructose-2,6-P_2, which activates the second enzyme and inhibits the third, currently attracts the most interest. The problem is that difficulties and potential errors in estimating the enzyme activities and levels of fructose-2,6-P_2 have been revealed (Podestá et al. 1994; Scott and Kruger 1994; Kruger 1995). Some past conclusions (see our earlier reviews) may therefore be in doubt. Two recent studies (Scott and Kruger 1995; Scott et al. 1995) in which in vivo fructose-2,6-P_2 levels were raised (by creating transgenic tobacco plants containing the mammalian enzyme that manufactures fructose-2,6-P_2) support the contention that, by inhibiting FBPase and activating PPi–PFK (Nielsen and Wischmann 1995), it can retard starch mobilisation and inhibit sucrose synthesis. Note, however, that fructose-1,6-P_2, a substrate for the latter enzyme, may substitute for fructose-2,6-P_2 in activating it (Nielsen 1995).

The question as to why plants need PPi–PFK has deepened. It catalyses a reversible reaction that functionally duplicates FBPase-plus-ATP-linked P-fructokinase. Its value in removing PPi (generated as UDP-glucose is formed on the way to sucrose) has been reiterated (Kruger 1995), but it is a worry that Paul et al. (1995b) found that tobacco plants genetically engineered to contain less than 3% of the normal level of PPi–PFK appeared to be totally unaffected by the loss.

f) Sucrose Movement and Deposition

Photosynthetically generated sucrose is directed toward the phloem for transport to sink tissues; a detailed review of this process in potato, tomato and tobacco (an odd trio, since the former two have sinks quantitatively far greater than the latter) is available (Frommer and Sonnewald 1995). Depending on the plant species, sucrose movement through the leaf is either symplastic or apoplastic. In the latter case, the sucrose is actively loaded into the phloem by a phloem-cell plasma-membrane sucrose-H^+ symporter (Truernit and Sauer 1995) powered by a plasma-membrane H^+-ATPase (Bouché-Pillon et al. 1994). It is the efficiency of this leaf process that drives the high sucrose accumulation in sugar-beet roots (Lohaus et al. 1994), where it is stored in vacuoles after crossing a tonoplast antiporter in exchange for a H^+ (Getz and Klein 1995). In the leaf, ATP to power phloem loading is presumably generated by diverting a portion of the sucrose entering companion cells and using it to fuel glycolysis and respiration in these cells (Fig. 2) – Lerchl et al. (1995) have evidence for this after crossing two types of transgenic tobacco.

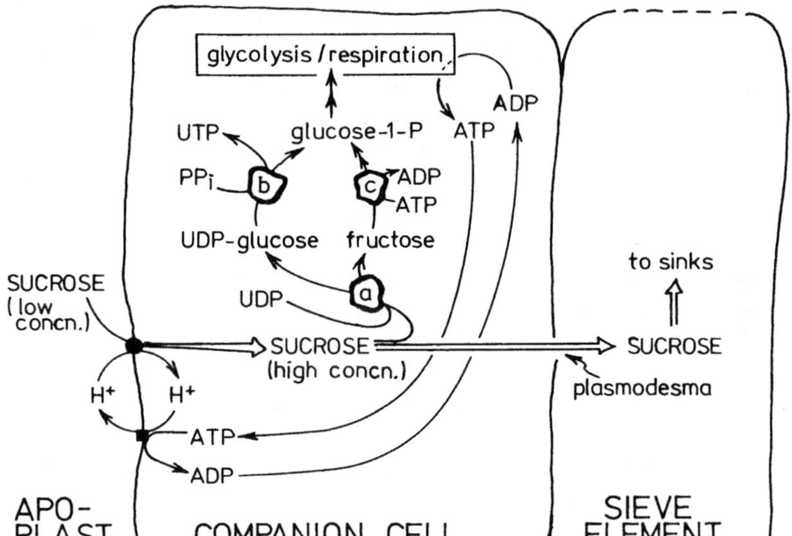

Fig. 2. Phloem loading. Respiratory ATP in the companion cell drives sucrose into the phloem against the concentration gradient. Support for the indicated metabolic pathway has been supplied by Lerchl et al. (1995) from whom this figure is adapted. Enzymes: *a* "sucrose synthase"; *b* UDP-glucose pyrophosphorylase; *c* hexokinase. If the complete respiration of one molecule of sucrose produces 72 ATP, and the import of each molecule of sucrose requires (as shown) 1 ATP, then only ca 1½% of the incoming sucrose would need to be respired

Details of the transport of sucrose from the sink end of phloem to sink cells are less clear. There are claims for the importance of simple sucrose transporters (Frommer and Sonnewald 1995), sucrose synthase (Zrenner et al. 1995), an apoplastic invertase (Roitsch et al. 1995), and even a source-type sucrose-H^+ symporter (Truernit and Sauer 1995) despite the fact that sucrose moves down its concentration gradient in most sink tissues. Incidentally, a new role for the poorly named sucrose synthase (that in vivo generates UDP-glucose and fructose from sucrose; see Fig. 2), has been revealed by Amor et al. (1995), who found that the enzyme can be tightly associated with the plasma membrane of cells busily synthesizing cellulose (e.g., cotton); the UDP-glucose product is quite possibly fed directly to the cellulose (β-glucan) synthase.

Some dramatic breakthroughs have finally been made in the long quest to explain how increased levels of soluble sugars in photosynthetic cells somehow exert a feedback control on photosynthesis, thereby reducing it. This phenomenon is well known for plants that have had a major sink organ removed (although in some grain-grasses it is not observed because alternative sinks are able to compensate for the loss; Nakano et al. 1995). A general idea of the mechanism involved came from Sheen (1994) and Krapp and Stitt (1994, 1995), who concluded that accumulated sugars acted at the level of gene transcription, specifically reducing the messenger RNAs for certain chloroplast enzymes involved in photosynthesis. Finally, a first hypothesis concerning details of the mechanism could be proposed by Jang and Sheen (1994): it seems likely that glucose and fructose are the critical soluble sugars, and that the enzyme hexokinase senses their concentrations and transmits this information, perhaps as an enzyme-substrate(hexose) complex, to the cell's DNA. Questions remain, such as how the information is passed across the nuclear envelope (directly or indirectly?), and how it operates when leaf vacuoles are able to actively accumulate glucose and fructose via a tonoplast hexose transporter (Heineke et al. 1994b). Nevertheless, it is fascinating to reflect that hexokinase, an enzyme almost neglected in the recent past, has suddenly been endowed with two important roles: the first cytosolic enzyme involved in the nocturnal mobilisation of leaf starch (Fig. 1C), and the messenger molecule in the soluble-sugars feedback regulation of photosynthesis.

g) Mitochondrial Respiration and Chlororespiration

It is generally believed that light at least partially inhibits mitochondrial respiration in photosynthetic cells, but that the portion remaining is important for sucrose synthesis and/or prevention of photoinhibition (Krömer 1995). In tree leaves the extent of inhibition is ca. 50% (Villar et al. 1995). No new information concerning the role of mitochondrial respiration in the light has appeared, except for an indication that it utilises some of the reducing equivalents generated by the thylakoid (Hurry et al. 1995a); these may be delivered as NADH, or even as NADPH, to the outer surface of the inner mitochondrial membrane with has separate dehydrogenases relatively specific for each of these two nucleotides (Roberts et al. 1995).

Three comments concerning the role of mitochondrial respiration in darkness are in order:

1. an initial higher rate in wheat leaves does not seem to be wasteful (Averill and ap Rees 1995);
2. almost one-third of it is calculated to be needed for the nocturnal mobilisation of chloroplast starch and export of the resultant sucrose (Bouma et al. 1995; see Fig. 2); and
3. a portion of it may terminate with the alternative, cyanide-insensitive oxidase which appears to undergo an NADPH-linked (and maybe thioredoxin-based) reductive activation in the mitochondrial matrix (Vanlerberghe et al. 1995), not unlike the light-mediated activation of chloroplast enzymes (Sect. 2.b). This alternative oxidase recently has been isolated from a unicellular green alga (Eriksson et al. 1995).

Minor attention continues to be devoted to the little-understood but potentially important process of chlororespiration. Recent research has centred on a possible thylakoid-based electron transport from NADH (Cuello et al. 1995) to oxygen (Büchel and Garab 1995) that involves plastoquinone (Gruszecki et al. 1994).

3. The "Stress" of Inadequate CO_2

It might be argued whether or not the ambient atmospheric CO_2 level of 360 ppm represents a stress to plants. We have adopted a liberal interpretation in claiming that it does: the level is so low that, for terrestrial C_3 plants,

1. it permits only ca. one-half of the photosynthetic capacity under full sunshine to be attained, and
2. it permits O_2 to compete for the active site of RuBP carboxylase, thereby initiating photorespiration (Sect. 3.c).

Certainly, C_3 plants growing at about half this CO_2 level, similar to levels that occurred 20 000 years ago during the Last Glacial Maximum, are barely able to grow (Dippery et al. 1995). However, we concede that C_4 and (perhaps) crassulacean acid metabolism (CAM) plants (Sects. 3.b and 4.d), and algae – including marine phytoplankton (Raven 1994) – have developed mechanisms for concentrating CO_2 in and around chloroplasts, and are therefore less likely to be stressed in this way.

a) Uptake of Inorganic Carbon by Eukaryotic Cells

Attention has swung from prokaryotic cyanobacteria (Espie and Kandasamy 1994; Yu et al. 1994) to eukaryotic algae in the study of inorganic carbon (Ci) uptake by aquatic photosynthetic cells grown in a "low CO_2" (i.e. equilibrated with air) medium. In green algae, several proteins, including special forms of carbonic anhydrase and ATPase, and a red-blood-cell-like anion antiporter, seem to be involved in taking up CO_2 and/or HCO_3^- (Sharkia et al. 1994; Gimmler and Slovik 1995; Matsuda and Colman 1995; Axelsson et al. 1995; Ramazanov et al. 1995). Establishment of the uptake system(s) is triggered by the low external Ci concentration ; neither the internal Ci concentration nor light seem to be important (Matsuda and Colman 1995; Rawat and Moroney 1995) although blue light may have a developmental role (Borodin et al. 1994) as it seems to have in brown algae (Forster and Dring 1992).

Much attention has been given to the enzyme carbonic anhydrase that accelerates the first of the $CO_2 + H_2O \rightleftharpoons \rightleftharpoons H^+ + HCO_3^-$ equilibrium reactions (Badger and Price 1994). The forms of the enzyme in *Chlamydomonas reinhardtii* have especially been targeted; at last count there were four: a periplasmic (between cell wall and plasmalemma) form, and three intracellular forms, two of which are associated with the chloroplast (Husic and Marcus 1994; Katzman et al. 1994; Karlsson et al. 1995; Sültemeyer et al. 1995). At least two of these forms are part of the Ci uptake process which (it should be recalled) can concentrate Ci inside cells to levels well in excess of the external concentration. An unusual twist to the links between carbonic anhydrase and the Ci concentrating mechanism is provided by the green alga *Coccomyxa* sp. that lacks the actual mechanism, but seems to compensate by developing a particularly high carbonic anhydrase activity (Hiltonen et al. 1995). Finally, the situation with chromophyte algae deserves some reconsideration: it seems that marine diatoms and coccolithophorids can possess Ci-concentrating mechanisms, but only with diatoms did carbonic anhydrase seem to play a role (Nimer et al. 1994; Sekino and Shiraiwa 1994; Colman and Rotatore 1995).

Sophisticated techniques have left little hope that a role for carbonic anhydrase in the chloroplasts of terrestrial C_3 plants will be found. Price et al. (1994) had trouble detecting any difference between normal tobacco plants and tobacco plants genetically engineered to contain only 2% of the usual level of carbonic anhydrase, while Peltier et al. (1995) found that the photosynthesis of *Commelina communis* was unaffected after its leaf carbonic anhydrase activity in vivo was inhibited by more than 90% (as detected by ^{18}O measurements). Nevertheless, the enzyme is not without effects; it promotes CO_2 solubilisation (Reising and Schreiber 1994; a new method for measuring this process has appeared, Oja and Laisk 1995), and it may have an important role in stomatal operation (Sharma et al. 1995).

b) C_4 Plants

C_4 plants avoid "low-CO_2 stress" by employing mesophyll cells to use a biochemical pump that soaks up the sparse supply of atmospheric CO_2 and delivers it to thick-walled bundle-sheath cells such that the concentration in the vicinity of RuBP carboxylase is sufficient to inhibit photorespiration (Dai et al. 1995) and allow this enzyme to operate closer to its V_{max}. All this is achieved at a cost: additional light energy is required to operate the pump, so that the quantum yield (molecules CO_2 fixed per photon absorbed) cannot match the higher values of ca. 0.10 measured with C_3 plants – not even after the external CO_2 level is increased 30-fold

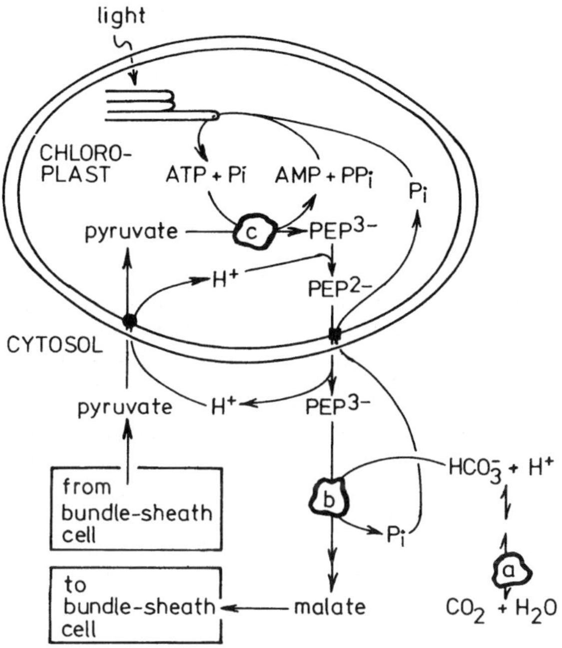

Fig. 3. Events in the leaf mesophyll cell of the C_4 plant *Zea mays*. Enzymes: *a* carbonic anhydrase; *b* PEP carboxylase; *c* pyruvate Pi dikinase. Details of transport processes on the chloroplast envelope are from Aoki and Kanai (1995)

(Lai and Edwards 1995). The CO_2 entering mesophyll cells must first be converted to HCO_3^- with the help of carbonic anhydrase (Sect. 3.a) because HCO_3^- is the substrate for the Ci-collecting enzyme phosphoenolpyruvate (PEP) carboxylase (Fig. 3). Consequently, inhibition of the in vivo carbonic anhydrase activity of C_4-plant leaves severely disrupts photosynthesis (Badger and Pfanz 1995), in contrast to the lack of effect with C_3-plant leaves (see above).

PEP carboxylation generates oxaloacetate that is reduced to malate. This malate enters the bundle-sheath cell (via plasmodesmata) where it is decarboxylated, thereby supplying the Calvin cycle with high CO_2. Normally, only about 10% of this CO_2 leaks back into mesophyll cells (Hatch et al. 1995), but an abnormality such as infection with a parasitic weed will increase this leakage (Smith et al. 1995). The other product of decarboxylation is pyruvate; this acid is returned to mesophyll cells where it is cotransported with Na^+ or H^+ into chloroplasts (Aoki et al. 1994) and then converted back to PEP in a reaction catalysed by pyruvate Pi dikinase. Furbank and Taylor (1995) have highlighted the importance of this enzyme by showing that *Flaveria bidentis* genetically manipulated to contain less than 20% of the normal activity cannot grow in air. Finally, in normal plants the PEP exits the chloroplast and goes to the cytosol to pick up more HCO_3^-. When pyruvate is cotransported into chloroplasts with H^+, stromal acidification is avoided by returning the H^+ as part of the PEP molecule when the latter moves to the cytosol (Aoki and Kanai 1995; Fig. 3).

C_4 plants are more plentiful in warmer, sunny climates, but a few efficient species have been found in cool (Beale and Long 1995) or shady (Smith and Wu 1994) habitats. More confusing is the existence of unusual variants of the "classical" C_4 plant, including the accepted "C_3–C_4 intermediate" category of plants that display reduced photorespiration (especially when adapted to warmer climates; Teese 1995) by restricting the photorespiratory glycine decarboxylase to bundle-sheath cells (Devi et al. 1995). There are both C_3-like and C_4-like C_3–C_4 intermediate species (Bruhl and Perry 1995; Reyes Fernández and Baker 1995), and there is an amphibious sedge, *Eleocharis baldwinii*, that is a C_4 plant on land but is a CAM-like C_3–C_4 intermediate in the water (Uchino et al. 1995). Finally, a touch of C_4 flavour has been given to tobacco, a typical C_3 plant, by genetically engineering it to contain chloroplast NADP-linked malate dehydrogenase from sorghum, a C_4 plant (Gallardo et al. 1995).

A central feature of typical C_4 plants is the restriction of the Calvin cycle to bundle-sheath cells. This concept has been challenged by Castrillo and Whatley (1994), who concluded from a differential mechanical grinding technique that the mesophyll cells of six species of *dicotyledonous* C_4 plants contained RuBP carboxylase. We eagerly await confirmation of this report.

Some recent studies have addressed the properties of two enzymes of PEP metabolism. PEP carboxylase is activated when phosphorylated by a protein kinase; the kinase from a CAM plant was strikingly similar to its counterpart from a C_4 plant (Li and Chollet 1994). Studies with sorghum have emphasised that the activation following phosphorylation is mainly a consequence of a decreased affinity for the allosteric inhibitor malate and an increased affinity for the allosteric activator glucose-6-P, especially when measured at physiological pH (Echevarria et al. 1994; Duff et al. 1995). The phosphorylative activation has again been demonstrated for a C_3-plant PEP carboxylase (Duff and Chollet 1995), but its translation into altered affinities for allosteric regulators is more subdued (Gupta et al. 1994), except perhaps in stomata (Zhang et al. 1994a). In CAM plants, the activation process is somehow favoured by a lower ambient temperature (Carter et al. 1995). On the other hand, Drilias et al. (1994) point out that the C_4 enzyme in vitro displays a higher (and more realistic) optimum temperature when assayed in the presence of a compatible solute such as glycerol. Finally, an interesting observation is the sixfold increase in PEP carboxylase that occurred in cucumber roots deprived of Fe^{2+}; presumably it could assist an increase in H^+ extrusion that would acidify the rhizosphere and promote the reduction of Fe^{3+} to Fe^{2+} (Rabotti et al. 1995).

PEP carboxykinase, an enzyme that catalyses a reaction comparable (but not identical) to PEP carboxylase, may have been poorly studied in the past due to its particular susceptibility to proteolytic degradation (Walker et al. 1995). The cucumber-cotyledon enzyme appears to undergo a phosphorylation/dephosphorylation cycle similar to that of PEP carboxylase (Walker and Leegood 1995); whether this also applies to the PEP carboxykinase in those C_4 plants that use this enzyme for releasing CO_2 in bundle-sheath cells remains to be seen.

c) Photorespiration

When RuBP carboxylase mistakenly accepts O_2, rather than CO_2, as its gaseous substrate (and thereby acts as an oxygenase; Sect. 2a), a cycle of reactions is initiated that results in loss of carbon as CO_2 and wasteful utilisation of photosynthetically-generated ATP and NADPH as the chloroplast endeavours to regenerate the RuBP. This, in summary, is the well-established process of photorespiration that normally occurs in C_3 plants. Total regeneration of RuBP cannot occur if the CO_2 concentration drops below a critical level (about one-seventh of the atmospheric level) where the straightforward competition between CO_2 and O_2 dictates that their utilisation by the carboxylase/oxygenase is in a 1:2 ratio; this break-even concentration is the so-called CO_2 compensation point (see Fig. 2 in our 1978 review). Of course, a similar ratio could be obtained by appropriately increasing the O_2 concentration while keeping the CO_2 at its atmospheric level. Tolbert et al. (1995) have pointed out that this increased O_2 level should be viewed as the "oxygen compensation point", and experimentally determined it to be 27% (cf. current atmospheric level is 21%). Therefore, if C_3 plants are exposed to 27% O_2, the O_2 taken up by the oxygenase activity and the O_2 released by photosynthetic water-splitting are equal.

In normal air, the CO_2-to-O_2 utilisation ratio is closer to the reverse of that at the compensation points, i.e. 2:1. This allows C_3 plants to regenerate RuBP *and* achieve net CO_2 fixation for photosynthate production; but concomitant photorespiratory CO_2 loss can be substantial; Valentini et al. (1995) recorded a rate of 56% of net assimilation for oak-tree leaves. This loss can be looked upon as a stress imposed by low CO_2; the likelihood that photorespiration is an important means of dissipating excess photosynthetic energy (by using ATP and NADPH) is low (Brestic et al. 1995; see our last review).

Further details of the photorespiratory metabolic pathway between glycolate (generated by the oxygenase reaction) and glycerate have been made available. The CO_2 release is catalysed by a mitochondrial multiprotein glycine decarboxylase complex (Oliver 1994) that is activated by ATP (Zhang and Wiskich 1995) and, incidentally, is inhibited by a toxin from the pathogenic fungus of oats, *Cochliobolus victoriae* (Navarre and Wolpert 1995). In contrast to some other mitochondrial proteins, the complex increases in amount slowly during leaf development, paralleling the attainment of photorespiratory activity (Lennon et al. 1995). Serine generated by the complex is directed to peroxisomes where it is deaminated and then reduced to glycerate; the reductant is cytosolic NADH delivered to the peroxisome via an unusual membrane-independent malate shuttle (Reumann et al. 1994).

Alternatives to the standard photorespiratory pathway have been suggested, based on the presence of the enzymes glycine oxidase and serine dehydratase in leaves (Igamberdiev and Bykova 1993), and enzymes of the glyoxylate pathway (usually associated with the conversion of fat to sucrose by germinating oil seeds) in diatom peroxisomes (Winkler and Stabenau 1995). However, the extent to which diatoms normally metabolise glycolate might be questioned:

1. like other unicellular algae, they are capable of actively accumulating Ci (Colman and Rotatore 1995) to a level that presumably could overwhelm the oxygenase, and
2. even if this Ci-pump fails to develop, as might happen in the Zn^{2+}-poor ocean (see our last review), glycolate simply might be excreted rather than metabolised – the detection of up to 89 µg glycolate per litre of Mediterranean seawater (Leboulanger et al. 1994) is evidence for this.

d) Experiments with High-CO_2 Atmospheres

One way to investigate the extent of stress imposed on plants by today's below-saturating (for RuBP carboxylase; Fig. 4A) level of atmospheric CO_2 is to study plants grown under higher levels of CO_2. A large volume of such studies is now available in the literature, prompted no doubt by the realisation that human activities are resulting in a steady annual addition of 1.5 ppm to the atmospheric CO_2 level. All these studies cannot be reviewed here. Many add to and refine the conclusions we noted in out 1990 review: for C_3 plants, increased CO_2 (e.g., a doubling to ca. 600 ppm) generally increases photosynthesis, growth, and biomass production, but with the caveats that, as might be expected, the extent of increase can be limited by a low availability of other plant resources

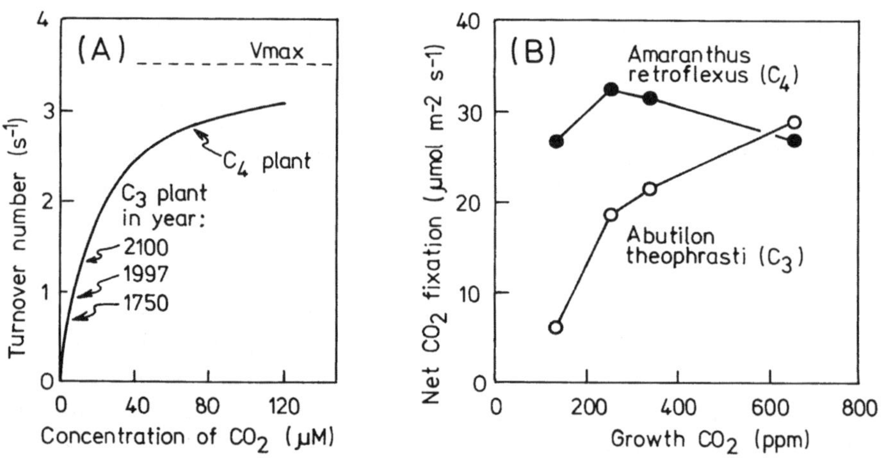

Fig. 4. A A calculated plot of RuBP carboxylase activity versus CO_2 concentration. The plot assumes Michaelis-Menten kinetics with $V_{max} = 3.5\ s^{-1}$ and $K_m\ (CO_2) = 18$ µM (von Caemmerer et al. 1994). Actual in vivo activities for C_3 plants under preindustrial (1750; 280 ppm CO_2), today's (1997; 360 ppm) and possible future (2100; 600 ppm) atmospheres are based on (i) a conversion factor of 12 µM = 360 ppm and (ii) the CO_2 concentration in the chloroplast stroma being 50% of that in the surrounding air (von Caemmerer and Evans 1991). The C_4-plant value is based upon a CO_2 concentration of 70 µM in bundle-sheath cells (Jenkins et al. 1989). **B** Photosynthesis by a C_3 plant and a C_4 plant grown and measured under CO_2 levels between one-half and twice the current atmospheric level of 360 ppm. (After Tissue et al. 1995)

(e.g., water, mineral nutrients), a low availability of plant sinks to accept photosynthate, or complex ecosystem interactions.

Plants with Ci pumps of one type or another are not expected to respond to increased CO_2 because they already develop it internally. Raven (1994) sees no reason to expect that marine phytoplankton (which have the added luxury of seawater HCO_3^- at 200-times the CO_2 level) should respond, but other reports are somewhat surprising: positive responses by the C_4 plants *Sorghum bicolor* (Amthor et al. 1994) and *Bouteloua gracilis* (Morgan et al. 1994) were recorded (although another C_4 plant, *Amaranthus retroflexus*, did not respond - Dippery et al. 1995; Fig. 4B), and a quite marked positive response by the CAM plant *Opuntia ficus-indica* was seen (Israel and Nobel 1994). In the latter case, stimulation was attributed to an enhanced collection of Ci during the early part of the night (see Sect. 4d).

For C_3 plants, further reports of high-CO_2-induced promotion of growth have appeared for crops (Amthor et al. 1994; Conroy et al. 1994; Thompson and Woodward 1994; Ziska and Bunce 1995), grasses (Jackson et al. 1994; Morgan et al. 1994; Greer et al. 1995), and trees (Gunderson and Wullschleger 1994; Idso and Kimball 1994). The extra growth of crops might benefit agricultural production, but for wheat this benefit might be at the expense of a reduction in grain quality (Conroy et al. 1994; Thompson and Woodward 1994). Two studies contrast with the rest, in that the responses of palnts grown under historically *low* CO_2 levels were examined. At 150 ppm CO_2 (less than half today's level), photosynthesis by the C_3 plants *Abutilon theophrasti* (Fig. 4B), oats, and wild mustard were reduced by a massive 70% (Polley et al. 1992; Tissue et al. 1995), but at the preindustrial level of ca. 270 ppm no noticeable difference to *A. theophrasti* grown under "today's" 350 ppm could be seen (Tissue et al. 1995).

The long-term responses of plants to high CO_2 are confusing. A number of studies (e.g. Idso and Kimball 1994) have recorded consistent long-term (years) higher rates of photosynthesis and growth under high CO_2, but in other studies initially high rates gradually decreased during a period of so-called acclimation (Reining 1994), presumably because of insufficient sinks for photosynthate. However, an acclimation-induced decrease can be sufficiently mild that plants grown under high CO_2 may still photosynthesize substantially faster than normal-CO_2-grown plants (Jacob et al. 1995). One case of a subsequent return to high rates has been noted by Ziska et al. (1995), and attributed to the development of additional sink capacity.

A symptom of insufficient sinks is the accumulation of leaf starch, and this symptom was patently obvious in a unique central Italian community of plants that always has been exposed to a naturally developed high-CO_2 atmosphere due to a chance combination of geological and geographical events (Körner and Miglietta 1994); the appearance and growth rates of these plants were otherwise similar to those of a nearby comparable plant community under a normal atmosphere, indicating that they were considerably acclimated to the high CO_2.

Physiological and biochemical experiments suggest that the term "insufficient sinks" should include, if present, any limited capacity to export GAP from chloroplasts, convert it to a translocatable soluble sugar such as sucrose, and export this sugar to sinks (Greiner de Mothes and Knoppik 1994). Both sucrose synthesis and phloem loading appear to be important: under 650 ppm CO_2 tomato plants genetically engineered to contain additional sucrose-P synthase grew better than control plants (Micallet et al. 1995), while a range of plant species with different (apoplastic or symplastic) types of phloem loading displayed the same absolute increase in soluble sugars and starch in leaves when grown under high CO_2 (Körner et al. 1995). It appears that, eventually, the accumulated soluble sugars may exert a feedback control on photosynthesis by inhibiting the transcription of genes that code far proteins central to photosynthesis (Sect. 2.f). This was found to be so for several tomato proteins, including chlorophyll-binding proteins, rubisco activase, and the small subunit of RuBP carboxylase (Van Oosten and Besford 1994; Van Oosten et al. 1994). However, it was not so for either subunit of RuBP carboxylase, of for several other Calvin cycle enzymes in wheat (Nie et al. 1995), but wheat is unusual in that its RuBP carboxylase content remains steady or even increases when plants are grown under high CO_2 (Makino 1994; Habash et al. 1995), in contrast to other species in which declines occur (Israel and Nobel 1994; Xu et al. 1994; Jacob et al. 1995).

4. Other Stresses, and Stress-Relief Mechanisms

a) Ascorbate, Zeaxanthin and Photoinhibition

We apologise for presenting a figure (Fig. 5) that is superficially similar to those dealing with ascorbate in two of our last three reviews; but we hasten to draw attention to a fundamental point of departure: we view the two schemes in Fig. 5 as representations of two relatively distinct plant physiological events, and it is noteworthy that neither scheme involves both of the enzymes ascorbate peroxidase and glutathione reductase. These two enzymes, together with superoxide dismutase and dehydroascorbate reductase, constitute the ascorbate-glutathione cycle (Fig. 2 in our last review) believed to be important for preventing the accumulation of toxic oxygen species (Foyer et al. 1994) and for dissipating excess photosynthetically generated reducing power (see Vácha 1995); but its role, and indeed that of photorespiration, in the latter dissipation process were questioned on the basis of these pathways' capacities

(Heber and Walker 1992). In particular, the relatively low activity of glutathione reductase (e.g., Mishra et al. 1995) was a concern, and the fruitless search by many investigators for *marked* increases in one or more of the enzyme activities following the imposition of a stress (see review by Foyer et al. 1994) puts the pathway's role in doubt. An additional observation was that leaves of tobacco genetically engineered to display a three-fold increase in superoxide dismutase activity also (via an unknown mechanism) showed a similar boost in ascorbate peroxidase activity, but neither of the activities of glutathione reductase or dehydroascorbate reductase showed any change (Gupta et al. 1993).

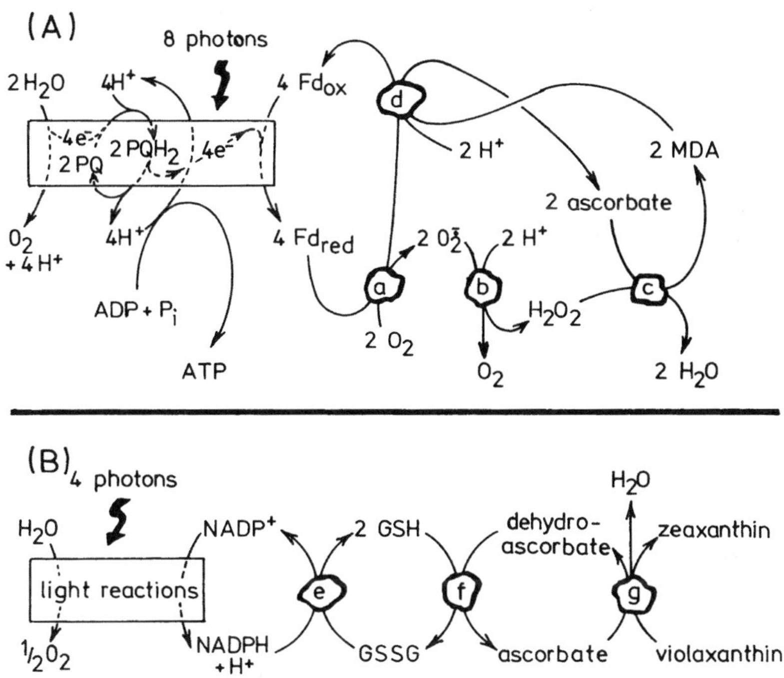

Fig. 5A, B. Two roles of ascorbate in plants. **A** Photosynthetic phosphorylation and detoxification of active oxygen. In the literature, this sequence of reactions is given the clumsy (and not altogether accurate) term "pseudocylic electron transport". It serves two purposes: (1) the processing of the products of oxygen reduction, and (2) the generation of ATP when the 4 H^+ moved into the thylakoid lumen by plastoquinone subsequently exit via the chloroplast coupling factor (Forti and Elli 1995). Note that this system, which is a representation of reactions 1, 2a, 3a and 4a in Forti and Elli (1995), is totally balanced with respect to the production and consumption of H_2O, O_2, H^+ and electrons. Enzymes: *a* ferredoxin-NADP$^+$ oxidoreductase (Goetze and Carpentier 1994); *b* superoxide dismutase; *c* ascorbate peroxidase; *d* monodehydroascorbate reductase (Miyake and Asada 1994). Abbreviations: *Fd* ferredoxin (oxidised or reduced); *MDA* monodehydroascorbate; *PQ* plastoquinone. **B** Generation of the photoprotective carotenoid zeaxanthin. Enzymes: *e* glutathione reductase; *f* dehydroascorbate reductase; *g* violaxanthin de-epoxidase. Abbreviations: *GSH* and *GSSG* glutathione (reduced and oxidised forms). (After Hager and Holocher 1994)

In the meantime, the fact that *mono*dehydroascorbate is the product of ascorbate peroxidase in the chloroplast has been reemphasised, and its very efficient reduction back to ascorbate by photosynthetically reduced ferredoxin on the chloroplast thylakoid has been confirmed (Miyake and Asada 1994). Thus, an abbreviated, glutathione-reductase-independent sequence of reactions can be drawn (Fig. 5A) that is identical to a sequence listed by Forti and Elli (1995) and shown by them to be capable of supplementing the supply of ATP for the Calvin cycle. This sequence has no net O_2 exchange, a fact supported by the observation that O_2 reduction is difficult to detect in non-stressed leaves (Ghashghaie and Cornic 1994). Overall, we suspect that the plant uses this sequence of reactions as its first line of attack against the superoxide anion, whether produced in low amounts during the course of "pseudocyclic" photophosphorylation, or in higher amounts when stressed (see below).

If Fig. 5A is accepted, then two questions arise:

1. does the ascorbate-glutathione cycle have any significant role in vivo, and
2. if not, does glutathione reductase have any role in preventing photoinhibition?

Possible answers to both questions are available. Firstly, it may be that the ascorbate-glutathione cycle serves a minor, backup role by scavenging any dehydroascorbate that is formed by disproportionation of any small amount of monodehydroascorbate (i.e. 2 monodehydroascorbate → ascorbate + dehydroascorbate) that may escape photoreduction on thylakoids. Correlations between glutathione reductase contents markedly altered by genetic engineering and suspectibility to oxidative stress (Aono et al. 1995; Foyer et al. 1995) support this suggestion.

Secondly, an alternative and probably more significant role for glutathione reductase is its participation in the three-reaction sequence by which photosynthetically generated NADPH is used to power the deepoxidation of violaxanthin to zeaxanthin in chloroplasts (Hager and Holocher 1994; Fig. 5B). This sequence may be a far more important mechanism than that in Fig. 5A for avoiding photoinhibition (Demmig-Adams and Adams 1992), in that it prevents reducing power from being generated in the first place. Any stess that reduces CO_2 fixation will reduce the demand for ATP, thus ADP + Pi will become unavailable to the thylakoid coupling factor which, in turn, will cease to operate, trapping plastoquinone-transported H^+ in the thylakoid lumen. The lower lumenal pH activates violaxanthin de-epoxidase by promoting its access to violaxanthin (Hagar and Holocher 1994) and increasing the proportion of ascorbate in the acid form, which appears to be this enzyme's true second substrate (Bratt et al. 1995). The generated zeaxanthin then initiates an as yet poorly defined sequence of photosystem-II membrane alterations and/or pigment interactions that result, ultimately, in excess harvested light energy being harmlessly converted to heat (Gilmore and Björkman 1994; Pfündel and Bilger 1994; Thiele and Krause 1994). It is noteworthy that this system does not require a continuous net flux of violaxanthin to zeaxanthin, and therefore might be less handicapped than the ascorbate-glutathione cycle by relatively low in vivo activities of glutathione reductase.

The alterations to photosystem II that promote this conversion of light to heat can persist for some hours after the reinstatement of conditions favourable to photosynthesis. Absorbed light energy is then partitioned between heat generation and CO_2 fixation, so that a low quantum yield (CO_2 fixed per photon absorbed) is detected. In this situation the photosynthetic tissue is said to be "photoinhibited", but this is a strong term to use because the low photosynthesis is merely a manifestation of a properly organised protective device at the molecular level, and not an indication of damage (Adams and Demmig-Adams 1995; Lovelock et al. 1995; see Thiele and Krause 1994). Admittedly, it can lead to reduced agricultural production (Laing et al. 1995). However, when the capacity to fix CO_2 is severely reduced by stresses, such as low-temperature-induced slow-down of the Calvin cycle enzymes or low-water-induced stomatal closure, the photoprotective devices in the thylakoid become overloaded. Real photoinhibition then develops as a surge of reduced oxygen species, apparently exceeding the capacity of the system in Fig. 5A, begins to oxidatively damage chloroplast components (Allen 1995; Wise 1995), eventually leading to plant death. Except, perhaps, in algae: the marine green alga *Ulva rigida* simply excretes H_2O_2 (Collén et al. 1995), while the marine microalga *Chattonella antiqua* releases superoxide anions into seawater, leading to the death of fish (Tanaka et al. 1994).

b) The Stress of Low Temperature

The simplified picture of low temperature slowing down the Calvin cycle and thereby initiating a chain of events leading the oxidative damage (because the light reactions are less affected by low temperature) needs to be viewed in the context of the wide range of temperature regimes to which different plant species are adapted. For example, zeaxanthin-induced energy dissipation was able to protect the Antarctic moss *Grimmia antarctici* (Lovelock et al. 1995) and two evergreen plants (Adams and Demmig-Adams 1995) from "real" photoinhibition at low temperature, but it could not protect pumpkin (which belongs to the group of so-called cold-sensitive plants), at least partly because, in pumpkin leaves, chilling retards the de-epoxidation of violaxanthin to zeaxanthin (Brüggemann and Koroleva 1995); but other enzymes may also be involved. Enlightening comparisons have been made between two species of tomato, the chilling-sensitive commercial *Lycopersicon esculentum* and the chilling-resistant *L. peruvianum.* Only in the latter were RuBP carboxylase and the FBPases stable after the plants were exposed to 10 °C (Brüggemann et al. 1994). In the former, a loss of one-half of the RuBP carboxylase activity coincided with the oxidation of enzyme sulfhydryl groups (Brüggemann 1995), and it is not inconceivable that this was somehow a consequence of damaged thylakoid membranes being a liability to the action of rubisco activase (Byrd et al. 1995).

Many studies have addressed the adaptation of plants to lower temperatures, and correlated success with increases in enzyme activities or stabilities. When rice, maize and red spruce were near their lower temperature limits, this applied to one or more of the oxygen-scavenging enzymes shown in Fig. 5 (Hausladen and Alscher 1994; Prasad et al. 1994; Saruyama and Tanida 1995; Massacci et al. 1995). In several agricultural plants, evidence for increased leaf contents or greater stabilities of RuBP carboxylase, FBPase (both chloroplast and cytosolic forms), sucrose-P synthase, glucose-6-P dehydrogenase, and PEP carboxykinase has been obtained (Brüggemann et al. 1994; Bredemeijer and Esselink 1995; Hurry et al. 1995b; Sáez-Vásquez et al. 1995; Gray et al. 1996). It is tempting to invoke some of these activity increases to explain a fundamental observation made by two groups and potentially of wide interest, viz. the photosynthesis rates (measured at 20 °C) by potato and rye leaves from plants grown at low temperatures were at least double those for leaves from plants grown at higher temperatures (Hurry et al. 1995a; Steffen et al. 1995; Table 1). Clearly, the cold-treated plants were not yet at temperatures low enough to cause stress. A similar result has been reported for the green alga *Chlorella vulgaris* (Maxwell et al. 1995).

Table 1. The greater photosynthetic rates of potato leaves, rye leaves and *Chlorella vulgaris* cells grown at low temperature

Species	Daytime growth temperature	Photosynthetic rate[a]
	(°C)	
Potato	12	118
	28	40
Rye	5	37
	24	18
C. vulgaris	5	770
	27	270

[a] Photosynthetic rates are in μmol mg^{-1} chl h^{-1} (potato and *C. vulgaris*) or μmol m^{-2} s^{-1} (rye), and were measured at 20 °C (the plants) or 27 °C (the alga). For each species, the light intensity during growth was the same at both temperatures. Data estimated from Maxwell et al. (1995), Hurry et al. (1995a) and Steffen et al. (1995).

An emerging concept is that, in cold-tolerant plants, the avoidance of photoinhibition is mediated by changes made, not in response to low temperature per se, but rather in response to the excitation pressure on photosystem II; identical changes occur when this pressure is increased by raising the light intensity rather than lowering the temperature. A fascinating (although not central) point is that different species adopt different changes: *Chlorella vulgaris* decreases its light-harvesting capacity and increases its capacity for zeaxanthin-based dissipation of energy as heat, but winter cereal grasses appear to concentrate on increasing their photosynthetic capacity, in particular by increasing their content of sucrose-P synthase (Maxwell et al. 1995; Gray et al. 1996). Potato appeared to adopt an intermediate approach (Steffen et al. 1995).

Finally, two points concerning *high* temperature:

1. the above-mentioned experiments with potato, rye and *Chlorella* (Table 1) indicate that, all other factors being optimal, relatively high (but not necessarily stressful) growth temperatures can decrease photosynthetic performance, and
2. one adaptive response of tree leaves to high temperature is to produce isoprene, which then constitutes 2% of the photosynthate produced by these leaves (Sharkey and Singsaas 1995).

c) The Stress of Low Water Availability

It is still unclear whether stomatal closure is the exclusive reason for lowered photosynthesis in drought-affected plants, or whether a disturbance to the metabolism of photosynthetic cells is also involved (see our 1992 review). Two recent studies point to stomatal closure as being the sole factor involved in French beans and white clover (Brestic et al. 1995; Grieu et al. 1995). In the former case, the avoidance of photoinhibition was attributed to effective light-to-heat conversion in the thylakoid; no significant role for photorespiration could be envisaged. However, the potential for photoinhibition following the diversion of reducing power to oxygen and consequent formation of reactive oxygen species such as H_2O_2 still exists (Allen 1995), but Moran et al. (1994) state that H_2O_2 did not accumulate in drought-stressed pea leaves because ascorbate peroxidase and monodehydroascorbate reductase removed it (cf. Fig. 5A). Four groups have examined changes to the activities of these two enzymes and other enzymes thought to be involved in metabolizing reactive oxygen species, in pea and wheat leaves before and after water stress (Zhang and Kirkham 1994; Baisak et al. 1994; Moran et al. 1994; Hernández et al. 1995). However, apart from a fairly consistent increase in superoxide dismutase, no clear picture emerged; some activities increased and then decreased, while glutathione reductase activity increased in wheat leaves but decreased in pea leaves.

Finally, it is fascinating to note that leaves treated with excess water also suffer stress: Ishibashi and Terashima (1995) report that bean leaves exposed to an artificial misty rain partially closed their stomates, resulting in reduced photosynthesis and some chronic damage to the photosynthetic apparatus.

d) Crassulacean Acid Metabolism (CAM)

CAM is one of the most dramatic adaptations of plants to stress. Faced with a low water potential due to either low water supply (e.g., *Kalanchoë daigremontiana*) or salinity (e.g., *Mesembryanthemum crystallinum*), CAM plants have totally reversed stomatal operation, opening stomates by night so that CO_2 may be collected by PEP carboxylase and then stored (as one of the carboxyl groups of malate) in vacuoles, and then closing stomates by day to block transpiration; however, photosynthesis can proceed behind these closed stomates because an ample supply of CO_2 can be released by decarboxylating the stored malate. The enigmatic reversal of stomatal operation may be explained, at least in part, by the inactivation of guard-cell photoreceptors (Mawson and Zaugg 1994).

The PEP carboxylase in CAM plants is cyclically activated/inhibited by phosphorylation/dephosphorylation (see Sect. 3.b), as recently demonstrated with the *M. crystallinum* enzyme (Weigand 1994). This regulation contributed to the circadian rhythm of CAM in *Graptopetalum paraguayense* (Kusumi et. al. 1994), but in other CAM plants additional enzyme properties, including alterations to the affinity for PEP (Conti and Smirnoff 1994) and abscisic acid-induced enzyme synthesis (Taybi et al. 1995), were found to be important.

Cytosolic PEP carboxylation leads to the formation of malate which is taken up by vacuoles, probably via a "carboxylate uniporter" protein that has a malate affinity and transport capacity sufficient to accommodate observed CAM activities (Ratajczak et al. 1994). These authors concluded that an electrical membrane potential could provide the driving force for malate uptake. The potential might, in turn, be developed by the combined action of two or three other tonoplast proteins: a Na^+/H^+ antiporter that can produce high concentrations of vacuolar Na^+ (at least in *M. crystallinum*), and an H^+-ATPase and/or H^+-pyrophosphatase (Barkla et al. 1995; Becker et al. 1995). Electron microscopy has permitted the visualisation of both of these last two proteins as "particles" on tonoplasts of *K. daigremontiana* and *M. crystallinum* (Mariaux et al. 1994; Rockel et al. 1994).

By day, malate is gradually returned to the cytosol where it is decarboxylated, either by an NADP-linked malic enzyme that shows a 12-fold quantitative increase when CAM is induced in *M. crystallinum* (Saitou et al. 1994), or by a mitochondrial NAD-linked malic enzyme that is more active diurnally than nocturnally (Cooke et al. 1995). The other product of the decarboxylation, pyruvate, is converted to PEP by a pyruvate Pi dikinase, whose biosynthesis in *M. crystallinum*, unlike that of PEP carboxylase, is only mildly responsive to whether or not the plant is in a C_3 mode (i.e. well-watered) or a CAM mode (i.e. poorly watered or salt-stressed; Fißlthaler et al. 1995). Finally, the PEP is converted to starch, which is degraded the following night via glycolysis to generate the next batch of PEP. The last enzyme in this sequence, enolase, experiences a fourfold boost in activity when *M. crystallinum* is salt stressed (Forsthoefel et al. 1995); it could be the next enzyme for which regulation by covalent modification (e.g., phosphorylation/dephosphorylation) is reported.

One last instance of CAM is odd, in that it occurs in the Australian *aquatic* macrophyte *Crassula helmsii*. In this case, CAM is not a response to water stress (obviously!), but possibly a response to low-CO_2 stress: nocturnal uptake and storage of CO_2 would extend the total period of CO_2 uptake, undoubtedly a benefit in an aquatic environment, where the rate of CO_2 diffusion is less than 0.01% of the rate in air (Newman and Raven 1995).

e) The Stress of Ultraviolet Radiation

Plants cannot avoid exposure to ultraviolet (UV) radiation, and thus have evolved protective and repair mechanisms to deal with it. Protection of terrestrial photosynthetic cells is afforded by a natural sunscreen of flavonoid compounds in leaf epidermal cells (Caldwell 1981). The vital importance of these flavonoids, and several other closely related phenolic compounds, as UV filters has been highlighted by the increased UV sensitivity of plants with reduced levels of these compounds (Reuber et al. 1993; Stapleton and Walbot 1994), and the drastic UV-induced damage done to *Arabidopsis thaliana* mutants deficient in their synthesis (Li et al. 1993; Lois and Buchanan 1994; Landry et al. 1995).

Two targets of UV-radiation damage are DNA and photosynthesis (Caldwell 1981). Damage and subsequent repair of DNA have been reviewed by Britt (1995). Here we briefly survey reported damage to photosynthesis. Several UV-induced effects have been noted, including marked alterations to chloroplast morphology, damage to photosystem II (including degradation of the D1 protein), and damage to the chloroplast coupling factor (He et al. 1994; Zhang et al. 1994b; Friso et al. 1995; Nogués and Baker 1995). In addition, there are further reports (cf. our 1992 review) that the activity of RuBP carboxylase is reduced (Rao et al. 1995), at least partly due to a photomodification of the enzyme's large subunit (Wilson et al. 1995). Finally, an indirect effect is that plants damaged by UV radiation suffer difficulties in moving photosynthate to sinks, or sinks (e.g., developing leaves) are damaged; in either case, photosynthate accumulates as chloroplast starch (Britz and Adamse 1994; He et al. 1994).

Global interest is paid to the possibility that plant productivity might be adversely affected by the increased amount of UV light that plants are now exposed to following the depletion of stratospheric ozone. The likelihood of this possibility deserves to be monitored, but conclusions must take into account the extent to which experiments, such as those referred to above, involved UV treatments in excess of projected future levels of UV radiation; hopefully, future levels will not be much greater than current levels now that action has been taken to curb further loss of stratospheric ozone. In a recent review, Fiscus and Booker (1995) are optimistic that actual UV stress to terrestrial plants will be minimal, and even phytoplankton in Antarctic waters, above which substantial ozone depletion occurs each austral spring, are not as stressed as popularly believed: annual primary productivity by these microalgae does not seem to be reduced by more than 1% (Holm-Hansen et al. 1993; Arrigo 1994).

f) The Stress of Pathogens

Plant pathology is largely beyond the scope of this chapter; but some interesting connections to photosynthesis deserve mention. One of these is the proposed involvement of reactive species of oxygen (e.g., superoxide anions and H_2O_2) in the defense mechanisms used by plants against pathogens; these same species of oxygen can be produced during other stress responses (Sect. 4.a–c). Their involvement in combating pathogens at infection sites is part of the so-called hypersensitive response (Mehdy 1994), which operates better in young plant tissue (Barna et al. 1993), and was improved in a potato plant genetically engineered to contain the H_2O_2-generating fungal enzyme glucose oxidase (Wu et al. 1995).

A more recent observation is that plant tissues away from the site of pathogen attack develop enhanced resistance to pathogens, presumably because one or more messenger molecules, produced at the site of attack, move through the phloem to other parts of the plant where changes that strengthen resistance are initiated, including the synthesis of several new proteins called pathogenesis-related proteins. These events are collectively termed "systemic acquired resistance" (Ryals et al. 1994). There is some evidence that one messenger molecule is the small phenolic compound salicylic acid, and an intriguing observation is that salicylic acid binds to and inhibits the enzyme catalase in a variety of plants (Sánchez-Casas and Klessig 1994). This might be expected to cause H_2O_2 (and other reactive oxygen species) to accumulate, and it has been proposed that this H_2O_2 might be the second messenger molecule that initiates the transcription of genes for pathogenesis-related proteins (Conrath et al. 1995; see Ryals et al. 1994). However, the evidence for this is not clear-cut (Neuenschwander et al. 1995), and we have to confess to some degree of awe when contemplating that a relatively dangerous and structurally simple molecule like H_2O_2 should be endowed with a second-messenger role. In addition, one wonders how H_2O_2 could effectively carry out such a role while simultaneously performing another duty, viz. assisting the generation of ATP for photosynthetic carbon fixation (Forti and Elli 1995; Fig. 5A). Finally, the suspected role of catalase is not without problems, given the notoriously low affinity of this enzyme for H_2O_2, and the report that other factors (e.g., light and temperature) cause fluctuations in its in vivo activity (Volk and Feierabend 1989).

The functions of only some of the pathogenesis-related proteins have been discovered. One recent surprise was that, in celery, one of them is a mannitol dehydrogenase capable of initiating the metabolism of mannitol, which is a major photosynthate in this plant (Williamson et al. 1995). In the future, celery (*Apium graveolens*) may provide real links between photosynthetic carbon metabolism and the mechanisms used by plants to fight off foreign organisms.

Acknowledgment. We are most grateful to Diagatha for excellent and cheerful assistance.

References

Adams WW, Demmig-Adams B (1995) Plant Cell Environ 18:117–127. **Allen RD** (1995) Plant Physiol 107:1049–1054. **Amor Y,** Haigler CH, Johnson S, Wainscott M, Delmer DP (1995) Proc Natl Acad Sci USA 92:9353–9357. **Amthor JS,** Mitchell RJ, Runion GB, Rogers HH, Prior SA, Wood CW (1994) New Phytol 128:443–450. **Anderson LE,** Goldhaber-Gordon IM, Li D, Tang X, Xiang M, Prakash N (1995) Planta 196:245–255. **Andralojc PJ,** Dawson GW, Parry MAJ, Keys AJ (1994) Biochem J 304:781–786. **Anwaruzzaman,** Sawada S, Usuda H, Yokota A (1995) Plant Cell Physiol 36:425–433. **Aoki N,** Kanai R (1995) Plant Cell Physiol 36:187–189. **Aoki N,** Ohnishi J, Kanai R (1994) Plant Cell Physiol 35:801–806. **Aono M,** Saji H, Fujiyama K, Sugita M, Kondo N, Tanaka K (1995) Plant Physiol 107:645–648. **Arrigo KR** (1994) Mar Ecol Prog Ser 114:1–12. **Averill RH,** ap Rees T (1995) Planta 196:344–349. **Axelsson L,** Ryberg H, Beer S (1995) Plant Cell Environ 18:439–445.

Baalmann E, Backhausen JE, Kitzmann C, Scheibe R (1994) Bot Acta 107:313–320. **Badger MR,** Pfanz H (1995) Aust J Plant Physiol 22:45–49. **Badger MR,** Price GD (1994) Annu Rev Plant Physiol Plant Mol Biol 45:369–392. **Baisak R,** Rana D, Acharya PBB, Kar M (1994) Plant Cell Physiol 35:489–495. **Ball K,** Preiss J (1994) J Biol Chem 269:24706–24711. **Ballicora MA,** Wolosiuk RA (1994) Eur J Biochem 222:467–474. **Ballicora MA,** Laughlin MJ, Fu Y, Okita TW, Barry GF, Preiss J (1995) Plant Physiol 109:245–251. **Barkla BJ,** Zingarelli L, Blumwald E, Smith JAC (1995) Plant Physiol 109:549–556. **Barna B,** Ádám AL, Király Z (1993) Naturwissenschaften 80:420–422. **Barnes SA,** Knight JS, Gray JC (1994) Plant Physiol 106:1123–1129. **Batz O,** Scheibe R, Neuhaus HE (1995) Planta 196:50–57. **Beale CV,** Long SP (1995) Plant Cell Environ 18:641–650. **Becker A,** Canut H, Lüttge U, Maeshima M, Marigo G, Ratajczak R (1995) J Plant Physiol 146:88–94. **Boldt R,** Pelzer-Reith B, Börner T, Schnarrenberger C (1994) J Plant Physiol 144:282–286. **Borodin V,** Gardeström P, Samuelsson G (1994) Physiol Plant 92:254–260. **Bouché-Pillon S,** Fleurat-Lessard P, Fromont J-C, Serrano R, Bonnemain J-L (1994) Plant Physiol 105:691–697. **Bouma TJ,** De Visser R, Van Leeuwen PH, De Kock MJ, Lambers H (1995) J Exp Bot 46:1185–1194. **Bratt CE,** Arvidsson P-O, Carlsson M, Åkerlund H-E (1995) Photosynth Res 45:169–175. **Bredemeijer GMM,** Esselink G (1995) J Plant Physiol 145:565–569. **Brestic M,** Cornic G, Fryer MJ, Baker NR (1995) Planta 196:450–457. **Britt AB** (1995) Plant Physiol 108:891–896. **Britz SJ,** Adamse P (1994) Photochem Photobiol 60:116–119. **Brüggemann W** (1995) Plant Cell Physiol 36:733–736. **Brüggemann W,** Koroleva OY (1995) Plant Physiol Biochem 33:251–259. **Brüggemann W,** Klaucke S, Maas-Kantel K (1994) Planta 194:160–168. **Bruhl JJ,** Perry S (1995) Aust J Plant Physiol 22:521–530. **Büchel C,** Garab G (1995) Planta 197:69–75. **Byrd GT,** Ort DR, Ogren WL (1995) Plant Physiol 107:585–591.

Caldwell MM (1981) In: Lange OL, Nobel PS, Osmond CB, Ziegler H (eds) Encyclopedia of Plant Physiology 12A. Springer, Berlin Heidelberg New York, pp 169–197. **Campbell WJ,** Ogren WL (1995) Plant Cell Physiol 36:215–220. **Carter PJ,** Wilkins MB, Nimmo HG, Fewson CA (1995) Planta 196:375–380. **Castrillo M,** Whatley FR (1994) Photosynthetica 30:269–277. **Clasper S,** Chelvarajan REL, Easterby JS, Powls R (1994) Biochim Biophys Acta 1209:101–106. **Collén J,** Del Rio MJ, García-Reina G, Pedersén M (1995) Planta 196:225–230. **Colman B,** Rotatore C (1995) Plant Cell Environ 18:919–924. **Conrath U,** Chen Z, Ricigliano JR, Klessig DF (1995) Proc Natl Acad Sci USA 92:7143–7147. **Conroy JP,** Seneweera S, Basra AS, Rogers G, Nissen-Wooller B (1994) Aust J Plant Physiol 21:741–758. **Conti S,** Smirnoff N (1994) J Exp Bot 45:1613–1621. **Cook RM,** Lindsay JG, Wilkins MB, Nimmo HG (1995) Plant Physiol 109:1301–1307. **Crafts-Brandner SJ,** Salvucci ME (1994) Planta 194:110–116. **Cuello J,** Quiles MJ, Albacete ME, Sabater B (1995) Plant Cell Physiol 36:265–271.

Dai Z, Ku MSB, Edwards GE (1995) Plant Physiol 107:815–825. **Delgado E,** Medrano H, Keys AJ, Parry MAJ (1995) J Exp Bot 46:1775–1777. **Demmig-Adams B,** Adams WW (1992) Annu Rev Plant Physiol Plant Mol Biol 43:599–626. **Devi MT,** Rajagopalan AV, Raghavendra AS (1995) Plant Cell Environ 18:589–594. **Dippery JK,** Tissue DT, Thomas RB, Strain BR (1995) Oecologia 101:13–20. **Drilias P,** Gousias H, Manetas Y, Gavalas NA (1994) Photosynthetica 30:225–232. **Duchateau N,** Bortlik K, Simmen U, Wiemken A, Bancal P (1995) Plant Physiol 107:1249–1255. **Duff SMG,** Chollet R (1995) Plant Physiol 107:775–782. **Duff SMG,** Andreo CS, Pacquit V, Lepiniec L, Sarath G, Condon SA, Vidal J, Gadal P, Chollet R (1995) Eur J Biochem 228:92–95.

Echevarria C, Pacquit V, Bakrim N, Osuna L, Delgado B, Arrio-Dupont M, Vidal J (1994) Arch Biochem Biophys 315:425–430. **Eichelmann H,** Laisk A (1994) Plant Physiol 106:679–687. **Eriksson M,** Gardeström P, Samuelsson G (1995) Plant Physiol 107:479–483. **Ernst M,** Bufler G (1994) New Phytol 128:403–406. **Espie GS,** Kandasamy RA (1994) Plant Physiol 104:1419–1428.

Farr TJ, Huppe HC, Turpin DH (1994) Plant Physiol 105:1037–1042. **Faske M,** Holtgrefe S, Ocheretina O, Meister M, Backhausen JE, Scheibe R (1995) Biochim Biophys Acta 1247:135–142. **Fiscus EL,** Booker FL (1995) Photosynth Res 43:81–92. **Fißlthaler B,** Meyer G, Bohnert HJ, Schmitt JM (1995) Planta 196:492–500. **Forster RM,** Dring MJ (1992) Plant Cell Environ 15:241–247. **Forsthoefel NR,** Cushman MAF, Cushman JC (1995) Plant Physiol 108:1185–1195. **Forti G,** Elli G (1995) Plant Physiol 109:1207–1211. **Foyer CH,** Parry MAJ (1995) J Exp Bot 46:1055–1060. **Foyer CH,** Descourvières P, Kunert KJ (1994) Plant Cell Environ 17:507–523. **Foyer CH,** Souriau N, Perret S, Lelandais M, Kunert K-J, Pruvost C, Jouanin L (1995) Plant Physiol 109:1047–1057. **Friso G,** Vass I, Spetea C, Barber J, Barbato R (1995) Biochim Biophys Acta 1231:41–46. **Frommer WB,** Sonnewald U (1995) J Exp Bot 46:587–607. **Furbank RT,** Taylor WC (1995) Plant Cell 7:797–807.

Gallardo F, Miginiac-Maslow M, Sangwan RS, Decottignies P, Keryer E, Dubois F, Bismuth E, Galvez S, Sangwan-Norreel B, Gadal P, Crétin C (1995) Planta 197:324–332. **Geiger DR,** Servaites JC (1994) Plant Physiol Biochem 32:173–183. **Geiger DR,** Shieh W-J, Yu X-M (1995) Plant Physiol 107:507–514. **Getz HP,** Klein M (1995) Plant Physiol 107:459–467. **Ghashghaie J,** Cornic G (1994) J Plant Physiol 143:643–650. **Gilmore AM,** Björkman O (1994) Planta 192:537–544. **Gimmler H,** Slovik S (1995) Plant Physiol Biochem 33:655–664. **Goetze DC,** Carpentier R (1994) Can J Bot 72:256–260. **Gotor C,** Hong S, Spreitzer RJ (1994) Planta 193:313–319. **Gray GR,** Savitch LV, Ivanov AG, Huner NPA (1996) Plant Physiol 110:61–71. **Greer DH,** Laing WA, Campbell BD (1995) Aust J Plant Physiol 22:713–722. **Greiner de Mothes MA,** Knoppik D (1994) Photosynthetica 30:435–445. **Grieu P,** Robin C, Guckert A (1995) Plant Physiol Biochem 33:19–24. **Gruszecki WI,** Bader KP, Schmid GH (1994) Biochim Biophys Acta 1188:335–338. **Gunderson CA,** Wullschleger SD (1994) Photosynth Res 39:369–388. **Gupta AS,** Webb RP, Holaday AS, Allen RD (1993) Plant Physiol 103:1067–1073. **Gupta SK,** Ku MSB, Lin J-H, Zhang D, Edwards GE (1994) Photosynth Res 42:133–143. **Gutteridge S,** Gatenby AA (1995) Plant Cell 7:809–819.

Habash DZ, Paul MJ, Parry MAJ, Keys AJ, Lawlor DW (1995) Planta 197:482–489. **Hager A,** Holocher K (1994) Planta 192:581–589. **Hatch MD,** Agostino A, Jenkins CLD (1995) Plant Physiol 108:173–181. **Hausladen A,** Alscher RG (1994) Plant Physiol 105:205–213. **He J,** Huang L-K, Whitecross MI (1994) Plant Cell Environ 17:771–775. **Heber U,** Walker D (1992) Plant Physiol 100:1621–1626. **Heineke D,** Kruse A, Flügge U-I, Frommer WB, Riesmeier JW, Willmitzer L, Heldt HW (1994a) Planta 193:174–180. **Heineke D,** Wildenberger K, Sonnewald U, Willmitzer L, Heldt HW (1994b) Planta 194:29–33. **Hernández JA,** Olmos E, Corpas FJ, Sevilla F, del Río LA (1995) Plant Sci 105:151–167. **Hill LM,** Smith AM (1995) J Plant Physiol 146:411–417. **Hiltonen T,** Karlsson J, Palmqvist K, Clarke AK,

Samuelsson G (1995) Planta 195:345–351. **Holbrook GP,** Campbell WJ, Rowland-Bamford A, Bowes G (1994) J Exp Bot 45:1119–1126. **Holm-Hansen O,** Helbling EW, Lubin D (1993) Photochem Photobiol 58:567–570. **Hurry V,** Tobiaeson M, Krömer S, Gardeström P, Öquist G (1995a) Plant Cell Environ 18:69–76. **Hurry VM,** Strand Å, Tibiaeson M, Gardeström P, Öquist G (1995b) Plant Physiol 109:697–706. **Husic HD,** Marcus CA (1994) Plant Physiol 105:133–139.

Idso SB, Kimball BA (1994) J Exp Bot 45:1669–1672. **Igamberdiev AU,** Bykova NV (1993) Photosynthetica 29:499–506. **Iglesias AA,** Charng Y, Ball S, Preiss J (1994) Plant Physiol 104:1287–1294. **Ishibashi M,** Terashima I (1995) Plant Cell Environ 18:431–438. **Israel AA,** Nobel PS (1994) Photosynth Res 40:223–229.

Jackson RB, Sala OE, Field CB, Mooney HA (1994) Oecologia 98:257–262. **Jacob J,** Greitner C, Drake BG (1995) Plant Cell Environ 18:875–884. **Jang J-C,** Sheen J (1994) Plant Cell 6:1665–1679. **Jeannette E,** Prioul J-L (1994) Plant Cell Physiol 35:869–878. **Jenkins CLD,** Furbank RT, Hatch MD (1989) Plant Physiol 91:1372–1381.

Kampfenkel K, Möhlmann T, Batz O, Van Montagu M, Inzé D, Neuhaus HE (1995) FEBS Lett 374:351–355. **Karlsson J,** Hiltonen T, Husic HD, Ramazanov Z, Samuelsson G (1995) Plant Physiol 109:533–539. **Katzman GL,** Carlson SJ, Marcus Y, Moroney JV, Togasaki RK (1994) Plant Physiol 105:1197–1202. **Kent SS,** Tomany MJ (1995) Plant Physiol Biochem 33:71–80. **Koops AJ,** Jonker HH (1994) J Exp Bot 45:1623–1631. **Körner C,** Miglietta F (1994) Oecologia 99:343–351. **Körner C,** Pelaez-Riedl S, Van Bel AJE (1995) Plant Cell Environ 18:595–600. **Kosegarten H,** Mengel K (1994) Physiol Plant 91:111–120. **Koßmann J,** Sonnewald U, Willmitzer L (1994) Plant J 6:637–650. **Krapp A,** Stitt M (1994) Plant Cell Environ 17:861–866. **Krapp A,** Stitt M (1995) Planta 195:313–323. **Krapp A,** Chaves MM, David MM, Rodriques ML, Pereira JS, Stitt M (1994) Plant Cell Environ 17:945–953. **Krömer S** (1995) Annu Rev Plant Physiol Plant Mol Biol 46:45–70. **Kruger NJ** (1995) Phytochemistry 38:1065–1071. **Kusumi K,** Arata H, Iwasaki I, Nishimura M (1994) Plant Cell Physiol 35:233–242.

Laing WA, Greer DH, Schnell TA (1995) Aust J Plant Physiol 22:511–520. **Lal A,** Edwards GE (1995) Plant Cell Physiol 36:1311–1317. **Landry LG,** Chapple CCS, Last RL (1995) Plant Physiol 109:1159–1166. **Leboulanger C,** Descolas-Gros C, Jupin H (1994) J Plankton Res 16:897–903. **Leidreiter K,** Heineke D, Heldt HW, Müller-Röber B, Sonnewald U, Willmitzer L (1995) Plant Cell Physiol 36:615–624. **Lennon AM,** Pratt J, Leach G, Moore AL (1995) Plant Physiol 107:925–932. **Lerchl J,** Geigenberger P, Stitt M, Sonnewald U (1995) Plant Cell 7:259–270. **Li B,** Chollet R (1994) Arch Biochem Biophys 314:247–254. **Li J,** Ou-Lee T-M, Raba R, Amundson RG, Last RL (1993) Plant Cell 5:171–179. **Lippard SJ** (1995) Science 268:996–997. **Lohaus G,** Burba M, Heldt HW (1994) J Exp Bot 45:1097–1101. **Lois R,** Buchanan BB (1994) Planta 194:504–509. **Lovelock CE,** Osmond CB, Seppelt RD (1995) Plant Cell Environ 18:1395–1402. **Lunn JE,** Hatch MD (1995) Planta 197:385–391. **Lüscher M,** Nelson CJ (1995) Plant Physiol 107:1419–1425.

Makino A (1994) J Plant Res 107:79–84. **Mariaux J-B,** Becker A, Kemna I, Ratajczak R, Fischer-Schliebs E, Kramer D, Lüttge U, Marigo G (1994) Bot Acta 107:321–327. **Massacci A,** Iannelli MA, Pietrini F, Loreto F (1995) J Exp Bot 46:119–127. **Matsuda Y,** Colman B (1995) Plant Physiol 108:253–260. **Mawson BT,** Zaugg MW (1994) J Plant Physiol 144:740–746. **Maxwell DP,** Falk S, Huner NPA (1995) Plant Physiol 107:687–694. **McMichael RW,** Bachmann M, Huber SC (1995) Plant Physiol 108:1077–1082. **Mehdy MC** (1994) Plant Physiol 105:467–472. **Micallef BJ,** Haskins KA, Vanderveer PJ, Roh K-S, Shewmaker CK, Sharkey TD (1995) Planta 196:327–334. **Mishra NP,** Fatma T, Singhal GS (1995) Physiol Plant 95:77–82. **Misra JB,** Daniel EV, Premchand (1994) Plant Physiol

Biochem 32:131–136. **Miyake C,** Asada K (1994) Plant Cell Physiol 35:539–549. **Möhlmann T,** Batz O, Maaß U, Neuhaus H-E (1995) Biochem J 307:521–526. **Moore BD,** Seemann JR (1994) Plant Physiol 105:731–737. **Moore BD,** Kürats O, Charlet T, Seemann JR (1995) Plant Cell Physiol 36:1097–1103. **Moran JF,** Becana M, Iturbe-Ormaetxe I, Frechilla S, Klucas RV, Aparicio-Tejo P (1994) Planta 194:346–352. **Morgan JA,** Hunt HW, Monz CA, Lecain DR (1994) Plant Cell Environ 17:1023–1033. **Morse D,** Salois P, Markovic P, Hastings JW (1995) Science 268:1622–1624.

Nakano H, Makino A, Mae T (1995) Plant Cell Physiol 36:653–659. **Navarre DA,** Wolpert TJ (1995) Plant Cell 7:463–471. **Neuenschwander U,** Vernooij B, Friedrich L, Uknes S, Kessmann H, Ryals J (1995) Plant J 8:227–233. **Neuhaus H-E,** Henrichs G, Scheibe R (1995) Planta 195:496–504. **Newman JR,** Raven JA (1995) Oecologia 101:494–499. **Nie G,** Hendrix DL, Webber AN, Kimball BA, Long SP (1995) Plant Physiol 108:975–983. **Nielsen TH** (1995) Plant Physiol 108:69–73. **Nielsen TH,** Wischmann B (1995) Plant Physiol 109:1033–1038. **Nimer NA,** Guan Q, Merrett MJ (1994) New Phytol 126:601–607. **Nogués S,** Baker NR (1995) Plant Cell Environ 18:781–787.

Ocheretina O, Scheibe R (1994) FEBS Lett 355:254–258. **Oja V,** Laisk A (1995) J Exp Bot 45:401–406. **Oliver DJ** (1994) Annu Rev Plant Physiol Plant Mol Biol 45:323–337.

Pacold ME, Stevens FJ, Li D, Anderson LE (1995a) Photosynth Res 43:125–130. **Pacold ME,** Anderson LE, Li D, Stevens FJ (1995b) J Phycol 31:297–301. **Paul MJ,** Knight JS, Habash D, Parry MAJ, Lawlor DW, Barnes SA, Loynes A, Gray JC (1995a) Plant J 7:535–542. **Paul M,** Sonnewald U, Hajirezaei M, Dennis D, Stitt M (1995b) Planta 196:277–283. **Peltier G,** Cournac L, Despax V, Dimon B, Fina L, Genty B, Rumeau D (1995) Planta 196:732–739. **Pelzer-Reith B,** Wiegand S, Schnarrenberger C (1994) Plant Physiol 106:1137–1144. **Penson SP,** Cairns AJ (1994) New Phytol 128:395–402. **Pfündel E,** Bilger W (1994) Photosynth Res 42:89–109. **Podestá FE,** Moorhead GBG, Plaxton WC (1994) Arch Biochem Biophys 313:50–57. **Polley HW,** Johnson HB, Mayeux HS (1992) Int J Plant Sci 153:453–461. **Portis AR,** Lilley RMcC, Andrews TJ (1995) Plant Physiol 109:1441–1451. **Prasad TK,** Anderson MD, Martin BA, Stewart CR (1994) Plant Cell 6:65–74. **Price GD,** von Caemmerer S, Evans JR, Yu J-W, Lloyd J, Oja V, Kell P, Harrison K, Gallagher A, Badger MR (1994) Planta 193:331–340. **Price GD,** Evans JR, von Caemmerer S, Yu J-W, Badger MR (1995) Planta 195:369–378.

Queiroz-Claret C, Meunier JC (1995) J Plant Physiol 145:45–49. **Quick WP,** Scheibe R, Neuhaus HE (1995) Plant Physiol 109:113–121.

Rabotti G, De Nisi P, Zocchi G (1995) Plant Physiol 107:1195–1199. **Ramazanov Z,** Sosa PA, Henk MC, del Rio MJ, Gómez-Pinchetti JL, Reina GG (1995) Planta 195:519–524. **Rao IM,** Terry N (1995) Plant Physiol 107:1313–1321. **Rao MV,** Paliyath G, Ormrod DP (1995) Photochem Photobiol 62:727–735. **Ratajczak R,** Kemna I, Lüttge U (1994) Planta 195:226–236. **Raven JA** (1994) Photosynth Res 39:259–273. **Rawat M,** Moroney JV (1995) Plant Physiol 109:937–944. **Read BA,** Tabita FR (1994) Arch Biochem Biophys 312:210–218. **Reimholz R,** Geigenberger P, Stitt M (1994) Planta 192:480–488. **Reining E** (1994) Photosynthetica 30:519–525. **Reising H,** Schreiber U (1994) Photosynth Res 42:65–73. **Reuber S,** Leitsch J, Krause GH, Weissenböck G (1993) Z Naturforsch 48c:749–756. **Reumann S,** Heupel R, Heldt HW (1994) Planta 193:167–173. **Reyes Fernández ME,** Baker DA (1995) Photosynthetica 31:379–388. **Riens B,** Lohaus G, Winter H, Heldt HW (1994) Planta 192:497–501. **Roberts TH,** Fredlund KM, Møller IM (1995) FEBS Lett 373:307–309. **Rockel B,** Ratajczak R, Becker A, Lüttge U (1994) J Plant Physiol 143:318–324. **Roitsch T,** Bittner M, Godt DE (1995) Plant Physiol 108:285–294. **Ryals J,** Uknes S, Ward E (1994) Plant Physiol 104:1109–1112.

Sáez-Vásquez J, Raynal M, Delseny M (1995) Plant Physiol 109:611–618. **Saitou K,** Agata W, Masui Y, Asakura M, Kubota F (1994) Plant Cell Physiol 35:1165–1171. **Sánchez-Casas P,** Klessig DF (1994) Plant Physiol 106:1675–1679. **Saruyama H,** Tanida M (1995) Plant Sci 109:105–113. **Sassenrath-Cole GF,** Pearcy RW (1994) Plant Physiol 105:1115–1123. **Sawada S,** Enomoto S, Tozu T, Kasai M (1995) Plant Cell Physiol 36:551–556. **Scheibe R** (1994) Naturwissenschaften 81:443–448. **Schnarrenberger C,** Pelzer-Reith B, Yatsuki H, Freund S, Jacobshagen S, Hori K (1994) Arch Biochem Biophys 313:173–178. **Schnarrenberger C,** Flechner A, Martin W (1995) Plant Physiol 108:609–614. **Schott K,** Borchert S, Müller-Röber B, Heldt HW (1995) Planta 196:647–652. **Schünemann D,** Borchert S (1994) Bot Acta 107:461–467. **Schwarz M,** Gross A, Steinkamp T, Flügge UI, Wagner R (1994) J Biol Chem 269:29481–29489. **Scott P,** Kruger NJ (1994) Planta 193:16–20. **Scott P,** Kruger NJ (1995) Plant Physiol 108:1569–1577. **Scott P,** Lange AJ, Pilkis SJ, Kruger NJ (1995) Plant J 7:461–469. **Sekino K,** Shiraiwa Y (1994) Plant Cell Physiol 35:353–361. **Shah N,** Bradbeer JW (1994) Planta 193:232–237. **Sharkey TD,** Singsaas EL (1995) Nature 374:769. **Sharkia R,** Beer S, Cabantchik ZI (1994) Planta 194:247–249. **Sharma PN,** Tripathi A, Bisht SS (1995) Plant Physiol 107:751–756. **Scheen J** (1994) Photosynth Res 39:427–438. **Smith LH,** Keys AJ, Evans MCW (1995) J Exp Bot 46:759–765. **Smith M,** Wu Y (1994) Plant Cell Environ 17:763–769. **Spreitzer RJ,** Thow G, Zhu G (1995) Plant Physiol 109:681–685. **Stapleton AE,** Walbot V (1994) Plant Physiol 105:881–889. **Steffen KL,** Wheeler RM, Arora R, Palta JP, Tibbitts TW (1995) Physiol Plant 94:51–56. **Stitt M,** Sonnewald U (1995) Annu Rev Plant Physiol Plant Mol Biol 46:341–368. **Stoop JMH,** Pharr DM (1994) Plant Physiol 106:503–511. **Sültemeyer D,** Amoroso G, Fock H (1995) Planta 196:217–224. **Sun Z,** Duke SH, Henson CA (1995) Plant Physiol 108:211–217. **Süss K-H,** Prokhorenko I, Adler K (1995) Plant Physiol 107:1387–1397.

Takeda T, Yokota A, Shigeoka S (1995) Plant Cell Physiol 36:1089–1095. **Tanaka K,** Muto Y, Shimada M (1994) J Plankton Res 16:161–169. **Tao R,** Uratsu SL, Dandekar AM (1995) Plant Cell Physiol 36:525–532. **Tarczynski MC,** Jensen RG, Bohnert HJ (1992) Proc Natl Acad Sci USA 89:2600–2604. **Tarczynski MC,** Jensen RG, Bohnert HJ (1993) Science 259:508–510. **Taybi T,** Sotta B, Gehrig H, Güçlü S, Kluge M, Brulfert J (1995) Bot Acta 108:240–246. **Teese P** (1995) Oecologia 102:371–376. **Tetlow IJ,** Blissett KJ, Emes MJ (1994) Planta 194:454–460. **Thiele A,** Krause GH (1994) J Plant Physiol 144:324–332. **Thom E,** Neuhaus HE (1995) Plant Physiol 109:1421–1426. **Thompson GB,** Woodward FI (1994) J Exp Bot 45:937–942. **Tissue DT,** Griffin KL, Thomas RB, Strain BR (1995) Oecologia 101:21–28. **Tolbert NE,** Benker C, Beck E (1995) Proc Natl Acad Sci USA 92:11230–11233. **Trethewey RN,** ap Rees T (1994a) Biochem J 301:449–454. **Trethewey RN,** ap Rees T (1994b) Planta 195:168–174. **Truernit E,** Sauer N (1995) Planta 196:564–570.

Uchino A, Samejima M, Ishii R, Ueno O (1995) Plant Cell Physiol 36:229–238.

Vácha F (1995) Photosynthetica 31:321–334. **Valentini R,** Epron D, de Angelis P, Matteucci G, Dreyer E (1995) Plant Cell Environ 18:631–640. **van der Meer IM,** Ebskamp MJM, Visser RGF, Weisbeek PJ, Smeekens SCM (1994) Plant Cell 6:561–570. **Vanlerberghe GC,** Day DA, Wiskich JT, Vanlerberghe AE, McIntosh L (1995) Plant Physiol 109:353–361. **Van Oosten J-J,** Besford RT (1994) J Plant Physiol 143:306–312. **Van Oosten J-J,** Wilkins D, Besford RT (1994) Plant Cell Environ 17:913–923. **Villar R,** Held AA, Merino J (1995) Plant Physiol 107:421–427. **Volk S,** Feierabend J (1989) Plant Cell Environ 12:701–712. **von Caemmerer S,** Evans JR (1991) Aust J Plant Physiol 18:287–305. **von Caemmerer S,** Evans JR, Hudson GS, Andrews TJ (1994) Planta 195:88–97.

Wadano A, Kamata Y, Iwaki T, Nishikawa K, Hirahashi T (1995) Plant Cell Physiol 36:1381–1385. **Walker D** (1992) Energy, plants and man. Oxygraphics, Brighton, 227 pp. **Walker RP,** Leegood RC (1995) FEBS Lett 362:70–74. **Walker RP,** Trevanion SJ, Leegood

RC (1995) Planta 196:58–63. **Weigend M** (1994) J Plant Physiol 144:654–660. **Weiner H** (1995) Plant Physiol 108:219–225. **Whitney SM,** Yellowlees D (1995) J Phycol 31:138–146. **Williamson JD,** Stoop JMH, Massel MO, Conkling MA, Pharr DM (1995) Proc Natl Acad Sci USA 92:7148–7152. **Wilson MI,** Ghosh S, Gerhardt KE, Holland N, Babu TS, Edelman M, Dumbroff EB, Greenberg BM (1995) Plant Physiol 109:221–229. **Winkler U,** Stabenau H (1995) Planta 195:403–407. **Winter H,** Robinson DG, Heldt HW (1994) Planta 193:530–535. **Wise RR** (1995) Photosynth Res 45:79–97. **Wu G,** Shortt BJ, Lawrence EB, Levine EB, Fitzsimmons KC, Shah DM (1995) Plant Cell 7:1357–1368.

Xu D-Q, Gifford RM, Chow WS (1994) Plant Physiol 106:661–671.

Yu J-W, Price GD, Badger MR (1994) Plant Physiol 104:605–611.

Zhang J, Kirkham MB (1994) Plant Cell Physiol 35:785–791. **Zhang Q,** Wiskich JT (1995) Arch Biochem Biophys 320:250–256. **Zhang SQ,** Outlaw WH, Chollet R (1994a) FEBS Lett 352:45–48. **Zhang J,** Hu X, Henkow L, Jordan BR, Strid Å (1994b) Biochim Biophys Acta 1185:295–302. **Ziska LH,** Bunce JA (1995) Physiol Plant 94:575–584. **Ziska LH,** Sicher RC, Kremer DF (1995) Physiol Plant 95:355–364. **Zrenner R,** Salanoubat M, Willmitzer L, Sonnewald U (1995) Plant J 7:97–107.

Dr. Grahame J. Kelly
Centre for Molecular Biotechnology
School of Life Science
Queensland University of Technology
Brisbane, Queensland 4000, Australia

Prof. Dr. Dr. Erwin Latzko
Institut für Allgemeine Chemie und Biochemie
Technische Universität München
85354 Freising-Weihenstephan, Germany

Notes added in proof (2005):

Rubisco activase (Sect. 2.a) does, after all, undergo light-mediated activation (Sect. 2.a, 2002 review).

Fig. 1A-D may be updated by showing ADP-glucose import in (B) because this occurs in cereal grains (Sect. 4.b, 1998 review), and maltose export in (C) (Sect. 4.1, 2004 review).

Photorespiration (Sect. 3.c) probably is, after all, an important means of dissipating excess photosynthetic energy (Sect. 5, 1998 review; Sect. 2.c, 2002 review).

Despite our awe (Sect. 4.f), the possibility that H_2O_2 acts as a messenger molecule is still being entertained [Apel K, Hirt H (2004) Annu Rev Plant Biol 55:373-399].

Photosynthesis. Carbon Metabolism: In and Beyond the Chloroplast

By Grahame J. Kelly

> *"It is far more probable that starch is only elaborated within the cell when the supply of nutriment is in excess of the cell requirements, and that most of the assimilated products never pass through the stage of starch at all. (Our experiments) .... point to the somewhat unexpected conclusion that, at any rate in the leaves of the Tropaeolum, cane sugar is the first sugar to be synthesised ..."* (Brown and Morris 1893).

1 Introduction

Views concerning the central sugars of photosynthesis have varied. Starch was popular in the late 1800s after Sachs (1862) found that light promoted its synthesis in Chlorophyllkörnern. The possibility that the flow of photosynthetically fixed carbon could bypass starch was not seriously entertained until the turn of the century, (see above quotation), after which cane sugar (the disaccharide sucrose) gradually achieved the reputation of being the end product of photosynthesis in the green leaf cell; but simplistic views seldom hold true in the biological world. The anticipation (see Walker 1997) and characterization (Heldt and Rapley 1970) of the chloroplast's phosphate translocator ultimately gave glyceraldehyde (as glyceraldehyde-3-P) the distinction of being *the* product of photosynthesis within the chloroplast. Still, glyceraldehyde-3-P (GAP) never accumulates in quantity, whereas sucrose (which is synthesized from GAP in the cytosol) does. Additionally, sucrose is a principal form in which photosynthetic product is translocated to other parts of the plant. Therefore, the triose phosphates (GAP and its partner dihydroxyacetone-P) and sucrose currently share the title of central sugars of photosynthesis; one in, and the other beyond, the chloroplast. Recent research on photosynthetic carbon metabolism reflects this distinction: more and more attention is being devoted to what happens beyond the chloroplast, while research on intrachloroplast events has been directed at the perennially popular enzyme Rubisco, and at the mechanism chloroplasts use to dissipate whatever portion of harvested light energy cannot be used to convert CO_2, via GAP, to sucrose. These topics are reviewed below, covering the literature that has appeared since our last chapter in Volume 58 of Progress in Botany.

G.J. Kelly and E. Latzko:
Thirty Years of Photosynthesis

Originally published in Progress in Botany 60

2 From CO_2 to GAP

a) Ribulose-Bisphosphate Carboxylase/Oxygenase (Rubisco)

The Calvin cylcle's CO_2 fixing enzyme, Rubisco, has continued to receive dedicated attention. Its reputation for being the major determinant of photosynthetic rate in nonstressed sunlit plants in the field has been extended to C_4 plants (Furbank et al. 1996). It also contributes to the establishment of the lower photosynthetic rates of stressed plants: stresses such as low water availability (Kanechi et al. 1996), rain treatment (Ishibashi et al. 1996), oxidative stress (Desimone et al. 1996; Stieger and Feller 1997; Ishida et al. 1997), and ultraviolet radiation (Ferreira et al. 1996; Greenberg et al. 1996; Allen et al. 1997) all cause Rubisco to be physically damaged and/or reduced in amount.

Research on the regulation of Rubisco has centered on the activity-enhancing protein Rubisco activase and the inhibitory sugar-P carboxyarabinitol-1-P (CA1P). Rubisco activase is a relatively heat-sensitive protein (Crafts-Brandner et al. 1997; Eckardt and Portis 1997) that slowly (Woodrow et al. 1996) promotes dissociation of inhibitory sugar-Ps from Rubisco, permitting the latter to display its full activity. Exactly how it does this is obscure (Salvucci and Ogren 1996), although there are hints that it first forms an oligomer reminiscent of the better-known microfilament-protein actin (Lilley and Portis 1997). Another intriguing parallel has been noted by Komatsu et al. (1996), who discovered an amino acid sequence homology between the activase and a gibberellin-binding protein. Whatever its mechanism, it is clear from antisense-gene experiments that the activase is vital for photosynthesis, the only point of contention is whether or not it is present in leaves in great excess (Mate et al. 1996; Eckardt et al. 1997).

Little progress, and some odd results, have appeared in relation to the CA1P inhibition of Rubisco. This sugar-P is presumably synthesized at night when it binds to any CO_2-plus-Mg^{2+}-activated Rubisco and stops it from working. However, this regulatory event occurs only in some plant species, e.g., French bean and tobacco, and perhaps in the green alga *Dunaliella tertiolecta* (MacIntyre et al. 1997), but not in others, e.g., spinach and wheat. In French bean leaves it is unclear whether it is restricted to chloroplasts (Moore et al. 1995) or whether some exists in unexpected locations such as leaf veins (Anwaruzzaman et al. 1996). Finally, its mode of synthesis remains unclear: it is quite possibly derived from fructose-1,6-P_2 via the seldom-met sugar hamamelose (Andralojc et al. 1996; Martindale et al. 1997), but it is annoying that a kinase needed to phosphorylate carboxyarabinitol to CA1P has not yet been detected. In addition, to complicate matters, recent research by the Rothamsted group has raised a suspicion that there exists another tight-binding inhibitor of Rubisco that, unlike CA1P, is present by day as well as by night (Keys et al. 1995; Parry et al. 1997), but there is so far no clue concerning the identity of this inhibitor.

The possibility that the *oxygenase* activity of Rubisco (and, consequently, photorespiration: Sect. 5) might be eliminated continues to be entertained. It has been argued that the oxygenase activity is inevitable in normal air, but the existence of certain aldolases that catalyze reactions involving similar chemistry but which nevertheless display no tendency to catalyze a similar oxygenase reaction bolster hopes that molecular biologists may some day create an oxygenase-free Rubisco (Hixon et al. 1996; Wildner et al. 1996). Slight success in this direction has been reported by Kostov et al. (1997) for a cyanobacterial Rubisco, but no recent efforts have been reported for higher plant Rubiscos; they all possess very similar ratios of carboxylase to oxygenase activities that are about double those of the cyanobacterial Rubiscos (Balaguer et al. 1996; Laisk and Loreto 1996; Uemura et al. 1996).

b) Remainder of the Calvin Cycle

Almost all attention to the remainder of the Calvin cycle has been focused on the mechanistic details by which certain enzymes undergo light-mediated activation, and on the existence of multienzyme complexes. Light-mediated activation involves the reduction of disulfide bonds by reduced thioredoxin. The cysteine residues between which the disulfide bonds are formed and broken have been more precisely identified in the higher plant and *Chlamydomonas reinhardtii* variants of glyceraldehyde-3-P (GAP) dehydrogenase, fructose-1,6-bisphosphatase (FBPase), and sedoheptulose-1,7-bisphosphatase (SBPase) (Anderson et al. 1996a; Jacquot et al. 1997a; Li et al. 1997). One of the cysteine residues is missing from a cyanobacterial GAP dehydrogenase, and indeed this enzyme probably is not light-activated (Tamoi et al. 1996). Reduced thioredoxin has attracted more attention because this small protein occurs not only in the chloroplast in two forms that participate in light-mediated enzyme activation (Reche et al. 1997), but also in mitochondria in two forms, one of which may have an as yet undefined role in photosynthesis, and in seeds, where it reduces the disulfide bonds of storage proteins prior to their mobilization during germination (Konrad et al. 1996; Lozano et al. 1996; Jacquot et al. 1997b). It may even have a role in protecting stressed plants by substituting for the enzyme dehydroascorbate reductase (Morell et al. 1997). Finally, it has attained importance in fatty acid biosynthesis in that it has been shown to participate in the light-mediated activation of the chloroplast's acetyl-CoA carboxylase (Sasaki et al. 1997).

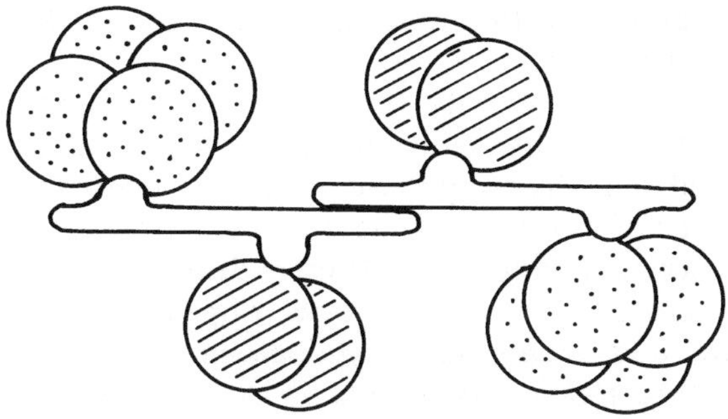

Fig. 1. An adaption of the model of Wedel et al. (1997) for the association of two molecules of GAP dehydrogenase (*dotted*, four-subunit enzyme), two molecules of Ru5P kinase (*cross-hatched*, two-subunit enzyme), and two molecules of a small linking protein (*thin, unshaded*). It is proposed that the enzymes are reductively activated (by reduced thioredoxin) while associated as shown, but that photosynthetically generated NADPH induces dissociation of the enzymes from the dimeric linking protein; the free, reduced enzymes are the fully active forms that participate in photosynthesis

Increased attention is being given to protein-protein interactions in the chloroplast. The simplest is aggregation of one enzyme, as evidenced by the recent demonstrations of the association of four molecules of either GAP dehydrogenase (Baalmann et al. 1995) or FBPase (Grotjohann 1997) into relatively inactive oligomers. Alternatively, Calvin cycle enzymes may associate with each other or with other proteins. Examples include an association between FBPase and thioredoxin that appears to regulate the FBPase activity in a nonreductive fashion (Häberlein and Vogeler 1995), the association of a major or minor portion of each of six Calvin cycle enzymes with thylakoid membranes (Anderson et al. 1996b), and associations of GAP dehydrogenase with glycerate-3-P kinase (Wang et al. 1996) and with ribulose-5-P (Ru5P) kinase (Avilan et al. 1997; Wedel et al. 1997). These protein-protein interactions may be expected in the chloroplast stroma, where the concentration of proteins is quite high, and they may benefit the Calvin cycle's operation by channeling intermediates from one enzyme to the next (Harris and Königer 1997); evidence for such chanelling in the *oxidative* pentose phosphate pathway in root nodules and in yeast has recently been presented (Debnam et al. 1997). Of special interest is the interpretation of Wedel et al. (1997) that there exists in chloroplasts a small protein with the specific task of mediating the tantalizing (Lebreton et al. 1997) GAP dehydrogenase/Ru5P kinase association and participating in the light-mediated activations of these enzymes (Fig. 1). Given that other Calvin cycle enzyme associations have been reported (see above, and our previous reviews), one wonders whether similar small proteins also exist. This one must be important if, as concluded by Fridlyand et al. (1997), GAP dehydrogenase has a part in determining the rate of CO_2-

saturated photosynthesis (cf. Rubisco is the major determinant with air levels of CO_2).

c) The C_4 Pathway

Photosynthetic carbon metabolism begins with the carboxylation of ribulose-1,5-P_2 (RuBP) in C_3 plants, but in C_4 plants, and in plants with crassulacean acid metabolism (CAM), preceding metabolic events designed to concentrate and/or indirectly store the CO_2 destined for RuBP carboxylation occur. The majority of these events occur beyond the chloroplast.

In C_4 plants, the well-known C_4 pathway uses the enzyme phosphoenolpyruvate (PEP) carboxylase in mesophyll cells to collect CO_2 (as HCO_3^-) and form oxaloacetate which, after being reduced to malate, is either directly or in a roundabout metabolic way transported to bundle-sheath cells and decarboxylated, thereby feeding CO_2 to the Calvin cycle, which is restricted to these cells (but see Castrillo et al. 1997). This CO_2 pump greatly suppresses photorespiration and increases water use efficiency. Its evolution was encouraged about 7 million years ago by a drop in the level of atmospheric CO_2; consequently, terrestrial C_4 plants have existed only during the last 2% of the period when vascular plants have been present on land (Cerling et al. 1997). However, a submersed monocot (*Hydrilla verticillata*) that displays C_4-type photosynthesis has been identified as a possible ancestor of terrestrial C_4 plants, and may have existed up to 100 million years ago (Magnin et al. 1997).

Several reports have emphasized the importance and properties of component enzymes of the C_4 pathway. The most studied is PEP carboxylase, partly because of its fascinating regulation via a phosphorylation (that activates it) catalyzed by another enzyme appropriately named PEP-carboxylase kinase (reviewed by Vidal and Chollet 1997). The mRNA for this kinase increases when plants are illuminated (Hartwell et al. 1996) via an as yet only partly elucidated signaling system that involves Ca^{2+} and glycerate-3-P (Giglioli-Guivarc'h et al. 1996). A similar, but not identical, system potentially operates in the cells of C_3-plant leaves (Li et al. 1996; Smith et al. 1996; but see Leport et al. 1996), and in other organs including germinating wheat seeds (Osuna et al. 1996), banana fruit (Law and Plaxton 1997), and soybean root nodules, in which photosynthate supply, rather than light, triggers the signaling system (Zhang and Chollet 1997a). Finally, Du et al. (1997) have added stomatal guard cells to the list of cells in which PEP carboxylase has a central role and is regulated by phosphorylation. In all of these cases, phosphorylation of the carboxylase seems to activate it primarily by reducing its susceptibility to inhibition by malate, although altered sensitivities to other regulatory metabolites could be important (Gao and Woo 1996); these would include the recently reported inhibitors phosphate and sulfate (Salahas et al. 1997) and, interestingly, shikimic acid, which is synthesized from PEP via a metabolic pathway that does not involve PEP carboxylation (Colombo et al. 1996).

PEP carboxylation produces oxaloacetate which is reduced to malate by a malate dehydrogenase shown to be essential for normal C_4 photosynthesis (Trevanion et al. 1997), and which undergoes light-mediated activation via reduction of two disulfide bonds per subunit (Issakidis et al. 1996). The malate is then processed by one of three possible metabolic routes (that designate the three subtypes of C_4 plants), each of which may vary subtly between plant species (Meister et al. 1996) and have special requirements such as an active 2-oxoglutarate/malate translocator (Taniguchi and Sugiyama 1997) or reduced capacity for respiration in bundle-sheath mitochondria (Agostino et al. 1996). Ultimately, CO_2 is released to the Calvins cycle's Rubisco, and the remaining 3-carbon pyruvate is returned to the mesophyll cells, where it is reconverted to PEP in the chloroplast by the remarkable enzyme pyruvate, P_i dikinase (PPDK). Molecular biologists managed to introduce this C_4 enzyme, and a functionally similar bacterial enzyme, into C_3 plants, but nothing spectacular happened (Ishimaru et al. 1997; Panstruga et al. 1997). PPDK produces, in addition to PEP, a good amount of AMP, which is returned to the ADP/ATP pool by adenylate kinase, an enzyme that is abundant in mesophyll chloroplasts where it seems to assemble into an inactive storage form overnight (Wild et al. 1997).

Further reports confirm that when the C_4 pathway concentrates CO_2 into bundle-sheath cells, photorespiration (Sect. 5) is largely, although not totally, inhibited, and although back diffusion of CO_2 is greater than earlier believed, the mesophyll cell's PEP carboxylase recaptures it so that the overall pathway remains effective (Dai et al. 1996; He and Edwards 1996; Saliendra et al. 1996; Lacuesta et al. 1997). However, without the C_4 pathway, C_4 plants can photorespire (Devi et al. 1996, Lacuesta et al. 1997).

Most C_4 plants are stressed by cool temperatures and, like all plants, by severe water stress. Sensitivity to cool temperatures appears to be due to the impairment of critical enzymes such as PEP carboxylase and PPDK, but a few species possess cold-resistant forms of these enzymes (Usami et al. 1995; Simon 1996; Matsuba et al. 1997) so that in cool climates C_4 plants, while rare, can be found (Beale et al. 1996). In the case of the enzyme PPDK, it was found that substitution of only 0.3% of the amino acid residues could confer cold tolerance (Ohta et al. 1996). PPDK is also implicated in water-stressed C_4 plants whose photosynthesis is decreased not only by a reduced supply of CO_2 to Rubisco (Lal and Edwards 1996), but also by a dramatic, over nine-fold drop in the PPDK activity (Du et al. 1996).

Characteristics of C_4 photosynthesis have again been reported in two aquatic plants, the leafless sedge *Eleocharis vivipara* (Ueno 1996) and the monocot *Hydrilla verticillata* (Spencer et al. 1996; Magnin et al. 1997), and in the ears of C_3 cereals and green tomato fruits that refix respiratory CO_2 (Bort et al. 1996; Xu et al. 1997). Of particular note is the predominant role being attributed to PPDK in the lemmas, paleae, and immature seeds of rice (Sadimantara et al. 1996; Imaizumi et al. 1997).

d) Crassulacean Acid Metabolism (CAM)

Like C_4 plants, CAM plants employ considerable metabolism beyond the chloroplast prior to RuBP carboxylation. Events within the cytosol and vacuole, and on the vacuolar membrane (tonoplast) are especially notable. In summary, PEP carboxylation in the cytosol at night produces malate that is delivered to and stored in the vacuole, from where it is dispensed back to the cytosol next day and decarboxylated, thereby providing CO_2 for Rubisco behind closed stomates. Stomate closure by day can help to protect CAM plants during drought (Güerere et al. 1996), but not if conditions become too extreme (Castillo 1996; Roberts et al. 1996), and indeed the expected superior water use efficiency of CAM plants (Helliker and Martin 1997) has recently been thrown into doubt (Eller and Ferrari 1997). It is also worth noting that CAM is not restricted to dry, sunny habitats: it is common among rainforest epiphytes (Zotz and Ziegler 1997), while some uncommon examples include species from the shaded rainforest understory, including *Aechmea magdalenae* (Skillman and Winter 1997; Zotz and Ziegler 1997). Two extreme examples are the tiny *Blossfeldia liliputana*, a unique poikilohydric cactus (Barthlott and Porembski 1996), and the brown marine macroalga *Ectocarpus siliculosus*, where a CAM-like mechanism helps to overcome the limiting CO_2 availability in seawater (Schmid and Dring 1996).

The mechanism of CAM in conventional CAM plants has been further investigated. The upside-down behavior of CAM-plant stomates (open at night and closed by day) has been correlated with a loss of the sensitivity of stomatal guard cells to blue light (Tallman et al. 1997) which is detected by the guard-cell chloroplasts of other plants (Quiñones et al. 1996). CO_2 that enters stomates at night is captured by a PEP carboxylase that is activated via a phosphorylation; in this case a circadian oscillator, rather than light (as in C_3 and C_4 plants), initiates the signaling system that culminates in enzyme phosphorylation (Hartwell et al. 1996). By day the PEP carboxylase is deactivated, more so in those species that display CAM more strongly (Borland and Griffiths 1997). Nocturnal PEP carboxylation leads to the production of the large quantity of malic acid that accumulates in the vacuole (Fig. 2); an accumulation of citric acid is also detected in some species, but does not seem to be a central component of CAM (Herppich et al. 1995; Borland et al. 1996). Malate is moved into the vacuole by a recently characterized tonoplast transporter (Steiger et al. 1997; Fig. 2); accumulation is powered by a tonoplast H^+-ATPase (Zhigang et al. 1996) that is possibly regulated by a H^+-

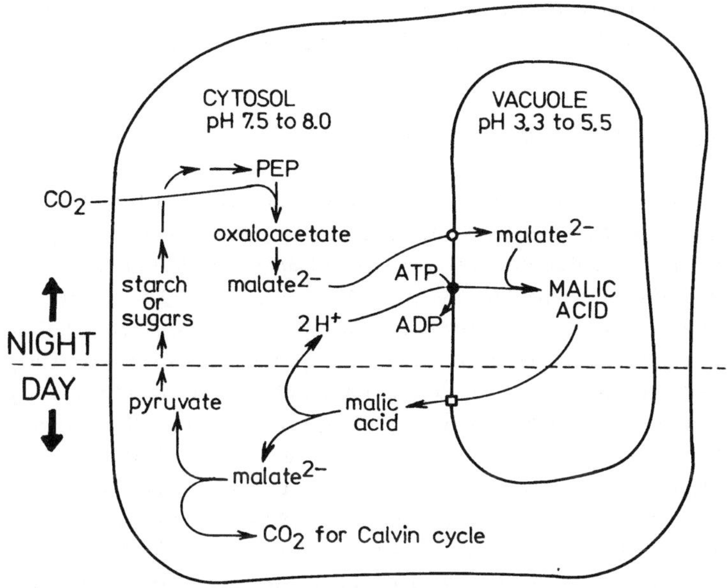

Fig. 2. The role of the tonoplast during CAM. By night, active transport of malate anions into the vacuole is driven by an ATPase that delivers H^+ into the same vacuole. Thus it is the acid form (malic acid) that accumulates. Next day, this malic acid diffuses passively (via a channel protein?) into the cytosol, where it first dissociates (returning the H^+, which is presumably accomodated by the buffering capacity of cytosolic metabolism), and then undergoes the decarboxylation that supplies CO_2 for the Calvin cycle and pyruvate for synthesis of the carbohydrate reserve needed to generate the next night's supply of PEP. (After Lüttge et al. 1995)

pyrophosphatase found on the same membrane (Lüttge et al. 1995). Next day, malic acid moves passively across the tonoplast and into the cytosol (Fig. 2). This process is notably sensitive to high or low temperature, indicating that the biophysical state of the membrane is important for its normal operation (Lüttge et al. 1995; Grams et al. 1996; Grams et al. 1997). Once in the cytosol, the malate is decarboxylated by one of three possible enzymes; one of these (PEP carboxykinase) is similar to PEP carboxylase in that its activity is potentially regulated by phosphorylation (Walker and Leegood 1996). The remaining 3C-pieces are converted gluconeogenically to stores of chloroplast starch or extrachloroplast sugars (Borland 1996; Christopher and Holtum 1996) from which PEP is derived the following night. It was thought that the storage form (starch or sugars) accumulated by a particular CAM species correlated with its particular decarboxylation enzyme, but it is now clear that this rule is invalid (Christopher and Holtum 1996).

3 The Multifaceted Fate of GAP

a) Transitory Starch in Chloroplasts

The conclusion of Brown and Morris (1893) that "... starch is only elaborated in the cell when the supply of nutriment is in excess ..." (see introductory quotation) has been confirmed for peach, a species that synthesizes sorbitol rather than cane sugar (sucrose) as the predominant translocated carbohydrate (Escobar-Gutiérrez and Gaudillère 1997), and for spinach and soybean deprived of nitrogen (Robinson 1997). However, no new details concerning the biochemistry of the synthesis of this chloroplast starch by day and its mobilization by night have been reported, except for further complications concerning the role of phosphorylase in the mobilization process: C_4 plant leaves contain even more nonchloroplastic forms of this enzyme than do C_3 plant leaves (Shatters and West 1996; Venkaiah and Kumar 1996), and Duwenig et al. (1997a) have indications that the (chloro)plastidic form might be more important for starch synthesis in fruits and seeds than for starch degradation in leaves. Thus, it seems to be even more certain that the nocturnal mobilization of starch in chloroplasts is hydrolytic (not phosphorylytic), producing maltose and glucose that are exported to the cytosol (Rost et al. 1996).

b) Three Notes About Chloroplast Lipids

The major role of the chloroplast in *mature* leaves is to act as a factory that manufactures soluble sugars (sucrose, or a sugar alcohol such as sorbitol, or one of the small raffinose type of oligosaccharides; Moore et al. 1997) for export to metabolic or storage sinks; but in *immature* leaves the chloroplast implements several other biosynthetic capabilities whereby the molecular building blocks of the proteins and membranes needed by a growing leaf are generated from Calvin cycle precursors (Fig. 3; Herbers and Sonnewald 1996). Three interesting pieces of information concerning lipid synthesis in chloroplasts have appeared:

1. The initial enzyme of fatty acid synthesis, acetyl-CoA carboxylase, is similar to several Calvin cycle enzymes in that it undergoes a reductive light-mediated activation (Sasaki et al. 1997).
2. The enzymes that synthesize long-chain fatty acids from acetyl-CoA are organized into a multienzyme complex adjacent to the thylakoid, thereby permitting substrate channeling and the consequent increase in rate of fatty acid synthesis for a given chloroplast content of each substrate (Roughan and Ohlrogge 1996; Roughan 1997).
3. Isopentenyl-P_2, the precursor of chloroplast isoprenoids such as the carotenoids and the phytol tail of chlorophyll, is not synthesized through the classical pathway (from acetate, and via mevalonate), but rather through a different pathway in which one molecule each of GAP and pyruvate are its precursors; in the 5-carbon sugar deoxyxylulose is an intermediate in this pathway (Schwender et al. 1996; Arigoni et al. 1997; Lichtenthaler et al. 1997; Fig. 3).

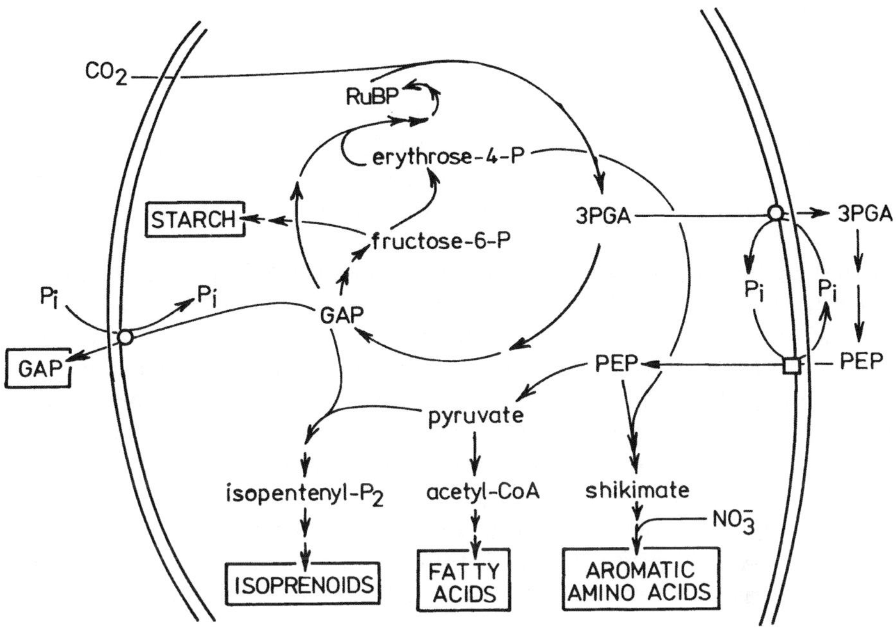

Fig. 3. Five products of photosynthesis (*boxed items*) in developing leaves. Based on research covered in our earlier reviews, and on Herbers and Sonnewald (1996), Fischer et al. (1997), and Lichtenthaler et al. (1997). *3PGA* Glycerate-3-P

c) Export from the Chloroplast

The now-classical phosphate translocator that exports net photosynthetic product, as GAP, from the chloroplast (in exchange for an incoming P_i; Fig. 3) is clearly not alone. It is now more certain that a separate, distinct envelope protein that favors the transport of glucose-6-P is present: such a transporter was identified on *Capsicum annuum* fruit chromoplasts (Quick and Neuhaus 1996) and on the chloroplasts of a CAM plant (Neuhaus and Schulte 1996; Kore-eda and Kanai 1997), where, presumably, it could participate in the nocturnal conversion of chloroplast starch to cytosolic PEP. In addition, a third phosphate translocator that exchanges PEP and P_i has been confirmed for chloroplasts and other types of plastids; it could be important for the synthesis of aromatic amino acids (Fischer et al. 1997; Fig. 3), and is almost certainly central to the C_4 pathway (Aoki and Kanai 1995). Finally, the presence of an ATP/ADP translocator has been more firmly established (Neuhaus et al. 1997), and the probability that there is a transporter for maltose has been raised (Rost et al. 1996). More surprises may be imminent, given the results of studies with other types of plastids (see below, Sect. 4.b).

d) Cytosolic Sucrose Synthesis and its Regulation

When GAP arrives in the cytosol of leaf cells, it is converted to soluble sugars ready for export from the leaf. Synthesis of the sugar sucrose has been by far the most intensively studied. Two regulatory enzymes have received recent attention: cytosolic FBPase that forms fructose-6-P from the FBP generated from two GAPs, and sucrose-P synthase (SPS) that forms sucrose-P from two fructose-6-Ps (one of which is first converted to UDP-glucose). The FBPase is potentially regulated by a reductive process (Anderson et al. 1997) similar to that of its chloroplast counterpart (Sect. 2.b), but the value of this is not yet established. Its activity is also controlled through inhibition by the regulatory sugar-P fructose-2,6-P_2 (Kruger and Scott 1995), especially during water stress (Reddy 1996). Plants containing low amounts of the FBPase clearly made less sucrose (Micallef and Sharkey 1996; Zrenner et al. 1996), but in one case (genetically engineered potato), plant growth was unaffected because photosynthate partitioning was appropriately adjusted, i.e., chloroplasts made more starch and then, by exporting hexose or hexose-P at night, the cytosolic FBPase was not required (Zrenner et al. 1996).

The well-established principal regulatory mechanism for SPS, viz. activation by dephosphorylation and inactivation by phosphorylation, catalyzed by a SPS phosphatase and a SPS kinase (respectively), has been reviewed (Huber and Huber 1996). The activity of the SPS phosphatase is, in turn, modulated by a circadian-rhythm-based signaling system (Jones and Ort 1997). An additional level of complexity was reported by Toroser and Huber (1997), who found that osmotic stress somehow activates a second form of SPS kinase that attaches a second phosphate group to the already phosphorylated (and therefore inactive) SPS. This, like total dephosphorylation, revives the enzyme's activity, in this case perhaps to promote the formation of sucrose for osmoregulatory purposes. There is also one report that SPS itself might occur in more than one form in leaves (Reimholz et al. 1997), and another that it forms a complex with the following (and final) enzyme of sucrose synthesis, viz. sucrose-P phosphatase (Echeverria et al. 1997). Interest in SPS will continue because there is evidence that it is a principal determinant of plant production: tomato plants genetically engineered to contain extra SPS photosynthesized faster (Galtier et al. 1995), although they tended to produce sweeter fruit rather than more fruit (Laporte et al. 1997).

A new topic, introduced in our last review, is the mechanism by which soluble sugars (that accumulate in actively photosynthesizing cells when the ability of the rest of the plant to use them is exceeded) somehow initiate a feedback inhibition of photosynthesis. Recent clear examples of the effect were wheat and sugar beet leaves in which the mRNA transcripts for Rubisco, FBPase, SBPase, and glycerate-3-P kinase were reduced when the leaves were fed 1–2% sucrose or glucose (Jones et al. 1996; Lee and Daie 1997). The current hypothesis is that the genes for these Calvin cycle enzymes are downregulated by a molecular signaling system that begins with the traditional glycolytic enzyme hexokinase and its substrates glucose and/or fructose, comparable mechanisms are known from animal and yeast cells (Jang and Sheen 1997; Smeekens and Rook 1997). Using transgenic *Arabidopsis* plants, Jang et al. (1997) have further strong evidence that hexokinase is the sugar sensor that links sugar-related metabolic activities to photosynthesis rates in plants, but other research implies that there is some explaining to be done:

1. Strand et al. (1997) found that the correlation between accumulation of sugars and downregulation of genes in *Arabidopsis* broke down when leaves developed at a low temperature (5 °C). Under this temperature the well-known accumulation of sugars as a component of cold tolerance took place, and was supported by an unrestricted photosynthesis. Similarly, spinach leaves (Martindale and Leegood 1997) and wheat and rye (Gray et al. 1996) acclimated to low temperatures developed increased capacities for photosynthesis, in the latter case this was deemed necessary to compensate for the slower metabolism at lower temperatures and therefore the greater difficulty in using harvested light energy at these temperatures (Huner et al. 1996). It seems that the success of this adaptation mechanism may not be total (Savitch et al. 1997).
2. Sugar beet, tobacco, and *Flaveria bidentis* supplied with sucrose in culture systems showed no evidence for photosynthetic repression (Kovtun and Daie 1995; Furbank et al. 1997).
3. More than 90% of photosynthetically generated hexoses (glucose and fructose) was compartmented into the vacuole of tobacco, snapdragon, and parsley leaves (Moore et al. 1997), whereas the hexokinase-based sugar-sensing mechanism would be expected to be in the cytosol. Possibly the concentration of hexoses in the cytosol increases at the end of the day (Moore et al. 1997) after invertase hydrolyzes the vacuolar store of sucrose (Scholes et al. 1996) and the resultant hexoses are moved across the tonoplast. The accumulation of hexoses in the cytosol appears to be possible: Schaffer and Petreikov (1997) measured 30 mM fructose in the cytosol of young tomato-fruit cells.

A new angle to the hexokinase story might develop if plant scientists follow up the report that rabbit erythrocyte hexokinase is inactivated by dehydroascorbate (Fiorani et al. 1996). Ascorbate and its oxidation products (e.g., monodehydroascorbate) have central roles in the responses of plants to stress.

4 Distant Fates of Photosynthetic Product

a) Translocation

The majority of photosynthate produced by mature leaves is exported as soluble sugars to metabolic or storage sink tissues via the phloem. A concept for the loading of sucrose into the phloem, driven by a H^+-ATPase and a H^+/sucrose symporter (see Fig. 2 in our last review), has received more experimental support (Stadler et al. 1995; Kühn et al. 1996, 1997), although it seems there is some flexibility as to whether the H^+/sucrose symporter is on the sieve-tube membrane (e.g., tobacco, potato, and tomato) or on the membrane of its associated companion cell (e.g., *Plantago major*). The presence of the symporter implies that sucrose is loaded from an apoplastic source; however, its arrival at this source, at least in monocots, seems to require good symplastic (i.e., plasmodesmatal) connections between preceeding parenchyma cells (Evert et al. 1996; Russin et al. 1996).

b) Storage

The ultimate fates of translocated sugars include metabolism to form the proteins, membranes, and cell walls of metabolic sinks (such as growing shoots and roots); conversion to storage materials such as protein, starch, fructans, and oils in fruits, seeds, and modified organs; direct storage as soluble sugars; and respiration to power these processes. The synthesis of storage starch and (briefly) storage sucrose and oils is reviewed below.

Starch, like photosynthesis, is located in plastids; storage starch is located in amyloplasts (cf. transitory starch in chloroplasts, Sect. 3.a). Its biosynthesis begins with translocated sucrose, which is cleaved by one or

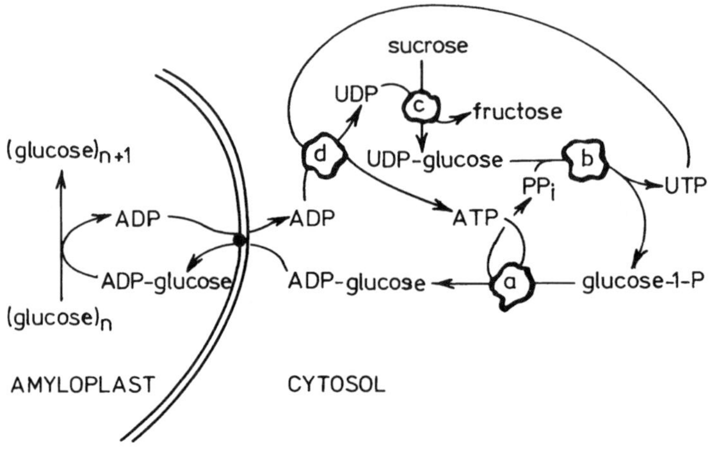

Fig. 4. Proposed pathway for the conversion of sucrose to starch in starch-storing tissues. The sum of the reactions is: sucrose + $(\text{glucose})_n \rightarrow$ fructose + $(\text{glucose})_{n+1}$. Enzymes: *a* ADP-glucose pyrophosphorylase; *b* UDP-glucose pyrophosphorylase; *c* sucrose synthase; *d* nucleoside-P_2 kinase. (After Kleczkowski 1996)

other of the isozymes (Guerin and Carbonero 1997) of the rather inappropriately named sucrose synthase (Fig. 4). This enzyme, which is thought to be a reliable marker for sink tissues (Déjardin et al. 1997), is yet another example of an enzyme whose activity is regulated by a phosphorylation/dephosphorylation mechanism; in this case, phosphorylation increases activity (Huber et al. 1996; Zhang and Chollet 1997b). The UDP-glucose generated by sucrose synthase has, according to recent reports, one surprise fate: conversion *in the cytosol* to ADP-glucose. This is a surprise because the enzyme required to do this, ADP-glucose pyrophosphorylase, has been traditionally thought of as the principal regulatory enzyme of starch synthesis *within plastids*, but now Denyer et al. (1996) and Thorbjørnsen et al. (1996) report that the majority of the ADP-glucose pyrophosphorylase in the cells of maize and barley endosperm is in the cytosol rather than in the amyloplast, and Rudi et al. (1997) report that this cytosolic enzyme is not allosterically regulated by glycerate-3-P and Pi like its chloroplast counterpart.

The second group of authors have been quick to point out (Thorbjørnsen et al. 1996) that there still seems to be enough of this enzyme in barley amyloplasts to accommodate starch synthesis, possibly from imported glucose-6-P (Hill and Smith 1995; Emes and Neuhaus 1997; Möhlmann et al. 1997), and indeed another study with transgenic potato indicated that there is plenty of the enzyme in starch-storing tissues (Sweetlove et al. 1996). However, indications that isolated amyloplasts can synthesize starch from exogenously supplied ADP-glucose and that they possess a translocator that recognizes this nucleotide sugar (Emes and Neuhaus 1997; Möhlmann et al. 1997; Naeem et al. 1997) support the possibility that at least some amyloplast starch is generated from the ADP-glucose synthesized in the cytosol (Kleczkowski 1996; Fig. 4). The fructose shown in Fig. 4 need not be wasted. It could be phosphorylated by hexokinase and then isomerized to glucose-6-P, which could be imported into the amyloplast for starch synthesis. Perhaps this could occur in one step by using the enzyme phosphorylase, which is thought by Duwenig et al. (1997a) to have a role in starch formation. The P_i released by this enzyme would need to be exported to properly balance all reactions, thus the design of the amyloplastic glucose-6-P translocator such that it preferentially exports P_i to the cytosol (when P_i is counterexchanged with glucose-6-P) (Neuhaus and Maaß 1996) makes it eminently suitable for this role.

The mobilization of stored starch has received little attention. Amylases (Rost et al. 1996) rather than phosphorylases probably catalyze the initial reactions. As noted above, plastid phosphorylase is believed to be more important for starch synthesis, while a role for the cytosolic enzyme is still unclear. Remarkably, when the activity of this cytosolic enzyme in the potato plant was reduced by antisense technology, the plants appeared normal but gave double the yield of tubers (Duwenig et al.

1997b). Incidentally, another remarkable aspect of phosphorylase is its potential longevity: active enzyme has been extracted from *Nelumbo nucifera* (lotus) seeds estimated to be more than 500 years old (Minamikawa et al. 1995).

Besides starch, sink tissues may also store sucrose or oils. Sucrose storage does not seem to involve the straightforward uptake and accumulation of translocated sucrose. Rather, the sucrose is first hydrolyzed by an apoplastic invertase and, in sugar-cane stem, resynthesis of sucrose by sucrose-P synthase occurs (Zhu et al. 1997). In tomato fruit, glucose released by the invertase appears to be pumped into cells by a cell membrane transporter powered by a H^+-ATPase (Brown et al. 1997). Tissues that store oil do so in modified plastids. In castor-bean endosperm, these plastids were found to possess a unique translocator that imports malate (the substrate for fatty acid synthesis) in exchange for an exported Pi (released when ATP is hydrolyzed to drive the acetyl-CoA carboxylase reaction) (Eastmond et al. 1997).

5 Photorespiration

Photorespiration (Leegood et al. 1995) consists of a cycle of reactions initiated by the oxygenase activity of Rubisco and interwoven with the Calvin cycle in such a way that, as clarified in our last review, it achieves nothing other than the consumption of photosynthetically generated ATP and NADPH when the CO_2 concentration drops to its compensation point, as might occur after leaf stomates shut on water-stressed plants. Theoretically, it could also occur if the oxygen evolved during intensive photosynthesis became trapped in the leaf (Tolbert et al. 1995), but Ligeza et al. (1997) believe that the normal leaf is too well ventilated for this to happen.

The fact that photorespiration consumes ATP and NADPH implies that it could be another approach whereby excess harvested light energy could be dissipated. This concept, which has had its ups and downs over the past years, is again on an "up" (Osmond et al. 1997). Firstly, experiments with the leaves of wheat, *Chenopodium bonus-henricus*, and pea all strongly supported it (Biehler and Fock 1996; Heber et al. 1996b; Park et al. 1996), and partial support came from another study with French bean leaves (Daniel 1997). Secondly, when the photorespiration of tobacco leaves was retarded by first identifying the rate-limiting enzyme (it turned out to be the chloroplast's glutamine synthetase) and then using antisense technology to reduce the amount of this enzyme in the leaf, the leaves' susceptibility to high-light-induced photoinhibition was notably increased (Kozaki and Takeba 1996).

Apart from an interesting report that glycolate may be oxidized in chloroplasts by a glycolate-quinone oxidoreductase system (Goyal and Tolbert 1996) rather than is "usual" (i.e., in peroxisomes by glycolate dehydrogenase), most recent enzyme studies have centered on the mitochondrial glycine decarboxylase complex that converts two molecules of glycine to serine, CO_2 and NH_4^+; also, an NAD^+ is reduced. The reassimilation of this NH_4^+, and another subsequently released from the serine, is the role of the rate-limiting glutamine synthase mentioned above. However, the glycine decarboxylase itself appears to be only a little less important with respect to rate limitation (Wingler et al. 1997), especially in the palisade layer of mesophyll cells, which has less of the enzyme compared to the spongy cells (Guinel and Ireland 1996). In the mitochondrion, the enzyme is intimately associated with the pyruvate dehydrogenase complex (Bourguignon et al. 1996) that produces acetyl-CoA for the Krebs cycle, and much of the NADH it produces may be oxidized through the alternative cyanide-resistant electron transport system (Igamberdiev et al. 1997). Finally, there is good evidence that plants obtain most of their serine by courtesy of this enzyme, rather than others that can synthesize serine from glycine and formate (Prabhu et al. 1996).

6 Conclusions and Prospects

The literature on photosynthetic carbon metabolism from the past 2 years has contained a few exciting items, although (in our opinion) not as many as in some previous 2-year periods. Nevertheless, there are several indications that the future will not be dull. Firstly, there are new views of chloroplasts. Evolutionary relics of these organelles have been discovered in protozoan parasites of humans, such as *Toxoplasma gondii* (McFadden et al. 1996). In plants such as *Chenopodium album* there is a vertical gradient of chloroplast abundance, with the greatest number of chloroplasts per mesophyll cell in the uppermost fully expanded leaves (Yamasaki et al. 1996); and, most amazingly, within the cell the chloroplasts can no longer be considered as discrete, relatively independent organelles. Molecular biologists have used their fascinating new tool (green fluorescent protein) to rediscover a structural feature first reported in the 1960s, viz. chloroplasts are sometimes interconnected by thin tubules through which not only small molecules, but even proteins, may pass from the stroma of one chloroplast to the stroma of another (Köhler et al. 1997).

Secondly, the aim of linking photosynthesis research to increased plant productivity (an aim that has been entertained for many decades) has been addressed from both traditional and unexpected quarters. One of the latter was an experiment showing that the productivity of wheat plants was reduced by 25% when they were grown in space (on board the space shuttle Discovery; Tripathy et al. 1996). Meanwhile, back on Earth, two groups have attained productively *increases* of ca. 22% by using advanced plant biotechnology and plant breeding and selection procedures, but the increases appeared to be due to factors such as the improved acquisition and maintenance of photosynthetic leaves rather than to improvements in photosynthetic carbon metabolism (Black et al.

1995; Lawlor 1995; Medrano et al. 1995). Finally, the ancient discipline of philosophy has been applied to photosynthesis research by raising the more fundamental question of just what good increased plant productivity would do. While possible benefits in the sectors of socioeconomics and sustainable agriculture (Lawlor 1995) are undeniable, Walker (1995) has emphasized that these benefits could do little more than buy a little time, while more robust solutions to the central facets of the human predicament (including the increasing population of *Homo sapiens*) are sought. Meanwhile, he implies (Walker 1995) that the simple objective of satisfying scientific curiosity is as legitimate a reason for researching photosynthesis as entertainment is a reason for funding sporting events and operatic performances. We hope that the imagination of photosynthesis researchers will in the future be allowed, at least sometimes, to wander unhindered, so that a cornucopia of exciting items on photosynthetic carbon metabolism will appear in our next reviews.

Acknowledgments. We are most grateful to Agatha and Helen for their ever-cheerful assistance.

References

Agostino A, Heldt HW, Hatch MD (1996) Mitochondrial respiration in relation to photosynthetic C_4 acid decarboxylation in C4 species. Aust J Plant Physiol 23:1–7

Allen DJ, McKee IF, Farage PK, Baker NR (1997) Analysis of limitations to CO_2 assimilation on exposure of leaves of two *Brassica napus* cultivars to UV-B. Plant Cell Environ 20:633–640

Anderson LE, Huppe HC, Li AD, Stevens FJ (1996a) Identification of a potential redox-sensitive interdomain disulfide in the sedoheptulose bisphosphatase of *Chlamydomonas reinhardtii.* Plant J 10:553–560

Anderson LE, Gibbons JT, Wang X (1996b) Distribution of ten enzymes of carbon metabolism in pea (Pisum sativum) chloroplasts. Int J Plant Sci 157:525–538

Anderson LE, Li AD, Nehrlich SC, Hill MH, Stevens FJ (1997) The cytosolic fructose bisphosphatase of *Brassica napus* contains a new potential regulatory disulfide and is redox-sensitive. Plant Sci 128:23–30

Andralojc PJ, Keys AJ, Martindale W, Dawson WG, Parry MAJ (1996) Conversion of D-hamamelose into 2-carboxy-D-arabinitol and 2-carboxy-D-arabinitol 1-phosphate in leaves of *Phaseolus vulgaris* L. J Biol Chem 271:26803–26809

Anwaruzzaman, Nakano Y, Yokota A (1996) Different location in dark-adapted leaves of *Phaseolus vulgaris* of ribulose-1,5-bisphosphate carboxylase/oxygenase and 2-carboxyarabinitol 1-phosphate. FEBS Lett 388:223–227

Aoki N, Kanai R (1995) The role of phosphoenolpyruvate in proton/pyruvate cotransport into mesophyll chloroplasts of maize. Plant Cell Physiol 36:187–189

Arigoni D, Sagner S, Latzel C, Eisenreich W, Bacher A, Zenk MH (1997) Terpenoid biosynthesis from 1-deoxy-D-xylulose in higher plants by intramolecular skeletal rearrangement. Proc Natl Acad Sci USA 94:10600–10605

Avilan L, Gontero B, Lebreton S, Ricard J (1997) Memory and imprinting effects in multienzyme complexes. Eur J Biochem 246:78–84

Baalmann E, Backhausen JE, Rak C, Vetter S, Scheibe R (1995) Reductive modification and nonreductive activation of purified spinach chloroplast NADP-dependent glyceraldehyde-3-phosphate dehydrogenase. Arch Biochem Biophys 324:201–208

Balaguer L, Afif D, Dizengremel P, Dreyer E (1996) Specifity factor of ribulose bisphosphate carboxylase/oxygenase of *Quercus robur.* Plant Physiol Biochem 34:879–883

Barthlott W, Porembski ST (1996) Ecology and morphology of *Blossfeldia liliputana* (Cactaceae): a poikilohydric and almost astomate succulent. Bot Acta 109:161–166

Beale CV, Bint DA, Long SP (1996) Leaf photosynthesis in the C_4-grass *Miscanthus x giganteus*, growing in the cool temperate climate of southern England. J Exp Bot 47:267–273

Biehler K, Fock H (1996) Evidence for the contribution of the Mehler-peroxidase reaction in dissipating excess electrons in drought-stressed wheat. Plant Physiol 112:265–272

Black CC, Tu Z-P, Counce PA, Yao P-F, Angelov MN (1995) An integration of photosynthetic traits and mechanisms that can increase crop photosynthesis and grain production. Photosynth Res 46:169–175

Borland AM (1996) A model for the partitioning of photosynthetically fixed carbon during the C_3-CAM transition in *Sedum telephium.* New Phytol 134:433–444

Borland AM, Griffiths H (1997) A comparative study on the regulation of C_3 and C_4 carboxylation processes in the constitutive crassulacean acid metabolism (CAM) plant *Kalanchoë daigremontiana* and the C_3-CAM intermediate *Clusia minor.* Planta 201:368–378

Borland AM, Griffiths H, Maxwell C, Fordham MC, Broadmeadow MSJ (1996) CAM induction in *Clusia minor* L. during the transition from wet to dry season in Trinidad: the role of organic acid speciation and decarboxylation. Plant Cell Environ 19:655-664

Bort J, Brown RH, Araus JL (1996) Refixation of respiratory CO_2 in the ears of C_3 cereals. J Exp Bot 47:1567–1575

Bourguignon J, Merand V, Rawsthorne S, Forest E, Douce R (1996) Glycine decarboxylase and pyruvate dehydrogenase complexes share the same dihydrolipoamide dehydrogenase in pea leaf mitochondria: evidence from mass spectrometry and primary-structure analysis. Biochem J 313:229–234

Brown HT, Morris GH (1893) A contribution to the chemistry and physiology of foilage leaves. J Chem Soc 63:604–677

Brown MM, Hall JL, Ho LC (1997) Sugar uptake by protoplasts isolated from tomato fruit tissues during various stages of fruit growth. Physiol Plant 101:533–539

Castillo FJ (1996) Antioxidative protection in the inducible CAM plant *Sedum album* L. following the imposition of severe water stress and recovery. Oecologia 107:469–477

Castrillo M, Aso P, Longart M, Vermehren A (1997) *In situ* immunofluorescent localisation of ribulose-1,5-bisphosphate carboxylase/oxygenase in mesophyll of C_4 dicotyledonous plants. Photosynthetica 33:39–50

Cerling TE, Harris JM, MacFadden BJ, Leakey MG, Quade J, Eisenmann V, Ehleringer JR (1997) Global vegetation change through the miocene/pliocene boundary. Nature 389:153–158

Christopher JT, Holtum JAM (1996) Patterns of carbon partitioning in leaves of crassulacean acid metabolism species during deacidification. Plant Physiol 112:393–399

Colombo SL, Pairoba CF, Andreo CS (1996) Inhibitory effect of shikimic acid on PEP carboxylase activity. Plant Cell Physiol 37:870–872

Crafts-Brandner SJ, van de Loo FJ, Salvucci ME (1997) The two forms of ribulose-1,5-bisphosphate carboxylase/oxygenase activase differ in sensitivity to elevated temperature. Plant Physiol 114:439–444

Dai Z, Ku MSB, Edwards GE (1996) Oxygen sensitivity of photosynthesis and photorespiration in different photosynthetic types in the genus *Flaveria.* Planta 198:563–571

Daniel E (1997) The temperature dependence of photoinhibition in leaves of *Phaseolus vulgaris* (L.), Influence of CO_2 and O_2 concentrations. Plant Sci 124:1–8

Debnam PM, Shearer G, Blackwood L, Kohl DH (1997) Evidence for channeling of intermediates in the oxidative pentose phosphate pathway by soybean and pea nodule extracts, yeast extracts, and purified yeast enzymes. Eur J Biochem 246:283–290

Déjardin A, Rochat C, Wuillème S, Boutin J-P (1997) Contribution of sucrose synthase, ADP-glucose pyrophosphorylase and starch synthase to starch synthesis in developing pea seeds. Plant Cell Environ 20:1421–1430

Denyer K, Dunlap F, Thorbjørnsen T, Keeling P, Smith AM (1996) The major form of ADP-glucose pyrophosphorylase in maize endosperm is extra-plastidal. Plant Physiol 112:779–785

Desimone M, Henke A, Wagner E (1996) Oxidative stress induces partial degradation of the large subunit of ribulose-1,5-bisphosphate carboxylase/oxygenase in isolated chloroplasts of barley. Plant Physiol 111:789–796

Devi MT, Rajagopalan AV, Raghavendra AS (1996) Purification and properties of glycolate oxidase from plants with different photosynthetic pathways: distinctness of C_4 enzyme from that of a C_3 species and a C_3–C_4 intermediate. Photosynth Res 47:231–238

Du YC, Kawamitsu Y, Nose A, Hiyane S, Murayama S, Wasano K, Uchida Y (1996) Effects of water stress on carbon exchange rate and activities of photosynthetic enzymes in leaves of sugarcane (*Saccharum* sp.). Aust J Plant Physiol 23:719–726

Du Z, Aghoram K, Outlaw WH (1997) *In vivo* phosphorylation of phospho*enol*pyruvate carboxylase in guard cells of *Vicia faba* L. is enhanced by fusicoccin and suppressed by abscisic acid. Arch Biochem Biophys 337:345–350

Duwenig E, Steup M, Kossmann J (1997a) Induction of genes encoding plastidic phosphorylase from spinach (*Spinacia oleracea* L.) and potato (*Solanum tuberosum* L.) by exogenously supplied carbohydrates in excised leaf discs. Planta 203:111–120

Duwenig E, Steup M, Willmitzer L, Kossmann J (1997b) Antisense inhibition of cytosolic phosphorylase in potato plants (*Solanum tuberosum* L.) affects tuber sprouting and flower formation with only little impact on carbohydrate metabolism. Plant J 12:323–333

Eastmond PJ, Dennis DT, Rawsthorne S (1997) Evidence that a malate/inorganic phosphate exchange translocator imports carbon across the leucoplast envelope for fatty acid synthesis in developing castor seed endosperm. Plant Physiol 114:851–856

Echeverria E, Salvucci ME, Gonzalez P, Paris G, Salerno G (1997) Physical and kinetic evidence for an association between sucrose-phosphate synthase and sucrose-phosphate phosphatase. Plant Physiol 115:223–227

Eckardt NA, Portis AR (1997) Heat denaturation profiles of ribulose-1,5-bisphosphate carboxylase/oxygenase (rubisco) and rubisco activase and the inability of rubisco activase to restore activity of heat-denaturated rubisco. Plant Physiol 113:243–248

Eckardt NA, Snyder GW, Portis AR, Ogren WL (1997) Growth and photosynthesis under high and low irradiance of *Arabidopsis thaliana* antisense mutants with reduced ribulose-1,5-bisphosphate carboxylase/oxygenase activase content. Plant Physiol 113:575–586

Eller BM, Ferrari S (1997) Water use efficiency of two succulents with contrasting CO_2 fixation pathways. Plant Cell Environ 20:93–100

Emes MJ, Neuhaus HE (1997) Metabolism and transport in non-photosynthetic plastids. J Exp Bot 48:1995–2005

Escobar-Gutiérrez AJ, Gaudillère J-P (1997) Carbon partitioning in source leaves of peach, a sorbitol-synthesizing species, is modified by photosynthetic rate. Physiol Plant 100:353–360

Evert RF, Russin WA, Botha CEJ (1996) Distribution and frequency of plasmodesmata in relation to photoassimilate pathways and phloem loading in the barley leaf. Planta 198:572–579

Ferreira RMB, Franco E, Teixeira ARN (1996) Covalent dimerization of ribulose bisphosphate carboxylase subunits by UV radiation. Biochem J 318:227–234

Fiorani M, Saltarelle R, De Sanctis R, Palma F, Ceccaroli P, Stocchi V (1996) Role of dehydroascorbate in rabbit erythrocyte hexokinase inactivation induced by ascorbic acid/Fe(II). Arch Biochem Biophys 334:357–361

Fischer K, Kammerer B, Gutensohn M, Arbinger B, Weber A, Häusler RE, Flügge U-F (1997) A new class of plastidic phosphate translocators: a putative link between primary and secondary metabolism by the phosphoenolpyruvate/phosphate antiporter. Plant Cell 9:453–462

Fridlyand LE, Backhausen JE, Holtgrefe S, Kitzmann C, Scheibe R (1997) Quantitative evaluation of the rate of 3-phosphoglycerate reduction in chloroplasts. Plant Cell Physiol 38:1177–1186

Furbank RT, Chitty JA, von Caemmerer S, Jenkins CLD (1996) Antisense RNA inhibition of *RbcS* gene expression reduces rubisco level and photosynthesis in the C_4 plant *Flaveria bidentis.* Plant Physiol 111:725–734

Furbank RT, Pritchard J, Jenkins CLD (1997) Effects of exogenous sucrose feeding on photosynthesis in the C_3 plant tobacco and the C_4 plant *Flaveria bidentis.* Aust J Plant Physiol 24:291–299

Galtier N, Foyer CH, Murchie E, Alred R, Quick P, Voelker TA, Thépenier C, Lascève G, Betsche T (1995) Effects of light and atmospheric carbon dioxide enrichment on photosynthesis and carbon partitioning in the leaves of tomato (*Lycopersicon esculentum* L.) plants over-expressing sucrose phosphate synthase. J Exp Bot 46:1335–1344

Gao Y, Woo KC (1996) Regulation of phospho*enol*pyruvate carboxylase in *Zea mays* by protein phosphorylation and metabolites and their roles in photosynthesis. Aust J Plant Physiol 23:25–32

Giglioli-Guivarc'h N, Pierre J-N, Brown S, Chollet R, Vidal J, Gadal P (1996) The light-dependent transduction pathway controlling the regulatory phosphorylation of C4 phospho*enol*pyruvate carboxylase in protoplasts from *Digitaria sanguinalis.* Plant Cell 8:573–586

Goyal A, Tolbert NE (1996) Association of glycolate oxidation with photosynthetic electron transport in plant and algal chloroplasts. Proc Natl Acad Sci USA 93:3319–3324

Grams TEE, Beck F, Lüttge U (1996) Generation of rhythmic and arrhythmic behaviour of crassulacean acid metabolism in *Kalanchoe daigremontiana* under continuous light by varying the irradiance or temperature: measurements in vivo and model stimulations. Planta 198:110–117

Grams TEE, Borland AM, Roberts A, Griffiths H, Beck F, Lüttge U (1997) On the mechanism of reinitiation of endogenous crassulacean acid metabolism rhythm by temperature changes. Plant Physiol 113:1309–1317

Gray GR, Savitch LV, Ivanov AG, Huner NPA (1996) Photosystem II excitation pressure and development of resistance to photoinhibition. Plant Physiol 110:61–71

Greenberg BM, Wilson MI, Gerhardt KE, Wilson KE (1996) Morphological and physiological responses of *Brassica napus* to ultraviolet-B radiation: photomodification of ribulose-1,5-bisphosphate carboxylase/oxygenase and potential acclimation processes. J Plant Physiol 148:78–85

Grotjohann N (1997) Activation of chloroplast fructose-1,6-bisphosphatase of *Pisum sativum* by mole mass change. Bot Acta 110:323–327

Güerere I, Tezara W, Herrera C, Fernández MD, Herrera A (1996) Recycling of CO_2 during induction of CAM by drought in *Talinum paniculatum* (Portulacaceae). Physiol Plant 98:471–476

Guerin J, Carbonero P (1997) The spatial distribution of sucrose synthase isozymes in barley. Plant Physiol 114:55–62

Guinel FC, Ireland RJ (1996) Immunogold localization and quantitative distribution of two proteins of the glycine decarboxylase complex in developing leaflets of pea (Pisum sativum L.). Int J Plant Sci 157:539–545

Häberlein I, Vogeler B (1995) Completion of the thioredoxin reaction mechanism: kinetic evidence for protein complexes between thioredoxin and fructose 1,6-bisphosphatase. Biochim Biophys Acta 1253:169–174

Harris GC, Königer M (1997) The 'high' concentrations of enzymes within the chloroplast. Photosynth Res 54:5–23

Hartwell J, Smith LH, Wilkins MB, Jenkins GI, Nimmo HG (1996) Higher plant phosphoenolpyruvate carboxylase kinase is regulated at the level of translatable mRNA in response to light or circadian rhythm. Plant J 10:1071–1078

He D, Edwards GE (1996) Estimation of diffusive resistance of bundle sheath cells to CO_2 from modeling of C_4 photosynthesis. Photosynth Res 49:195–208

Heber U, Bligny R, Streb P, Douce R (1996) Photorespiration is essential for the protection of the photosynthetic apparatus of C3 plants against photoinactivation under sunlight. Bot Acta 109:307–315

Heldt HW, Rapley L (1970) Specific transport of inorganic phosphate, 3-P-glycerate and dihydroxyacetone-P, and of dicarboxylates across the inner membrane of spinach chloroplasts. FEBS Lett 10:143–148

Helliker BR, Martin CE (1997) Comparative water-use efficiences of three species of *Peperomia* (Piperaceae) having different photosynthetic pathways. J Plant Physiol 150:259–263

Herbers K, Sonnewald U (1996) Manipulating metabolic partitioning in transgenic plants. Trends Biotechnol 14:198–205

Herppich M, von Willert DJ, Herppich WB (1995) Diurnal rhythm in citric acid content preceded the onset of nighttime malic acid accumulation during metabolic changes form C_3 to CAM in salt-stressed plants of *Mesembryanthemum crystallinum*. J Plant Physiol 147:38–42

Hill LM, Smith AM (1995) Coupled movements of glucose 6-phosphate and triose phosphate through the envelopes of plastids from developing embryos of pea (*Pisum sativum* L.). J Plant Physiol 146:411–417

Hixon M, Sinerius G, Schneider A, Walter C, Fessner W-D, Schloss JV (1996) Quo vadis photorespiration: a tale of two aldolases. FEBS Lett 392:281–284

Huber SC, Huber JL (1996) Role and regulation of sucrose-phosphate synthase in higher plants. Annu Rev Plant Physiol Plant Mol Biol 47:431–444

Huber SC, Huber JL, Liao P-C, Gage DA, McMichael RW, Chourey PS, Hannah LC, Koch K (1996) Phosphorylation of serine-15 of maize leaf sucrose synthase. Plant Physiol 112:793–802

Huner NPA, Maxwell DP, Gray GR, Savitch LV, Krol M, Ivanov AG, Falk S (1996) Sensing environmental temperature change through imbalances between energy supply and energy consumption: redox state of photosystem II. Physiol Plant 98:358–364

Igamberdiev AU, Bykova NV, Gardeström P (1997) Involvement of cyanide-resistant and rotenone-intensitive pathways of mitochondrial electron transport during oxidation of glycine in higher plants. FEBS Lett 412:265–269

Imaizumi N, Samejima M, Ishihara K (1997) Characteristics of photosynthetic carbon metabolism of spikelets in rice. Photosynth Res 52:75–82

Ishibashi M, Usuda H, Terashima I (1996) The loss of ribulose-1,5-bisphosphate carboxylase/oxygenase caused by 24-hour rain treatment fully explains the decrease in the photosynthetic rate in bean leaves. Plant Physiol 111:635–640

Ishida H, Nishimori Y, Sugisawa M, Makino A, Mae T (1997) The large subunit of ribulose-1,5-bisphosphate carboxylase/oxygenase is fragmented into 37-kDa and 16-kDa polypeptides by active oxygen in the lysates of chloroplasts from primary leaves of wheat. Plant Cell Physiol 38:471–479

Ishimaru K, Ichikawa H, Matsuoka M, Ohsugi R (1997) Analysis of C4 maize pyruvate, orthophosphate dikinase expressed in C3 transgenic *Arabidopsis* plants. Plant Sci 129:57–64

Issakidis E, Lemaire M, Decottignies P, Jacquot J-P, Migniac-Maslow M (1996) Direct evidence for the different roles of the N- and C-terminal regulatory disulfides of sorghum leaf NADP-malate dehydrogenase in its activation by reduced thioredoxin. FEBS Lett 392:121–124

Jacquot J-P, Lopez-Jaramillo J, Miginiac-Maslow M, Lemaire S, Cherfils J, Chueca A, Lopez-Gorge J (1997a) Cysteine-153 is required for redox regulation of pea chloroplast fructose-1,6-bisphosphatase. FEBS Lett 401:143–147

Jacquot J-P, Lancelin J-M, Meyer Y (1997b) Thioredoxins: structure and function in plant cells. New Phytol 136:543–570

Jang J-C, Sheen J (1997) Sugar sensing in higher plants. Trends Plant Sci 2:208–214

Jang J-C, León P, Zhou L, Sheen J (1997) Hexokinase as a sugar sensor in higher plants. Plant Cell 9:5–19

Jones TL, Ort DR (1997) Circadian regulation of sucrose phosphate synthase activity in tomato by protein phosphatase activity. Plant Physiol 113:1167–1175

Jones PG, Lloyd JC, Raines CA (1996) Glucose feeding of intact wheat plants represses the expression of a number of Calvin cycle genes. Plant Cell Environ 19:231–236

Kanechi M, Uchida N, Yasuda T, Yamaguchi T (1996) Non-stomatal inhibition associated with inactivation of rubisco in dehydrated coffee leaves under unshaded and shaded conditions. Plant Cell Physiol 37:455–460

Kelly GJ, Latzko E (1996) Photosynthesis. Carbon metabolism: the carbon metabolisms of unstressed and stressed plants. Progress Bot 58:187–220

Keys AJ, Major I, Parry MAJ (1995) Is there another player in the game of rubisco regulation? J Exp Bot 46:1245–1251

Kleczkowski LA (1996) Back to the drawing board: redefining starch synthesis in cereals. Trends Plant Sci 1:363–364

Köhler RH, Cao J, Zipfel WR, Webb WW, Hanson MR (1997) Exchange of protein molecules through connections between higher plant plastids. Science 276:2039–2042

Komatsu S, Masuda T, Hirano H (1996) Rice gibberellin-binding phosphoprotein structurally related to ribulose-1,5-bisphosphate carboxylase/oxygenase activase. FEBS Lett 384:167–171

Konrad A, Banze M, Follmann H (1996) Mitochondria of plant leaves contain two thioredoxins. Completion of the thioredoxin profile of higher plants. J Plant Physiol 149:317–321

Kore-eda S, Kanai R (1997) Induction of glucose 6-phosphate transport activity in chloroplasts of *Mesembryanthemum crystallinum* by the C_3-CAM transition. Plant Cell Physiol 38:895–901

Kostov RV, Small CL, McFadden BA (1997) Mutations in a sequence near the N-terminus of the small subunit alter the CO_2/O_2 specificity factor for ribulose bisphosphate carboxylase/oxygenase. Photosynth Res 54:127–134

Kovtun Y, Daie J (1995) End-product control of carbon metabolism in culture-grown sugar beet plants. Plant Physiol 108:1647–1656

Kozaki A, Takeba G (1996) Photorespiration protects C3 plants from photooxidation. Nature 384:557–560

Kruger NJ, Scott P (1995) Integration of cytosolic and plastidic carbon metabolism by fructose 2,6-bisphosphate. J Exp Bot 46:1325–1333

Kühn C, Quick WP, Schulz A, Riesmeier JW, Sonnewald U, Frommer WB (1996) Companion cell-specific inhibition of the potato sucrose transporter SUT1. Plant Cell Environ 19:1115–1123

Kühn C, Franceschi VR, Schulz A, Lemoine R, Frommer WB (1997) Macromolecular trafficking indicated by localization and turnover of sucrose transporters in enucleate sieve elements. Science 275:1298–1300

Lacuesta M, Dever LV, Muñoz-Rueda A, Lea PJ (1997) A study of photorespiratory ammonia production in the C_4 plant *Amaranthus edulis*, using mutants with altered photosynthetic capacities. Physiol Plant 99:447–455

Laisk A, Loreto F (1996) Determining photosynthetic parameters from leaf CO_2 exchange and chlorophyll fluorescence. Plant Physiol 110:903–912

Lal A, Edwards GE (1996) Analysis of inhibition of photosynthesis under water stress in the C4 species *Amaranthus cruentus* and *Zea mays*: electron tansport, CO_2 fixation and carboxylation capacity. Aust J Plant Physiol 23:403–412

Laporte MM, Galagan JA, Shapiro JA, Boersig MR, Shewmaker CK, Sharkey TD (1997) Sucrose-phosphate synthase activity, and yield analysis of tomato plants transformed with maize sucrose-phosphate synthase. Planta 203:253–259

Law RD, Plaxton WC (1997) Regulatory phosphorylation of banana fruit phospho*enol*pyruvate carboxylase by a copurifying phospho*enol*pyruvate carboxylasekinase. Eur J Biochem 247:642–651

Lawlor DW (1995) Photosynthesis, productivity and environment. J Exp Bot 46:1449–1461

Lebreton S, Gontero B, Avilan L, Ricard J (1997) Information transfer in multienzyme complexes. 1. Thermodynamics of conformational constaints and memory effects in the bienzyme glyceraldehyde-3-phosphate-dehydrogenase-phosphoribulokinase complex of *Chlamydomonas reinhardtii* chloroplasts. Eur J Biochem 250:286–295

Lee S-J, Daie J (1997) End-product repression of genes involving carbon metabolism in photosynthetically active leaves of sugarbeet. Plant Cell Physiol 38:887–894

Leegood RC, Lea PJ, Adcock MD, Häusler RE (1995) The regulation and control of photorespiration. J Exp Bot 46:1397–1414

Leport L, Kandlbinder A, Baur B, Kaiser WM (1996) Diurnal modulation of phosphoenolpyruvate carboxylation in pea leaves and roots as related to tissue malate concentrations and to the nitrogen source. Planta 198:495–501

Li B, Zhang X-Q, Chollet R (1996) Phospho*enol*pyruvate carboxylase kinase in tobacco leaves is activated by light in a similar but not identical way as in maize. Plant Physiol 111:497–505

Li AD, Stevens FJ, Huppe HC, Kersanach R, Anderson LE (1997) *Chlamydomonas reinhardtii* NADP-linked glyceraldehyde-3-phosphate dehydrogenase contains the cysteine residues identified as potentially domain-locking in the higher plant enzyme and is light activated. Photosynth Res 51:167–177

Lichtenthaler HK, Schwender J, Disch A, Rohmer M (1997) Biosynthesis of isoprenoids in higher plant chloroplasts proceeds via a mevalonate-independent pathway. FEBS Lett 400:271–274

Ligeza A, Tikhonov AN, Subczynski WK (1997) In situ measurements of oxygen production and consumption using paramagnetic fusinite particles injected into a bean leaf. Biochim Biophys Acta 1319:133–137

Lilley RMcC, Portis AR (1997) ATP hydrolysis activity and polymerization state of ribulose-1,5-bisphosphate carboxylase oxygenase activase. Plant Physiol 114:605–613

Lozano RM, Wong JH, Yee BC, Peters A, Kobrehel K, Buchanan BB (1996) New evidence for a role for thioredoxin *h* in germination and seedling development. Planta 200:100–106

Lüttge U, Fischer-Schliebs E, Ratajczak R, Kramer D, Berndt D, Kluge M (1995) Functioning of the tonoplast in vacuolar C-storage and remobilization in crassulacean acid metabolism. J Exp Bot 46:1377–1388

MacIntyre HL, Sharkey TD, Geider RJ (1997) Activation and deactivation of ribulose-1,5-bisphosphate carboxylase/oxygenase (rubisco) in three marine microalgae. Photosynth Res 51:93–106

Magnin NC, Cooley BA, Reiskind JB, Bowes G (1997) Regulation and localization of key enzymes during the induction of Kranz-less, C_4-type photosynthesis in *Hydrilla verticillata*. Plant Physiol 115:1681–1689

Martindale W, Leegood RC (1997) Acclimation of photosynthesis to low temperature in *Spinacia oleraceae* L. 1. Effects of acclimation on CO_2 assimilation and carbon partitioning. J Exp Bot 48:1865–1872

Martindale W, Parry MAJ, Andralojc PJ, Keys AJ (1997) Synthesis of 2'-carboxy-D-arabinitol-1-phosphate in french bean (*Phaseolus vulgaris* L.): a search for precursors. J Exp Bot 48:9–14

Mate CJ, von Caemmerer S, Evans JR, Hudson GS, Andrews TJ (1996) The relationship between CO_2-assimilation rate, rubisco carbamylation and rubisco activase content in activase-deficient transgenic tobacco suggests a simple model of activase action. Planta 198:604–613

Matsuba K, Imaizumi N, Kaneko S, Samejima M, Ohsugi R (1997) Photosynthetic responses to temperature of phosphoenolpyruvate carboxykinase type C_4 species differing in cold sensitivity. Plant Cell Environ 20:268–274

McFadden GI, Reith ME, Munholland J, Lang-Unnasch N (1996) Plastid in human parasites. Nature 381:482

Medrano H, Keys AJ, Lawlor DW, Parry MAJ, Azcon-Bieto J, Delgado E (1995) Improving plant production by selection for survival at low CO_2 concentrations. J Exp Bot 46:1389–1396

Meister M, Agostino A, Hatch MD (1996) The roles of malate and aspartate in C_4 photosynthetic metabolism of *Flaveria bidentis* (L.). Planta 199:262–269

Micallef BJ, Sharkey TD (1996) Genetic and physiological characterization of *Flaveria linearis* plants having a reduced activity of cytosolic fructose-1,6-bisphosphatase. Plant Cell Environ 19:1–9

Minamikawa T, Maeda Y, Xu B-M (1995) Ancient lotus seeds preserve activity of starch phosphorylase. Naturwissenschaften 82:375–376

Möhlmann T, Tjaden J, Henrichs G, Quick WP, Häusler R, Neuhaus HE (1997) ADP-glucose drives starch synthesis in isolated maize endosperm amyloplasts: characterization of starch synthesis and transport properties across the amyloplast envelope. Biochem J 324:503–509

Moore BD, Sharkey TD, Seemann JR (1995) Intracellular localization of CA1P and CA1P phosphatase activity in leaves of *Phaseolus vulgaris* L. Photosynth Res 45:219–224

Moore BD, Palmquist DE, Seemann JR (1997) Influence of plant growth at high CO_2 concentrations on leaf content of ribulose-1,5-bisphosphate carboxylase/oxygenase and intracellular distribution of soluble carbohydrates in tobacco, snapdragon and parsley. Plant Physiol 115:241–248

Morell S, Follmann H, De Tullio M, Häberlein I (1997) Dehydroascorbate and dehydroascorbate reductase are phantom indicators of oxidative stress in plants. FEBS Lett 414:567–570

Naeem M, Tetlow IJ, Emes MJ (1997) Starch synthesis in amyloplasts purified from developing potato tubers. Plant J 11:1095–1103

Neuhaus H-E, Maaß U (1996) Unidirectional transport of orthophosphate across the envelope of isolated cauliflower-bud amyloplasts. Planta 198:542–548

Neuhaus HE, Schulte N (1996) Starch degradation in chloroplasts isolated from C_3 or CAM (crassulacean acid metabolism)-induced *Mesembryanthemum crystallinum* L. Biochem J 318:945–953

Neuhaus HE, Thom E, Möhlmann T, Steup M, Kampfenkel K (1997) Characterization of a novel eukaryotic ATP/ADP translocator located in the plastid envelope of *Arabidopsis thaliana* L. Plant J 11:73–82

Ohta S, Usami S, Ueki J, Kumashiro T, Komari T, Burnell JN (1996) Identification of the amino acid residues responsible for cold tolerance in *Flaveria brownii* pyruvate, orthophosphate dikinase. FEBS Lett 396:152–156

Osmond B, Badger M, Maxwell K, Björkman O, Leegood R (1997) Too many photons: photorespiration, photoinhibition and photooxidation. Trends Plant Sci 2:119–121

Osuna L, González M-C, Cejudo FJ, Vidal J, Echevarría C (1996) In vivo and vitro phosphorylation of the phospho*enol*pyruvate carboxylase from wheat seeds during germination. Plant Physiol 111:551–558

Panstruga R, Hippe-Sanwald S, Lee Y-K, Lataster M, Lipka V, Fischer R, Liao YC, Häusler RE, Kreuzaler F, Hirsch H-J (1997) Expression and chloroplast-targeting of active phosphoenolpyruvate synthetase from *Escherichia coli* in *Solanum tuberosum*. Plant Sci 127:191–205

Park Y-I, Chow WS, Osmond CB, Anderson JM (1996) Electron transport to oxygen mitigates against the photoinactivation of photosystem II *in vivo*. Photosynth Res 50:23–32

Parry MAJ, Andralojc PJ, Parmar S, Keys AJ, Habash D, Paul MJ, Alred R, Quick WP, Servaites JC (1997) Regulation of rubisco by inhibitors in the light. Plant Cell Environ 20:528–534

Prabhu V, Chatson KB, Abrams GD, King J (1996) ^{13}C nuclear magnetic resonance detection of interactions of serine hydroxymethyltransferase with C1-tetrahydrofolate synthase and glycine decarboxylase complex activities in Arabidopsis. Plant Physiol 112:207–216

Quick WP, Neuhaus HE (1996) Evidence of two types of phosphate translocators in sweet-pepper (*Capsicum annuum* L.) fruit chromoplasts. Biochem J 320:7–10

Quiñones MA, Lu Z, Zeiger E (1996) Close correspondence between the action spectra for the blue light responses of the guard cell and coleoptile chloroplasts, and the spectra for blue light-dependent stomatal opening and coleoptile phototropism. Proc Natl Acad Sci USA 93:2224–2228

Reche A, Lázaro JJ, Hermoso R, Chueca A, López Gorgé J (1997) Binding and activation of pea chloroplast fructose-1,6-bisphosphatase by homologous thioredoxins *m* and *f*. Physiol Plant 101:463–470

Reddy AR (1996) Fructose 2,6-bisphosphate-modulated photosynthesis in sorghum leaves grown under low water regimes. Phytochemistry 43:319–322

Reimholz R, Geiger M, Haake V, Deiting U, Krause K-P, Sonnewald U, Stitt M (1997) Potato plants contain multiple forms of sucrose phosphate synthase, which differ in their tissue distributions, their levels during development, and their responses to low temperature. Plant Cell Environ 20:291–305

Roberts A, Griffiths H, Borland AM, Reinert F (1996) Is crassulacean acid metabolism activity in sympatric species of hemi-epiphytic stranglers such as *Clusia* related to carbon cycling as a photoprotective process? Oecologia 106:28–38

Robinson JM (1997) Nitrogen limitation of spinach plants causes a simultaneous rise in foliar levels of orthophosphate sucrose and starch. Int J Plant Sci 158:432–441

Rost S, Frank C, Beck E (1996) The chloroplast envelope is permeable for maltose but not for maltodextrins. Biochim Biophys Acta 1291:221–227

Roughnan PG (1997) Stromal concentrations of coenzyme A and its esters are insufficient to account for rates of chloroplast fatty acid synthesis: evidence for substrate channelling within the chloroplast fatty acid synthase. Biochem J 327:267–273

Roughan PG, Ohlrogge JB (1996) Evidence that isolated chloroplasts contain an integrated lipid-synthesizing assembly that channels acetate into long-chain fatty acids. Plant Physiol 110:1239–1247

Rudi H, Doan DNP, Olsen O-A (1997) A $(his)_6$-tagged recombinant barley (*Hordeum vulgare* L.) endosperm ADP-glucose pyrophosphorylase expressed in the baculovirus-insect cell system is insensitive to allosteric regulation by 3-phosphoglycerate and inorganic phosphate. FEBS Lett 419:124–130

Russin WA, Evert RF, Vanderveer PJ, Sharkey TD, Briggs SP (1996) Modification of a specific class of plasmodesmata and loss of sucrose export ability in the *sucrose export defective1* maize mutant. Plant Cell 8:645–658

Sachs J (1862) Über den Einfluß des Lichtes auf die Bildung des Amylums in den Chlorophyllkörnern. Bot Zeit 20:365–373

Sadimantara GR, Abe T, Suzuki J, Hirano H, Sasahara T (1996) Characterization and partial amino acid sequence of a high molecular weight protein from rice seed en-

dosperm: homology to pyruvate orthophosphate dikinase. J Plant Physiol 149:285–289
Salahas G, Hatzidimitrakis K, Georgiou CD, Angelopoulos K, Gavalas NA (1997) Phosphate and sulfate activate the phosphoenolpyruvate carboxylase from the C_4 plant *Cynodon dactylon* L. Bot Acta 110:309–313
Saliendra NZ, Meinzer FC, Perry M, Thom M (1996) Associations between partitioning of carboxylase activity and bundle sheath leakiness to CO_2, carbon isotope discrimination, photosynthesis, and growth in sugarcane. J Exp Bot 47:907–914
Salvucci ME, Ogren WL (1996) The mechanism of rubisco activase: insights from studies of the properties and structure of the enzyme. Photosynth Res 47:1–11
Sasaki Y, Kozaki A, Hatano M (1997) Link between light and fatty acid synthesis: thioredoxin-linked reductive activation of plastidic acetyl-CoA carboxylase. Proc Natl Acad Sci USA 94:11096–11101
Savitch LV, Gray GR, Huner NPA (1997) Feedback-limited photosynthesis and regulation of sucrose-starch accumulation during cold acclimation and low-temperature stress in a spring and winter wheat. Planta 201:18–26
Schaffer AA, Petreikov M (1997) Inhibition of fructokinase and sucrose synthase by cytosolic levels of fructose in young tomato fruit undergoing transient starch synthesis. Physiol Plant 101:800–806
Schmid R, Dring MJ (1996) Influence of carbon supply on the circadian rhythmicity of photosynthesis and its stimulation by blue light in *Ectocarpus siliculosus*: clues to the mechanism of inorganic carbon acquisition in lower brown algae. Plant Cell Environ 19:373–382
Scholes J, Bundock N, Wilde R, Rolfe S (1996) The impact of reduced vacuolar invertase activity on the photosynthetic and carbohydrate metabolism of tomato leaves. Planta 200:265–272
Schwender J, Seemann M, Lichtenthaler HK, Rohmer M (1996) Biosynthesis of isoprenoids (carotenoids, sterols, prenyl side-chains of chlorophylls and plastoquinone) via a novel pyruvate/glyceraldehyde 3-phosphate non-mevalonate pathway in the green alga *Scenedesmus obliquus*. Biochem J 316:73–80
Shatters RG, West SH (1996) Purification and characterization of non-chloroplastic α-1,4-glucan phosphorylases from leaves of *Digitaria eriantha* Stent. J Plant Physiol 149:501–509
Simon J-P (1996) Molecular forms and kinetic properties of pyruvate P_i dikinase from two populations of barnyard grass (*Echinochloa crus-galli*) from sites of contrasting climates. Aust J Plant Physiol 23:191–199
Skillman JB, Winter K (1997) High photosynthetic capacity in a shade-tolerant crassulacean acid metabolism plant. Plant Physiol 113:441-450
Smeekens S, Rook F (1997) Sugar sensing and sugar-mediated signal transduction in plants. Plant Physiol 115:7–13
Smith LH, Lillo C, Nimmo HG, Wilkins MB (1996) Light regulation of phosphoenolpyruvate carboxylase in barley mesophyll protoplasts is modulated by protein synthesis and calcium, and not necessarily correlated with phosphoenolpyruvate carboxylase kinase activity. Planta 200:174–180
Spencer WE, Wetzel RG, Teeri J (1996) Photosynthetic phentoype plasticity and the role of phosphoenolpyruvate carboxylase in *Hydrilla verticillata*. Plant Sci 118:1–9
Stadler R, Brandner J, Schulz A, Gahrtz M, Sauer N (1995) Phloem loading by the PmSUC2 sucrose carrier from *Plantago major* occurs into companion cells. Plant Cell 7:1545–1554
Steiger S, Pfeifer T, Ratajczak R, Martinoia E, Lüttge U (1997) The vacuolar malate transporter of *Kalanchoe daigremontiana*: A 32 kDa polypeptide? J Plant Physiol 151:137–141
Stieger PA, Feller U (1997) Degradation of stromal proteins in pea (*Pisum sativum* L.) chloroplasts under oxidising conditions. J Plant Physiol 151:556–562

Strand Å, Hurry V, Gustafsson P, Gardeström P (1997) Development of *Arabidopsis thaliana* leaves at low temperatures releases the suppression of photosynthesis and photosynthetic gene expression despite the accumulation of soluble carbohydrates. Plant J 12:605–614

Sweetlove LJ, Burrell MM, ap Rees T (1996) Starch metabolism in tubers of transgenic potato (*Solanum tuberosum*) with increased ADP glucose pyrophosphorylase. Biochem J 320:493–498

Tallman G, Zhu J, Mawson BT, Amodeo G, Nouhi Z, Levy K, Zeiger E (1997) Induction of CAM in *Mesembryanthemum crystallinum* abolishes the stomatal response to blue light and light-dependent zeaxanthin formation in guard cell chloroplasts. Plant Cell Physiol 38:236–242

Tamoi M, Ishikawa T, Takeda T, Shigeoka S (1996) Enzymic and molecular characterization of NADP-dependent glyceraldehyde-3-phosphate dehydrogenase from *Synechococcus* PCC 7942: resistance of the enzyme to hydrogen peroxide. Biochem J 316:685–690

Taniguchi M, Sugiyama T (1997) The expression of 2-oxoglutarate/malate translocator in the bundle-sheath mitochondria of *Panicum miliaceum*, a NAD-malic enzyme-type C_4 plant, is regulated by light and development. Plant Physiol 114:285–293

Thorbjørnsen T, Villand P, Denyer K, Olsen O-A, Smith AM (1996) Distinct isoforms of ADPglucose pyrophosphorylase occur inside and outside the amyloplasts in barley endosperm. Plant J 10:243–250

Tolbert NE, Benker C, Beck E (1995) The oxygen and carbon dioxide compensation points of C_3 plants: possible role in regulating atmospheric oxygen. Proc Natl Acad Sci USA 92:11230-11233

Toroser D, Huber SC (1997) Protein phosphorylation as a mechanism for osmotic-stress activation of sucrose-phosphate synthase in spinach leaves. Plant Physiol 114:947–955

Trevanion SJ, Furbank RT, Ashton AR (1997) NADP-malate dehydrogenase in the C_4 plant *Flaveria bidentis*. Plant Physiol 113:1153–1165

Tripathy BC, Brown CS, Levine HG, Krikorian AD (1996) Growth and photosynthetic responses of wheat plants grown in space. Plant Physiol 110:801–806

Uemura K, Suzuki Y, Shikanai T, Wadano A, Jensen RG, Ohmara W, Yokota A (1996) A rapid sensitive method for determination of relative specificity of rubisco from various species by anion-exchange chromatography. Plant Cell Physiol 37:325–331

Ueno O (1996) Immunocytochemical localization of enzymes involved in the C_3 and C_4 pathways in the photosynthetic cells of an amphibious sedge, *Eleocharis vivipara*. Planta 199:394–403

Usami S, Ohta S, Komari T, Burnell JN (1995) Cold stability of pyruvate, orthophosphate dikinase of *Flaveria brownii*. Plant Mol Biol 27:969–980

Venkaiah B, Kumar A (1996) Multiple forms of starch phosphorylase from *Sorghum* leaves. Phytochemistry 41:713–717

Vidal J, Chollet R (1997) Regulatory phosphorylation of C_4 PEP carboxylase. Trends Plant Sci 2:230–237

Walker DA (1995) Manipulating photosynthetic metabolism to improve crops: an inversion of ends and means. J Exp Bot 46:1253–1259

Walker DA (1997) Tell me where all past years are. Photosynth Res 51:1–26

Walker RP, Leegood RC (1996) Phosphorylation of phosphoenolpyruvate carboxykinase in plants. Biochem J 317:653–658

Wang X, Tang X-y, Anderson LE (1996) Enzyme-enzyme interaction in the chloroplast: physical evidence for association between phosphoglycerate kinase and glyceraldehyde-3-phosphate dehydrogenase in vitro. Plant Sci 117:45–53

Wedel N, Soll J, Paap BK (1997) CP12 provides a new mode of light regulation of Calvin cycle activity in higher plants. Proc Natl Acad Sci USA 94:10479–10484

Wild K, Grafmüller R, Wagner E, Schulz GE (1997) Structure, catalysis and supramolecular assembly of adenylate kinase from maize. Eur J Biochem 250:326–331

Wildner GF, Schlitter J, Müller M (1996) Rubisco, an old challenge with new perspectives. Z Naturforsch 51c:263–276

Wingler A, Lea PJ, Leegood RC (1997) Control of photosynthesis in barley plants with reduced activities of glycine decarboxylase. Planta 202:171–178

Woodrow IE, Kelly ME, Mott KA (1996) Limitation of the rate of ribulosebisphosphate carboxylase activation by carbamylation and ribulosebisphosphate carboxylase activase activity: development and tests of a mechanistic model. Aust J Plant Physiol 23:141–149

Xu H-L, Gauthier L, Desjardins Y, Gosselin A (1997) Photosynthesis in leaves, fruit, stem and petioles of greenhouse-grown tomato plants. Photosynthetica 33:113–123

Yamasaki T, Kudoh T, Kamimura Y, Katoh S (1996) A vertical gradient of the chloroplast abundance among leaves of *Chenopodium album*. Plant Cell Physiol 37:43–48

Zhang X-Q, Chollet R (1997a) Phospho*enol*pyruvate carboxylase protein kinase from soybean root nodules: partial purification, characterization, and up/down-regulation by photosynthate supply from the shoots. Arch Biochem Biophys 343:260–268

Zhang X-Q, Chollet R (1997b) Seryl-phosphorylation of soybean nodule sucrose synthase (nodulin-100) by a Ca^{2+}-dependent protein kinase. FEBS Lett 410:126–130

Zhigang A, Löw R, Rausch T, Lüttge U, Ratajczak R (1996) The 32 kDa tonoplast polypeptide D_i associated with the V-type H^+-ATPase of *Mesembryanthemum crystallinum* L. in the CAM state: A proteolytically processed subunit B? FEBS Lett 389:314–318

Zhu YJ, Komor E, Moore PE (1997) Sucrose accumulation in the sugarcane stem is regulated by the difference between the activities of soluble acid invertase and sucrose phosphate synthase. Plant Physiol 115:609–616

Zotz G, Ziegler H (1997) The occurrence of crassulacean acid metabolism among vascular epiphytes from central Panama. New Phytol 137:223–229

Zrenner R, Krause K-P, Apel P, Sonnewald U (1996) Reduction of the cytosolic fructose-1,6-bisphosphatase in transgenic potato plants limits photosynthetic sucrose biosynthesis with no impact on plant growth and tuber yield. Plant J 9:671–681

Dr. Grahame J. Kelly
Centre for Molecular Biotechnology
School of Life Sciences
Queensland University of Technology
Brisbane, Queensland 4000, Australia
e-mail g.kelly@qut.edu.au

Edited by
U. Lüttge

Note added in proof (2005):
The second-last sentence of Sect. 4b should read: "Tissues that store oil synthesise fatty acids in modified plastids." Oils are stored in oleosomes, not in plastids.

Photosynthesis: Carbon Metabolism from DNA to Deoxyribose

Grahame J. Kelly

"One of the first eukaryotic organisms that will be completely sequenced is the small mustard species *Arabidopsis thaliana* .... about 59% of the genome sequence is currently available in public databases and .... it is currently anticipated that the complete genome sequence .... will be available by the end of the year 2000."
(Somerville and Somerville 1999).

1 Introduction

Life, like the Calvin cycle, is a revolution that returns to its beginning. Even photosynthetic carbon metabolism is a part of this revolving life. Traditionally, photosynthesis is seen as life's energy transducer, whereby solar light energy is collected and changed into the potential energy in the covalent bonds of sugar molecules. Indeed, the fate of a large proportion of these sugar molecules is respiration so that this energy can become available to non-photosynthetic cells or to photosynthetic cells in darkness. However, a small fraction of these sugars are metabolized to form the monomers (including deoxyribose; Debnam and Emes 1999) that are linked to make the new DNA of newly emerging photosynthetic cells. This DNA will be transcribed to initiate the synthesis of the Calvin-cycle enzymes in these cells so that they too can contribute to the conversion of CO_2 to more sugars and, ultimately, more DNA. We should reflect on the photosynthesis–DNA link because, as the above quote indicates, we can expect to be able to view the entire nucleotide sequences of the five chromosomes of the small, photosynthetic, flowering plant *Arabidopsis thaliana* within the next year or two. We will also be able to determine the position (or positions) on these chromosomes where each of the 11 enzymes of the Calvin cycle are encoded. Due to the imminence of these events, this review of photosynthetic carbon metabolism concentrates on the properties of all of the proteins that make up the machinery of CO_2 fixation. However, it must be remembered that these proteins are synthesized from instructions on DNA via a process that may be influenced by the status of CO_2 fixation and that may be manipulated using the tools of modern molecular biology. The literature covered in this review has predominantly been published since the last review in volume 60 of *Progress in Botany.*

G.J. Kelly and E. Latzko:
Thirty Years of Photosynthesis

Originally published in Progress in Botany 62

2 The Chloroplast

One impressive advance and one surprise concerning chloroplasts have recently occurred. The advance was the successful isolation of functional chloroplasts from the diatoms *Odontella sinensis* and *Coscinodiscus granii*. These chloroplasts are unusual in that they have four envelope membranes; at least one of these must have remained intact, because the chloroplasts photosynthesized and appeared to possess a phosphate (P) translocator (Wittpoth et al. 1998). The surprise is the report that chloroplasts, and not phytochrome, mediate the photo-induction of spore germination in the liverwort *Marchantia polymorpha* (Nakazato et al. 1999). Such observations are reminders that there is substantial diversity in the cellular and molecular characteristics of photosynthetic organisms and that the genome of shade-loving *A. thaliana* will only be a guide to the genomes of other photosynthetic organisms.

a) Ribulose-Bisphosphate Carboxylase/Oxygenase

The focus of research on ribulose-bisphosphate (P_2) carboxylase/oxygenase (Rubisco), the most popular of the Calvin-cycle enzymes, has shifted toward the two DNA sequences that encode its two subunits. The higher-plant enzyme consists of eight chloroplast-encoded large subunits that contain catalytic sites, and eight nuclear-encoded small subunits whose function is poorly defined (Spreitzer 1999). Molecular approaches to the study of the enzyme have been hampered by this complex arrangement, but recent successes in producing a directed mutation in the large subunit of the tobacco enzyme (Whitney et al. 1999) suggest that there will be more of this type of research in the future. The creation of hybrid enzymes has been reported; both *Arabidopsis* large subunit/pea small subunit and sunflower large subunit/tobacco small subunit enzymes have been constructed (Getzoff et al. 1998; Kanevski et al. 1999). Both hybrid Rubiscos were active and, in fact, were hardly different from their normal counterparts.

Investigations into the movement of entire Rubiscos from one species to another for practical purposes have also been performed. In particular, the observation (Uemura et al. 1997) that certain red algae contain a form of Rubisco that is more than twice as efficient as the higher-plant Rubisco at distinguishing between CO_2 and O_2 (thereby favoring photosynthesis over photorespiration) led to a suggestion that agricultural productivity might be increased if the Rubisco in crop plants is replaced by this red-algal enzyme (Mann 1999). However, the overall lower activity of the red-algal enzyme is a drawback to this proposal (Portis and Zhang 1999). In addition, while the activity of Rubisco may be the principal determinant of the photosynthetic rate, the photosynthetic rate is certainly not the principal determinant of the growth rate of crops in the field (Hansen et al. 1999). In related research, Ramage et al. (1998) attempted to improve the ability of a cyanobacterial Rubisco to distinguish between CO_2 and O_2 using site-directed mutagenesis to alter its amino acid sequence so that it mimics that of the efficient red-algal enzyme. The attempt was not successful.

Nevertheless, study of the relatively unique algal Rubiscos is enriching our understanding of the enzyme. In the chloroplasts of many green micro-algae, most of the enzyme is in the proteinaceous structure called the pyrenoid and performs CO_2 fixation there (Borkhsenious et al. 1998; Moroney and Chen 1998). In dinoflagellates, it is a strange enzyme; it is composed only of large subunits and has not displayed its catalytic ability in vitro (Whitney and Andrews 1998).

Considerable attention has been given to the relatively complex regulation of Rubisco activity. Initially, the inactive protein is activated when CO_2 and Mg^{2+} bind to an activation site, an event that seems to be partially reversed near midday, when the well-known midday depression of photosynthesis is evident (Sinha et al. 1997). Secondly, the enzyme fully activated by CO_2 and Mg^{2+} can be inactivated again by the binding of various sugar-Ps, such as xylulose-P_2 and, as recently reported by Zhu et al. (1998), 3-ketoarabinitol-P_2; both of these are formed when the chemical reaction follows a false route. Another recently discovered inactivator is glycero-2,3-pentodiulose-P_2, which can occur as a contaminant in stored preparations of the enzyme's substrate, ribulose-P_2 (RuBP; Kane et al. 1998). However, the most physiologically relevant inactivator is 2-carboxy-arabinitol-1-P (CA1P), which appears in the leaves of some (but not all) plant species at night, binding to the enzyme and keeping it inactive. A previous report that this CA1P might not be in the same part of the leaf as Rubisco has been enthusiastically challenged by Parry et al. (1999).

Of course, removal of CA1P must occur if photosynthesis is to be possible during the next day. Dissociation of CA1P from the enzyme (and dissociation of all other inhibitory sugar-Ps) is deemed to be the task of a second protein, called Rubisco activase (Hammond et al. 1998a). In a new twist to the topic of light-mediated regulation of Calvin-cycle enzymes (our earlier reviews), Zhang and Portis (1999) have discovered that one form of this protein undergoes a thioredoxin-f-mediated reduction of disulfide bonds. This reduction turns on its ability to interact with the Rubisco and rid the latter of the inhibitory CA1P. Subsequent dissolution of the CA1P also seems to be enhanced by illumination, because the activity of the participating phosphatase has been found (Heo and Holbrook 1999) to be markedly stimulated when the chloroplast pool of glutathione becomes totally reduced, as expected in light.

The notable temperature sensitivity of the Rubisco activase, and experiments with plants engineered to contain reduced levels of the activase, have been used to provide further evidence that the activase is intimately involved in the overall light-mediated activation of Rubisco (Feller at al. 1998; Hammond et al. 1998b). However, there are tantalizing indications that the influence of the activase extends beyond this role (He et al. 1997; Li et al. 1999). This suspicion and the data of Morales et al. (1999), which imply that the productivity of maize is correlated with the content of Rubisco activase in its leaves (an observation that conflicts somewhat with the above-mentioned dogma that the photosynthetic rate is not a primary determinant of growth rate), will ensure that this mysterious protein will receive more attention from researchers during the coming years.

b) Other Calvin-Cycle Enzymes

A relative resurgence of interest in the remaining ten enzymes of the Calvin cycle has occurred during the past 2 years. This should continue, because the *Arabidopsis* genome sequence will soon be available (Sect. 1), and a complete collection of clones of complimentary DNAs for all of the Calvin-cycle enzymes is now available; the last to be added to the list was transketolase (Flechner et al. 1996). Another stimulus is the increasingly sophisticated view of the in vivo status of the enzymes. The association of glyceraldehyde-3-P (GAP) dehydrogenase and ribulose-5-P (Ru5P) kinase with a small linking peptide, reported for higher plants in our previous review, has now been shown to be widespread. It occurs in mosses, green algae and even cyanobacteria (Wedel and Soll 1998). Several of the Calvin-cycle enzymes, including Ru5P kinase, transketolase and Ru5P 3-epimerase, seem to be arranged in a thin layer on thylakoid membrane surfaces. They are surrounded by more soluble stromal proteins, including Rubisco and the enigmatic carbonic anhydrase (Jebanathirajah and Coleman 1998; Teige et al. 1998). The 3-epimerase was also studied by Chen et al. (1998b), who found that, in spinach leaves, it is an eight-subunit protein that is extremely active but very labile. The well-known light-mediated reductive activation of several of the Calvin-cycle enzymes, including fructose-1,6-bisphosphatase (FBPase), sedoheptulose-1,7-bisphosphatase (SBPase; Dunford et al. 1998) and Ru5P kinase, have been reviewed by Ruelland and Miginiac-Maslow (1999), who suggested that a central feature of these enzymes is a loosening of the structure of the enzymes on reduction of the disulfide bonds, thereby increasing the flexibility of the active site and, consequently, the likelihood of catalysis. Nicotinamide adenine dinucleotide phosphate (NADP)-linked malate dehydrogenase from sorghum provides an excellent example of this phenomenon (Goyer et al. 1999).

Most of the Calvin cycle's GAP dehydrogenase exists in chloroplasts as a tetramer of two "A" subunits and two slightly different "B" subunits, although a minor portion (~20%) occurs as a homotetramer of four "A" subunits (Scagliarini et al. 1998). Surprisingly, some of the GAP dehydrogenase is located in the nucleus, where it apparently has a totally different catalytic function, namely the excision of the base uracil from DNA (Wang et al. 1999b). Another surprising enzyme is aldolase (a protein responsible for two of the 13 reactions in the Calvin cycle; Flechner et al. 1999). When Haake et al. (1998) used anti-sense DNA technology to lower the aldolase activity in potato leaves, the inhibition of photosynthesis and growth was unexpectedly dramatic, given that the enzyme catalyzes a reversible reaction and possesses "no known regulatory properties" (authors' quote). It should be recalled, however, that this enzyme is susceptible to inhibition by RuBP when operating photosynthetically (i.e., catalyzing fructose-1,6-P_2 formation; our 1992 review).

Unlike the two aldolase reactions, the Calvin cycle's two phosphatase reactions in higher-plant chloroplasts require two enzyme proteins: FBPase and SBPase. Only SBPase is specific for its reaction; the FBPase can act as an SBPase (Ashton 1998). However, the specific SBPase appears to be vital for the plant because, when its level in tobacco leaves was reduced to less than 20% of the normal level, the plants exhibited chlorosis and reduced growth (Harrison et al. 1998b). These conclusions may not pertain to cyanobacteria, however, because Tamoi et al. (1999) found that the most important FBPase in cyanobacteria is a strange, bifunctional (FBPase/SBPase) protein not found in land plants; it is quite resistant to inactivation by H_2O_2. Finally, marine micro-algae have again provided an example of a Calvin cycle enzyme with relatively unique properties: an Ru5P kinase that is a tetramer of especially large subunits, from the chromophyte *Heterosigma carterae* (Hariharan et al. 1998).

Continuous operation of the Calvin cycle requires that five-sixths of the GAP pool be used to regenerate RuBP and that no more than one-sixth be diverted to sucrose or starch synthesis. It is well recognized that the chloroplast envelope's P translocator ensures that the quota for sucrose is not exceeded (Fig. 1), but the mechanism restricting starch synthesis within the chloroplast has been poorly defined. Isotope studies by Schleucher et al. (1999) have revealed that the reaction representing the

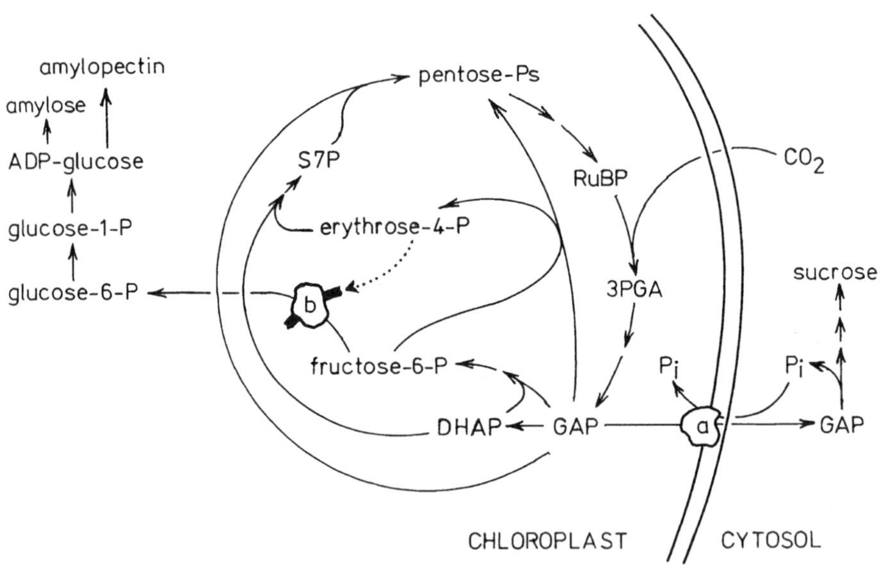

Fig. 1. Maintaining the Calvin cycle's pool of ribulose-P_2. The phosphate translocator (*a*) prevents excessive export of glyceraldehyde-3-P. The pool of the Calvin cycle's erythrose-4-P may be sufficient to appreciably inhibit hexose-P isomerase (*b*) so that the synthesis of transitory starch (amylose and amylopectin) is not excessive. *3PGA* 3-P-glycerate, *DHAP* dihydroxyacetone-P, *S7P* sedoheptulose-7-P

first step out of the Calvin cycle and toward starch (the conversion of fructose-6-P to glucose-6-P) is not at equilibrium in the chloroplast. This indicates a restriction of the activity of the enzyme hexose-P isomerase, presumably to ensure that the quota of photosynthate for starch is not exceeded. The restriction may be effected through the strong inhibition of this enzyme by erythrose-4-P (our 1982 review; Backhausen et al. 1997; Fig. 1).

One last point of interest is the fascinating disappearance of enzymes specific to the Calvin cycles of tomato and sweet-pepper fruits as these fruits ripen and the population of plastids changes from predominantly chloroplasts to predominantly chromoplasts. The NADP-linked GAP dehydrogenase disappears from sweet-pepper fruits (Backhausen et al. 1998), and the plastidic FBPase disappears from tomato fruits (Büker et al. 1998). Unexpectedly, the latter enzyme remains in ripe sweet-pepper fruits (Thom et al. 1998).

c) Transitory Starch Metabolism

There is currently a burgeoning interest in the enzymology of the biosynthesis of starch in plant plastids, partly due to the commercial value of starch products and the perceived possibility that molecular-biology techniques may be used to increase this value by altering the proportions and/or characteristics of the two principal components, amylose (a linear glucose polymer) and amylopectin (a branched glucose polymer). In chloroplasts, starch is synthesized and decomposed diurnally. Synthesis by day represents an important transient sink of photosynthate; without it, the rate of photosynthesis may be reduced (Sun et al. 1999a). As noted above, synthesis begins when some of the Calvin cycle's fructose-6-P is diverted and isomerized to glucose-6-P; this glucose-6-P is then converted, in two steps, to adenosine diphosphate (ADP)-glucose, the glucose donor for polymer manufacture (Fig. 1). The second of these two steps is catalyzed by ADP-glucose pyrophosphorylase, which is restricted to chloroplasts in leaves (Chen et al. 1998a). This pyrophosphorylase is a principal regulatory enzyme of starch synthesis (Greene and Hannah 1998; Kleczkowski 1999), but the importance of the preceding enzyme, P-glucomutase, should not be overlooked. Its indispensability has been demonstrated in experiments with pea leaves (Harrison et al. 1998a), and the possibility that it may contribute to regulation has been raised by Hattenbach and Heineke (1999), who found that both RuBP and FBP inhibit its activity.

A collection of starch synthases catalyzes the glucosyl transfer from ADP-glucose to the growing starch molecules that constitute the insoluble starch granule in the chloroplast. Some of these are totally or predominantly soluble enzymes, and these forms add glucosyl units to amylopectin (Tomlinson et al. 1998; Kossmann et al. 1999). Branching of the amylopectin molecule is achieved by specialized branching enzymes (Jobling et al. 1999). There are indications that the degree of branching is initially excessive, resulting in a highly branched, soluble, glycogen-like molecule, but this is apparently rectified by the indirect intercession of a de-branching enzyme (Greene and Hannah 1998; Zeeman et al. 1998a). The other component of starch, amylose, is synthesized within the amylopectin matrix by a special form of starch synthase that is bound to the starch granule and is therefore referred to as the granule-bound starch synthase (Tomlinson et al. 1998; Tatge et al. 1999; Wang et al. 1999a).

A marine alga has provided another surprise. Nyvall et al. (1999), having noted that red algae synthesize starch in the cytosol rather than in the chloroplast, investigated *Gracilaria tenuistipitata* and found that its starch synthase prefers uridine diphosphate–glucose rather than ADP–glucose.

The dissimulation of chloroplast starch proceeds at night. Experiments with *Arabidopsis* indicate that the intact starch granule is initially attacked by a novel endoamylase, whose activity is induced by the onset of darkness and then mysteriously disappears after 2 h (Kakefuda and Preiss 1997). By then, however, the granule is sufficiently eroded, so other endoamylases, including one specifically directed at starch (Zeeman et al. 1998b), can continue the degradation process, releasing small, soluble sugars. New evidence has increased the likelihood that glucose is the sugar exported to the cytosol, where it is phosphorylated by a hexokinase as a first step toward sucrose manufacture (Schleucher et al. 1998; Veramendi et al. 1999). It seems unlikely that either glucose-1-P or glucose-6-P could be exported, because chloroplasts lack envelope transporter proteins for these glucose-Ps, despite the fact that non-photosynthetic plastids (amyloplasts, chromoplasts, embryo plastids) have them (Eastmond and Rawsthorne 1998; Harrison et al. 1998a; Kammerer et al. 1998; Thom et al. 1998). Amyloplasts even possess an ADP–glucose transporter for the *import* of this substrate of starch synthesis, coincident with the previously reported unusual cytosolic location of ADP–glucose pyrophosphorylase in cereal grain endosperm (Shannon et al. 1998).

3 The Photosynthetic Cell

a) Uptake of Inorganic Carbon

The collection of CO_2 by photosynthetic cells and the delivery of this CO_2 to Rubisco is intriguing. Cyanobacteria and algae seem to engage in this process, but land plants usually do not. Much recent research has been devoted to the elucidation of the molecular mechanisms used by algae to collect and concentrate CO_2 within themselves. Because this research has been recently reviewed (Kaplan and Reinhold 1999; Moroney and Somanchi 1999), no more than a succinct summary is presented here.

The central feature of the inorganic carbon (C_i)-uptake mechanism of these algae is the possession of a CO_2 transporter protein and/or a bicarbonate transporter protein on the plasma membrane. Recent evidence from experiments with marine red, green and brown macro-algae, a marine diatom, and a marine dinoflagellate indicate that CO_2 transport is the preferred or only option. In that case, the transporter is assisted by a carbonic anhydrase delivered to the extracellular, periplasmic space, where it catalyzes the generation of CO_2 from bicarbonate, the most abundant C_i species in seawater (Flores-Moya and Fernández 1998; Nimer et al. 1998, 1999; Andría et al. 1999; Axelsson et al. 1999). A similar situation exists with freshwater green micro-algae, such as *Chlorella* and *Chlamydomonas* (see above reviews). Experiments with another member of this group, *Eremosphaera viridis*, indicate that the CO_2 transporter also performs the pumping task; i.e., it hydrolyzes cytosolic adenosine triphosphate (ATP) and uses the liberated energy to pump CO_2 into the cell against the concentration gradient (Deveau et al. 1998).

However, the C_i-concentration process does not stop there. Additional proteins involved include CO_2 and bicarbonate transporters on the chloroplast envelopes of the green microalgae *Chlamydomonas reinhardtii* and *Dunaliella tertiolecta* (Amoroso et al. 1998), and several intracellular carbonic anhydrases, (including forms associated with pyrenoids, thylakoids and even mitochondria; Eriksson et al. 1998; Villarejo et al. 1998; Park et al. 1999). The large proteinaceous pyrenoids common in microalgae are generally considered to be a part of (and perhaps essential for) the C_i-concentrating mechanism, but Morita et al. (1998) have questioned this by reporting that two of four examined species of the green-algae genus *Chloromonas* lack pyrenoids but have the ability to concentrate C_i.

However, the observation of pyrenoids has led to the first report of an algae-like, C_i-concentrating mechanism in a land plant. Primitive bryophytes in the class Anthocerotae (hornworts) possess pyrenoids and a C_i-concentrating mechanism (Smith and Griffiths 1996).

Exactly why this pyrenoid-based C_i-concentrating mechanism was lost as land plants evolved is a mystery (Smith and Griffiths 1996). However, its loss may explain another mystery: the presence of carbonic anhydrase in land plants. Perhaps the loss was not complete, and this enzyme is simply a relic of the ancestral mechanism. It may even exist in a loose association with Rubisco in the chloroplast stroma (a relic pyrenoid?; Jebanathirajah and Coleman 1998), and plasmalemma and thylakoid-bound forms are also reported to exist (Stemler 1997; Ignatova et al. 1998). Some researchers have been tempted to conclude that it has no role in the leaves of land plants (our 1996 review). However, this might not apply to rice plants; Sasaki et al. (1998) recently proposed that rice plants suffer from Zn deficiency because of the consequent drop in the activity of carbonic anhydrase, which is a Zn metalloprotein. Incidentally, a very recent report that marine micro-algae can use Cd instead of Zn to assemble this enzyme (Cullen et al. 1999) may be relevant to the observation (Siedlecka et al. 1999) that Cd benefits the activation of Rubisco in bean leaves through a process involving carbonic anhydrase.

Carbonic anhydrase may gain notoriety if the conclusions of Tchernov et al. (1998) become widely known. These authors found that the enzyme in marine phytoplankton is so effective that a large excess of CO_2 is produced; because it is not assimilated, it is simply evolved, making phytoplankton a source (rather than a sink, as popularly believed) of this greenhouse gas.

b) Sucrose Biosynthesis

Rather surprisingly, research focussed directly on sucrose biosynthesis and its regulation in photosynthetic cells has been almost non-existent during the past 2 years. The only point of interest has been the entry of a new regulator of the penultimate enzyme, sucrose-P synthase. It seems that the regulatory proteins named "14-3-3"s, well known in mammalian cell biology, also exist in photosynthetic cells and bind to phosphorylated proteins, including sucrose-P synthase (and nitrate reductase; Toroser et al. 1998; Moorhead et al. 1999). It seems likely that the inactivation of these two enzymes by phosphorylation is not complete until a 14-3-3 protein has attached to the phosphorylated site.

c) The Enigma of Glucose and DNA

The downregulation of photosynthesis that occurs if soluble sugars accumulate in photosynthetic cells and tissues is a relatively new topic that has become both controversial and complicated. Hexokinase is thought to be the enzyme that detects the level of sugars (as hexoses) and then initiates a signal-transduction pathway that leads to a reduction in the transcription of DNA and a consequent reduction in the biosynthesis of enzymes required for CO_2 fixation. Alteration of the cell's content of hexokinase and experiments with alternative substrates of the enzyme that are phosphorylated but not further metabolized support this role. However, it is unknown whether hexokinase performs the role directly or indirectly, through perturbations that its altered activity causes in the metabolism (Dai et al. 1999; Halford et al. 1999; Moore et al. 1999). Added to this uncertainty is the complication that more than one sugar-sensing/signaling system might exist; there is evidence that hexokinase-independent systems influence the expression of other genes not directly related to CO_2 fixation (Chiou and Bush 1998; Cheng et al. 1999; Lalonde et al. 1999). There are also indications that environmental parameters, such as low temperature or an abundant supply of nitrogen, can reduce or cancel the system that links sugar levels to the production of photosynthetic enzymes (Huner et al. 1998; Nielsen et al. 1998).

Another query concerning the hexokinase system is how the over-accumulation of sucrose (a disaccharide of glucose and fructose) could result in an increased involvement of this enzyme, for which the substrates are hexoses (glucose or fructose, but not sucrose). A solution has been supplied by Moore et al. (1999), who propose that the sucrose is hydrolyzed to the hexoses and then re-synthesized from them in a futile cycle that involves hexokinase but achieves nothing metabolically, other than ATP hydrolysis (Fig. 2). In this scheme, sucrose is hydrolyzed by the enzyme invertase, an event that might explain why photosynthetic

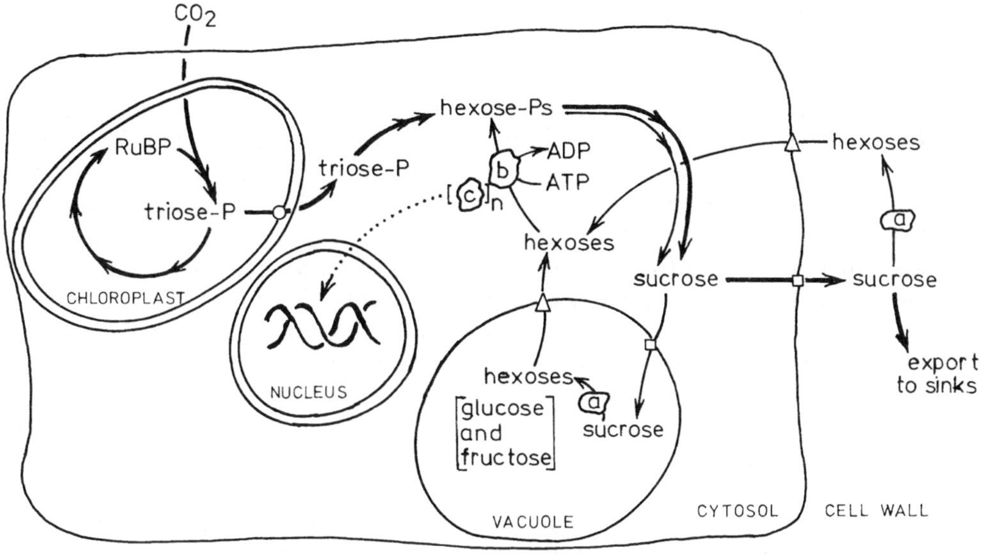

Fig. 2. Sucrose cycling and the sugar-mediated regulation of genes encoding enzymes of photosynthesis. The *thick arrows* represent the standard overview of photosynthesis, with CO_2 entering chloroplasts and sucrose being exported to sink tissues. No free hexoses are involved in this pathway. However, if sucrose accumulates in the vacuoles or cell wall, it may be cycled (*thin arrows*). Initial hydrolysis by invertase (*a*) releases free hexoses, which are transported to the cytosol, where hexokinase (*b*) converts them into hexose-Ps, which can be used to remanufacture the sucrose via the standard pathway. The hexokinase, on becoming more active, somehow initiates a signal pathway involving other proteins ($[c]_n$) and culminating in the repression of the DNA transcription that encodes enzymes of the photosynthetic carbon metabolism. *Open circles, squares and triangles* represent membrane transporter proteins (Moore et al. 1999)

cells possess ample amounts of this enzyme (Kingston-Smith et al. 1999). The scheme might even be elaborated to include the enigmatic cytosolic invertase which, unlike those in the vacuole and cell wall, operates at neutral pH and is more specific for sucrose (Sturm 1999). Lastly, one wonders about the unknown molecules in the signal-transduction pathway subsequent to the enlivened hexokinase. An intriguing possibility is that trehalose (a disaccharide of two glucose units linked C1-to-C1), common in fungi and invertebrates but seldom detected in plants, is involved. By revealing the presence in plants of enzymes of trehalose metabolism, plant-cell DNA and protein analysis have indicated that this might be so (Goddijn and Smeekens 1998; Moorhead et al. 1999; Müller et al. 1999). Indeed, a link to hexokinase might be involved if trehalose-6-P strongly inhibits this enzyme in plants, as it does in the fungus *Aspergillus niger* (Panneman et al. 1998).

d) Mitochondrial Respiration and the Oxidative Pentose-P Pathway

There is current consensus, recently reviewed by Hoefnagel et al. (1998), that an intricate metabolic interaction occurs between mitochondria and chloroplasts in illuminated photosynthetic cells. This includes continued respiration by mitochondria in light, such that 10–15% of newly produced photosynthate is re-oxidized to CO_2 (Lawlor 1995). This respiration is considered essential for the generation of carbon skeletons (via the Krebs cycle) destined for use in the biosynthesis of amino acids. The respiration may also be essential for the oxidation of excess reducing equivalents (via the respiratory electron-transport chain; Hoefnagel et al. 1998). The latter role may be especially pertinent at the onset of illumination when, during the so-called induction phase, the Calvin cycle lacks the ability to use the available reducing power supplied by the photochemistry (Igamberdiev et al. 1998; Padmasree and Raghavendra 1999).

A second route by which photosynthate may be re-oxidized to CO_2 is the oxidative pentose-P pathway, in which glucose-6-P is oxidatively decarboxylated to Ru5P by two enzymes. Six molecules of this Ru5P may then be metabolized back to five molecules of glucose-6-P by eight enzymes. A report that this pathway is restricted to plastids (our 1996 review) has been supported in general; nevertheless, it has been shown

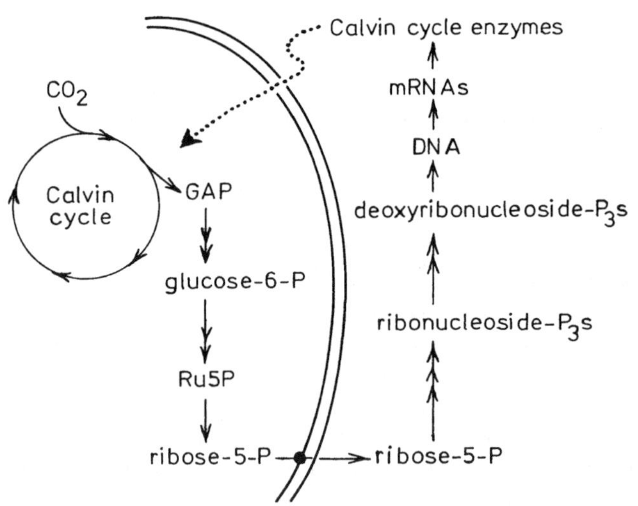

Fig. 3. From DNA to deoxyribose: a conceptual overview. The DNA-directed synthesis of Calvin-cycle enzymes leads to the photosynthetic generation of sugar-Ps, including ribose-5-P, which is the immediate precursor of the deoxyribose in the sugar-phosphate backbone of DNA. A plastid-envelope pentose-P translocator (Debnam and Emes 1999) may be a component of this system. In reality, the events shown would predominantly occur in different cells, with photosynthesis in mature cells and DNA synthesis in meristematic cells

that it is not universal. It applies to spinach, maize and pea plastids, but not to tobacco and carrot plastids (Krook et al. 1998; Debnam and Emes 1999). A common feature, highlighted by Debnam and Emes (1999), is that the initial two enzymes that catalyze the oxidative de-carboxylation of the glucose-6-P are always found in both the plastid and the cytosol, but the subsequent two enzymes, which begin the metabolism of the Ru5P, are often restricted to the plastid. This implies that a plastid-envelope pentose-P translocator must exist to permit the metabolism of any Ru5P generated in the cytosol. It is not inconceivable that such a translocator may be critical for the synthesis of the deoxyribose component of the DNA of future photosynthetic cells (Fig. 3).

e) Photorespiration

A third process by which photosynthate may be re-oxidized to CO_2 is photorespiration. It is initiated when Rubisco (Sect. 2.a) accepts O_2 rather than CO_2 as its second substrate; these two small, gaseous molecules act in a straightforward competitive manner for access to the enzyme's catalytic site. When O_2 is successful, the products are one molecule of glycerate-3-P (a Calvin-cycle intermediate) and one molecule of glycolate-P (not a Calvin-cycle intermediate). Subsequently, two molecules of glycolate-P are metabolized (via seven reactions distributed in four cellular compartments) to one molecule of glycerate-P and one molecule of CO_2; the latter represents a loss to plant growth. Part of this pathway was recently reconstituted in vitro from mitochondria and peroxisomes isolated from spinach leaves (Raghavendra et al. 1998). There may be some flexibility in the pathway; where the glycine-to-serine conversion was blocked in mutant plants, an alternative route involving formate was observed (Wingler et al. 1999a).

Researchers have long been attracted by the concept that photorespiratory CO_2 loss could be avoided (and plant growth consequently increased) if CO_2's competitive advantage at the active site of Rubisco could be improved. Recent attention has focussed on C_4 plants, which achieve this naturally by using the C_4 pathway as a CO_2 pump. Based on this phenomenon, Ishimaru et al. (1998) genetically engineered potato to express the C_4 enzyme pyruvate orthophosphate (Pi) dikinase from maize, Ku et al. (1999) engineered the addition of the maize gene for P-enol-pyruvate (PEP) carboxylase into rice to achieve an increased activity of that enzyme in the leaves, and Lipka et al. (1999) genetically engineered potato to simultaneously achieve expression of a bacterial PEP carboxylase and expression of a *Flaveria pringlei* NADP-linked malic enzyme. In the latter two cases, there was a limited indication that some reduction in photorespiration occurred, but a notable increase in plant growth rate has not been seen. With rice, increased growth may not occur, because this plant already seems to be at its capacity for utilizing the photosynthate that its leaves generate. Any extra production resulting from a reduction in photorespiration may only result in feedback inhibition of photosynthesis (Winder et al. 1998; Sun et al. 1999b), perhaps via the hexokinase system reviewed above (Sect. 3.c).

f) C_4 Photosynthesis

The most impressive aspect of C_4 photosynthesis is the near-total absence of concomitant photorespiration. The well-known, PEP-carboxylase-based CO_2 pump in the mesophyll cells of C_4 plants raises the CO_2 concentration around the Calvin cycle's Rubisco in bundle-sheath cells 10- to 20-fold (Jenkins 1997), so the competition between CO_2 and O_2 in bundle-sheath cells heavily favors CO_2. Nevertheless, recent intensive investigations (Maroco et al. 1997, 1998; Laisk and Edwards 1998) have revealed that O_2 still influences C_4 photosynthesis. It initiates a minor amount of photorespiration in C_4-plant leaves in normal air but, if these leaves are exposed to air with progressively reduced levels of O_2 (so that photorespiration becomes non-existent), another role emerges. This role is that of electron acceptor in the so-called pseudo-cyclic electron transport system, whereby the electron flow from water splitting to O_2 boosts the supply of ATP. This event is of special benefit to C_4 photosynthesis, which has a higher requirement for ATP than C_3 photosynthesis. When these two effects of O_2 (one positive, the other negative) are balanced, an optimum O_2 concentration of approximately 7% is revealed (Maroco et al. 1997).

Modern DNA technology has been applied to sorghum leaves to confirm the differential gene expression expected to produce the typical compartmentation of enzymes between mesophyll and bundle-sheath cells of C_4-plant leaves (Wyrich et al. 1998). Two surprises have arisen. The first is from Ueno (1998a), who found that pyruvate Pi dikinase, which is classically a mesophyll-cell enzyme, is also present in the bundle-sheath cells of some C_4 species. The second surprise is from Wingler et al. (1999b), who have destroyed the long-standing status of maize as the traditional NADP–malic enzyme-type C_4 plant by demonstrating that the release of CO_2 in the bundle-sheath cells of maize is not catalyzed exclusively by NADP-malic enzyme. Instead, it is catalyzed by a partnership of this enzyme and PEP carboxykinase.

Another study with NADP–malic enzyme has shown that it is a marker for the evolution of C_4 photosynthesis. Forms of the enzyme in C_3, C_3–C_4 intermediate, C_4-like and true C_4 species of *Flaveria* were examined, and it was observed that slightly smaller forms of the enzyme with distinctive kinetic properties became more significant as C_4 photosynthesis became more prominent (Drincovich et al. 1998; Casati et al. 1999). The other C_4 enzyme that has received considerable attention has been PEP carboxylase. Light-mediated activation by phosphorylation of this enzyme is apparently independent of any Calvin-cycle activity (Smith et al. 1998). The activation is theoretically vital, because the enzyme is judged to be well below saturation with respect to its substrate (Mg–PEP) in cells, thereby making it more responsive to effector molecules (Tovar-Méndez et al. 1998). The phosphorylated enzyme is, therefore, more influenced by these effectors (Bakrim et al. 1998). Interestingly, it has been reported that the enzyme's other substrate, bicarbonate, also influences these regulatory phenomena when its concentration fluctuates around physiological levels (Parvathi et al. 1998).

The emergence of C_4 plants some seven million years ago (see our preceding review) has been blamed on increased aridity at that time and on lowered atmospheric CO_2 levels (Pagani et al. 1999). The present-day interplay between the CO_2 level, water supply and temperature has been analyzed by Collatz et al. (1998), who conclude that, with the current CO_2 level of 0.035%, a warmest-month temperature greater than 22 °C favors C_4 plants, whereas C_3 plants are favored below this temperature. An intriguing link with the physiology of water stress has also emerged from studies of the strange amphibious sedge *Eleocharis vivipara*, which possesses C_4-like traits (including Kranz anatomy) when growing out of water, but not when submerged. Ueno (1998b) has shown that the C_4-like traits develop in submerged plants treated with abscisic acid. The author consequently proposes that this plant hormone might be central in the differentiation of C_4 photosynthetic tissue. Another link involves the C_3 plant rice; pyruvate Pi dikinase, traditionally a C_4 chloroplast enzyme, appears in the cytosol of root cells of this plant when they are water stressed or exposed to abscisic acid (Moons et al. 1998).

g) Crassulacean Acid Metabolism

Unlike C_4 plants, crassulacean acid metabolism (CAM) plants do everything in one cell. At night, with stomates open, two enzymes (PEP carboxylase and malate dehydrogenase) collect CO_2 and reductively combine it with PEP to generate malate. The malate is delivered to (and stored in) the large central vacuole that gives CAM plant tissues their succulent, fleshy character. During the day, with stomates shut, the reserve of malate is used to release the CO_2 needed to operate the Calvin cycle and permit plant growth. The remaining pyruvate is metabolized to a carbohydrate reserve, from which the next night's supply of PEP is obtained. Clearly, CAM is revolutionary in two senses: it entails the Calvin cycle supported by the diurnal/nocturnal PEP–malate–pyruvate–PEP cycle, and it enables these plants to survive in hot, dry habitats by being unconventional with respect to stomatal operation.

The majority of recent research regarding CAM has highlighted the flexibility of this type of photosynthetic carbon metabolism and the diversity of hot dry habitats in which CAM plants can be found. In a recent

Table 1. A comparison of C_3 photosynthesis, classic crassulacean-acid metabolism (CAM), and CAM variants (Cushman and Bohnert 1999). Plants engaged in CAM cycling or CAM idling accumulate smaller pools of organic acids nocturnally compared with plants performing classic CAM; such plants synthesize these organic acids from internally respired CO2

	C_3		CAM cycling		CAM (classic)		CAM idling	
	Night	Day	Night	Day	Night	Day	Night	Day
Stomates open	–	+	–	+	+	–	–	–
Net CO_2 uptake	–	+	–	+	+	–	–	–
Pool of organic acid	–	–	+	–	+	–	+	–
Net reduction of CO_2 to sugars	–	+	–	+	–	+	–	+

review, Cushman and Bohnert (1999) clarified the possible permutations of CAM (Table 1). In addition, Cockburn (1998) has predicted that a more subtle, high-frequency and low-amplitude organic acid fluctuation termed rapid-cycling CAM may be a C_i-concentrating mechanism in some plants. However, Cockburn has also predicted that it will be difficult to detect if it indeed exists. Extensive attention has been given to the neo-tropical genus *Clusia* [which includes species that are obligatory CAM, C_3–CAM intermediate, weak-CAM inducible and (perhaps) non-CAM (i.e., C_3)], because comparisons between morphologically similar species with different potentials for CAM permit better analyses of the true value of CAM. Notably, species with C_3/CAM plasticity seemed to have the advantage of a greater ability to adapt their photosynthetic carbon metabolism to changes in environmental parameters, such as light availability and water supply (Borland et al. 1998; Grams et al. 1998; Herzog et al. 1999a,b; Lüttge 1999). A similar conclusion has been used to explain the rather unexpected occurrence of the CAM plant *Aechmea magdalenae* in the shaded understory of rainforests. It appears to use brief inputs of high light more efficiently compared with similar C_3 species and (as expected) grew better during the dry season (Skillman et al. 1999). There is another indication (Baattrup-Pedersen and Madsen 1999) that flexibility in CAM serves its purpose. Its intuitively unexpected occurrence in some fresh-water plants is related to the acquisition of the scarce supply of CO_2; i.e., night-time collection and storage (as malate) of CO_2 becomes possible. Incidentally, it appears that CO_2 supply can also be a problem *inside* the leaves of classic CAM plants; their succulence raises mesophyll resistance to CO_2 diffusion (Maxwell et al. 1998).

Comparison of two species of *Peperomia*, one CAM and the other C_3, led Woerner and Martin (1999) to propose that CAM plants use water more efficiently not only because they open their stomates when the atmospheric evaporative demand is lower (i.e., at night), but also because they have fewer stomates. Some plants only display CAM when they are water-stressed, e.g., *Mesembryanthemum crystallinum*. An ingenious experiment by Eastmond and Ross (1997) has indicated that the water stress is primarily detected by the roots, which presumably send a signal to the leaves to switch from C_3 to CAM. However, the nature of the signal is not yet known, and Taybi and Cushman (1999) have shown that the imposition of water stress on detached leaves can initiate a Ca^{2+}-dependent signaling pathway that promotes transcription of the gene for PEP carboxylase.

A fundamental feature of CAM is the diurnal rhythm of its components (Blasius et al. 1998). Rhythmic oscillations can be detected at all levels, from whole-plant features (such as CO_2 exchange and leaf malate levels) to molecules (such as enzymes). A fine recent example is the day-night oscillation of one of the subunits of the vacuolar ATPase in a photoautotrophic cell-suspension culture from *M. crystallinum* (Rockel et al. 1998).

4 The Whole Plant

An impressive description of whole-plant photosynthesis has recently been provided by Shishido et al. (1999). These authors prepared an account of the entire carbon balance of a young tomato plant and recorded that 42% of fixed carbon was subsequently respired, 28% remained in the source leaves and 29% went into sinks (8% to the shoot apex, 12% to the stem and 9% to the roots).

a) Translocation

Fascinating details concerning the movement of photosynthetically generated sucrose from leaves to sink tissues via the phloem continue to emerge. There is some contradiction concerning the capacity of this process. It appears to be saturated during the day in castor beans, as evidenced by its unchanged rate when leaf photosynthesis is stimulated 40% on doubling the CO_2 supply (Grimmer and Komor 1999). However, in an *Arabidopsis* mutant deficient in the ability to synthesize transitory starch, the rate of sucrose translocation to roots was doubled, apparently to compensate for the loss of the starch sink (Zeeman and ap Rees 1999).

Excellent research during recent years has established that, in plants such as carrot and potato (Shakya and Sturm 1998; Kühn et al. 1999), sucrose is actively loaded into the phloem by a distinctive sucrose–H^+

symporter of the phloem cell's plasma membrane. Subsequently, the sucrose is moved to sinks; otherwise, it is capable of initiating a signal-transduction pathway that culminates in a reduced activity of this protein (Chiou and Bush 1998), possibly caused by its phosphorylation (Roblin et al. 1998). Clearly, this method of translocation is sophisticated, but it is not universal. It has long been known that cucurbits (such as pumpkin) translocate much of their photosynthetic product as the trisaccharide raffinose or the tetra-saccharide stachyose (sucrose extended by the addition of one or two galactose units). These two sugars have been shown to constitute 85% of the translocated sugars in the culinary herb basil (Büchi et al. 1998). More startling is the discovery (Wang and Nobel 1998) that fructans (sucrose extended by the addition of one to three fructose units) can be the translocated sugars. Wang and Nobel reported the first example of this in the CAM plant *Agave deserti*. In both these cases, extension of the sucrose molecule was determined to occur in the vicinity of the phloem rather than in the photosynthetic cells. There is a report that the translocation performed by willow (*Salix babylonica*) and possibly most other temperate forest trees is not sophisticated. Apparently, sucrose simply diffuses through the abundant plasmodesmatal connections that exist between the parenchyma and phloem cells of the leaf (Turgeon and Medville 1998).

b) CO_2 Fixation by Stressed Plants

Chilling-resistant plants, such as wheat, rye and *Arabidopsis*, acclimate to cold temperatures (5 °C) by synthesizing extra Calvin-cycle and sucrose-biosynthesis enzymes so that their ability to utilize the harvested light energy becomes similar to that of plants grown at moderate temperatures (23 °C; Huner et al. 1998; Strand et al. 1999). Studies with rice indicate that this response may occur to some extent when a relatively minor drop in temperature takes place (Ohashi et al. 1998). Chilling-sensitive plants, such as tomato and grapevine, have photosynthesis curtailed by low temperatures in the light. Even low temperatures in the dark (i.e., at night) decrease daytime photosynthesis, due (at least in part) to interference in the activation of sucrose-P synthase and a proportional increase in photorespiration (Jones et al. 1998b; Flexas et al. 1999).

Water stress reduces photosynthesis both directly (by inducing stomatal closure) and indirectly (by promoting reductions in the activities of certain photosynthetic enzymes; Saccardy et al. 1996; Sánchez-Rodríguez et al. 1999; Wingler et al. 1999c). Stomatal closure has been favored as the more significant factor, but Tezara et al. (1999) have recently added the chloroplast thylakoid's ATP synthase (coupling factor) to the list of enzymes whose activities are lowered during water stress.

They argue that this loss of activity is of prime importance. An odd pair of results relates to the C_4 enzyme pyruvate Pi dikinase. Water stress causes its activity to plummet in leaves of the C_4 plant sugarcane but causes it to unexpectedly appear in the roots of the C_3 plant rice (Du et al. 1998; Moons et al. 1998).

Two reports relate to current environmental issues. In the first, Jones et al. (1998a) consider that high-temperature damage to the Rubisco system of dinoflagellate micro-algae is the primary reason for the bleaching of coral reefs. In the second, Allen et al. (1998) review the likelihood that ozone depletion and the concurrent rise in ultraviolet-B irradiance might reduce the photosynthesis of crops and natural vegetation; they conclude that this effect is minimal.

c) CO_2 Fixation when CO_2 is Supply Abundant

The consistent increase (0.015% per year) in the concentration of CO_2 in the atmosphere due to human activities has led to voluminous research on the photosynthetic carbon metabolism of plants growing in the presence of CO_2-enriched air. Several reviews of this research have appeared recently (our 1996 review; Makino and Mae 1999; Norby et al. 1999); thus, a detailed coverage of the literature will not be made here. In the most recent work, large-scale (and technologically impressive) field experiments indicate that most agricultural species (including wheat, rice and radish) and most forest trees in their usual environments display a consistent and substantial (~40%) increase in their rates of photosynthesis when the concentration of CO_2 in the surrounding air is approximately doubled (Garcia et al. 1998; Monje and Bugbee 1998; Usuda and Shimogawara 1998; Vu et al. 1998; Norby et al. 1999). Acclimation (i.e., downregulation) of photosynthesis involves the sugar-sensing transduction pathway that culminates in the reduction of gene transcription for photosynthetic enzymes (most notably Rubisco) and the consequent reduction in photosynthetic capacity (Moore et al. 1999; Sims et al. 1999; Fig. 2). This acclimation, often observed in previous experiments involving glasshouse and/or potted plants exposed to elevated CO_2, appears to be minor. Therefore, it is seldom likely to interfere with the CO_2 stimulation of photosynthesis except in plants, such as grasses (Iceland's *Nardus stricta*), that lack relatively large sink organs (Bryant et al. 1998; Cook et al. 1998). Apparently, humanity is slowly but surely accelerating photosynthetic CO_2 fixation of the world's plants; we are doing so inadvertently and without recourse to DNA manipulation.

References

Allen DJ, Nogués S, Baker NR (1998) Ozone depletion and increased UV-B radiation: is there a real threat to photosynthesis? J Exp Bot 49:1775–1788

Amoroso G, Sültemeyer D, Thyssen C, Fock HP (1998) Uptake of HCO_3^- and CO_2 in cells and chloroplasts from the microalgae *Chlamydomonas reinhardtii* and *Dunaliella tertiolecta*. Plant Physiol 116:193–201

Andría JR, Pérez-Lloréns JL, Vergara JJ (1999) Mechanisms of inorganic carbon acquisition in *Gracilaria gaditana* nom. prov. (Rhodophyta). Planta 208:564–573

Ashton AR (1998) Sedoheptulose-1,7-bisphosphate phosphatase activity of chloroplast fructose-1,6-bisphosphatase: identification of enzymes hydrolysing fructose-1,6-bisphosphate and sedoheptulose-1,7-bisphosphate in stromal extracts from chloroplasts of spinach (*Spinacia oleracea*). Aust J Plant Physiol 25:531–537

Axelsson L, Larsson C, Ryberg H (1999) Affinity, capacity and oxygen sensitivity of two different mechanisms for bicarbonate utilization in *Ulva lactuca* L. (Chlorophyta). Plant Cell Environ 22:969–978

Baattrup-Pedersen A, Madsen TV (1999) Interdependence of CO_2 and inorganic nitrogen on crassulacean acid metabolism and efficiency of nitrogen use by *Littorella uniflora* (L.) Aschers. Plant Cell Environ 22:535–542

Backhausen JE, Jöstingmeyer P, Scheibe R (1997) Competitive inhibition of spinach leaf phosphoglucose isomerase isoenzymes by erythrose 4-phosphate. Plant Sci 130:121–131

Backhausen JE, Vetter S, Baalmann E, Kitzmann C, Scheibe R (1998) NAD-dependent malate dehydrogenase and glyceraldehyde 3-phosphate dehydrogenase isoenzymes play an important role in dark metabolism of various plastid types. Planta 205:359–366

Bakrim N, Nhiri M, Pierre J-N, Vidal J (1998) Metabolite control of *Sorghum* C_4 phosphoenol-pyruvate carboxylase catalytic activity and phosphorylation state. Photosynth Res 58:153–162

Blasius B, Beck F, Lüttge U (1998) Oscillatory model of crassulacean acid metabolism: structural analysis and stability boundaries with a discrete hysteresis switch. Plant Cell Environ 21:775–784

Borkhsenious ON, Mason CB, Moroney JV (1998) The intracellular localization of ribulose-1,5-bisphosphate carboxylase/oxygenase in *Chlamydomonas reinhardtii*. Plant Physiol 116:1585–1591

Borland AM, Técsi LI, Leegood RC, Walker RP (1998) Inducibility of crassulacean acid metabolism (CAM) in *Clusia* species: physiological/biochemical characterisation and intercellular localization of carboxylation and decarboxylation processes in three species which exhibit different degrees of CAM. Planta 205:342–351

Bryant J, Taylor G, Frehner M (1998) Photosynthetic acclimation to elevated CO_2 is modified by source:sink balance in three component species of chalk grassland swards grown in a free air carbon dioxide enrichment (FACE) experiment. Plant Cell Environ 21:159–168

Büchi R, Bachmann M, Keller F (1998) Carbohydrate metabolism in source leaves of sweet basil (*Ocimum basilicum* L.), a starch-storing and stachyose-translocating labiate. J Plant Physiol 153:308–315

Büker M, Schünemann D, Borchert S (1998) Enzymic properties and capacities of developing tomato (*Lycopersicon esculentum* L.) fruit plastids. J Exp Bot 49:681–691

Casati P, Fresco AG, Andreo CS, Drincovich MF (1999) An intermediate form of NADP-malic enzyme from the C_3–C_4 intermediate species *Flaveria floridana*. Plant Sci 147:101–109

Chen B-Y, Wang Y, Janes HW (1998a) ADP-glucose pyrophosphorylase is localized to both the cytoplasm and plastids in developing pericarp of tomato fruit. Plant Physiol 116:101–106

Chen Y-R, Hartman FC, Lu T-YS, Larimer FW (1998b) D-ribulose-5-phosphate 3-epimerase: cloning and heterologous expression of the spinach gene, and purification and characterization of the recombinant enzyme. Plant Physiol 118:199–207
Cheng W-H, Taliercio EW, Chourey PS (1999) Sugars modulate an unusual mode of control of the cell-wall invertase gene (*Incw* 1) through its 3' untranslated region in a cell suspension culture of maize. Proc Natl Acad Sci USA 96:10512–10517
Chiou T-J, Bush DR (1998) Sucrose is a signal molecule in assimilate partitioning. Proc Natl Acad Sci USA 95:4784–4788
Cockburn W (1998) Rapid-cycling CAM; an hypothetical variant of photosynthetic metabolism. Plant Cell Environ 21:845–848
Collatz GJ, Berry JA, Clark JS (1998) Effects of climate and atmospheric CO_2 partial pressure on the global distribution of C_4 grasses: present, past, and future. Oecologia 114:441–454
Cook AC, Tissue DT, Roberts SW, Oechel WC (1998) Effects of long-term elevated [CO_2] from natural CO_2 springs on *Nardus stricta*: photosynthesis, biochemistry, growth and phenology. Plant Cell Environ 21:417–425
Cullen JT, Lane TW, Morel FMM, Sherrell RM (1999) Modulation of cadmium uptake in phytoplankton by seawater CO_2 concentration. Nature 402:165–167
Cushman JC, Bohnert HJ (1999) Crassulacean acid metabolism: molecular genetics. Annu Rev Plant Physiol Plant Mol Biol 50:305–332
Dai N, Schaffer A, Petreikov M, Shahak Y, Giller Y, Ratner K, Levine A, Granot D (1999) Overexpression of *Arabidopsis* hexokinase in tomato plants inhibits growth, reduces photosynthesis, and induces rapid senescence. Plant Cell 11:1253–1266
Debnam PM, Emes MJ (1999) Subcellular distribution of enzymes of the oxidative pentose phosphate pathway in root and leaf tissues. J Exp Bot 50:1653–1661
Deveau JST, Khosravani H, Lew RR, Colman B (1998) CO_2 uptake mechanism in *Eremosphaera viridis*. Can J Bot 76:1161–1164
Drincovich MF, Casati P, Andreo CS, Chessin SJ, Franceschi VR, Edwards GE, Ku MSB (1998) Evolution of C_4 photosynthesis in *Flaveria* species. Plant Physiol 117:733–744
Du Y-C, Nose A, Wasano K, Uchida Y (1998) Responses to water stress of enzyme activities and metabolite levels in relation to sucrose and starch synthesis, the Calvin cycle and the C_4 pathway in sugarcane (*Saccharum* sp.) leaves. Aust J Plant Physiol 25:253–260
Dunford RP, Durrant MC, Catley MA, Dyer TA (1998) Location of the redox-active cysteines in chloroplast sedoheptulose-1,7-bisphosphatase indicate that its allosteric regulation is similar but not identical to that of fructose-1,6-bisphosphatase. Photosynth Res 58:221–230
Eastmond PJ, Rawsthorne S (1998) Comparison of the metabolic properties of plastids isolated from developing leaves or embryos of *Brassica napus* L. J Exp Bot 49:1105–1111
Eastmond PJ, Ross JD (1997) Evidence that the induction of crassulacean acid metabolism by water stress in *Mesembryanthemum crystallinum* (L.) involves root signalling. Plant Cell Environ 20:1559–1565
Eriksson M, Villand P, Gardeström P, Samuelsson G (1998) Induction and regulation of expression of a low-CO_2-induced mitochondrial carbonic anhydrase in *Chlamydomonas reinhardtii*. Plant Physiol 116:637–641
Feller U, Crafts-Brandner SJ, Salvucci ME (1998) Moderately high temperatures inhibit ribulose-1,5-bisphosphate carboxylase/oxygenase (Rubisco) activase-mediated activation of Rubisco. Plant Physiol 116:539–546
Flechner A, Dressen U, Westhoff P, Henze K, Schnarrenberger C, Martin W (1996) Molecular characterization of transketolase (EC 2.2.1.1) active in the Calvin cycle of spinach chloroplasts. Plant Mol Biol 32:475–484

Flechner A, Gross W, Martin WF, Schnarrenberger C (1999) Chloroplast class I and class II aldolases are bifunctional for fructose-1,6-bisphosphate and sedoheptulose-1,7-bisphosphate cleavage in the Calvin cycle. FEBS Lett 447:200–202
Flexas J, Badger M, Chow WS, Medrano H, Osmond CB (1999) Analysis of the relative increase in photosynthetic O_2 uptake when photosynthesis in grapevine leaves is inhibited following low night temperatures and/or water stress. Plant Physiol 121:675–684
Flores-Moya A, Fernández JA (1998) The role of external carbonic anhydrase in the photosynthetic use of inorganic carbon in the deep-water alga *Phyllariopsis purpurascens* (Laminariales, Phaeophyta). Planta 207:115–119
Garcia RL, Long SP, Wall GW, Osborne CP, Kimball BA, Nie GY, Pinter PJ, Lamorte RL, Wechsung F (1998) Photosynthesis and conductance of spring-wheat leaves: field response to continuous free-air atmospheric CO_2 enrichment. Plant Cell Environ 21:659–669
Getzoff TP, Zhu G, Bohnert HJ, Jensen RG (1998) Chimeric *Arabidopsis thaliana* ribulose-1,5-bisphosphate carboxylase/oxygenase containing a pea small subunit protein is compromised in carbamylation. Plant Physiol 116:695–702
Goddijn O, Smeekens S (1998) Sensing trehalose biosynthesis in plants. Plant J 14:143–146
Goyer A, Decottignies P, Lemaire S, Ruelland E, Issakidis-Bourguet E, Jacquot J-P, Miginiac-Maslow M (1999) The internal cys-207 of sorghum leaf NADP-malate dehydrogenase can form mixed disulphides with thioredoxin. FEBS Lett 444:165–169
Grams TEE, Herzog B, Lüttge U (1998) Are there species in the genus *Clusia* with obligate C_3-photosynthesis? J Plant Physiol 152:1–9
Greene TW, Hannah LC (1998) Adenosine diphosphate glucose pyrophosphorylase, a rate-limiting step in starch biosynthesis. Physiol Plant 103:574–580
Grimmer C, Komor E (1999) Assimilate export by leaves of *Ricinus communis* L. growing under normal and elevated carbon dioxide concentrations: the same rate during the day, a different rate at night. Planta 209:275–281
Haake V, Zrenner R, Sonnewald U, Stitt M (1998) A moderate decrease of plastid aldolase activity inhibits photosynthesis, alters the levels of sugars and starch, and inhibits growth of potato plants. Plant J 14:147–157
Halford NG, Purcell PC, Hardie DG (1999) Is hexokinase really a sugar sensor in plants? Trends Plant Sci 4:117–120
Hammond ET, Andrews TJ, Woodrow IE (1998a) Regulation of ribulose-1,5-bisphosphate carboxylase/oxygenase by carbamylation and 2-carboxyarabinitol 1-phosphate in tobacco: insights from studies of antisense plants containing reduced amounts of rubisco activase. Plant Physiol 118:1463–1471
Hammond ET, Andrews TJ, Mott KA, Woodrow IE (1998b) Regulation of rubisco activation in antisense plants of tobacco containing reduced levels of rubisco activase. Plant J 14:101–110
Hansen LD, Smith BN, Criddle RS (1999) Future food. Science 283:1115–1116
Hariharan T, Johnson PJ, Cattolico RA (1998) Purification and characterization of phosphoribulokinase from the marine chromophytic alga *Heterosigma carterae*. Plant Physiol 117:321–329
Harrison CJ, Hedley CL, Wang TL (1998a) Evidence that the *rug3* locus of pea (*Pisum sativum* L.) encodes plastidial phosphoglucomutase confirms that the imported substrate for starch synthesis in pea amyloplasts is glucose-6-phosphate. Plant J 13:753–762
Harrison EP, Willingham NM, Lloyd JC, Raines CA (1998b) Reduced sedoheptulose-1,7-bisphosphatase levels in transgenic tobacco lead to decreased photosynthetic capacity and altered carbohydrate accumulation. Planta 204:27–36
Hattenbach A, Heineke D (1999) On the role of chloroplastic phosphoglucomutase in the regulation of starch turnover. Planta 207:527–532

He Z, von Caemmerer S, Hudson GS, Price GD, Badger MR, Andrews TJ (1997) Ribulose-1,5-bisphosphate carboxylase/oxygenase activase deficiency delays senescence of ribulose-1,5-bisphosphate carboxylase/oxygenase but progressively impairs its catalysis during tobacco leaf development. Plant Physiol 115:1569–1580

Heo J, Holbrook GP (1999) Regulation of 2-carboxy-D-arabinitol 1-phosphate phosphatase: activation by glutathione and interaction with thiol reagents. Biochem J 338:409–416

Herzog B, Hübner C, Ball E, Bastos RdoN, Franco AC, Scarano FR, Lüttge U (1999a) Comparative study of the C_3/CAM intermediate species *Clusia parviflora* Saldanha et Engl. and the obligate CAM species *Clusia hilariana* Schlecht. growing sympatrically exposed and shaded in the coastal restinga of Brazil. Plant Biol 1:453–459

Herzog B, Hoffmann S, Hartung W, Lüttge U (1999b) Comparison of photosynthetic responses of the sympatric tropical C_3 species *Clusia multiflora* H.B.K. and the C_3-CAM intermediate species *Clusia minor* L. to irradiance and drought stress in a phytotron. Plant Biol 1:460–470

Hoefnagel MHN, Atkin OK, Wiskich JT (1998) Interdependence between chloroplasts and mitochondria in the light and the dark. Biochim Biophys Acta 1366:235–255

Huner NPA, Öquist G, Sarhan F (1998) Energy balance and acclimation to light and cold. Trends Plant Sci 3:224–230

Igamberdiev AU, Hurry V, Krömer S, Gardeström P (1998) The role of mitochondrial electron transport during photosynthetic induction. A study with barley (*Hordeum vulgare*) protoplasts incubated with rotenone and oligomycin. Physiol Plant 104:431–439

Ignatova LK, Moskvin OV, Romanova AK, Ivanov BN (1998) Carbonic anhydrases in the C_3-plant leaf cell. Aust J Plant Physiol 25:673–677

Ishimaru K, Ohkawa Y, Ishige T, Tobias DJ, Ohsugi R (1998) Elevated pyruvate orthophosphate dikinase (PPDK) activity alters carbon metabolism in C_3 transgenic potatoes with a C_4 maize PPDK gene. Physiol Plant 103:340–346

Jebanathirajah JA, Coleman JR (1998) Association of carbonic anhydrase with a Calvin cycle enzyme complex in *Nicotiana tabacum*. Planta 204:177–182

Jenkins CLD (1997) The CO_2 concentrating mechanism of C_4 photosynthesis: bundle sheath cell CO_2 concentration and leakage. Aust J Plant Physiol 24:543–547

Jobling SA, Schwall GP, Westcott RJ, Sidebottom CM, Debet M, Gidley MJ, Jeffcoat R, Safford R (1999) A minor form of starch branching enzyme in potato (*Solanum tuberosum* L.) tubers has a major effect on starch structure: cloning and characterisation of multiple forms of *SBE A*. Plant J 18:163–171

Jones RJ, Hoegh-Guldberg O, Larkum AWD, Schreiber U (1998a) Temperature-induced bleaching of corals begins with impairment of the CO_2 fixation mechanism in zooxanthellae. Plant Cell Environ 21:1219–1230

Jones TL, Tucker DE, Ort DR (1998b) Chilling delays circadian pattern of sucrose phosphate synthase and nitrate reductase activity in tomato. Plant Physiol 118:149–158

Kakefuda G, Preiss J (1997) Partial purification and characterization of a diurnally fluctuating novel endoamylase from *Arabidopsis thaliana* leaves. Plant Physiol Biochem 35:907–913

Kammerer B, Fischer K, Hilpert B, Schubert S, Gutensohn M, Weber A, Flügge U-I (1998) Molecular characterization of a carbon transporter in plastids from heterotrophic tissues: the glucose 6-phosphate/phosphate antiporter. Plant Cell 10:105–117

Kane HJ, Wiklen JM, Portis AR, Andrews TJ (1998) Potent inhibition of ribulose-bisphosphate carboxylase by an oxidized impurity in ribulose-1,5-bisphosphate. Plant Physiol 117:1059–1069

Kanevski I, Maliga P, Rhoades DF, Gutteridge S (1999) Plastome engineering of ribulose-1,5-bisphosphate carboxylase/oxygenase in tobacco to form a sunflower large subunit and tobacco small subunit hybrid. Plant Physiol 119:133–141

Kaplan A, Reinhold L (1999) CO_2 concentrating mechanisms in photosynthetic microorganisms. Annu Rev Plant Physiol Plant Mol Biol 50:539–570

Kingston-Smith AH, Walker RP, Pollock CJ (1999) Invertase in leaves: conundrum or control point? J Exp Bot 50:735–743

Kleczkowski LA (1999) A phosphoglycerate to inorganic phosphate ratio is the major factor in controlling starch levels in chloroplasts via ADP-glucose pyrophosphorylase regulation. FEBS Lett 448:153–156

Kossmann J, Abel GJW, Springer F, Lloyd JR, Willmitzer L (1999) Cloning and functional analysis of a cDNA encoding a starch synthase from potato (*Solanum tuberosum* L.) that is predominantly expressed in leaf tissue. Planta 208:503–511

Krook J, Vreugdenhil D, Dijkema C, van der Plas LHW (1998) Sucrose and starch metabolism in carrot (*Daucus carota* L.) cell suspensions analysed by ^{13}C-labelling: indications for a cytosol and a plastid-localized oxidative pentose phosphate pathway. J Exp Bot 49:1917–1924

Ku MSB, Agarie S, Nomura M, Fukayama H, Tsuchida H, Ono K, Hirose S, Toki S, Miyao M, Matsuoka M (1999) High-level expression of maize phospho-enol-pyruvate carboxylase in transgenic rice plants. Nature Biotech 17:76–80

Kühn C, Barker L, Bürkle L, Frommer W-B (1999) Update on sucrose transport in higher plants. J Exp Bot 50:935–953

Laisk A, Edwards GE (1998) Oxygen and electron flow in C_4 photosynthesis: Mehler reaction, photorespiration and CO_2 concentration in the bundle sheath. Planta 205:632–645

Lalonde S, Boles E, Hellmann H, Barker L, Patrick JW, Frommer WB, Ward JM (1999) The dual function of sugar carriers: transport and sugar sensing. Plant Cell 11:707–726

Lawlor DW (1995) Photosynthesis, productivity and environment. J Exp Bot 46:1449–1461

Li L-A, Zianni MR, Tabita FR (1999) Inactivation of the monocistronic *rca* gene in *Anabaena variabilis* suggests a physiological ribulose bisphosphate carboxylase/oxygenase activase-like function in heterocystous cyanobacteria. Plant Mol Biol 40:467–478

Lipka V, Häusler RE, Rademacher T, Li J, Hirsch HJ, Kreuzaler F (1999) *Solanum tuberosum* double transgenic expressing phosphoenolpyruvate carboxylase and NADP-malic enzyme display reduced electron requirement for CO_2 fixation. Plant Sci 144:93–105

Lüttge U (1999) One morphotype, three physiotypes: sympatric species of *Clusia* with obligate C_3 photosynthesis, obligate CAM and C_3-CAM intermediate behaviour. Plant Biol 1:138–148

Makino A, Mae T (1999) Photosynthesis and plant growth at elevated levels of CO_2. Plant Cell Physiol 40:999–1006

Mann CC (1999) Genetic engineers aim to soup up crop photosynthesis. Science 283:314–316

Maroco JP, Ku MSB, Edwards GE (1997) Oxygen sensitivity of C_4 photosynthesis: evidence from gas exchange and chlorophyll fluorescence analyses with different C_4 subtypes. Plant Cell Environ 20:1525–1533

Maroco JP, Ku MSB, Lea PJ, Dever LV, Leegood RC, Furbank RT, Edwards GE (1998) Oxygen requirement and inhibition of C_4 photosynthesis. Plant Physiol 116:823–832

Maxwell K, Badger MR, Osmond CB (1998) A comparison of CO_2 and O_2 exchange patterns and the relationship with chlorophyll fluorescence during photosynthesis in C_3 and CAM plants. Aust J Plant Physiol 25:45–52

Monje O, Bugbee B (1998) Adaptation to high CO_2 concentration in an optimal environment: radiation capture, canopy quantum yield and carbon use efficiency. Plant Cell Environ 21:315–324

Moons A, Valcke R, Van Montagu M (1998) Low-oxygen stress and water deficit induce cytosolic pyruvate orthophosphate dikinase (PPDK) expression in roots of rice, a C_3 plant. Plant J 15:89–98

Moore BD, Cheng S-H, Sims D, Seemann JR (1999) The biochemical and molecular basis for photosynthetic acclimation to elevated atmospheric CO_2. Plant Cell Environ 22:567–582

Moorhead G, Douglas P, Cotelle V, Harthill J, Morrice N, Meek S, Deiting U, Stitt M, Scarabel M, Aitken A, MacKintosh C (1999) Phosphorylation-dependent interactions between enzymes of plant metabolism and 14-3-3 proteins. Plant J 18:1–12

Morales A, Ortega-Delgado ML, Molina-Galán J, Sánchez de Jiménez E (1999) Importance of Rubisco activase in maize productivity based on mass selection procedure. J Exp Bot 50:823–829

Morita E, Abe T, Tsuzuki M, Fujiwara S, Sato N, Hirata A, Sonoike K, Nozaki H (1998) Presence of the CO_2-concentrating mechanism in some species of the pyrenoid-less free-living algal genus *Chloromonas* (Volvocales, Chlorophyta). Planta 204:269–276

Moroney JV, Chen Z-Y (1998) The role of the chloroplast in inorganic carbon uptake by eukaryotic algae. Can J Bot 76:1025–1034

Moroney JV, Somanchi A (1999) How do algae concentrate CO_2 to increase the efficiency of photosynthetic carbon fixation? Plant Physiol 119:9–16

Müller J, Wiemken A, Aeschbacher R (1999) Trehalose metabolism in sugar sensing and plant development. Plant Sci 147:37–47

Nakazato T, Kadota A, Wada M (1999) Photoinduction of spore germination in *Marchantia polymorpha* L. is mediated by photosynthesis. Plant Cell Physiol 40:1014–1020

Nielsen TH, Krapp A, Röper-Schwarz U, Stitt M (1998) The sugar-mediated regulation of genes encoding the small subunit of Rubisco and the regulatory subunit of ADP glucose pyrophosphorylase is modified by phosphate and nitrogen. Plant Cell Environ 21:443–454

Nimer NA, Warren M, Merrett MJ (1998) The regulation of photosynthetic rate and activation of extracellular carbonic anhydrase under CO_2-limiting conditions in the marine diatom *Skeletonema costatum*. Plant Cell Environ 21:805–812

Nimer NA, Brownlee C, Merrett MJ (1999) Extracellular carbonic anhydrase facilitates carbon dioxide availability for photosynthesis in the marine dinoflagellate *Prorocentrum micans*. Plant Physiol 120:105–111

Norby RJ, Wullschleger SD, Gunderson CA, Johnson DW, Ceulemans R (1999) Tree responses to rising CO_2 in field experiments: implications for the future forest. Plant Cell Environ 22:683–714

Nyvall P, Pelloux J, Davies HV, Pedersén M, Viola R (1999) Purification and characterisation of a novel starch synthase selective for uridine 5'-diphosphate glucose from the red alga *Gracilaria tenuistipitata*. Planta 209:143–152

Ohashi K, Makino A, Mae T (1998) Gas exchange characteristics in rice leaves grown under the conditions of physiologically low temperature and irradiance. Plant Cell Physiol 39:1384–1387

Padmasree K, Raghavendra AS (1999) Prolongation of photosynthetic induction as a consequence of interference with mitochondrial oxidative metabolism in mesophyll protoplasts of the pea (*Pisum sativum* L.). Plant Sci 142:29–36

Pagani M, Freeman KH, Arthur MA (1999) Late miocene atmospheric CO_2 concentrations and the expansion of C_4 grasses. Science 285:876–879

Panneman H, Ruijter GJG, van den Broeck HC, Visser J (1998) Cloning and biochemical characterisation of *Aspergillus niger* hexokinase. The enzyme is strongly inhibited by physiological concentrations of trehalose 6-phosphate. Eur J Biochem 258:223–232

Park Y-I, Karlsson J, Rojdestvenski I, Pronina N, Klimov V, Öquist G, Samuelsson G (1999) Role of a novel photosystem-II-associated carbonic anhydrase in photosynthetic carbon assimilation in *Chlamydomonas reinhardtii*. FEBS Lett 444:102–105

Parry MAJ, Andralojc PJ, Lowe HM, Keys AJ (1999) The localisation of 2-carboxy-D-arabinitol 1-phosphate and inhibition of Rubisco in leaves of *Phaseolus vulgaris* L. FEBS Lett 444:106–110

Parvathi K, Bhagwat AS, Raghavendra AS (1998) Modulation by bicarbonate of catalytic and regulatory properties of C_4 phospho-enol-pyruvate carboxylase from *Amaranthus hypochondriacus*: desensitization to malate and glucose 6-phosphate and sensitization to Mg^{2+}. Plant Cell Physiol 39:1294–1298

Portis AR, Zhang X-H (1999) Green revolutions. Science 283:1267

Raghavendra AS, Reumann S, Heldt HW (1998) Participation of mitochondrial metabolism in photorespiration. Plant Physiol 116:1333–1337

Ramage RT, Read BA, Tabita FR (1998) Alteration of the α helix region of cyanobacterial ribulose 1,5-bisphosphate carboxylase/oxygenase to reflect sequences found in high substrate specificity enzymes. Arch Biochem Biophys 349:81–88

Roblin G, Sakr S, Bonmort J, Delrot S (1998) Regulation of a plant plasma membrane sucrose transporter by phosphorylation. FEBS Lett 424:165–168

Rockel B, Jia C, Ratajczak R, Lüttge U (1998) Day-night changes of the amount of subunit-c transcript of the V-ATPase in suspension cells of *Mesembryanthemum crystallinum* L. J Plant Physiol 152:189–193

Ruelland E, Miginiac-Maslow M (1999) Regulation of chloroplast enzyme activities by thioredoxins: activation or relief from inhibition? Trends Plant Sci 4:136–141

Saccardy K, Cornic G, Brulfert J, Reyss A (1996) Effect of drought stress on net CO_2 uptake by *Zea* leaves. Planta 199:589–595

Sánchez-Rodríguez J, Pérez P, Martínez-Carrasco R (1999) Photosynthesis, carbohydrate levels and chlorophyll fluorescence-estimated intercellular CO_2 in water-stressed *Casuarina equisetifolia* Forst. & Forst. Plant Cell Environ 22:867–873

Sasaki H, Hirose T, Watanabe Y, Ohsugi R (1998) Carbonic anhydrase activity and CO_2-transfer resistance in Zn-deficient rice leaves. Plant Physiol 118:929–934

Scagliarini S, Trost P, Pupillo P (1998) The non-regulatory isoform of NAD(P)-glyceraldehyde-3-phosphate dehydrogenase from spinach chloroplasts. J Exp Bot 49:1307–1315

Schleucher J, Vanderveer PJ, Sharkey TD (1998) Export of carbon from chloroplasts at night. Plant Physiol 118:1439–1445

Schleucher J, Vanderveer P, Markley JL, Sharkey TD (1999) Intramolecular deuterium distributions reveal disequilibrium of chloroplast phosphoglucose isomerase. Plant Cell Environ 22:525–533

Shakya R, Sturm A (1998) Characterization of source- and sink-specific sucrose/H^+ symporters from carrot. Plant Physiol 118:1473–1480

Shannon JC, Pien F-M, Cao H, Liu K-C (1998) Brittle-1, an adenylate translocator, facilitates transfer of extraplastidial synthesized ADP-glucose into amyloplasts of maize endosperms. Plant Physiol 117:1235–1252

Shishido Y, Kumakura H, Nishizawa T (1999) Carbon balance of a whole tomato plant and the contribution of source leaves to sink growth using the $^{14}CO_2$ steady-state feeding method. Physiol Plant 106:402–408

Siedlecka A, Gardeström P, Samuelsson G, Kleczkowski LA, Krupa Z (1999) A relationship between carbonic anhydrase and rubisco in response to moderate cadmium stress during light activation of photosynthesis. Z Naturforsch C 54:759–763

Sims DA, Cheng W, Luo Y, Seemann JR (1999) Photosynthetic acclimation to elevated CO_2 in a sunflower canopy. J Exp Bot 50:645–653

Sinha AK, Shirke PA, Pathre U, Sane PV (1997) Midday depression in photosynthesis: Effect on sucrose-phosphate synthase and ribulose-1,5-bisphosphate carboxylase in leaves of *Prosopis juliflora* (Swartz) DC. Photosynthetica 34:115–124

Skillman JB, Garcia M, Winter K (1999) Whole-plant consequences of crassulacean acid metabolism for a tropical forest understory plant. Ecology 80:1584–1593

Smith EC, Griffiths H (1996) A pyrenoid-based carbon-concentrating mechanism is present in terrestrial bryophytes of the class Anthocerotae. Planta 200:203–212

Smith LH, Langdale JA, Chollet R (1998) A functional Calvin cycle is not indispensable for the light activation of C_4 phospho-enol-pyruvate carboxylase kinase and its target enzyme in the maize mutant *bundle sheath defective2-mutable1*. Plant Physiol 118:191–197

Somerville C, Somerville S (1999) Plant functional genomics. Science 285:380–383

Spreitzer RJ (1999) Questions about the complexity of chloroplast ribulose-1,5-bisphosphate carboxylase/oxygenase. Photosynth Res 60:29–42

Stemler AJ (1997) The case for chloroplast thylakoid carbonic anhydrase. Physiol Plant 99:348–353

Strand Å, Hurry V, Henkes S, Huner N, Gustafsson P, Gardeström P, Stitt M (1999) Acclimation of *Arabidopsis* leaves developing at low temperatures. Increasing cytoplasmic volume accompanies increased activities of enzymes in the Calvin cycle and in the sucrose-biosynthesis pathway. Plant Physiol 119:1387–1397

Sturm A (1999) Invertases. Primary structures, functions, and roles in plant development and sucrose partitioning. Plant Physiol 121:1–7

Sun J, Okita TW, Edwards GE (1999a) Modification of carbon partitioning, photosynthetic capacity, and O_2 sensitivity in *Arabidopsis* plants with low ADP-glucose pyrophosphorylase activity. Plant Physiol 119:267–276

Sun J, Edwards GE, Okita TW (1999b) Feedback inhibition of photosynthesis in rice measured by O_2 dependent transients. Photosynth Res 59:187–200

Tamoi M, Takeda T, Shigeoka S (1999) Functional analysis of fructose-1,6-bisphosphatase isozymes (*fbp-I* and *fbp-II* gene products) in cyanobacteria. Plant Cell Physiol 40:257–261

Tatge H, Marshall J, Martin C, Edwards EA, Smith AM (1999) Evidence that amylose synthesis occurs within the matrix of the starch granule in potato tubers. Plant Cell Environ 22:543–550

Taybi T, Cushman JC (1999) Signaling events leading to crassulacean acid metabolism induction in the common ice plant. Plant Physiol 121:545–555

Tchernov D, Hassidim M, Vardi A, Luz B, Sukenik A, Reinhold L, Kaplan A (1998) Photosynthesizing marine microorganisms can constitute a source of CO_2 rather than a sink. Can J Bot 76:949–953

Teige M, Melzer M, Süss K-H (1998) Purification, properties and in situ localization of the amphibolic enzymes D-ribulose 5-phosphate 3-epimerase and transketolase from spinach chloroplasts. Eur J Biochem 252:237–244

Tezara W, Mitchell VJ, Driscoll SD, Lawlor DW (1999) Water stress inhibits plant photosynthesis by decreasing coupling factor and ATP. Nature 401:914–917

Thom E, Möhlmann T, Quick WP, Camara B, Neuhaus H-E (1998) Sweet pepper plastids: enzymic equipment, characterisation of the plastidic oxidative pentose-phosphate pathway, and transport of phosphorylated intermediates across the envelope membrane. Planta 204:226–233

Tomlinson K, Craig J, Smith AM (1998) Major differences in isoform composition of starch synthase between leaves and embryos of pea (*Pisum sativum* L.). Planta 204:86–92

Toroser D, Athwal GS, Huber SC (1998) Site-specific regulatory interaction between spinach leaf sucrose-phosphate synthase and 14-3-3 proteins. FEBS Lett 435:110–114

Tovar-Méndez A, Rodríguez-Sotres R, López-Valentín DM, Muñoz-Clares RA (1998) Re-examination of the roles of PEP and Mg^{2+} in the reaction catalysed by the phosphorylated and non-phosphorylated forms of phosphoenolpyruvate carboxylase from leaves of *Zea mays*. Biochem J 332:633–642

Turgeon R, Medville R (1998) The absence of phloem loading in willow leaves. Proc Natl Acad Sci USA 95:12055–12060

Uemura K, Anwaruzzaman, Miyachi S, Yokota A (1997) Ribulose-1,5-bisphosphate carboxylase/oxygenase from thermophilic red algae with a strong specificity for CO_2 fixation. Biochem Biophys Res Commun 233:568–571
Ueno O (1998a) Immunogold localization of photosynthetic enzymes in leaves of various C_4 plants, with particular reference to pyruvate orthophosphate dikinase. J Exp Bot 49:1637–1646
Ueno O (1998b) Induction of Kranz anatomy and C_4-like biochemical characteristics in a submerged amphibious plant by abscisic acid. Plant Cell 10:571–583
Usuda H, Shimogawara K (1998) The effects of increased atmospheric carbon dioxide on growth, carbohydrates, and photosynthesis in radish, *Raphanus sativus*. Plant Cell Physiol 39:1–7
Veramendi J, Roessner U, Renz A, Willmitzer L, Trethewey RN (1999) Antisense repression of hexokinase 1 leads to an overaccumulation of starch in leaves of transgenic potato plants but not to significant changes in tuber carbohydrate metabolism. Plant Physiol 121:123–133
Villarejo A, Orús MI, Ramazanov Z, Martínez F (1998) A 38-kilodalton low-CO_2-inducible polypeptide is associated with the pyrenoid in *Chlorella vulgaris*. Planta 206:416–425
Vu JCV, Baker JT, Pennanen AH, Allen LH, Bowes G, Boote KJ (1998) Elevated CO_2 and water deficit effects on photosynthesis, ribulose bisphosphate carboxylase–oxygenase, and carbohydrate metabolism in rice. Physiol Plant 103:327–339
Wang N, Nobel PS (1998) Phloem transport of fructans in the crassulacean acid metabolism species *Agave deserti*. Plant Physiol 116:709–714
Wang S-J, Yeh K-W, Tsai C-Y (1999a) Molecular characterization and expression of starch granule-bound starch synthase in the sink and source tissues of sweet potato. Physiol Plant 106:253–261
Wang X, Sirover MA, Anderson LE (1999b) Pea chloroplast glyceraldehyde-3-phosphate dehydrogenase has uracil glycosylase activity. Arch Biochem Biophys 367:348–353
Wedel N, Soll J (1998) Evolutionary conserved light regulation of Calvin cycle activity by NADPH-mediated reversible phosphoribulokinase/CP12/glyceraldehyde-3-phosphate dehydrogenase complex dissociation. Proc Natl Acad Sci USA 95:9699–9704
Whitney SM, Andrews TJ (1998) The CO_2/O_2 specificity of single-subunit ribulose-bisphosphate carboxylase from the dinoflagellate *Amphidinium carterae*. Aust J Plant Physiol 25:131–138
Whitney SM, Von Caemmerer S, Hudson GS, Andrews TJ (1999) Directed mutation of the Rubisco large subunit of tobacco influences photorespiration and growth. Plant Physiol 121:579–588
Winder TL, Sun J, Okita TW, Edwards GE (1998) Evidence for the occurrence of feedback inhibition of photosynthesis in rice. Plant Cell Physiol 39:813–820
Wingler A, Lea PJ, Leegood RC (1999a) Photorespiratory metabolism of glyoxylate and formate in glycine-accumulating mutants of barley and *Amaranthus edulis*. Planta 207:518–526
Wingler A, Walker RP, Chen Z-H, Leegood RC (1999b) Phospho-enol-pyruvate carboxykinase is involved in the decarboxylation of aspartate in the bundle sheath of maize. Plant Physiol 120:539–545
Wingler A, Quick WP, Bungard RA, Bailey KJ, Lea PJ, Leegood RC (1999c) The role of photorespiration during drought stress: an analysis utilizing barley mutants with reduced activities of photorespiratory enzymes. Plant Cell Environ 22:361–373
Wittpoth C, Kroth PG, Weyrauch K, Kowallik KV, Strotmann H (1998) Functional characterization of isolated plastids from two marine diatoms. Planta 206:79–85
Woerner AC, Martin CE (1999) Mechanistic basis of differences in water-use efficiency between a CAM and a C_3 species of *Peperomia* (Piperaceae). New Phytol 144:307:312
Wyrich R, Dressen U, Brockmann S, Streubel M, Chang C, Qiang D, Paterson AH, Westhoff P (1998) The molecular basis of C_4 photosynthesis in sorghum: isolation, charac-

terization and RFLP mapping of mesophyll- and bundle-sheath-specific cDNAs obtained by differential screening. Plant Mol Biol 37:319–335

Zeeman SC, ap Rees T (1999) Changes in carbohydrate metabolism and assimilate export in starch-excess mutants of *Arabidopsis*. Plant Cell Environ 22:1445–1453

Zeeman SC, Umemoto T, Lue W-L, Au-Yeung P, Martin C, Smith AM, Chen J (1998a) A mutant of *Arabidopsis* lacking a chloroplastic isoamylase accumulates both starch and phytoglycogen. Plant Cell 10:1699–1711

Zeeman SC, Northrop F, Smith AM, ap Rees T (1998b) A starch-accumulating mutant of *Arabidopsis thaliana* deficient in a chloroplastic starch-hydrolysing enzyme. Plant J 15:357–365

Zhang N, Portis AR (1999) Mechanism of light regulation of Rubisco: a specific role for the larger Rubisco activase isoform involving reductive activation by thioredoxin-f. Proc Natl Acad Sci USA 96:9438–9443

Zhu G, Bohnert HJ, Jensen RG, Wildner GF (1998) Formation of the tight-binding inhibitor, 3-ketoarabinitol-1,5-bisphosphate by ribulose-1,5-bisphosphate carboxylase/oxygenase is O_2-dependent. Photosynth Res 55:67–74

Communicated by
U. Lüttge

Dr. Grahame J. Kelly
Centre for Molecular Biotechnology
School of Life Sciences
Queensland University of Technology
Brisbane, Queensland 4000, Australia
e-mail: g.kelly@qut.edu.au

Photosynthesis. Carbon Metabolism: Quantification and Manipulation

Grahame J. Kelly

"And, he gave it for his opinion, that whoever could make two ears of corn, or two blades of grass to grow upon a spot of ground where only one grew before, would deserve better of mankind, and do more essential service to his country, than the whole race of politicians put together."

Jonathan Swift in Gulliver's Travels, ca. 1727

This review is dedicated to Professor Martin Gibbs on the occasion of his 80th birthday

1 Introduction

Two philosophies have always guided research on photosynthetic carbon metabolism. The first is based on the desire to satisfy scientific curiosity by elucidating the chemical reactions and metabolic pathways involved, and the metabolic interconnections between chloroplasts, other parts of the cell, and the rest of the photosynthetic organism. An impressive overview in this regard has just been published (Cen et al. 2001). The second is the expectation that a thorough understanding of photosynthesis might in some way lead to benefits for humanity, the most obvious contender being faster growing crop plants that provide more food and fibre. As the years and decades have passed, the pendulum of emphasis has swung slowly but consistently away from the first and toward the second of these philosophies, and is perhaps now midway between the two. In this review, the literature relating to chemical reactions and metabolic pathways is given the most attention, but it will be noted that the latest research contains few new reactions and no new pathways. Is it possible that, finally, they have all been discovered? On the other hand, the literature relating to the second philosophy is given some attention because it deals with those topics that are linking photosynthetic carbon metabolism to recurring public news items of the type that have "GM foods" and "Kyoto Protocol" in their titles. The business of genetic manipulation of photosynthesis for increased productivity of food crops has been initiated (Sect. 5), but it is wise to point out from the

G.J. Kelly and E. Latzko:
Thirty Years of Photosynthesis

Originally published in Progress in Botany 64

start that the potential of this business is still largely unknown. The business of quantifying the biosphere's total photosynthesis, in both today's atmosphere and a future atmosphere containing extra CO_2, and of quantifying the sinks in which photosynthetically fixed carbon is sequestered on long-term scales, is also proceeding actively (Malhi and Grace 2000; Field 2001; Schlesinger and Lichter 2001), because these values are factored into the rules and calculations of the Kyoto Protocol (Schiermeier 2001). It is to be hoped, however, that readers of this review will find items that are fascinating not because of their relevance to these applied matters, but purely because they bring to light new tantalising facets of the wonder that is photosynthesis. The literature covered is predominantly that which has appeared since the last review in Volume 62 of Progress in Botany.

2 Carbon Metabolism Common to All Photosynthetic Cells

a) The Ribulose Bisphosphate (RuBP) Carboxylation System

Attention given to the Calvin cycle's CO_2-fixing enzyme ribulose-1,5-bisphosphate carboxylase/oxygenase (Rubisco) has been less than normal during the past 2 years, and an almost equal amount of attention has been given to the protein (Rubisco activase) that assists its activation (see below). Concerning Rubisco itself, Schlitter and Wildner (2000) have hypothesised that the enzyme's greater specificity for CO_2 (leading to photosynthesis) compared with that for O_2 (leading to photorespiration) is based on the rapidity with which the enzyme undergoes the conformational change that opens the niche in which the gaseous substrate binds. The content and activity of Rubisco in tissues are often difficult to estimate, so new approaches using tritium-labelled RuBP (Ferreira et al. 2000) and a trichloroacetic acid based procedure for determining the amount of enzyme lost during extraction (Rogers et al. 2001) may be welcomed by some researchers, including those giving the enzyme a major place in models directed toward quantifying the global carbon cycle (Bernacchi et al. 2001).

Algal and bacterial Rubiscos have become of special interest because their study has shown that the evolutionary diversity of the enzyme is much broader than indicated by studies of the plant enzyme. Some algae and photosynthetic bacteria contain an eight-large-subunit/eight-small-subunit Rubisco similar to the plant enzyme (termed the "green form I"), others contain a "red form I" in which the amino acid sequences of the two types of subunits are considerably different to those of the green form I, and some (but not all) dinoflagellates contain a "form II" Rubisco composed of eight large subunits only (Horken and Tabita 1999; Jenks and Gibbs 2000). Another difference, first observed in the 1980s, was that the two genes for the two types of subunit are both in the chloroplast in algae such as *Olisthodiscus luteus* (see Jenks and Gibbs 2000), in contrast to higher plants where the large subunit is encoded in the chloroplast but the small subunit is encoded in the nucleus. Whitney and Andrews (2001) recently attempted to create the *O. luteus* situation in a higher plant by relocating the Rubisco small subunit gene to the chloroplast in tobacco. They were moderately successful in that totally chloroplast-encoded Rubisco was obtained, but only at a level of about 1% of the total Rubisco; the tobacco synthesised the remaining 99% conventionally. A final note on algal Rubiscos relates to a field observation: in the dinoflagellate alga *Gonyaulax* sp., Rubisco moves into and out of pyrenoids in a circadian fashion (Nassoury et al. 2001), thereby contributing to this alga's earlier detected circadian rhythm in photosynthesis.

Recent research on the regulation of Rubisco activity in higher plants has centred almost exclusively on the protein Rubisco activase. The activase catalyses the early-morning dissociation of inhibitory sugar-Ps, such as RuBP (when previously bound to the inactive form of the enzyme) and, in some species, carboxyarabinitol-1-P, from Rubisco. These inhibitors maintain the enzyme in an inactive form during the night, and, as recently observed by Khan et al. (1999), may protect it against proteolytic breakdown. The activase presumably becomes effective only after sunrise because it undergoes a light-mediated activation very similar to that experienced by the Calvin cycle's fructose-1,6-bisphosphatase (Zhang et al. 2001). However, it seldom manages to fully activate all the Rubisco in leaves because its capacity is finite, and any synthesis of more of it would entail a trade-off of less Rubisco synthesis such that photosynthesis might decline rather than increase (Mott and Woodrow 2000). Clearly, all of a leaf's activase is needed to maintain photosynthesis, and this is dramatically demonstrated when the notable high-temperature sensitivity of the activase is examined. At temperatures around 40–45 °C (which are not unrealistic: it was 42 °C in Brisbane on Christmas Day last year!), the activase is thermally denatured (Salvucci et al. 2001) and also induced to associate with thylakoid membranes (Rokka et al. 2001). Consequent declines of about 50% in photosynthetic rates are observed (Crafts-Brandner and Law 2000; Crafts-Brandner and Salvucci 2000; Sharkey et al. 2001).

b) Other Enzymes of Chloroplast Carbon Metabolism

Recent research reports focused directly on the properties of the remaining ten enzymes of the Calvin cycle and other enzymes of chloroplast carbon metabolism have been few in number; one detailed study of spinach ribose-5-P isomerase has been provided by Jung et al. (2000). Two studies of the ubiquitous enzyme aldolase relate to stresses: the enzyme occurs as slightly different isozymes in chloroplasts, effectively indistinguishable except that certain isozymes become more predominant when the plant is stressed by salinity or high temperature (Michelis and Gepstein 2000; Yamada et al. 2000). A similar situation exists for chloroplast fructose-1,6-bisphosphatase (FBPase), in that the rice enzyme is salt-resistant when isolated from a salt-tolerant variety, but not when isolated from a salt-sensitive variety (Ghosh et al. 2001). It is well known that this chloroplast FBPase undergoes light-mediated reductive activation mediated by reduced thioredoxin *f*. The FBPase, initially inactive because a disulfide bridge severely disrupts the catalytic site (Chiadmi et al. 1999), transiently forms a mixed disulfide with the thioredoxin during the process in which the disulfide bridge is cleaved and reduced to two sulfhydryl groups (Balmer and Schürmann 2001). Thioredoxin is an incredible protein with many other roles, as pointed out in our 1998 review; a new addition is that it has a vital role in the complex interplay between stigma and pollen proteins that mediate self-incompatibility during sexual reproduction in many plant species (Cabrillac et al. 2001). Returning to leaves, another enzyme that undergoes light-mediated activation is P-ribulokinase, but it seems that this activation contributes to the regulation of photosynthesis only in low-light grown plants suddenly exposed to high light (Paul et al. 2000). Finally, three studies relating to the first two enzymes in the oxidative pentose-P pathway in chloroplasts have appeared. A new isoform of plastidic glucose-6-P dehydrogenase has been discovered by Wendt et al. (2000), while the delicate 6-P-gluconate dehydrogenase in chloroplasts has finally been purified from spinach by Krepinsky et al. (2001). A principal function of these two enzymes is to generate NADPH, the supply of which can be inadequate in situations such as low nitrogen availability (Robinson 2000).

c) The Three Respirations of Photosynthetic Cells: Mitochondrial Respiration, Chlororespiration, and Photorespiration

Photosynthetic cells are unique in possessing not one, but three respiratory activities. This makes their metabolism interesting, but also difficult to analyse. Haupt-Herting et al. (2001) have developed a new approach that may help to differentiate mitochondrial and photorespiratory CO_2 releases. During photosynthesis, mitochondrial respiration is reduced to a variable degree (Atkin et al. 2000), but it must continue, at least to some extent, because its effect on the cell's metabolite profile and redox status seems to be important for maintaining proper light-mediated regulation of enzymes and sucrose biosynthesis (Padmasree and Raghavendra 1999, 2001) and because the Krebs cycle is the immediate source of some of the carbon skeletons for amino acid biosynthesis following nitrate reduction (Cen et al. 2001). Not all of the CO_2 evolved from the Krebs cycle is necessarily lost from the plant: some may be recaptured by photosynthesis, especially in drought-stressed leaves (Haupt-Herting et al. 2001) and the glumes of cereal ears (Gebbing and Schnyder 2001).

Chlororespiration, like mitochondrial respiration, involves the oxidation of organic molecules and consequent generation of NAD(P)H which then feeds reducing equivalents into a membrane-located electron transport chain that delivers them finally to O_2. The major difference is that the membrane is the chloroplast's thylakoid membrane, rather than the mitochondrion's inner membrane (Fig. 1). Of course, being the thylakoid membrane, it is not difficult to envisage that some of the electron-transporting components might be the same as those used in photosynthesis. The best-established example is plastoquinone, and the central version of chlororespiration involves little more than two enzymes, the first of which catalyses the transfer of the reducing equivalents from NAD(P)H to this plastoquinone (see Carol and Kuntz 2001) while the second, a plastoquinol oxidase, catalyses a reaction between the resultant plastoquinol and O_2, regenerating the plastoquinone and releasing water (Cournac et al. 2000). A possible ancestor of the latter enzyme has been detected in a cyanobacterium by Büchel et al. (1998). Improved evidence for chlororespiration in leaf chloroplasts in darkness has become available (Feild et al. 1998), but of greater interest are three variants of the chlororespiratory process (Fig. 1). In the first, the organic molecule oxidised is not a carbohydrate (as in conventional respiration),

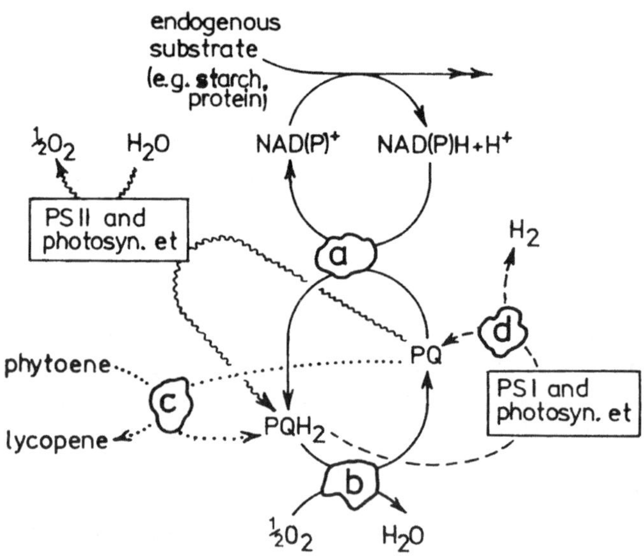

Fig. 1. Chlororespiration (*solid lines*) on the *Chlamydomonas* thylakoid membrane, and its connections with carotenoid biosynthesis (*dotted lines*), anaerobiosis-induced H_2 evolution (*dashed lines*), and a water-to-water cycle when photosystem I is missing (*squiggly lines*). *a* NAD(P)H-plastoquinone oxidoreductase; *b* plastid terminal oxidase; *c* phytoene desaturase; *d* hydrogenase. *photosyn. et* photosynthetic electron transport; *PQ* plastoquinone; *PS* photosystem. (Adapted from Cournac et al. 2000; Melis et al. 2000; Carol and Kuntz 2001)

but rather the carotenoid precursor phytoene, which undergoes four cycles of desaturation to generate lycopene and (after ring cyclisation) β-carotene (Bennoun 2001; Carol and Kuntz 2001). In the second, the reducing equivalents do not come from an organic molecule (and it is, therefore, strictly speaking, not a form of respiration), but rather from water which is photosynthetically split by photosystem II in the light (Grotjohann et al. 1999). Clearly, this process must be of low magnitude in normal cells (or else there would be no photosynthesis), but its capacity is considerable, as indicated by a *Chlamydomonas* mutant that is deficient in photosystem I and thus has nowhere else to deposit the reducing equivalents (Cournac et al. 2000). The third variant appears in certain green microalgae (and also certain cyanobacteria) that are capable of synthesising the enzyme hydrogenase. When these cells are placed in an anaerobic environment (thus precluding the transfer of the reducing equivalents from the plastoquinol to O_2; again, strictly speaking, the process is no longer a true respiration), they simply synthesise hydrogenase and use it to recombine the electron and proton parts of the reducing equivalents into H_2 gas (which is evolved). When this metabolism is linked in some way to the original release of hydrogen from water splitting by the photochemical apparatus of these cells (not an easy task because the hydrogenase is quite sensitive to inactivation by O_2, the other product of water splitting), then the application of photosynthesis to the production of a clean and renewable fuel (H_2) for humanity can be envisaged. Such an objective, long championed by Martin Gibbs, has recently come one step closer to reality (Melis et al. 2000).

The third form of respiration in photosynthetic cells, photorespiration, differs from the above two in that there is minimal involvement of cell membranes. As described in our earlier reviews, it is composed of the glycolate cycle (a series of eight reactions distributed in the soluble phases of chloroplasts, mitochondria and peroxisomes) connected to the Calvin cycle at the Rubisco reaction. A potential role of photorespiration is the consumption of excess ATP and NADPH and consequent avoidance of photoinhibition. More evidence supporting this role has come from a mutant barley with low glycine decarboxylase activity (Igamberdiev et al. 2001a), from a tobacco transgenic for low glycine decarboxylase activity (Yamaguchi and Nishimura 2000), and from plants under stress situations: tobacco exposed for 1 week to a low O_2 atmosphere (Migge et al. 1999), winter wheat exposed to high light at 20 °C (Savitch et al. 2000), and the rainforest plant *Arisaema heterophyllum* exposed to high light (Muraoka et al. 2000). A different facet of stress vis-à-vis photorespiration has been raised by Sivakumar et al. (2000) who found that the proline accumulated by plants in response to salt stress interacts with Rubisco in such a way as to counteract a salt-induced enhancement of the oxygenase activity of that enzyme.

Enzymes of the glycolate cycle are still attracting interest. Iwamoto and Ikawa (2000) remind us that two types of glycolate-oxidising enzymes (oxidase or dehydrogenase) exist in algae, sometimes even within the same algal group. Liepman and Olsen (2001) have DNA-based evidence that just one protein may accommodate the two transamination reactions required in the peroxisome. Finally, the tremendously high concentration and activity (five-times that of the Krebs cycle) of the four-enzyme complex that decarboxylates two glycines to one serine in the mitochondrial matrix has been re-emphasised by Douce et al. (2001). The sophistication of the complex is such that it contributes to the fractionation between ^{12}C and ^{13}C in photosynthetic cells (Igamberdiev et al. 2001b). This apparently happens despite the fact that most of the released CO_2 never gets to exit the leaf, but is returned to Rubisco (Loreto et al. 1999), an event that is intuitively intriguing but, of course, does not alter the basic premise that photorespiration can be viewed as a wasteful loss (as CO_2) of previously fixed CO_2.

d) Starch

The diurnal diversion of some of the product of the Calvin cycle into transitory starch in the chloroplast is achieved by the sequential action of hexose-P isomerase, P-glucomutase, ADP-glucose pyrophosphorylase, and a collection of starch synthases and associated enzymes that sculpture starch molecules. Indirect evidence for the indispensability of P-glucomutase (Tauberger et al. 2000; Fernie et al. 2001) has been complemented by the conclusion of Yu et al. (2000) that the hexose-P isomerase is also essential. Of course, the ADP-glucose pyrophosphorylase is essential because it forms the glucose-donor molecule (ADP-glucose). In leaves, all of this enzyme is in the chloroplast. It almost certainly influences the rate of starch synthesis through its regulatory properties (see our earlier reviews) and the overall amount present that, in turn, may be influenced by the mysterious sugar trehalose (Wingler et al. 2000). Another recent surprise has been the detection by Rodriguez-López et al. (2000) of an enzyme that destroys ADP-glucose by hydrolysing it into AMP and glucose-1-P, thereby potentially hindering starch synthesis. The role of this new enzyme is unclear. It might be envisioned that it could prevent starch biosynthesis from becoming excessive (Kleczkowski 2001), but its capacity to do this might be questioned given that chloroplasts are well known for their proclivity to accumulate excessive amounts of starch, e.g. after removal of sink organs (Nakano et al. 2000), or when the supply of CO_2 to the leaf is luxurious (Sawada et al. 2001). Such starch accumulation is also hard to reconcile with the proposal that the growth of the starch granule is kept under control by a refined regulatory system orchestrated by certain 14-3-3 proteins in the granule (Sehnke et al. 2001). [Note: "14-3-3" is a somewhat obscure term derived from the loci of migration of these proteins in a double chromatography system; the proteins are ubiquitous in nature, and usually enmeshed in intracellular molecular signalling and regulatory systems that involve the phosphorylation and dephosphorylation of other proteins (Voigt et al. 2001)].

Two points concerning the earlier surprise (see our 1998 review) that most ADP-glucose pyrophosphorylase is located in the cytosol, rather than the plastid, of non-photosynthetic starch-storing organs have come to light. The first is that this phenomenon is restricted to the seed endosperm of graminaceous species (Beckles et al. 2001a). An earlier report that it applies to developing tomato fruit has been challenged (Beckles et al. 2001b). The second is that, while this cytosolic form of the enzyme commonly lacks (Doan et al. 1999) the allosteric regulatory properties typical of the chloroplast, and particularly cyanobacterial (Gómez Casati et al. 2000) enzyme, this feature may not be universal: Sikka et al. (2001) report that the rice-endosperm enzyme is subject to allosteric regulation.

The manufacture of amylose and amylopectin by starch synthases and branching enzyme, and the reshaping of amylopectin by debranching enzyme, were described in our last review. The fate of the short chains of glucoses (termed maltooligosaccharides) released by the debranching enzyme can now be considered. One role is that they can be used as a start-up substrate by the granule-bound starch synthase, for elongation into amylose molecules (Denyer at al. 1999). A second, somewhat fascinating, role has been considered by Takaha et al. (1998), who propose that yet another enzyme, named disproportionating enzyme or "D-enzyme", can remove a piece of one maltooligosaccharide and add it to another. The latter (lengthened at the expense of the former) is then long enough to be recognised by a well-known enzyme of old, starch phosphorylase, that proceeds to cleave glucose-1-Ps from it. These are then able to re-enter starch synthesis via the ADP-glucose pyrophosphorylase reaction. The beauty of this scheme is that it explains how starch phosphorylase might participate in starch biosynthesis in chloroplasts and amyloplasts, an event long entertained by investigators of this enzyme (Albrecht et al. 2001; Yu et al. 2001). The likelihood of these events are strong, given that D-enzyme has been shown to be essential for the synthesis of normal starch in *Chlamydomonas reinhardtii* (Colleoni et al. 1999).

The nocturnal degradation of chloroplast starch begins with hydrolysis by endoamylases of the α-type, and possibly also of the β-type (Lao et al. 1999). Any released maltooligosaccharides too small to be recognised by the amylases are manipulated by D-enzyme (as described above) into suitably larger molecules (Critchley et al. 2001). Branch points in amylopectin are cleaved by debranching enzyme which, it seems, is regulated by the *h* form of thioredoxin (Cho et al. 1999).

e) Sucrose

The product of the chloroplast's CO_2 fixation is exported as triose-P to the cytosol where it is manufactured into sucrose (for subsequent translocation) in most species. Two challenges to this conventional scheme have appeared, one in which it is claimed that fumaric acid may have a role similar to that of sucrose (Chia et al. 2000), and the other reviving the old debate of whether or not sucrose might occur in chloroplasts as well as in the cytosol: Gerrits et al. (2001) provide evidence that substantial amounts of sucrose can be pulled into the chloroplast if it is then metabolised to an end product such as fructan.

Triose-P export is mediated by the well-known triose-P/Pi translocator ("phosphate translocator") of the chloroplast envelope's inner membrane. This translocator helps to determine how much photosynthetic product is directed to sucrose synthesis versus the alternative fates of transitory storage (as starch) in the chloroplast, and respiration by mitochondria (Häusler et al. 2000a), but it may not be the only player in this game: the chloroplast envelope's *outer* membrane has suddenly achieved prominence because it has been found to possess an anion-selective channel protein with the potential to control the direction of triose-P flow (Bölter et al. 1999), and a hexokinase (Wiese et al. 1999) that might energise glucose arriving from the inner membrane's glucose translocator during the night when the mobilisation of the transitory starch reserve takes place (Weber et al. 2000). Incidentally, a recent astonishing report concerning such glucose transporter proteins was that the marine alga *Phaeodactylum tricornutum* could be converted from photosynthesis to heterotrophy simply by inserting a single gene encoding the glucose transporter from human red blood cells (Zaslavskaia et al. 2001).

The roles of the cytosolic FBPase and sucrose-P synthase have been further investigated by decreasing the expression of their corresponding genes in *Arabidopsis thaliana*. In each case, sucrose synthesis was hindered, but only with lowered FBPase did a reciprocal rise in chloroplast starch occur (Strand et al. 2000). Confirmatory evidence that the FBPase, and also the enigmatic PPi-linked P-fructokinase, are subject to control by the regulatory sugar-P fructose-2,6-P_2 is provided by Scott et al. (2000) and Theodorou and Kruger (2001), and is complemented by a demonstration with transgenic *Arabidopsis* that this sugar-P has an influence on whether photosynthetic product is partitioned into cytosolic sucrose or chloroplast starch (Draborg et al. 2001). Finally, a full description of the ultimate enzyme of sucrose synthesis, sucrose-P phosphatase, has appeared (Lunn et al. 2000).

The movement of sucrose from mesophyll photosynthetic cells to phloem sieve tubes is a topic currently receiving intense scrutiny. At least five different sucrose transporter proteins associated with phloem-cell membranes have been identified, varying in their relative affinities for sucrose and capacities for transporting it. It is envisaged that most are engaged in loading sucrose into the phloem (Gottwald et al. 2000;

Meyer et al. 2000; Weise et al. 2000), but at least one is considered to be a sucrose sensor as well (Barker et al. 2000). In a unique report, Ludwig et al. (2000) describe how one, and possibly more, of these sucrose transporters can also transport the vitamin biotin. This may appear mystifying, but for those of us lucky enough to have had the opportunity to prepare culture media for growing marine microalgae, it is less mystical because biotin (and also thiamine and vitamin B_{12}) is included in the recipes for these media. Could an essential microalgal biotin transporter be the ancestor of the plant sucrose transporters? And to what extent are these transporters related to the mannitol transporter more recently identified (Noiraud et al. 2001) in the phloem of celery that translocates most photosynthetic product as mannitol rather than sucrose?

3 C_4 Photosynthesis and Crassulacean Acid Metabolism (CAM)

a) C_4 Photosynthesis

It has been more than 30 years since C_4 photosynthesis was discovered and, indeed, this topic may have passed its climax; during the past 2 years there has been little direct research on this marvellous metabolic extension of photosynthetic carbon metabolism. Great effort is being devoted to an indirect aspect, viz., the roles of DNA regulatory elements and associated proteins in bringing about the differential gene expression and distribution of enzymes between mesophyll and bundle-sheath cells, but this topic is beyond the scope of this review.

The tenant that C_4 photosynthesis is always associated with distinctive mesophyll and bundle-sheath cells (Kranz anatomy) is being shaken from three directions. Firstly, two aquatic angiosperms that display variants of C_4 photosynthesis lack Kranz anatomy (Casati et al. 2000; van Ginkel et al. 2001). Secondly, an obscure C_4 member of the Chenopodiaceae, *Borszczowia aralocaspica*, that has a strange leaf anatomy of one-layered chlorenchyma (Freitag and Stichler 2000), is being used by Voznesenskaya et al. (2001) to support the notion that Kranz anatomy is not essential for terrestrial C_4 photosynthesis. They describe how the two types of chloroplasts (and their associated enzymes) typical of mesophyll and bundle-sheath cells in conventional C_4 plants are positioned into discrete proximal and distal sections of the single type of photosynthetic cell in this plant. Finally, enzyme and ^{14}C-labelling experiments with the unicellular diatom *Thalassiosira weissflogii* under conditions that retard its CO_2-concentrating mechanism led Reinfelder et al. (2000) to claim that, under such conditions, the diatom performed a "unicellular C_4 photosynthesis". However, two types of chloroplasts were not identified, and the degree to which the data support C_4 photosynthesis in its traditional sense has been questioned by Johnston et al.

(2001); perhaps diatoms perform C_3 photosynthesis with a "C_4ish" tendency, akin to the aquatic angiosperms mentioned above.

More details concerning the activation upon sunrise of the C_4 pathway's initial enzyme, phosphoenolpyruvate (PEP) carboxylase, have emerged. As reviewed previously, the activation (now detected in one of the submersed aquatic species that perform C_4-like photosynthesis; Lara et al. 2001) results from a phosphorylation of the enzyme catalysed by a second enzyme appropriately named PEP carboxylase kinase (PEPCK). This PEPCK appears to be absent in darkness, or present in quite low amounts, with activity suppressed by yet a third protein, an inhibitor that binds to it (G.A. Nimmo et al. 2001). However, large amounts of PEPCK appear upon illumination when the light initiates an intracellular signalling system that culminates in the expression of its gene (H.G. Nimmo et al. 2001). The signalling system involves the ubiquitous second messengers inositol-P_3 (Coursol et al. 2000) and Ca^{2+} (Ogawa et al. 1998), but the nature of the (coloured?) molecule that detects light and sets off this chain of events is still elusive. The end result is an activated PEP carboxylase with, among other properties, an increased affinity for its substrate bicarbonate (Parvathi et al. 2000). This could be valuable because, at physiological levels of bicarbonate and Mg^{2+}, the enzyme operates below its potential (Tovar-Méndez et al. 2000).

b) Crassulacean Acid Metabolism

If C_4 photosynthesis has passed its 30th year, then crassulacean acid metabolism (CAM) must have passed its 40th. Nevertheless, the list of plants with CAM continues to lengthen, with more members of the Portulacaceae (Guralnick and Jackson 2001), and tropical epiphytic and lithophytic ferns (Holtum and Winter 1999) recently added. The principal benefit of CAM is that plant water is conserved when stomates are open only at night; CAM achieves nocturnal storage (as malate) of the CO_2 that enters these stomates. Next day this store is drawn upon for photosynthesis behind closed stomates. A not unexpected, but still impressive, observation concerning the CAM plant's water economy is that the plasma membrane and tonoplast of the CAM plant *Graptopetalum paraguayense* possess remarkably low levels (20 and 1% respectively) of the water-transporter protein, aquaporin, when compared to a representative C_3 plant, radish (Ohshima et al. 2001).

The distinctive nocturnal and diurnal metabolic activities of CAM plants are set into a circadian rhythm in the cells of these plants; wave fronts of the rhythm can be seen to propagate across leaves (Rascher et al. 2001). The master switch for the rhythm has not yet been confirmed, but a biophysical switch involving altered tonoplast tension is a strong contender (Lüttge 2000), especially since the activation of PEP carboxylase, which follows the turning-on of a PEPCK gene as described above for C_4 plants, follows a circadian rhythm determined by an oscillator that may be at the level of the tonoplast (Hartwell et al. 1999; Nimmo 2000; Taybi et al. 2000). Other points concerning PEP carboxylase relate to those plants for which CAM is an inducible phenomenon accompanied by an increase in the amount of the enzyme. Sometimes the extra enzyme is of a special form (Mazen 2000). The well-known induction by salinity in *Mesembryanthemum crystallinum* occurs only under high light (Miszalski et al. 2001). There is a surprise report that foliar application of gibberellic acid to this plant induces, like salinity, CAM and increased PEP carboxylase activity (Guralnick et al. 2001). Incidentally, another surprise arose when Malda et al. (1999) grew the CAM plant *Coryphantha minima* in vitro using plant tissue culture techniques. It showed net CO_2 uptake continuously in the light, as well as in the dark, and grew seven-fold larger than control plants.

Several other enzymes and membrane-located transporter proteins of CAM plants have been examined, including an unusual cytosolic NAD-linked malic enzyme with some function in night-time acidification (rather than daytime deacidification as established for its mitochondrial counterpart; Cuevas and Podestá 2000), a pyruvate Pi dikinase unexpectedly located in the cytosol (rather than in the chloroplast; Kondo et al. 2000), a putative tonoplast malate transporter (Lüttge et al. 2000), and chloroplast-envelope PEP/Pi and glucose-6-P/Pi translocators (Häusler et al. 2000b), all three of which are more plentiful when CAM is more prevalent. A notable conclusion from the latter study was that CAM chloroplasts might be more similar to non-photosynthetic plastids (Tauberger at al. 2000) than to other chloroplasts in that they can import glucose-6-P for starch synthesis.

4 Brief Notes on Carbon and Kyoto

Because the Kyoto Protocol makes reference to carbon sequestration in forests and soils, the current status of these systems and their responses in a world with extra CO_2 in its atmosphere are under intense investigation. The well-established stimulation of photosynthesis and growth by increased CO_2 is now even more strongly believed to be a characteristic and persistent attribute of trees (Curtis and Wang 1998; Saxe et al. 1998; Stylinski et al. 2000; Herrick and Thomas 2001; Idso and Kimball 2001) despite some down-regulation of the capacities of photosynthetic electron transport and Rubisco by some species (Medlyn et al. 1999; Griffin et al. 2000), and retardation of photosynthesis by excess starch accumulated in chloroplasts by others (Würth et al. 1998; see Sawada et al. 2001; Sicher and Bunce 2001). Another recent change in emphasis relates to respiration: it now seems that increased CO_2 does not reduce mitochondrial respiration, but rather has no effect on it (Amthor et al. 2001; Hamilton et al. 2001) when care is taken with measurement devices (Jahnke 2001). Of course, in quantifying the global carbon cycle, respiration below ground must also be factored in. This parameter may not be receiving proper attention in experiments with trees exposed to increased CO_2 (Luo 2001), and its importance is highlighted by the conclusion of Ekblad and Högberg (2001) that carbon photosynthetically fixed by tree leaves can be respired just 1–4 days later by the tree's roots or nearby soil microbes.

5 Genetic Manipulation of Photosynthetic Carbon Metabolism

The dream of bigger and/or faster growing crop plants has existed at least as far back as the times when the wonderful story of Gulliver's Travels and the impossible fantasy of Jack and the Beanstalk were penned (see quote above, and that preceding our 1994 review). Farmers, crop breeders and plant physiologists over the years have slowly but surely brought that dream to fruition (Richards 2000). However, dreams never die, and the prospect that the modern tools of molecular biology might be used to genetically engineer plants to be still bigger or still faster growing has encouraged many researchers in the past decade. Finally, there has appeared "the first report of the transgenic improvement of plant photosynthetic capacity" (Wood 2002). Encouraged, no doubt, by indications that the Calvin cycle enzyme sedoheptulose bisphosphatase (SBPase) plays a powerful part in determining the rate of photosynthetic carbon assimilation (Poolman et al. 2000; Harrison et al. 2001), and having on hand the gene for an unusual cyanobacterial SBPase that, among other odd properties, also has FBPase activity (Tamoi et al. 1998), Miyagawa et al. (2001) inserted this gene into tobacco chloroplasts and obtained transgenic plants with doubled SBPase activity and a photosynthetic rate increased by almost 25% (Fig. 2). One wonders whether an additional boost might be possible if extra transketolase could also be engineered into the chloroplast, given that Henkes et al. (2001) have discovered that the activity of this enzyme also has a dramatic influence on the rate of photosynthesis.

Whether or not the above was indeed the first case of a plant genetically engineered for faster photosynthesis might be debated. Three other cases deserve mention. In our 1998 review, we noted that Galtier et al. (1995) coaxed tomato plants to photosynthesise faster (the increase was about 20%) by transforming them with a maize sucrose-P synthase gene. More recently, Migge et al. (2000) reported a 20–30% increase in fresh weight of young tobacco transformed to contain a doubled activity of the chloroplast enzyme glutamine synthetase. This enzyme participates in the assimilation of ammonium, including that released by the mitochondrial four-enzyme complex that converts two glycines to one serine during photorespiration (Sect. 2.c). Finally, Paul et al. (2001) make reference to a relatively obscure publication pertaining to tobacco with photosynthesis increased by up to 30% following transformation with a

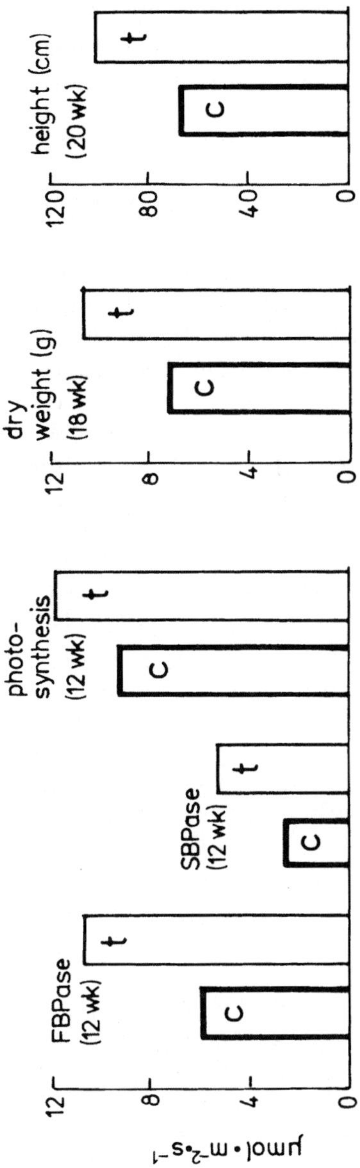

Fig. 2. Increased photosynthesis and plant size of tobacco transformed with a cyanobacterial FBPase/SBPase. The enzyme activities are also shown. *c* Control; *t* transformant TpFS-3. (Data from Miyagawa et al. 2001)

bacterial gene encoding trehalose-P synthase. Presumably, the trehalose alters sugar-signalling pathways such that the expression of the tobacco-leaf genes is altered.

Of course, the obliteration of photorespiration has been another goal of genetic engineers aiming to increase photosynthesis. Four approaches have been tried. One has been to alter the plant's native Rubisco so that it's oxygenase activity is removed, or at least altered. However, as noted in our earlier reviews, and as Lorimer (2001) has eloquently pointed out, this has come to naught after 30 years of effort (not to mention 2 billion years of evolution). A second approach recently considered and debated (see our last review) is to transform plants with an algal Rubisco that recognises O_2 (instead of CO_2) less often as a substrate. Whitney et al. (2001) now report that, when this was attempted, the two different subunits of the algal Rubisco were synthesised in the leaf chloroplasts in abundant amounts, but they failed to assemble into functional multi-subunit enzyme molecules. Unfortunately, the accumulated unassembled subunits clogged the chloroplast and photosynthesis consequently declined. A third idea from Barata et al. (2000) is to introduce the gene for the O_2-binding protein of legume nodules (leghaemoglobin) into chloroplasts so that the Rubisco no longer has access to free O_2. These authors attempted this, but so far gene expression has been too low to significantly lower the O_2 level. Finally, a popular approach to reduce photorespiration is to add characteristics of C_4 plants into C_3 plants (C_4 plants naturally avoid photorespiration by pumping CO_2 into bundle-sheath cells). To this end one, or sometimes two, of the genes for the C_4 enzymes PEP carboxylase, NADP-linked malate dehydrogenase, NADP-linked malic enzyme, PEP carboxykinase, and pyruvate Pi dikinase have been engineered into tobacco, potato or rice (Suzuki et al. 2000; Beaujean et al. 2001; Fukayama et al. 2001; Häusler et al. 2001). Gene expression was sometimes substantial (Fukayawa et al. 2001). However, in no case was a significant beneficial effect on the photosynthesis of the recipient plant observed. It could be argued that additional transformation with genes that direct the differentiation of cells into mesophyll and bundle-sheath types (possibly a daunting task) might be needed before a beneficial effect could ensue, but Voznesenskaya et al. (2001) argue that this is not necessarily so since they have discovered a C_4 plant that works well with only one type of cell (Sect. 3.a).

The only certainty concerning the genetic modification of photosynthesis to increase production is that a good deal more literature on the topic will be available for inclusion in the next review of this series in 2004. Meantime, optimism might be tempered by the comments from Horton (2000): "without extreme good fortune, manipulation of photosynthesis to increase yield is at present an unlikely prospect for the near future"; and from Richards (2000): "....photosynthesis does not limit grain yield and current crops have sufficient surplus photosynthetic capacity to respond to increases in yield potential". Nevertheless, we should be prepared for the unexpected. An example just reported in the media (but unconfirmed at the time of writing) is the production of a pig engineered to contain spinach genes. This first example of a mammal with plant genes will no doubt cause some of we older researchers of photosynthesis to squirm!

References

Albrecht T, Koch A, Lode A, Greve B, Schneider-Mergener J, Steup M (2001) Plastidic (Pho1-type) phosphorylase isoforms in potato (*Solanum tuberosum* L.) plants: expression analysis and immunochemical characterization. Planta 213:602–613

Amthor JS, Koch GW, Willms JR, Layzell DB (2001) Leaf O_2 uptake in the dark is independent of coincident CO_2 partial pressure. J Exp Bot 52:2235–2238

Atkin OK, Evans JR, Ball MC, Lambers H, Pons TL (2000) Leaf respiration of snow gum in the light and dark. Interactions between temperature and irradiance. Plant Physiol 122:915–923

Balmer Y, Schürmann P (2001) Heterodimer formation between thioredoxin *f* and fructose 1,6-bisphosphatase from spinach chloroplasts. FEBS Lett 492:58–61

Barata RM, Chaparro A, Chabregas SM, González R, Labate CA, Azevedo RA, Sarath G, Lea PJ, Silva-Filho MC (2000) Targeting of the soybean leghemoglobin to tobacco chloroplasts: effects on aerobic metabolism in transgenic plants. Plant Sci 155:193–202

Barker L, Kühn C, Weise A, Schulz A, Gebhardt C, Hirner B, Hellmann H, Schulze W, Ward JM, Frommer WB (2000) SUT2, a putative sucrose sensor in sieve elements. Plant Cell 12:1153–1164

Beaujean A, Issakidis-Bourguet E, Catterou M, Dubois F, Sangwan RS, Sangwan-Norreel BS (2001) Integration and expression of *Sorghum* C_4 phosphoenolpyruvate carboxylase and chloroplastic NADP$^+$-malate dehydrogenase separately or together in C_3 potato plants. Plant Sci 160:1199–1210

Beckles DM, Smith AM, ap Rees T (2001a) A cytosolic ADP-glucose pyrophosphorylase is a feature of graminaceous endosperms, but not of other starch storing organs. Plant Physiol 125:818–827

Beckles DM, Craig J, Smith AM (2001b) ADP-glucose pyrophosphorylase is located in the plastid in developing tomato fruit. Plant Physiol 126:261–266

Bennoun P (2001) Chlororespiration and the process of carotenoid biosynthesis. Biochim Biophys Acta 1506:133–142

Bernacchi CJ, Singsaas EL, Pimentel C, Portis AR, Long SP (2001) Improved temperature response functions for models of Rubisco-limited photosynthesis. Plant Cell Environ 24:253–259

Bölter B, Soll J, Hill K, Hemmler R, Wagner R (1999) A rectifying ATP-regulated solute channel in the chloroplastic outer envelope from pea. EMBO J 18:5505–5516

Büchel C, Zsíros O, Gareb G (1998) Alternative cyanide-sensitive oxidase interacting with photosynthesis in *Synechocystis* PCC6803. Ancestor of the terminal oxidase of chlororespiration? Photosynthetica 35:223–231

Cabrillac D, Cock JM, Dumas C, Gaude T (2001) The *S*-locus receptor kinase is inhibited by thioredoxins and activated by pollen coat proteins. Nature 410:220–223

Carol P, Kuntz M (2001) A plastid terminal oxidase comes to light: implications for carotenoid biosynthesis and chlororespiration. Trends Plant Sci 6:31–36

Casati P, Lara MV, Andreo CS (2000) Induction of a C_4-like mechanism of CO_2 fixation in *Egeria densa*, a submersed aquatic species. Plant Physiol 123:1611–1621

Cen Y-P, Turpin DH, Layzell DB (2001) Whole-plant gas exchange and reductive biosynthesis in white lupin. Plant Physiol 126:1555–1565

Chia DW, Yoder TJ, Reiter W-D, Gibson SI (2000) Fumaric acid: an overlooked form of fixed carbon in *Arabidopsis* and other plant species. Planta 211:743–751

Chiadmi M, Navaza A, Miginiac-Maslow M, Jacquot J-P, Cherfils J (1999) Redox signalling in the chloroplast: structure of oxidized pea fructose-1,6-bisphosphate phosphatase. EMBO J 18:6809–6815

Cho M-J, Wong JH, Marx C, Jiang W, Lemaux PG, Buchanan BB (1999) Overexpression of thioredoxin *h* leads to enhanced activity of starch debranching enzyme (pullulanase) in barley grain. Proc Natl Acad Sci USA 96:14641–14646

Colleoni C, Dauvillée D, Mouille G, Buléon A, Gallant D, Bouchet B, Morell M, Samuel M, Delrue B, d'Hulst C, Bliard C, Nuzillard J-M, Ball S (1999) Genetic and biochemical evidence for the involvement of α-1,4 glucanotransferases in amylopectin synthesis. Plant Physiol 120:993–1003

Cournac L, Redding K, Ravenel J, Rumeau D, Josse E-M, Kuntz M, Peltier G (2000) Electron flow between photosystem II and oxygen in chloroplasts of photosystem I-deficient algae is mediated by a quinol oxidase involved in chlororespiration. J Biol Chem 275:17256–17262

Coursol S, Giglioli-Guivarc'h N, Vidal J, Pierre J-N (2000) An increase in phosphoinositide-specific phospholipase C activity precedes induction of C_4 phosphoenolpyruvate carboxylase phosphorylation in illuminated and NH_4Cl-treated protoplasts from *Digitaria sanguinalis*. Plant J 23:497–506

Crafts-Brandner SJ, Law RD (2000) Effect of heat stress on the inhibition and recovery of the ribulose-1,5-bisphosphate carboxylase/oxygenase activation state. Planta 212:67–74

Crafts-Brandner SJ, Salvucci ME (2000) Rubisco activase constrains the photosynthetic potential of leaves at high temperature and CO_2. Proc Natl Acad Sci USA 97:13430–13435

Critchley JH, Zeeman SC, Takaha T, Smith AM, Smith SM (2001) A critical role for disproportionating enzyme in starch breakdown is revealed by a knock-out mutation in *Arabidopsis*. Plant J 26:89–100

Cuevas IC, Podestá FE (2000) Purification and physical and kinetic characterization of an NAD^+-dependent malate dehydrogenase from leaves of pineapple (*Ananas comosus*). Physiol Plant 108:240–248

Curtis PS, Wang X (1998) A meta-analysis of elevated CO_2 effects on woody plant mass, form, and physiology. Oecologia 113:299–313

Denyer K, Waite D, Motawia S, Møller BL, Smith AM (1999) Granule-bound starch synthase I in isolated starch granules elongates malto-oligosaccharides processively. Biochem J 340:183–191

Doan DNP, Rudi H, Olsen O-A (1999) The allosterically unregulated isoform of ADP-glucose pyrophosphorylase from barley endosperm is the most likely source of ADP-glucose incorporated into endosperm starch. Plant Physiol 121:965–975

Douce R, Bourguignon J, Neuburger M, Rébeillé F (2001) The glycine decarboxylase system: a fascinating complex. Trends Plant Sci 6:167–176

Draborg H, Villadsen D, Nielsen TH (2001) Transgenic *Arabidopsis* plants with decreased activity of fructose-6-phosphate-2-kinase/fructose-2,6-bisphosphatase have altered carbon partitioning. Plant Physiol 126:750–758

Ekblad A, Högberg P (2001) Natural abundance of ^{13}C in CO_2 respired from forest soils reveals speed of link between tree photosynthesis and root respiration. Oecologia 127:305–308

Feild TS, Nedbal L, Ort DR (1998) Nonphotochemical reduction of the plastoquinone pool in sunflower leaves originates from chlororespiration. Plant Physiol 116:1209–1218

Fernie AR, Roessner U, Trethewey RN, Willmitzer L (2001) The contribution of plastidial phosphoglucomutase to the control of starch synthesis within the potato tuber. Planta 213:418–426

Ferreira RB, Esquível MG, Teixeira AR (2000) An accurate method to quantify ribulose bisphosphate carboxylase content in plant tissue. Plant Cell Environ 23:1329–1340

Field CB (2001) Plant physiology of the "missing" carbon sink. Plant Physiol 125:25–28

Freitag H, Stichler W (2000) A remarkable new leaf type with unusual photosynthetic tissue in a central Asiatic genus of Chenopodiaceae. Plant Biol 2:154–160

Fukayama H, Tsuchida H, Agarie S, Nomura M, Onodera H, Ono K, Lee B-H, Hirose S, Toki S, Ku MSB, Makino A, Matsuoka M, Miyao M (2001) Significant accumulation of C_4-specific pyruvate, orthophosphate dikinase in a C_3 plant, rice. Plant Physiol 127:1136–1146

Galtier N, Foyer CH, Murchie E, Alred R, Quick P, Voelker TA, Thépenier C, Lascève G, Betsche T (1995) Effects of light and atmospheric carbon dioxide enrichment on photosynthesis and carbon partitioning in the leaves of tomato (*Lycopersicon esculentum* L.) plants over-expressing sucrose phosphate synthase. J Exp Bot 46:1335–1344

Gebbing T, Schnyder H (2001) ^{13}C labeling kinetics of sucrose in glumes indicates significant refixation of respiratory CO_2 in the wheat ear. Aust J Plant Physiol 28:1047–1053

Gerrits N, Turk SCHJ, van Dun KPM, Hulleman SHD, Visser RGF, Weisbeek PJ, Smeekens SCM (2001) Sucrose metabolism in plastids. Plant Physiol 125:926–934

Ghosh S, Bagchi S, Majumder AL (2001) Chloroplast fructose-1,6-bisphosphatase from *Oryza* differs in salt tolerance property from the *Porteresia* enzyme and is protected by osmolytes. Plant Sci 160:1171–1181

Gómez Casati DF, Aon MA, Iglesias AA (2000) Kinetic and structural analysis of the ultrasensitive behaviour of cyanobacterial ADP-glucose pyrophosphorylase. Biochem J 350:139–147

Gottwald JR, Krysan PJ, Young JC, Evert RF, Sussman MR (2000) Genetic evidence for the *in planta* role of phloem-specific plasma membrane sucrose transporters. Proc Natl Acad Sci USA 97:13979–13984

Griffin KL, Tissue DT, Turnbull MH, Whitehead D (2000) The onset of photosynthetic acclimation to elevated CO_2 partial pressure in field-grown *Pinus radiata* D. Don. after 4 years. Plant Cell Environ 23:1089–1098

Grotjohann N, Messdaghi D, Kowallik W (1999) Oxygen uptake during photosynthesis of isolated pea chloroplasts. Z Naturforsch 54c:209–219

Guralnick LJ, Jackson MD (2001) The occurrence and phylogenetics of crassulacean acid metabolism in the Portulacaceae. Int J Plant Sci 162:257–262

Guralnick LJ, Ku MSB, Edwards GE, Strand D, Hockema B, Earnest J (2001) Induction of PEP carboxylase and crassulacean acid metabolism by gibberellic acid in *Mesembryanthemum crystallinum*. Plant Cell Physiol 42:236–239

Häusler RE, Schlieben NH, Flügge U-I (2000a) Control of carbon partitioning and photosynthesis by the triose phosphate /phosphate translocator in transgenic tobacco plants (*Nicotiana tabacum*). II. Assessment of control coefficients of the triose phosphate/phosphate translocator. Planta 210:383–390

Häusler RE, Baur B, Scharte J, Teichmann T, Eicks M, Fischer KL, Flügge U-I, Schubert S, Weber A, Fischer K (2000b) Plastidic metabolite transporters and their physiological functions in the inducible crassulacean acid metabolism plant *Mesembryanthemum crystallinum*. Plant J 24:285–296

Häusler RE, Rademacher T, Li J, Lipka V, Fischer KL, Schubert S, Kreuzaler F, Hirsch H-J (2001) Single and double overexpression of C_4-cycle genes had differential effects on the pattern of endogenous enzymes, attenuation of photorespiration and on contents of UV protectants in transgenic potato and tobacco plants. J Exp Bot 52:1785–1803

Hamilton JG, Thomas RB, DeLucia EH (2001) Direct and indirect effects of elevated CO_2 on leaf respiration in a forest ecosystem. Plant Cell Environ 24:975–982

Harrison EP, Olcer H, Lloyd JC, Long SP, Paines CA (2001) Small decreases in SBPase cause a linear decline in the apparent RuBP regeneration rate, but do not affect Rubisco carboxylation capacity. J Exp Bot 52:1779–1784

Hartwell J, Gill A, Nimmo GA, Wilkins MB, Jenkins GI, Nimmo HG (1999) Phosphoenolpyruvate carboxylase kinase is a novel protein kinase regulated at the level of expression. Plant J 20:333–342

Haupt-Herting S, Klug K, Fock HP (2001) A new approach to measure gross CO_2 fluxes in leaves. Gross CO_2 assimilation, photorespiration, and mitochondrial respiration in the light in tomato under drought stress. Plant Physiol 126:388–396

Henkes S, Sonnewald U, Badur R, Flachmann R, Stitt M (2001) A small decrease of plastid transketolase activity in antisense tobacco transformants has dramatic effects on photosynthesis and phenylpropanoid metabolism. Plant Cell 13:535–551

Herrick JD, Thomas RB (2001) No photosynthetic down-regulation in sweetgum trees (*Liquidamber styraciflua* L.) after three years of CO_2 enrichment at the Duke Forest FACE experiment. Plant Cell Environ 24:53–64

Holtum JAM, Winter K (1999) Degrees of crassulacean acid metabolism in tropical epiphytic and lithophytic ferns. Aust J Plant Physiol 26:749–757

Horken KM, Tabita FR (1999) The "green" form I ribulose 1,5-bisphosphate carboxylase/oxygenase from the nonsulfur purple bacterium *Rhodobacter capsulatus.* J Bacteriol 181:3935–3941

Horton P (2000) Prospects for crop improvement through the genetic manipulation of photosynthesis: morphological and biochemical aspects of light capture. J Exp Bot 51:475–485

Idso SB, Kimball BA (2001) CO_2 enrichment of sour orange trees: 13 years and counting. Environ Exp Bot 46:147–153

Igamberdiev AU, Bykova NV, Lea PJ, Gardeström P (2001a) The role of photorespiration in redox and energy balance of photosynthetic plant cells: a study with a barley mutant deficient in glycine decarboxylase. Physiol Plant 111:427–438

Igamberdiev AU, Ivlev AA, Bykova NV, Threlkeld CN, Lea PJ, Gardeström P (2001b) Decarboxylation of glycine contributes to carbon isotope fractionation in photosynthetic organisms. Photosyn Res 67:177–184

Iwamoto K, Ikawa T (2000) A novel glycolate oxidase requiring flavin mononucleotide as the cofactor in the prasinophycean alga *Mesostigma viride.* Plant Cell Physiol 41:988–991

Jahnke S (2001) Atmospheric CO_2 concentration does not directly affect leaf respiration in bean or poplar. Plant Cell Environ 24:1139–1151

Jenks A, Gibbs AP (2000) Immunolocalization and distribution of form II Rubisco in the pyrenoid and chloroplast stroma of *Amphidinium carterae* and form I Rubisco in the symbiont-derived plastids of *Peridinium foliaceum* (Dinophyceae). J Phycol 36:127–138

Johnston AM, Raven JA, Beardall J, Leegood RC (2001) Photosynthesis in a marine diatom. Nature 412:40–41

Jung C-H, Hartman FC, Lu T-YS, Larimer FW (2000) D-Ribose-5-phosphate isomerase from spinach: heterologous overexpression, purification, characterization, and site-directed mutagenesis of the recombinant enzyme. Arch Biochem Biophys 373:409–417

Khan S, Andralojc PJ, Lea PJ, Parry MAJ (1999) 2′-Carboxy-D-arabinitol 1-phosphate protects ribulose 1,5-bisphosphate carboxylase/oxygenase against proteolytic breakdown. Eur J Biochem 266:840–847

Kleczkowski LA (2001) A new player in the starch field. Plant Physiol Biochem 39:759–761

Kondo A, Nose A, Yuasa H, Ueno O (2000) Species variation in the intracellular localization of pyruvate, Pi dikinase in leaves of crassulacean-acid-metabolism plants: an immunogold electron-microscope study. Planta 210:611–621

Krepinsky K, Plaumann M, Martin W, Schnarrenberger C (2001) Purification and cloning of chloroplast 6-phosphogluconate dehydrogenase from spinach. Eur J Biochem 268:2678–2686

Lao NT, Schoneveld O, Mould RM, Hibberd JM, Gray JC, Kavanagh TA (1999) An *Arabidopsis* gene encoding a chloroplast-targeted β-amylase. Plant J 20:519–527

Lara MV, Casati P, Andreo CS (2001) In vivo phosphorylation of phosphoenolpyruvate carboxylase in *Egeria densa*, a submersed aquatic species. Plant Cell Physiol 42:441–445

Liepman AH, Olsen LJ (2001) Peroxisomal alanine:glyoxylate aminotransferase (AGT1) is a photorespiratory enzyme with multiple substrates in *Arabidopsis thaliana*. Plant J 25:487–498

Loreto F, Delfine S, Di Marco G (1999) Estimation of photorespiratory carbon dioxide recycling during photosynthesis. Aust J Plant Physiol 26:733–736

Lorimer G (2001) Letter to the editor. Plant Physiol 127:3

Ludwig A, Stolz J, Sauer N (2000) Plant sucrose-H^+ symporters mediate the transport of vitamin H. Plant J 24:503–509

Lüttge U (2000) The tonoplast functioning as the master switch for circadian regulation of crassulacean acid metabolism. Planta 211:761–769

Lüttge U, Pfeifer T, Fischer-Schliebs E, Ratajczak R (2000) The role of vacuolar malate-transport capacity in crassulacean acid metabolism and nitrate nutrition. Higher malate-transport capacity in ice plant after crassulacean acid metabolism-induction and in tobacco under nitrate nutrition. Plant Physiol 124:1335–1347

Lunn JE, Ashton AR, Hatch MD, Heldt HW (2000) Purification, molecular cloning, and sequence analysis of sucrose-6^F-phosphate phosphohydrolase from plants. Proc Natl Acad Sci USA 97:12914–12919

Luo Y (2001) Transient ecosystem responses to free-air CO_2 enrichment (FACE): experimental evidence and methods of analysis. New Phytol 152:3–8

Malda G, Backhaus RA, Martin C (1999) Alterations in growth and crassulacean acid metabolism (CAM) activity of in vitro cultured cactus. Plant Cell Tissue Org Cult 58:1–9

Malhi Y, Grace J (2000) Tropical forests and atmospheric carbon dioxide. Trends Ecol Evol 15:332–337

Mazen AMA (2000) Changes in properties of phosphoenolpyruvate carboxylase with induction of crassulacean acid metabolism (CAM) in the C_4 plant *Portulaca oleracea*. Photosynthetica 38:385–391

Medlyn BE, Badeck F-W, De Pury DGG, Barton CVM, Broadmeadow M, Ceulemans R, De Angelis P, Forstreuter M, Jach ME, Kellomäki S, Laitat E, Marek M, Philippot S, Rey A, Strassemeyer J, Laitinen K, Liozon R, Portier B, Roberntz P, Wang K, Jarvis PG (1999) Effects of elevated [CO_2] on photosynthesis in European forest species: a meta-analysis of model parameters. Plant Cell Environ 22:1475–1495

Melis A, Zhang L, Forestier M, Ghirardi ML, Seibert M (2000) Sustained photobiological hydrogen gas production upon reversible inactivation of oxygen evolution in the green alga *Chlamydomonas reinhardtii*. Plant Physiol 122:127–135

Meyer S, Melzer M, Truernit E, Hümmer C, Besenbeck R, Stadler R, Sauer N (2000) *AtSUC3*, a gene encoding a new *Arabidopsis* sucrose transporter, is expressed in cells adjacent to the vascular tissue and in a carpel cell layer. Plant J 24:869–882

Michelis R, Gepstein S (2000) Identification and characterization of a heat-induced isoform of aldolase in oat chloroplast. Plant Mol Biol 44:487–498

Migge A, Kahmann U, Fock HP, Becker TW (1999) Prolonged exposure of tobacco to a low oxygen atmosphere to suppress photorespiration decreases net photosynthesis and results in changes in plant morphology and chloroplast structure. Photosynthetica 36:107–116

Migge A, Carrayol E, Hirel B, Becker TW (2000) Leaf-specific overexpression of plastidic glutamine synthetase stimulates the growth of transgenic tobacco seedlings. Planta 210:252–260

Miszalski Z, Niewiadomska E, Slesak I, Lüttge U, Kluge M, Ratajczak R (2001) The effect of irradiance on carboxylating/decarboxylating enzymes and fumarase activities in *Mesembryanthemum crystallinum* L. exposed to salinity stress. Plant Biol 3:17–23

Miyagawa Y, Tamoi M, Shigeoka S (2001) Overexpression of a cyanobacterial fructose-1,6-/sedoheptulose-1,7-bisphosphatase in tobacco enhances photosynthesis and growth. Nat Biotech 19:965–969

Mott KA, Woodrow IE (2000) Modelling the role of Rubisco activase in limiting non-steady-state photosynthesis. J Exp Bot 51:399–406

Muraoka H, Tang Y, Terashima I, Koizumi H, Washitani I (2000) Contributions of diffusional limitation, photoinhibition and photorespiration to midday depression of photosynthesis in *Arisaema heterophyllum* in natural high light. Plant Cell Environ 23:235–250

Nakano H, Muramatsu S, Makino A, Mae T (2000) Relationship between the suppression of photosynthesis and starch accumulation in the pod-removed bean. Aust J Plant Physiol 27:167–173

Nassoury N, Fritz L, Morse D (2001) Circadian changes in ribulose-1,5-bisphosphate carboxylase/oxygenase distribution inside individual chloroplasts can account for the rhythm in dinoflagellate carbon fixation. Plant Cell 13:923–934

Nimmo GA, Wilkins MB, Nimmo HG (2001) Partial purification and characterization of a protein inhibitor of phosphoenolpyruvate carboxylase kinase. Planta 213:250–257

Nimmo HG (2000) The regulation of phosphoenolpyruvate carboxylase in CAM plants. Trends Plant Sci 5:75–80

Nimmo HG, Fontaine V, Hartwell J, Jenkins GI, Nimmo GA, Wilkins MB (2001) PEP carboxylase kinase is a novel protein kinase controlled at the level of expression. New Phytol 151:91–97

Noiraud N, Maurousset L, Lemoine R (2001) Identification of a mannitol transporter, AgMaT1, in celery phloem. Plant Cell 13:695–705

Ogawa N, Yabuta N, Ueno Y, Izui K (1998) Characterization of a maize Ca^{2+}-dependent protein kinase phosphorylating phosphoenolpyruvate carboxylase. Plant Cell Physiol 39:1010–1019

Ohshima Y, Iwasaki I, Suga S, Murakami M, Inoue K, Maeshima M (2001) Low aquaporin content and low osmotic water permeability of the plasma and vacuolar membranes of a CAM plant *Graptopetalum paraguayense*: comparison with radish. Plant Cell Physiol 42:1119–1129

Padmasree K, Raghavendra AS (1999) Response of photosynthetic carbon assimilation in mesophyll protoplasts to restriction on mitochondrial oxidative metabolism: metabolites related to the redox status and sucrose biosynthesis. Photosyn Res 62:231–239

Padmasree K, Raghavendra AS (2001) Consequence of restricted mitochondrial oxidative metabolism on photosynthetic carbon assimilation in mesophyll protoplasts: decrease in light activation of four chloroplastic enzymes. Physiol Plant 112:582–588

Parvathi K, Bhagwat AS, Ueno Y, Izui K, Raghavendra AS (2000) Illumination increases the affinity of phosphoenolpyruvate carboxylase to bicarbonate in leaves of a C_4 plant, *Amaranthus hypochondriacus*. Plant Cell Physiol 41:905–910

Paul M, Pellny T, Goddijn O (2001) Enhancing photosynthesis with sugar signals. Trends Plant Sci 6:197–200

Paul MJ, Driscoll SP, Andralojc PJ, Knight JS, Gray JC, Lawlor DW (2000) Decrease of phosphoribulokinase activity by antisense RNA in transgenic tobacco: definition of the light environment under which phosphoribulokinase is not in large excess. Planta 211:112–119

Poolman MG, Fell DA, Thomas S (2000) Modelling photosynthesis and its control. J Exp Bot 51:319–328
Rascher U, Hütt M-T, Siebke K, Osmond B, Beck F, Lüttge U (2001) Spatiotemporal variation of metabolism in a plant circadian rhythm: the biological clock as an assembly of coupled individual oscillators. Proc Natl Acad Sci USA 98:11801–11805
Reinfelder JR, Kraepiel AML, Morel FMM (2000) Unicellular C_4 photosynthesis in a marine diatom. Nature 407:996–999
Richards RA (2000) Selectable traits to increase crop photosynthesis and yield of grain crops. J Exp Bot 51:447–458
Robinson JM (2000) Dark and daylight activity of glucose-6-phosphate dehydrogenase and 6-phosphogluconate dehydrogenase in the leaves of nitrogen-limited spinach and soybean plants. Int J Plant Sci 161:651–657
Rodriguez-López M, Baroja-Fernández E, Zandueta-Criado A, Pozueta-Romero J (2000) Adenosine diphosphate glucose pyrophosphatase: a plastidial phosphodiesterase that prevents starch biosynthesis. Proc Natl Acad Sci USA 97:8705–8710
Rogers A, Ellsworth DS, Humphries SW (2001) Possible explanation of the disparity between the in vitro and in vivo measurements of Rubisco activity: a study in loblolly pine grown in elevated pCO_2. J Exp Bot 52:1555–1561
Rokka A, Zhang L, Aro E-M (2001) Rubisco activase: an enzyme with a temperature-dependent dual function? Plant J 25:463–471
Salvucci ME, Osteryoung KW, Crafts-Brandner SJ, Vierling E (2001) Exceptional sensitivity of Rubisco activase to thermal denaturation in vitro and in vivo. Plant Physiol 127:1053–1064
Savitch LV, Massacci A, Gray GR, Huner NPA (2000) Acclimation to low temperature or high light mitigates sensitivity to photoinhibition: roles of the Calvin cycle and the Mehler reaction. Aust J Plant Physiol 27:253–264
Sawada S, Kuninaka M, Watanabe K, Sato A, Kawamura H, Komine K, Sakamoto T, Kasai M (2001) The mechanism to suppress photosynthesis through end-product inhibition in single-rooted soybean leaves during acclimation to CO_2 enrichment. Plant Cell Physiol 42:1093–1102
Saxe H, Ellsworth DS, Heath J (1998) Tree and forest functioning in an enriched CO_2 atmosphere. New Phytol 139:395–436
Schiermeier Q (2001) Cycle studies see carbon sinks rise to prominence. Nature 414:385
Schlesinger WH, Lichter J (2001) Limited carbon storage in soil and litter of experimental forest plots under increased atmospheric CO_2. Nature 411:466–469
Schlitter J, Wildner GF (2000) The kinetics of conformation change as determinant of Rubisco's specificity. Photosyn Res 65:7–13
Scott P, Lange AJ, Kruger NJ (2000) Photosynthetic carbon metabolism in leaves of transgenic tobacco (*Nicotiana tabacum* L.) containing decreased amounts of fructose 2,6-bisphosphate. Planta 211:864–873
Sehnke PC, Chung H-J, Wu K, Ferl RJ (2001) Regulation of starch accumulation by granule-associated plant 14-3-3 proteins. Proc Natl Acad Sci USA 98:765–770
Sharkey TD, Badger MR, von Caemmerer S, Andrews TJ (2001) Increased heat sensitivity of photosynthesis in tobacco plants with reduced Rubisco activase. Photosyn Res 67:147–156
Sicher RC, Bunce JA (2001) Adjustments of net photosynthesis in *Solanum tuberosum* in response to reciprocal changes in ambient and elevated growth CO_2 partial pressures. Physiol Plant 112:55–61
Sikka VK, Choi S-B, Kavakli IH, Sakulsingharoj C, Gupta S, Ito H, Okita TW (2001) Subcellular compartmentation and allosteric regulation of the rice endosperm ADPglucose pyrophosphorylase. Plant Sci 161:461–468
Sivakumar P, Sharmila P, Saradhi PP (2000) Proline alleviates salt-stress-induced enhancement in ribulose-1,5-bisphosphate oxygenase activity. Biochem Biophys Res Commun 279:512–515

Strand Å, Zrenner R, Trevanion S, Stitt M, Gustafsson P, Gardeström P (2000) Decreased expression of two key enzymes in the sucrose biosynthesis pathway, cytosolic fructose-1,6-bisphosphatase and sucrose phosphate synthase, has remarkably different consequences for photosynthetic carbon metabolism in transgenic *Arabidopsis thaliana*. Plant J 23:759–770

Stylinski CD, Oechel WC, Ganon JA, Tissue DT, Miglietta F, Raschi A (2000) Effects of lifelong [CO_2] enrichment on carboxylation and light utilization of *Quercus pubescens* Willd. examined with gas exchange, biochemistry and optical techniques. Plant Cell Environ 23:1353–1362

Suzuki S, Murai N, Burnell JN, Arai M (2000) Changes in photosynthetic carbon flow in transgenic rice plants that express C4-type phosphoenolpyruvate carboxykinase from *Urochloa panicoides*. Plant Physiol 124:163–172

Takaha T, Critchley J, Okada S, Smith SM (1998) Normal starch content and composition in tubers of antisense potato plants lacking D-enzyme (4-α-glucanotransferase). Planta 205:445–451

Tamoi M, Murakami A, Takeda T, Shigeoka S (1998) Acquisition of a new type of fructose-1,6-bisphosphatase with resistance to hydrogen peroxide in cyanobacteria: molecular characterization of the enzyme from *Synechocystis* PCC 6803. Biochim Biophys Acta 1383:232–244

Tauberger E, Fernie AR, Emmermann M, Renz A, Kossmann J, Willmitzer L, Trethewey RN (2000) Antisense inhibition of plastidial phosphoglucomutase provides compelling evidence that potato tuber amyloplasts import carbon from the cytosol in the form of glucose-6-phosphate. Plant J 23:43–53

Taybi T, Patil S, Chollet R, Cushman JC (2000) A minimal serine/threonine protein kinase circadianly regulates phosphoenolpyruvate carboxylase activity in crassulacean acid metabolism-induced leaves of the common ice plant. Plant Physiol 123:1471–1481

Theodorou ME, Kruger NJ (2001) Physiological relevance of fructose 2,6-bisphosphate in the regulation of spinach leaf pyrophosphate:fructose 6-phosphate 1-phosphotransferase. Planta 213:147–157

Tovar-Méndez A, Mújica-Jiménez C, Muñoz-Clares RA (2000) Physiological implications of the kinetics of maize leaf phosphoenolpyruvate carboxylase. Plant Physiol 123:149–160

van Ginkel LC, Bowes G, Reiskind JB, Prins HBA (2001) A CO_2-flux mechanism operating via pH-polarity in *Hydrilla verticillata* leaves with C_3 and C_4 photosynthesis. Photosyn Res 68:81–88

Voigt J, Liebich I, Kieß M, Frank R (2001) Subcellular distribution of 14-3-3 proteins in the unicellular green alga *Chlamydomonas reinhardtii*. Eur J Biochem 268:6449–6457

Voznesenskaya EV, Franceschi VR, Kiirats O, Freitag H, Edwards GE (2001) Kranz anatomy is not essential for terrestrial C_4 plant photosynthesis. Nature 414:543–546

Weber A, Servaites JC, Geiger DR, Kofler H, Hille D, Gröner F, Hebbeker U, Flügge U-I (2000) Identification, purification, and molecular cloning of a putative plastidic glucose transporter. Plant Cell 12:787–801

Weise A, Barker L, Kühn C, Lalonde S, Buschmann H, Frommer WB, Ward JM (2000) A new subfamily of sucrose transporters, SUT4, with low affinity/high capacity localized in enucleate sieve elements of plants. Plant Cell 12:1345–1355

Wendt UK, Wenderoth I, Tegeler A, von Schaewen A (2000) Molecular characterization of a novel glucose-6-phosphate dehydrogenase from potato (*Solanum tuberosum* L.). Plant J 23:723–733

Whitney SM, Andrews TJ (2001) The gene for the ribulose-1,5-bisphosphate carboxylase/oxygenase (Rubisco) small subunit relocated to the plastid genome of tobacco directs the synthesis of small subunits that assemble into Rubisco. Plant Cell 13:193–205

Whitney SM, Baldet P, Hudson GS, Andrews TJ (2001) Form I Rubiscos from non-green algae are expressed abundantly but not assembled in tobacco chloroplasts. Plant J 26:535–547

Wiese A, Gröner F, Sonnewald U, Deppner H, Lerchl J, Hebbeker U, Flügge U-I, Weber A (1999) Spinach hexokinase I is located in the outer envelope membrane of plastids. FEBS Lett 461:13–18

Wingler A, Fritzius T, Wiemken A, Boller T, Aeschbacher RA (2000) Trehalose induces the ADP-glucose pyrophosphorylase gene, *ApL3*, and starch synthesis in arabidopsis. Plant Physiol 124:105–114

Wood NT (2002) Cyanobacterial transgene paves the way to increased crop productivity. Trends Plant Sci 7:9

Würth MKR, Winter K, Körner C (1998) Leaf carbohydrate responses to CO_2 enrichment at the top of a tropical forest. Oecologia 116:18–25

Yamada S, Komori T, Hashimoto A, Kuwata S, Imaseki H, Kubo T (2000) Differential expression of plastidic aldolase genes in *Nicotiana* plants under salt stress. Plant Sci 154:61–69

Yamaguchi K, Nishimura M (2000) Reduction to below threshold levels of glycolate oxidase activities in transgenic tobacco enhances photoinhibition during irradiation. Plant Cell Physiol 41:1397–1406

Yu T-S, Lue W-L, Wang S-M, Chen J (2000) Mutation of *Arabidopsis* plastid phosphoglucose isomerase affects leaf starch synthesis and floral initiation. Plant Physiol 123:319–325

Yu Y, Mu HH, Wasserman BP, Carman GM (2001) Identification of the maize amyloplast stromal 112-kD protein as a plastidic starch phosphorylase. Plant Physiol 125:351–359

Zaslavskaia LA, Lippmeier JC, Shih C, Ehrhardt D, Grossman AR, Apt KE (2001) Trophic conversion of an obligate photoautotrophic organism through metabolic engineering. Science 292:2073–2075

Zhang N, Schürmann P, Portis AR (2001) Characterization of the regulatory function of the 46-kDa isoform of Rubisco activase from *Arabidopsis*. Photosyn Res 68:29–37

Dr. Grahame J. Kelly
Centre for Molecular Biotechnology
School of Life Sciences
Queensland University of Technology
Brisbane
Queensland 4000, Australia

e-mail: g.kelly@qut.edu.au

Notes added in proof (2005):

The transformation of tobacco with a bacterial gene encoding trehalose-P synthase (Sect. 5) has now been reported [Pellny et al. (2004) Plant Biotech J 2:71-82]. Photosynthesis was increased, but neither relative growth rate nor biomass was increased.

The transformation of a pig with a spinach gene (Sect. 5) has now been described by Saeki et al. [(2004) Proc Natl Acad Sci USA 101:6361-6366].

Photosynthesis. Carbon Metabolism: The Calvin Cycle's Golden Jubilee*

Grahame J. Kelly

'A crucial question was how the two molecules of 3PGA might arise from RuBP and CO_2...An ingenious solution to the question soon occurred to Melvin Calvin...I recall him saying that the answer...came to him as he was stopped in traffic waiting for a stoplight to change.'

Bassham (2003)

1 Personal Reflections

With this chapter, I have completed 30 years of reviewing photosynthetic carbon metabolism in *Progress in Botany*. Thirty years ago this series was produced as *Fortschritte der Botanik*, a formidable title for a young man from the Australian bush, freshly arrived in Germany. However, the task of reviewing in those early years was a pleasure: I had Erwin Latzko for a wise counsellor and coauthor, and since many journals were not available in our library at Weihenstephan I was permitted to enjoy a trip on the S-Bahn to the beautiful Botanic Gardens in Munich every now and then to peruse the literature in the magnificent old library there. With the latest research assembled, our task was to connect the newly discovered pathways of C_4 photosynthesis and photorespiration to the Calvin cycle, a challenge that we enthusiastically accepted over coffee in the laboratory, or over beer in the nearby 'oldest brewery in the world'.

Time has moved on, and the science of photosynthetic carbon metabolism has changed; thus with this chapter I will 'call it a day'. Erwin Latzko deserted me in 1998, no doubt because, being wiser than me, he could see that the likelihood of the discovery of new metabolic pathways was slim, while the influence of the emerging era of the gene and molecular biology could only be wondered upon. Indeed, no new pathways have been found,

* This review is dedicated to Professor Dr Erwin Latzko on the occasion of his 80th birthday.

G.J. Kelly and E. Latzko:
Thirty Years of Photosynthesis

Originally published in Progress in Botany 66

but molecular biology is calling into question the comfortable simplicity of the established pathways by showing us that there are often more forms of most enzymes (and other proteins) than we would have deemed necessary. For example, there are five genes encoding five sucrose transporters in rice (Aoki et al. 2003; Kühn 2003), and in *Arabidopsis* there are four genes encoding four phosphoenolpyruvate carboxylases (Sánchez and Cejudo 2003), six genes encoding six hexokinases (Frommer et al. 2003), six genes encoding eight ADP-glucose pyrophosphorylases (but four are faulty; Crevillén et al. 2003), nine genes encoding nine β-amylases (Laby et al. 2001), and 14 genes encoding 14 carbonic anhydrases (Moroney et al. 2001). The business of retrieving the literature reporting all of this has also changed. Hard copies of journals are being replaced by virtual forms; thus one is more-and-more tied to one's desk and a computer screen rather than walking to the library or enjoying an S-Bahn ride which, it must be admitted, is an opportunity for contemplation and reflection (even better than that provided while waiting in traffic for a stoplight to change?)

2 The Calvin Cycle's Golden Jubilee

This year (2004) can be considered the 50th anniversary of one of the most satisfying discoveries in science: the elucidation of the cycle of biochemical reactions that photosynthetically convert CO_2 to the sugars that support all life on Earth. Properly called the Calvin–Benson–Bassham cycle in recognition of the scientists who contributed so much to its discovery, its details gradually became apparent during several years of research [principally with haploid, clonal unicellular green algae (Raven and Girard-Bascou 2001)], but it was the publication 50 years ago by Bassham et al. (1954) that finalised and established it, as emphasised in the informative and entertaining historical accounts recently published by Benson (2002) and Bassham (2003). The cycle as presented by Bassham (2003), and reproduced here in 'word' form (Fig. 1), emphasises the substance of its discovery: most of the work involved investigations of sugar phosphates (sugar-Ps), while little attention needed to be given to assimilatory power (NADPH and ATP) or enzymes, because the former was known to be available, and for most of the contemplated reactions an appropriate enzyme was already described, or was being described, from research into the carbohydrate biochemistry of all types of organisms in other laboratories. It is remarkable that no alterations have been deemed necessary during the ensuing 50 years, and that the cycle has turned out to be universal in plants, algae and cyanobacteria. Only in a few relatively obscure photosynthetic prokaryotes has evidence for alternative cycles been detected, including a reductive form of the tricar-

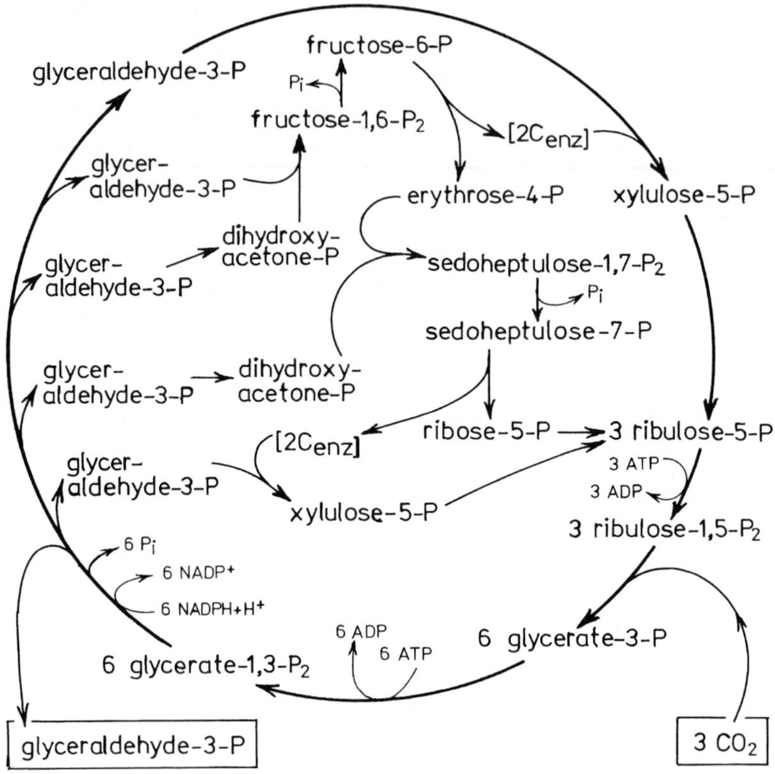

Fig. 1. The Calvin cycle, with emphasis on the sugar-P biochemistry. The sum of the reactions is: $3CO_2 + 6\ (NADPH + H^+) + 9\ ATP \rightarrow$ glyceraldehyde-3-P $+ 6\ NADP^+ + 9\ ADP + 8\ Pi$ [$2C_{enz}$], 2C-piece/enzyme (transketolase) complex. (Adapted from Bassham 2003)

boxylic acid cycle (see our 1978 review) and a bicyclic 3-hydroxypropionate cycle (Herter et al. 2002). Research into the critical carboxylating enzymes of these cycles is progressing (Kanao et al. 2002; Hügler et al. 2003).

3 The Calvin Cycle's Enzymology

3.1 Ribulose-1,5-Bisphosphate Carboxylase/Oxygenase (Rubisco)

Rubisco, the Calvin cycle's CO_2-fixing enzyme, is the protein that helps two molecules of glycerate-3-P (3PGA) to arise from ribulose-1,5-P_2 (RuBP) and CO_2 (see Bassham 2003 quote above). Its discovery was concomitant with that of the Calvin cycle 50 years ago (see Benson 2002). It has been, without doubt, the most studied of the 11 enzymes of the Calvin cycle, and the current perceived potential to manipulate it so that it works better and/or faster (Spreitzer and Salvucci 2002) is maintaining its popularity. Rubisco is not fast. In fact, it is distinctive in that it is a large and cumbersome enzyme. Its catalytic sites struggle to fix more than one molecule of CO_2 per second (von Caemmerer 2003). Thus, to keep global photosynthesis running there needs to be a lot of it. It is the most abundant protein on Earth.

The desire to fully explain the catalytic and regulatory properties of Rubisco has led to more detailed and impressive three-dimensional descriptions of the molecule (Okano et al. 2002; see Spreitzer and Salvucci 2002; Du et al. 2003). Some include its intimate association with a second protein, Rubisco activase, which acts in a three-fold chaperone capacity by dissociating inhibitory sugar-Ps from the enzyme (Portis 2003), by mediating its light activation (Zhang et al. 2002), and by indirectly determining its sensitivity (and therefore the sensitivity of photosynthesis) to high temperature, as recently shown for cotton (Law et al. 2001) and maize (Crafts-Brandner and Salvucci 2002). A realistic picture of Rubisco at sunrise would therefore show not only the 16 subunits of the enzyme (i.e. eight copies of the chloroplast-encoded large subunit and eight copies of the nuclear-encoded small subunit), but also the associated Rubisco activase. An adaptation of such a supramolecular complex shown by Portis (2003) is presented in Fig. 2. Incidentally, there is now stronger evidence that one of the more notable inhibitory sugar-Ps, viz. carboxyarabinitol-P (which binds to Rubisco and prevents it from being active at night), is synthesised from the Calvin cycle intermediate fructose-1,6-P_2 via the sugar hamamelose (Andralojc et al. 2002).

As mentioned above, much of the impetus for current research on Rubisco comes from the desire to apply the tools of modern biotechnology to improve this 'notoriously inefficient' (Spreitzer and Salvucci 2002) enzyme, and thereby increase agricultural production. This latter benefit presupposes that, under good (i.e. not water- or inorganic-nutrient-limited) field conditions, Rubisco activity is the ultimate determinate of photosynthetic rate and production. It certainly is a major (Jiang and Xu 2001; Sage 2002),

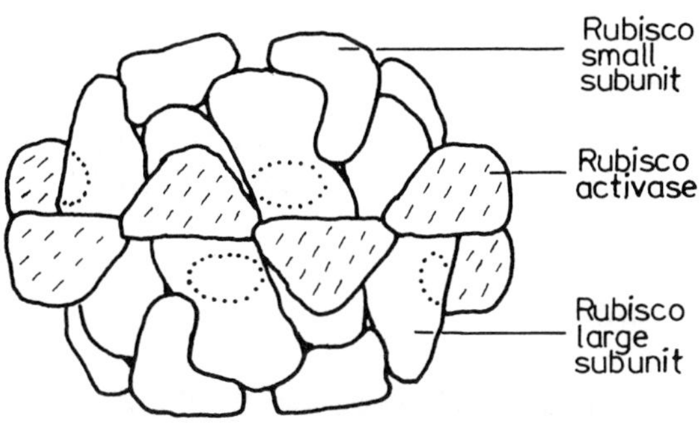

Fig. 2. The Rubisco complex at sunrise. One molecule of Rubisco activase is bound to each of the eight large subunits of Rubisco, facilitating dissociation of any inhibitory sugar-Ps from the enzyme's eight active sites (*dotted ellipses*) which are on these subunits. Having completed this task, the activase molecules themselves dissociate, liberating the active Rubisco for its diurnal chore of CO_2 fixation. Note that not all of the Rubisco subunits and activase molecules are in view in this 3D picture. (Adapted from Portis 2003)

although not exclusive (Bernacchi et al. 2002), determinate of photosynthetic rate in C_3 plants, but whether this translates into any important control function for photosynthesis (Raines 2003) is debatable. Furthermore, as emphasised in our 2000 review and now supported by the studies of Murchie et al. (2002) with rice, there is little evidence that Rubisco activity is a principal determinant of plant production in the field.

Rubisco could be improved by accelerating its activity, and/or by coaxing it to ignore O_2 (Parry et al. 2003). When the enzyme accepts O_2, rather than CO_2, as substrate the 'wasteful' photorespiratory pathway is initiated (see Section 5.5). Spreitzer and Salvucci (2002) are optimistic that some day a Rubisco better than any natural version will be created in a laboratory, but so far this has not happened, and they remind researchers that any alteration to the enzyme that improves one of the above two characteristics will be useless if it causes a depreciation of the other.

Research on Rubiscos from primitive autotrophic prokaryotes is providing intriguing insights. A study by Robinson et al. (2003) reminds us that Rubisco, and indeed the entire Calvin cycle, is not restricted to the sunlit parts of Earth; these authors have isolated Rubisco from chemoautotrophic (as opposed to photoautotrophic) bacteria that exist as endosymbionts in tubeworms colonising hydrothermal vents in the dark depths of the ocean. Rubisco presumably evolved in primitive prokaryotes, and indeed Ashida et al. (2003) have made the startling discovery that the unusually simple two-large-subunit Rubisco from the photosynthetic bacterium *Rhodospirillum rubrum* is so closely related to a similar protein in the common non-photosynthetic bacterium *Bacillus subtilis* that it can perform that protein's function (catalysis of a reaction in a pathway that salvages the amino acid methionine), and thereby rescue cells that have lost the protein but have had the *R. rubrum* Rubisco gene genetically engineered into them. Another impressive event involving *R. rubrum* Rubisco was the creation by Whitney and Andrews (2001) of a tobacco plant lacking its own Rubisco, but transformed to contain *R. rubrum* Rubisco in its chloroplasts. These plants grew well so long as they were treated like *R. rubrum*, i.e. supplied with a relatively high concentration of CO_2 (Whitney and Andrews 2003). Finally, Wang et al. (2001) manufactured prokaryotic/eukaryotic hybrid Rubiscos with cyanobacterial (*Synechococcus*) large subunits and either tobacco, wheat or rice small subunits. None of the hybrids was better than the native *Synechococcus* enzyme.

3.2 The Enzymes That Regenerate RuBP from 3PGA

In the Calvin cycle, 10 enzymes catalyse the 12 reactions that regenerate RuBP from the initial product of CO_2 fixation, 3PGA. The first (which converts 3PGA to glycerate-1,3-P_2) receives scant attention, but the second, NADP-linked glyceraldehyde-3-P dehydrogenase (GAPDH), which reductively forms glyceraldehyde-3-P (GAP) from the glycerate-1,3-P_2, is well researched. The majority of the recent studies on GAPDH also include the cycle's tenth enzyme which finally generates the RuBP, viz. ribulose-5-P kinase (Ru5PK), because these two enzymes come together with a third small protein (named CP12) to form a complex, and it is within this complex that many of the earlier-documented regulatory events, such as aggregation, light-mediated reductive activation and NADP-linked activation, take place and (additionally) interact. Consequences include lowering of the reactions' activation energies (Graciet et al. 2002), indirect reductive activation of GAPDH via intra-complex conformational changes following the reduction of a disulfide bond (to two sulfhydryl groups) in the Ru5PK (Lebreton et al. 2003) and the existence of multiple aggregation states (Scheibe et al. 2002). The plant GAPDH, unlike the algal enzyme, is a heterotetramer of two A-subunits and two B-subunits, and there is evidence that a unique disulfide bond in the C-terminus of the B subunit has critical regulatory functions (Qi et al. 2001; Sparla et al. 2002). While on the topic of molecular complexes, two further items are worth noting. Firstly, the possibility has again been raised (this time from experiments with a cyanobacterium) that the Calvin cycle enzymes may exist as ordered supramolecular complexes in vivo, perhaps with the help of thylakoid membranes (Sainis et al. 2003). Secondly, Giegé et al. (2003) have surprisingly announced that a dollop of the ten enzymes of the classic old pathway of glycolysis is associated with the outside surface of the cell's mitochondria, thereby enhancing the conversion of cytosolic glucose to mitochondrial-matrix pyruvate.

Returning to the Calvin cycle enzymes, four of the remaining seven have received some attention during the past 2 years. Two enzymes, viz. aldolase and triose-P isomerase, were recently observed by Ito et al. (2003) to be susceptible to in vivo modification by the covalent attachment of glutathione. Given the importance of glutathione to life, the significance of this observation will be eagerly awaited. The enzyme that hydrolyses fructose-1,6-P_2 to fructose-6-P [fructose1,6-bisphosphatase (FBPase)] remains popular; it has been the subject of a review (Chueca et al. 2002), and is considered an important factor in explanations of why the photosynthetic capacity of wheat leaves appears better than that of rice leaves (Sudo et al. 2003), and why photosynthesis runs poorly in cooled soybean leaves (van Heerden et al. 2003). Lastly, there is a report addressing transketolase. Maize transketolase has been cloned, crystallised and shown to be a homodimer, similar to its yeast counterpart (Gerhardt et al. 2003).

Regulation of the Calvin cycle and its constituent enzymes is taking some new twists. While it is clear that, upon sunrise, the well-established reductive activation of enzymes such as FBPase (Ocón et al. 2001; Chueca et al. 2002) and Ru5PK (Lebreton et al. 2003) and the increase in chloroplast-stromal Mg^{2+} concentration from 0.5–2 mM (Ishijima et al. 2003) must assist the start-up of photosynthesis, the view is still held that other enzymes such as aldolase and transketolase may be more important than these light-regulated enzymes (and even more important than Rubisco) in controlling steady-state carbon fixation (Raines 2003).

On the other hand, a fourth Calvin cycle enzyme susceptible to reductive activation, viz. sedoheptulose-1,7-bisphosphatase (SBPase), does seem to be an important determinant of the steady-state photosynthetic rate (Raines 2003); thus the reductive activation of enzymes will continue to be closely monitored. It is a burgeoning field. Enzyme reduction is mediated by the ubiquitous small protein thioredoxin (Tdx), and genomics has revealed that our old impression of chloroplasts containing two Tdxs (Tdx *f* and Tdx *m*) must be modified to one in which there are two forms of Tdx *f*, four forms of Tdx *m* and also a Tdx *x* and a Tdx *y* (Collin et al. 2003; Lemaire et al. 2003). Furthermore, many more proteins than the Calvin cycle's GAPDH, FBPase, SBPase and Ru5PK are potentially reduced by Tdx. Several others, including glucose-6-P dehydrogenase and NADP-linked malate dehydrogenase, have been mentioned in our earlier reviews, and now the list can be extended by the addition of two protein kinases [one that phosphorylates maize phosphoenolpyruvate carboxylase (Saze et al. 2001) and the other a thylakoid light-harvesting complex (Martinsuo et al. 2003)], and the initial enzyme of the shikimate pathway [3-deoxy-D-arabino-heptulosonate-P synthase (Entus et al. 2002)]. However, the list could explode, because Balmer et al. (2003) have used genomics to predict that up to 35 chloroplast proteins may be targets for reduced Trx.

4 The Calvin Cycle's Products

When the Calvin cycle is spinning in plant-leaf cells, two main products are generated: starch and sucrose. Starch is synthesised in the vicinity of the cycle, within the chloroplast, by diverting some of the cycle's fructose-6-P into the pathway: fructose-6-P → glucose-6-P → glucose-1-P → ADP-glucose → starch. Sucrose, on the other hand, is synthesised away from the cycle, in the cytosol, using GAP supplied by the cycle but exported into the cytosol via the chloroplast envelope's well-known, and still investigated (Schneider et al. 2002), 'phosphate translocator' protein. Ultimately, the sucrose is translocated to so-called sinks (growing shoots, roots, fruits, seeds and storage organs such as tubers). The starch accumulated in chloroplasts is later hydrolysed to soluble sugars at night, after the Calvin cycle has stopped spinning, and these are moved via chloroplast-envelope transporter proteins into the cytosol and used to continue the process of sucrose biosynthesis and translocation. Aspects of these metabolisms are reviewed below.

4.1 Starch

Research into the biosynthesis and structure of starch, both as a transitory reserve in chloroplasts (see above) and as a storage reserve in sink tissues, continues to flourish. In much of this research, properties of the participating enzymes have been addressed, especially those of the notable regulatory enzyme ADP-glucose pyrophosphorylase (ADPG-PPase) which generates the glucose donor molecule (ADP-glucose) utilised by the starch synthetases. In cyanobacteria ADPG-PPase is a relatively simple, highly regulated (Gómez-Casati et al. 2003) tetrameric protein composed of four identical subunits, whereas in plant chloroplasts it is a tetramer of two pairs of subunits, one slightly larger than the other (Salamone et al. 2002), and furthermore there are four genes encoding four versions of the larger subunit and two genes encoding two versions of the smaller subunit. Differential expression of these genes in different plant tissues possibly fine-tunes the enzyme's regulatory properties (especially relating to the well-established allosteric activation by 3PGA and inhibition by inorganic phosphate) in a way that is appropriate to each tissue (Crevillén et al. 2003). A recent observation indicates that regulation of the enzyme is more complex, in that both light and increased levels of soluble sugars can indirectly activate the enzyme by promoting the reduction of a disulfide bridge that interconnects the pair of smaller subunits when the enzyme is in its less active state (Hendriks et al. 2003). Intriguingly, no prominent mention of thioredoxin is included in this report, and similar regulation of the enzyme in potato tubers (but obviously not involving light!) has also been detected (Tiessen et al. 2002). Finally, Baroja-Fernández et al. (2001) are entertaining the possibility that ADPG-PPase might to some extent be side-stepped when ADP-glucose is synthesised in the cytosol by the enzyme sucrose synthase catalysing a reaction between ADP (instead of the usual UDP) and sucrose, followed by import of the ADP-glucose into the chloroplast. However, these authors have had to propose a complex cycling of carbohydrates between chloroplasts and the cytosol, possibly for control purposes, to explain the very existence of ADPG-PPase. This cycling would entail turnover of starch during the day, but Zeeman et al. (2002) report that this does not occur.

The existence of ADPG-PPase is currently viewed as being most significant, because one of the most successful advances in genetically engineering crop plants to increase productivity has centred on this enzyme. Smidansky et al. (2002, 2003) introduced a mutant form of the gene for the enzyme's larger subunit into wheat and rice. The mutant form of the subunit resulted in the enzyme being less sensitive to inhibition by orthophosphate, and the resultant increased activity effectively increased the rate of starch synthesis such that the wheat yield increased by 38% and the rice yield increased by 20%. Clearly, the plant leaves had the capacity to photosynthesise faster when called upon to do so. A similar success has been reported, rather unexpectedly, for another enzyme: when Regierer et al. (2002) genetically engineered potato to have a reduced activity of adenylate kinase in plastids, the yield of tubers doubled when compared to one plot of wild-type plants, although the increase was only 22% when compared to a second plot of these plants.

Concepts concerning the diurnal biosynthesis of starch granules from ADP-glucose in chloroplasts, and its subsequent nocturnal dissimulation, remain essentially as outlined in our preceding two reviews, but the discovery of two new proteins requires that these concepts be remodelled. The first protein (Ritte et al. 2002) is a granule-bound, enigmatic new enzyme that, during biosynthesis, phosphorylates starch via a reaction in which the central (rather than terminal) phosphate group of ATP is transferred to approximately 0.3% of the glucose moieties in the amylopectin component of starch; the ATP's terminal phosphate is simultaneously transferred to water (i.e. hydrolysed). Entitled 'α-glucan water dikinase' (an odd name, but I cannot think of a better one – hopefully, however, ATPases will not be renamed 'water kinases'!), this new enzyme is believed to be vital for the subsequent dissimulation of the starch, perhaps because the addition of phosphate groups assists starch degradation. The second new protein relates to starch degradation. Niittylä et al. (2004) have very recently discovered a previously unknown maltose transporter protein in chloroplast envelopes. Combined with preceding evidence that a chloroplast β-amylase is quite important for starch degradation (Scheidig et al. 2002), it now seems likely that chloroplasts at night are initially brewery-like in that most of the starch hydrolysis produces maltose. This maltose is then transported into the cytosol and used there to manufacture sucrose. The previous popular pathway of starch hydrolysis to glucose and transport of this sugar (Servaites and Geiger 2002) might be a minor one. However, one question remains: does the enzyme maltase exist in the cytosol of leaf cells?

4.2 Sucrose and Other Soluble Sugars

It is sometimes good to stand back, look at photosynthesis and wonder. When Salerno and Curatti (2003) did this, they found great difficulty in suggesting why most plants have adopted sucrose (a disaccharide of glucose and fructose) to be the major form of translocated photosynthate when many other soluble sugars may have been similarly suitable. Admittedly, some plants have chosen instead to translocate sugar alcohols such as sorbitol; the final enzyme in the biosynthesis of sorbitol in apple leaves has now been characterized (Zhou et al. 2003). A few other plants add one or two galactose or fructose units onto the sucrose before translocating it (see our 2000 review), but these are isolated exceptions. Sucrose easily predominates; thus it is little wonder that its biosynthesis and the regulation of this biosynthesis receive much attention.

One regulatory step receiving renewed attention is the interconversion between fructose-6-P and fructose-1,6-P_2 which involves three enzymes [viz. FBPase, ATP-linked phosphofructokinase (PFK) and pyrophosphate (PPi)-linked PFK] and the odd regulatory sugar-P, fructose-2,6-P_2. This sugar-P is both synthesised and degraded by a single, appropriately controlled enzyme (Markham and Kruger 2002). A clear description of the interplay between these enzymes is still emerging (Nielsen and Stitt 2001; Trevanion 2002). Meantime, a new role for the PPi-linked PFK has been proposed: Costa dos Santos et al. (2003) have evidence that, in maize roots, its reversible reaction operates in the direction of PPi synthesis to supply PPi to the tonoplast H^+-pyrophosphatase that pumps protons into the vacuole, thereby generating a pH in the vacuole 2 units lower than that in the cytosol. An extension of this proposal, involving the inclusion of the old classical ATP-linked PFK to supply the necessary fructose-1,6-P_2, is entertained in Fig. 3. If this system operated in leaves (and there is no reason why it should not), the acidification of the vacuole might therein promote hydrolysis of sucrose by acid invertase, releasing hexoses that exit and interact with cytosolic hexokinase in the newly emerging signal transduc-

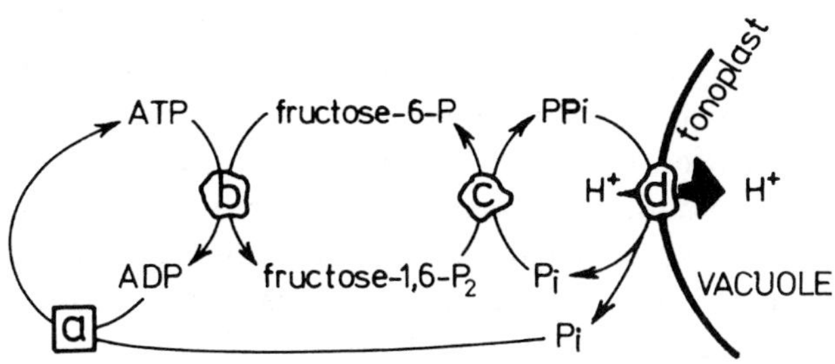

Fig. 3. Possible utilisation of PFKs for vacuolar acidification. Photosynthetically generated ATP (*a*), delivered either directly or indirectly (via a sugar-P shuttle system) to the cytosol, could be used by ATP-linked PFK (*b*) to supply fructose-1,6-P_2 to the PPi-linked PFK (*c*) which, by catalysing its reverse reaction, generates the PPi used by the tonoplast H^+-pyrophosphatase (*d*). Section involving *c* and *d* is based on Costa dos Santos et al. (2003)

tion pathways (see below). The central steps of sucrose synthesis would then be intertwined with the signal transduction system that influences the transcription of genes, including those of the Calvin cycle enzymes.

The remainder of the sucrose biosynthetic pathway involves the conversion of two molecules of fructose-6-P to one of sucrose. A cytosolic P-glucomutase normally operative in this pathway has been tenuously considered to be beneficial (Lytovchenko et al. 2002), but not necessarily essential since its chloroplast counterpart may be able to indirectly substitute for it in an emergency (Lytovchenko and Fernie 2003). Considerable attention is still given to the enzyme two steps further along the pathway, viz. sucrose-P synthase. The phosphorylation/dephosphorylation-based regulation of this enzyme, like so many others, now seems to be modulated by light (Pagnussat et al. 2002). It is still popular to genetically engineer plants to boost their sucrose-P synthase content. Recently, *Arabidopsis* became more freezing tolerant (Strand et al. 2003), old leaves of tobacco photosynthesised faster (Baxter et al. 2003) and tomato plants experienced a 58% increase in photosynthetic capacity (Lunn et al. 2003) when this was done.

Another popular topic relating to sucrose is the ability of this sugar, and other soluble sugars, to initiate signal transduction pathways (which terminate in the regulation of gene expression) when they accumulate in cells. Both sucrose and glucose, and possibly other sugars, can initiate these pathways, of which there are probably several (Xiao et al. 2000; Sinha et al. 2002; Tiessen et al. 2003). Most attention is given to the ability of the ubiquitous enzyme hexokinase to 'sense' the glucose concentration and, when this is substantial, to initiate a signalling pathway into the nucleus where the transcription of genes for Calvin cycle enzymes is reduced (see Fig. 2 in our 2000 review), while the transcription of genes for other proteins such as sucrose synthase (Ciereszko and Kleczkowski 2002), and possibly cytosolic glucose-6-P dehydrogenase (Hauschild and von Schaewen 2003), and a protein involved in ADPG-PPase activation (Tiessen et al. 2003) is increased. There has been some debate about how hexokinase does this, but an ingenious experiment by Moore et al. (2003) in which mutant hexokinases were introduced into mutant *Arabidopsis* plants demonstrated that the glucose-sensing function of the hexokinase was independent of this protein's catalytic ability (see Frommer et al. 2003; Harrington and Bush 2003). Clearly, the renewed attention being given to this famous old enzyme is worthwhile, and the discovery of a new form of hexokinase in the stroma of moss chloroplasts (Olsson et al. 2003) can only enliven the attention it receives in the future.

Before leaving soluble sugars, we should note that the mysterious trehalose (a disaccharide of two glucose units linked C1 to C1, not to be confused with the Weihenstephan brewery's maltose which is a disaccharide of two glucose units linked C1 to C4) is still arousing curiosity. It exists in fungi and some invertebrate animals, but no more than trace amounts have been detected in any plant except the primitive, desiccation-tolerant lycophyte *Selaginella lepidophylla* (resurrection plant). Nevertheless, genes encoding the two enzymes required for its synthesis from UDP-glucose and glucose-6-P, via trehalose-6-P (Penna 2003), appear to be widespread in plants, and investigations have demonstrated that the first, detected in *Arabidopsis* (Vogel et al. 2001), is essential for the development of embryos (Eastmond et al. 2002; Schluepmann et al. 2003). Presumably, trehalose is hydrolysed by a trehalase (Wingler 2002) as fast as it is formed in all plants except *S. lepidophylla*. In this plant, and other organisms where it accumulates, it confers tolerance to abiotic stresses such as drought; thus Garg et al. (2002) decided to introduce the two trehalose biosynthetic genes from a bacterium into rice and look for a similar situation. Indeed, their plants had elevated levels of trehalose, were more resistant to drought, salinity and cold, and even displayed an elevated capacity for photosynthesis. Meantime, trehalose still remains mysterious, especially since electronic database searches give no reports of it in any moss or alga except *Euglena gracilis* (Porchia et al. 1999) (but *E. gracilis* is equally mysterious, being considered by systematists to be more closely related to certain protozoa than to algae). Such database searches also give no reports of it in seeds, yet one might expect maturing seeds, destined to survive long periods of desiccation, to benefit by synthesising some.

4.3 Isoprene

A somewhat overlooked extension of photosynthetic carbon metabolism, alluded to in our 1996 review, is the manufacture of isoprene (2-methyl-1,3-butadiene) by forest plants, including ferns and many trees. This gas, which has value as a cellular thermoprotective agent (Brüggemann and Schnitzler 2002), possibly in association with photorespiration (Peñuelas and Llusià 2002), is easily emitted into the atmosphere. It is synthesised from the same 5-carbon precursor molecule as that used to build isoprenoids such as carotenoids and steroids, and most of this precursor is, in turn, synthesised in the chloroplast (Rosenstiel et al. 2002) in eight steps (Fellermeier et al. 2003) that begin with two molecules of GAP supplied by the Calvin cycle (one of the GAPs is first converted to a pyruvate). Sharkey (1996) earlier reported that some trees direct up to 15% of their assimilated CO_2 into isoprene, and since most of the Earth's photosynthesis is performed by forests, isoprene should clearly be included with starch and sucrose as an important end-product of the Calvin cycle.

5 Pathways and Processes That Assist the Calvin Cycle

5.1 Uptake of Inorganic Carbon

Terrestrial plants take up inorganic carbon (Ci) as CO_2, while aquatic plants, algae and cyanobacteria have the additional option of taking up bicarbonate (HCO_3^-). If HCO_3^- is selected it must, sooner or later, be converted to CO_2 because this is the form of Ci used by the Calvin cycle's Rubisco. The interconversion between CO_2 and HCO_3^- involves the two reversible reactions: $CO_2 + H_2O \rightarrow H_2CO_3 \rightarrow H^+ + HCO_3^-$, the first of which is slow, but is accelerated by the enzyme carbonic anhydrase (CA) if present. Algae and cyanobacteria, both in laboratories and in the ocean, variously use intracellular and extracellular (periplasmic) forms of this enzyme, plus plasmalemma-located, ATP-powered (directly or indirectly) transporter proteins for either CO_2 or HCO_3^- to not only deliver Ci as CO_2 to Rubisco, but also to actively pump the Ci into the cell so that the concentration of CO_2 in the vicinity of Rubisco is much higher than that in the surrounding water (or, for that matter, much higher than that in the air above) (Ghoshal et al. 2002; Tortell and Morel 2002; Rost et al. 2003). The concepts and research on this topic, last covered in our 2000 review, have been more recently massively reviewed (see Price and Badger 2002).

The ATP that powers Ci uptake is ultimately derived from photosynthetic electron transport [although in the odd case of the eustigmatophyte alga *Nannochloropsis gaditana* mitochondrial respiration also contributes (Huertas et al. 2002)]. In an intriguing report, Tchernov et al. (2003) partially reversed the functions of ATP and Ci uptake by providing evidence that many microalgae may take up much more Ci than they need, such that excess Ci leaks back out, but the cell still benefits because the associated ATP consumption represents yet another avenue for the dissipation of excess harvested light energy. Meanwhile, from other studies there are indications that extracellular structures, as well as extracellular CA, might assist Ci uptake. There is tenuous (Buitenhuis et al. 1999) but not universal (Herfort et al. 2002) support for an early proposal that the manufacture of $CaCO_3$ coccoliths by the widespread marine coccolithophorid microalgae promotes the conversion of seawater HCO_3^- to CO_2 for Rubisco by providing the necessary H^+ (Fig. 4). Now, a role for the elegant frustule of diatoms is also under consideration, because Milligan and Morel (2002) have evidence that the frustule's biosilica appropriately buffers the pH of the microenvironment of the extracellular CA.

Moving up to plants, it seems that those that have moved to aquatic environments, such as the fresh-water *Egeria densa* (Lara et al. 2002) and the marine seagrass *Ruppia cirrhosa*

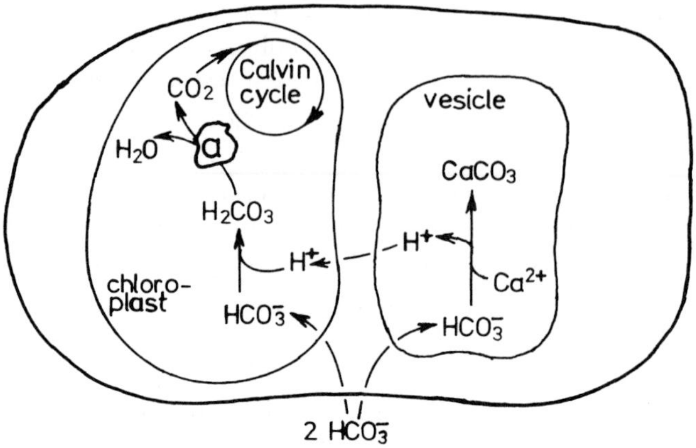

Fig. 4. Dual use of HCO_3^- by coccolithophorid microalgae such as *Emiliania huxleyi*. Coccoliths are synthesised from Ca^{2+} and HCO_3^- intracellularly in Golgi vesicles (before being transported to the cell surface). The H^+ released conceivably joins HCO_3^- in chloroplasts, forming H_2CO_3, the substrate of carbonic anhydrase (*a*). (Adapted from Buitenhuis et al. 1999)

(Hellblom and Axelsson 2003), have copied algae to some extent in that they can utilise HCO_3^- by acidifying the solution adjacent to the leaf surface and (in the case of the seagrass) producing an extracellular CA, thereby enhancing the conversion of HCO_3^- to CO_2, which then simply diffuses into the cells. In fact, the seagrass mechanism is little different to that used by some macroalgae that succeed with simple CO_2 diffusion as the final step (Snoeijs et al. 2002).

Most plants, however, are terrestrial; thus CO_2 is their only source of Ci. Most are 'C_3 plants', and a role for the CA that their leaves contain is, as ever, temptingly enigmatic. It may be important (Bernacchi et al. 2002) because, in CO_2's journey to the chloroplast, it must traverse the cytosol where at least some of it will be converted to HCO_3^-. However, we are currently distracted by a new group of proteins: the odd, membrane-located transporter proteins that facilitate the movement of water across membranes. Designated 'aquaporins', they have now been found to facilitate the entry of CO_2 into the photosynthetic cells of bean and tobacco (Terashima and Ono 2002; Uehlein et al. 2003), thereby assisting its journey to Rubisco. One now wonders whether chloroplast-envelope membranes also possess aquaporins.

5.2 Concentrating CO_2 Near Terrestrial Rubisco: C_4 Photosynthesis

The universality of the Calvin cycle was briefly challenged 22 years after its discovery when Hatch and Slack (1966) reported that sugar-cane leaves appeared to utilise an alternative cycle centred on the carboxylation of P-enolpyruvate (PEP) rather than RuBP. However, it soon became apparent that these leaves do contain the Calvin cycle in tough, photosynthetic bundle-sheath cells, and that they need it to grow (see our first, 1974 review). The high capacity for PEP carboxylation, located in cooperating adjacent mesophyll cells that are distinctive in lacking the Calvin cycle, was soon recognised to be the first component in a CO_2-collection-and-concentration system appropriately named 'the C_4 pathway of photosynthesis'.

It is sobering to be reminded that the sum of the reactions that constitute the C_4 pathway is nothing more than the hydrolysis of two ATP (Miyao 2003). The consequence, however, is the substantial elevation of the concentration of CO_2 in the vicinity of Rubisco (von Caemmerer and Furbank 2003), and this profoundly alters the plant's physiology: photorespiration is deterred (Kiirats et al. 2002), Rubisco works faster and water use efficiency is heightened. The latter allows C_4 plants to succeed well in relatively warm, dry habitats. By contrast, they do poorly in cool habitats, at least in part because their content of Rubisco, which is less than that of C_3 plants, is insufficient when it is slowed by low temperature (Kubien et al. 2003). The tendency for the C_4 plant's PEP carboxylase to lose activity in cooled leaves (Chinthapalli et al. 2003) may also contribute. Incidentally, further studies on the proteins that mediate the activation of this enzyme by phosphorylating it at sunrise (Bakrim et al. 2001; Echevarría et al. 2001; Saze et al. 2001) and dephosphorylating it at sunset (Dong et al. 2001) have been reported.

Exceptions to the above include a species of *Miscanthus* that maintains a higher content of Rubisco when cool (Naidu et al. 2003), and the high-altitude grass *Muhlenbergia richardsonis* that grows flat in order to take advantage of the much higher temperature of the thin boundary layer of air at ground level when it is calm in such localities (Sage and Sage 2002). C_4 plants also do poorly in shade because much of the photochemical activity is restricted to the bundle-sheath cells surrounding veins. Ogle (2003) has calculated that veins would need to be uncommonly close together before C_4 plants could venture into the shade.

Strange C_4 plants in the family Chenopodiaceae are being discovered in Asia. Following the surprise description of single-celled C_4 photosynthesis in *Borszczowia aralocaspica*, in which Calvin-cycle-containing chloroplasts are at one end of an elongated cell while Calvin-cycle-lacking, PEP-regenerating chloroplasts are at the opposite end (see our last review), there is now *Bienertia cycloptera*, also with single-celled C_4 photosynthesis, but in this plant the Calvin-cycle chloroplasts are embedded in a 'bubble' of cytoplasm surrounded by a large central vacuole, but in contact (via cytoplasmic strands) with a layer of peripheral cytoplasm in which the PEP-regenerating chloroplasts are located (Voznesenskaya et al. 2002). Finally, there are species of *Salsola* that are C_4 plants in which C_4 photosynthesis even occurs in green cotyledons following seed germination (Voznesenskaya et al. 2003).

The discovery of plants with single-celled C4 photosynthesis has created excitement because such plants reflect the vision of genetic engineers planning to improve C3 crop plants by transforming their single type of photosynthetic cell with one or more of the genes that encode enzymes of the C4 pathway. However, such excitement is probably premature for two reasons, as pointed out by von Caemmerer (2003): firstly, these two odd plants have quite low photosynthetic rates, even lower than C3 crop plants, and secondly, C3-plant cells tend to have chloroplasts appressed to the plasma membrane facing intercellular air spaces, thereby negating any possibility of locally concentrating CO_2. There is no impressive arrangement of two types of chloroplasts as in the two chenopods. Reports of C3 plants transformed with C4 enzymes continue to appear, but no clear-cut benefit has yet emerged.

A major reorientation of our perception of C4 photosynthesis is warranted following the recent demonstration by Hibberd and Quick (2002) that the sheath of green cells surrounding vascular strands in the petioles of typical C3 plants such as tobacco and celery (the latter obvious to all culinary chefs slicing celery stalks into fresh salads) performs photosynthesis with strong C4 characteristics (Fig. 5). Remarkably, these C3 bundle-

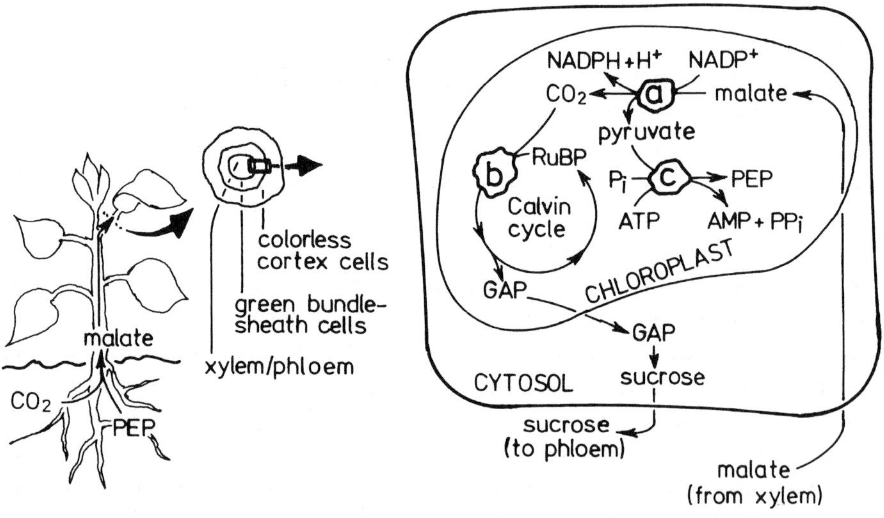

Fig. 5. CO_2 for photosynthesis in the petioles of C3 plants is obtained by decarboxylation of malate delivered from the xylem, and previously synthesised in lower plant parts (e.g. roots) from respiratory CO_2. The biochemical and transport processes that balance the PEP produced in the bundle-sheath chloroplast and the PEP consumed in the plant root are not shown. *a* NADP-linked malic enzyme; *b* Rubisco; *c* pyruvate Pi dikinase

sheath cells contain high activities of malate-decarboxylating enzymes that release CO_2 (for Rubisco) from malate, and of the 'C_4 enzyme' pyruvate Pi dikinase that generates PEP from pyruvate (the other product of decarboxylation), and which is probably regulated by the odd phosphorylation system known from C_4 plants (Chastain et al. 2002). Quite possibly, these C_3 bundle-sheath cells are the evolutionary ancestors of the bundle-sheath cells of genuine C_4 plants (Raven 2002). It is considered that the C_3-plant petiole cells retrieve respiratory CO_2 evolved in the rhizosphere and lower parts of the plant (and, somewhat ironically, this CO_2 must therefore have already met Rubisco during conventional C_3 photosynthesis by a leaf of the same plant earlier in its life!). The system envisaged is one whereby three root enzymes generate a malate from the respired CO_2 and a PEP, and this malate is then transported through the xylem to the petiole. Differences from traditional photosynthesis are: (1) the task of mesophyll cells is performed by a distant part of the plant, (2) PEP regeneration is in the same chloroplast as the Calvin cycle, and (3) no boost to the concentration of CO_2 near Rubisco has been mentioned. Nevertheless, it is not inconceivable that, in the future, petioles will be looked upon as 'C_4 organs within C_3 plants'.

5.3 Night-Time Preparations for the Calvin Cycle: Crassulacean Acid Metabolism (CAM)

CAM plants, as is well known, achieve the intuitively impossible task of organising the daytime delivery of CO_2 to the Calvin cycle behind closed stomata. Plants regularly subjected to severe water stress such as desert cacti, rainforest epiphytes (Pierce et al. 2002) and halophytes (Broetto et al. 2002) do this. Their sleuth is to collect CO_2 during the preceding night through open stomata; this CO_2 is joined to PEP (catalysed by PEP carboxylase), the oxaloacetate produced is reduced to malate, and this malate is then transported into the vacuole and accumulated there. Next day the malate is moved out into the cytosol and decarboxylated to supply the Calvin cycle with CO_2 while, simultaneously, the stomata are closed to conserve water. Later in the day they open their stomata a little and complete their day's photosynthesis by directly fixing atmospheric CO_2, so that, overall, typical CAM plants collect three-quarters of their CO_2 at night, and the remaining quarter during late afternoons (Winter and Holtum 2002) when Rubisco is maximally activated (Griffiths et al. 2002). Both carboxylases, i.e. Rubisco and PEP carboxylase, are important components in the molecular machinery that establishes the endogenous circadian rhythm which, in turn, sets in motion the day/night phases of CAM (Wyka and Lüttge 2003).

Active membrane-located transporter proteins are critical for CAM, and are being anxiously sought. The nocturnal entry of malate into vacuoles could conceivably be mediated by the tonoplast dicarboxylate transporter recently discovered by Emmerlich et al. (2003), but these authors made their discovery with the non-CAM plant *Arabidopsis*, and noted that the transporter was eerily similar to a human Na^+/dicarboxylate cotransporter. A different and possibly more serious candidate transporter protein is the voltage-gated malate-ion channel found in tonoplasts of the CAM plant *Kalanchoë daigremontiana* by Hafke et al. (2003). It was recorded to be present at a density of one channel per 5 μm^2, sufficient to accommodate observed rates of vacuolar malate uptake at night. Another tonoplast transporter recently proposed (McRae et al. 2002) is a sucrose uniporter that permits the diurnal uptake of this end-product of photosynthesis into vacuoles where it is hydrolysed into glucose and fructose; CAM plants such as pineapple prefer to accumulate these soluble sugars rather than starch in chloroplasts.

The conquest of stressful habitats by CAM plants does not mean that these plants are immune from stress. Their CAM needs to be accompanied by other biophysical and biochemical mechanisms that deal with potential oxidative stress and photoinhibition (Broetto et al. 2002; Lu et al. 2003). Nevertheless, the success of CAM is unquestionable, and has fascinated plant physiologists to such an extent that four reviews have recently appeared, dealing with the 407 CAM plants investigated to date (Sayed 2001), the consequences of the high CO_2 concentration and often-overlooked high O_2 concentration that develop in leaves during the day (Lüttge 2002), a Congress (Holtum 2002), and a brief history of CAM (that includes an unfortunate choice of words incorrectly implying that CAM occurs in nutshells!) (Black and Osmond 2003).

5.4 Mitochondrial Metabolism

The other famous cycle of carbon metabolism, viz. the respiratory Krebs cycle, must operate to some extent in photosynthetic cells because there are beneficial interactions with the Calvin cycle (Raghavendra and Padmasree 2003) and because the Krebs cycle is the source of some of the carbon skeletons for amino acid biosynthesis. In fact, Peisker and Apel (2001) estimate that 71% of mitochondrial respiration is sustained by C_3 plant leaves in the light. Loreto et al. (2001) imply that an even greater percentage may be sustained in C_4 plant leaves, but (as expected) most of the CO_2 released by the Krebs cycle is recaptured by PEP carboxylase. Even in C_3 plants a good portion of this CO_2 is caught by Rubisco before escaping from the leaf (Pinelli and Loreto 2003). Meanwhile, examination of a mutant wild tomato (*Lycopersicon pennellii*) depleted in aconitase (the second enzyme of the Krebs cycle) now indicates that these Krebs cycle activities may be excessive. The mutant tomato exhibited a reduced flux through the Krebs cycle, but nevertheless photosynthesised faster and yielded more fruit (Carrari et al. 2003).

5.5 Photorespiration

The Calvin cycle not only faced a challenge 22 years after its discovery (see Sect. 5.2), but also suffered a humbling experience 5 years later when it was revealed (see Ogren 2003) that its CO_2-fixing enzyme Rubisco was not only slow, but also unfaithful: on about one occasion in three it accepted O_2, rather than CO_2, as substrate, thereby producing glycolate-P and initiating photorespiration. Schloss (2002) declares that this discovery 'stunned the biochemical community', in part because an oxygenase-enzyme devoid of any organic cofactor was unheard of, but Benson (2002), on the other hand, states that he 'failed to get excited' by the discovery because he knew how labile RuBP is. Recently, Lilley et al. (2003) have brought Rubisco's oxygenase activity to light by showing that, in the presence of Mn^{2+} instead of Mg^{2+}, it displays far-red chemiluminescence, quite a feat for an enzyme classically considered to be one of the catalysts of 'the dark reactions' of photosynthesis.

The glycolate-P produced by Rubisco is hydrolysed to glycolate and Pi within the chloroplast by a specific phosphatase which finally has been purified (Mamedov et al. 2001). The glycolate is exported to the cytosol, but the Pi is not, thereby making the chloroplast independent of cytosolic Pi for the photorespiratory component of its carbon metabolism (Holbrook and Keys 2003) [this is in contrast to its delicate dependency on cytosolic Pi when GAP is being generated (Walker 2003)]. The exported glycolate is processed by reactions in peroxisomes (extensively reviewed by Igamberdiev and Lea 2002) and mitochondria, such that one molecule each of CO_2 and glycerate are derived from two molecules of glycolate. The CO_2 represents the photorespiratory CO_2 loss, which is difficult to measure and may only be about one-quarter of the initially fixed CO_2 in C_3 plants (Yoshimura et al. 2001), rather than close to one-half as generally suspected.

Much attention is still being given to the roles of the mitochondrion in photorespiration. It is home to the multi-enzyme complex (reviewed by Bauwe and Kolukisaoglu 2003) that converts two glycine (derived from the two glycolate) to one serine in reactions releasing one CO_2, one NH_3 and reducing one NAD^+. The resultant NADH is possibly reoxidised by the plant-specific mitochondrial NADH dehydrogenase located on the matrix side of the inner membrane; unlike other mitochondrial membrane proteins, this one is encoded by a gene whose expression is somehow (but appropriately) light activated (see Møller 2002). Reducing equivalents might then be funnelled to O_2 by the alternative oxidase (also specific to plant mitochondria) that is (also appropriately) activated by the photorespiratory intermediates glyoxylate and hydroxypyruvate (Pastore et al. 2001), and would best serve the purpose of linking photorespiration to the dissipation of excess harvested light energy (Millenaar and Lambers 2003) when leaves are stressed by high light and low water supply, as is common for savannah trees (Franco and Lüttge 2002). Such energy dissipation would occur because this electron transport pathway is 'energy wasteful', i.e. it is not linked to ATP synthesis (Møller 2002). Finally, a remarkable observation is that a *variant* of one of the enzymes in the multi-enzyme complex mentioned above is used by plants for a totally unrelated task: that of catalysing the synthesis of the signalling molecule nitric oxide (NO) which participates in defence against pathogens (Chandok et al. 2003), and which has skyrocketed into significance in mammalian biochemistry during the past decade.

6 A Few Departing Comments About Chloroplasts, the Calvin Cycle's Organelle

Somehow our excitement over chloroplasts in recent years has been less than in days gone by, such as 12 years post-Calvin cycle when the hunt for the first photosynthetically competent isolated chloroplast was on, and the word 'envelope' was adopted as the name for the organelle's pair of surrounding membranes (see Walker 2003). Much has been learnt since that time, including the often-forgotten statistic that chloroplasts occupy a greater volume than the cytosol in a typical leaf cell (Winter et al. 1993), the fact that many biosynthetic pathways (including those for fatty acids, some isoprenoids, and the shikimate pathway) reside beside the Calvin cycle in the chloroplast stroma (see Fig. 3 in our 1998 review), and the existence on the envelope of critical transporter proteins that interconnect the chloroplast's metabolism with the rest of the cell; most recently a transporter for xylulose-5-P has been revealed (Eicks et al. 2002). Now we find ourselves concluding by noting that there is an unusual algal chloroplast that still possesses a cyanobacterial-like peptidoglycan wall (Raven 2003), and that there are regular chloroplasts with odd functions or in unexpected places. Chloroplasts in the outer layer of stems and twigs of woody plants photosynthesise (Schmidt et al. 2000), but the benefit may not so much be sugar synthesis, but rather O_2 supply to the respiration of underlying non-photosynthetic tissues (Pfanz et al. 2002). Perhaps the photosynthesis by these stems and twigs is C_4-like (Hibberd and Quick 2002) and uses CO_2 delivered from root respiration (Fig. 5), in which case (ironically) root respiration would be indirectly helping to supply O_2 for shoot respiration. Finally, another case of chloroplasts being beneficial by photosynthetically generating O_2 for respiration is the developing embryo of the legume seed (Rolletschek et al. 2003).

The most bizarre chloroplasts are algal chloroplasts absconded by marine molluscs. A long-term (months) endosymbiotic association is established between the intracellular photosynthesising chloroplasts (which colour the molluscs green) and the mollusc (Green et al. 2000), with the latter obtaining more than half of its 'dietary' carbohydrate from the former (Raven et al. 2001). More extreme is the recently discovered, non-symbiotic acquisition of diatom chloroplasts by the common benthic foraminifer *Nonionella stella*, in the dark depths of the ocean, probably for the purpose of adding the chloroplast's nitrate reductase to its repertoire of nitrogen-assimilation enzymes (Grzymski et al. 2002).

It may seem strange to finalise these reviews with chloroplasts caught in continuous darkness, but it is symbolic of the bright/dark dichotomy with which we conclude. How absolutely wonderful it must have been, 50 years ago, for the Calvin cycle discoverers to have elucidated the biochemical outline of Earth's CO_2-to-sugars support system. Is it not somewhat naïve, just 50 years later, to connive to twist it to do better? The potential is not great. Walker (1995) has questioned the wisdom of such plans (see our 1998 review), and his wisdom has not been contested. Can we not be pleased with the Calvin cycle as it is, and be content with simply discovering more about its place in the chloroplast, the cell, the photosynthetic organism and the cycles of life on Earth?

Acknowledgement. We thank Vivienne Sofia Kelly (born in Münster) for her exceptionally fastidious typing of the list of 173 references.

References

Andralojc JP, Keys AJ, Kossmann J, Parry MAJ (2002) Elucidating the biosynthesis of 2-carboxyarabinitol 1-phosphate through reduced expression of chloroplastic fructose 1,6-bisphosphate phosphatase and radiotracer studies with $^{14}CO_2$. Proc Natl Acad Sci USA 99:4742–4747

Aoki N, Hirose T, Scofield GN, Whitfeld PR, Furbank RT (2003) The sucrose transporter gene family in rice. Plant Cell Physiol 44:223–232

Ashida H, Saito Y, Kojima C, Kobayashi K, Ogasawara N, Yokota A (2003) A functional link between RuBisCO-like protein of *Bacillus* and photosynthetic RuBisCO. Science 302:286–290

Bakrim N, Brulfert J, Vidal J, Chollet R (2001) Phospho*enol*pyruvate carboxylase kinase is controlled by a similar signaling cascade in CAM and C_4 plants. Biochem Biophys Res Commun 286:1158–1162

Balmer Y, Koller A, del Val G, Manieri W, Schürmann P, Buchanan BB (2003) Proteomics gives insight into the regulatory function of chloroplast thioredoxins. Proc Natl Acad Sci USA 100:370–375

Baroja-Fernández E, Muñoz FJ, Akazawa T, Pozueta-Romero J (2001) Reappraisal of the currently prevailing model of starch biosynthesis in photosynthetic tissues: a proposal involving the cytosolic production of ADP-glucose by sucrose synthase and occurrence of cyclic turnover of starch in the chloroplast. Plant Cell Physiol 42:1311–1320

Bassham JA (2003) Mapping the carbon reduction cycle: a personal retrospective. Photosyn Res 76:35–52

Bassham JA, Benson AA, Kay LD, Harris AZ, Wilson AT, Calvin M (1954) The path of carbon in photosynthesis. 21. The cyclic regeneration of carbon dioxide acceptor. J Am Chem Soc 76:1760–1770

Bauwe H, Kolukisaoglu Ü (2003) Genetic manipulation of glycine decarboxylation. J Exp Bot 54:1523–1535

Baxter CJ, Foyer CH, Turner J, Rolfe SA, Quick WP (2003) Elevated sucrose-phosphate synthase activity in transgenic tobacco sustains photosynthesis in older leaves and alters development. J Exp Bot 54:1813–1820

Benson AA (2002) Following the path of carbon in photosynthesis: a personal story. Photosyn Res 73:29–49

Bernacchi CJ, Portis AR, Nakano H, von Caemmerer S, Long SP (2002) Temperature response of mesophyll conductance. Implications for the determination of Rubisco enzyme kinetics and for limitations to photosynthesis in vivo. Plant Physiol 130:1992–1998

Black CC, Osmond CB (2003) Crassulacean acid metabolism photosynthesis: 'working the night shift'. Photosyn Res 76:329–341

Broetto F, Lüttge U, Ratajczak R (2002) Influence of light intensity and salt-treatment on mode of photosynthesis and enzymes of the antioxidative response system of *Mesembryanthemum crystallinum*. Funct Plant Biol 29:13–23

Brüggemann N, Schnitzler J-P (2002) Comparison of isoprene emission, intercellular isoprene concentration and photosynthetic performance in water-limited oak (*Quercus pubescens* Willd. and *Quercus robur* L.) saplings. Plant Biol 4:456–463

Buitenhuis ET, de Baar HJW, Veldhuis MJW (1999) Photosynthesis and calcification by *Emiliania huxleyi* (Prymnesiophyceae) as a function of inorganic carbon species. J Phycol 35:949–959

Carrari F, Nunes-Nesi A, Gibon Y, Lytovchenko A, Loureiro ME, Fernie AR (2003) Reduced expression of aconitase results in an enhanced rate of photosynthesis and marked shifts in carbon partitioning in illuminated leaves of wild species tomato. Plant Physiol 133:1322–1335

Chandok MR, Ytterberg AJ, van Wijk KJ, Klessig DF (2003) The pathogen-inducible nitric oxide synthase (iNOS) in plants is a variant of the P protein of the glycine decarboxylase complex. Cell 113:469–482

Chastain CJ, Fries JP, Vogel JA, Randklev CL, Vossen AP, Dittmer SK, Watkins EE, Fiedler LJ, Wacker SA, Meinhover KC, Sarath G, Chollet R (2002) Pyruvate, orthophosphate dikinase in leaves and chloroplasts of C_3 plants undergoes light-/dark-induced reversible phosphorylation. Plant Physiol 128:1368–1378

Chinthapalli B, Murmu J, Raghavendra AS (2003) Dramatic difference in the responses of phospho*enol*pyruvate carboxylase to temperature in leaves of C_3 and C_4 plants. J Exp Bot 54:707–714

Chueca A, Sahrawy M, Pagano EA, López Gorge J (2002) Chloroplast fructose-1,6-bisphosphatase: structure and function. Photosyn Res 74:235–249

Ciereszko I, Kleczkowski LA (2002) Glucose and mannose regulate the expression of a major sucrose synthase gene in *Arabidopsis* via hexokinase-dependent mechanisms. Plant Physiol Biochem 40:907–911

Collin V, Issakidis-Bourguet E, Marchand C, Hirasawa M, Lancelin J-M, Knaff DB, Miginiac-Maslow M (2003) The *Arabidopsis* plastidial thioredoxins. New functions and new insights into specificity. J Biol Chem 278:23747–23752

Costa dos Santos A, da-Silva WS, de Meis L, Galina A (2003) Proton transport in maize tonoplasts supported by fructose-1,6-bisphosphate cleavage. Pyrophosphate-dependent phosphofructokinase as a pyrophosphate-regenerating system. Plant Physiol 133:885–892

Crafts-Brandner SJ, Salvucci ME (2002) Sensitivity of photosynthesis in a C_4 plant, maize, to heat stress. Plant Physiol 129:1773–1780

Crevillén P, Ballicora MA, Mérida A, Preiss J, Romero JM (2003) The different large subunit isoforms of *Arabidopsis thaliana* ADP-glucose pyrophosphorylase confer distinct kinetic and regulatory properties to the heterotetrameric enzyme. J Biol Chem 278:28508–28515

Dong L, Ermolova NV, Chollet R (2001) Partial purification and biochemical characterization of a heteromeric protein phosphatase 2A holoenzyme from maize (*Zea mays* L.) leaves that dephosphorylates C_4 phospho*enol*pyruvate carboxylase. Planta 213:379–389

Du Y-C, Peddi SR, Spreitzer RJ (2003) Assessment of structural and functional divergence far from the large subunit active site of ribulose-1,5-bisphosphate carboxylase/oxygenase. J Biol Chem 278:49401–49405

Eastmond PJ, van Dijken AJH, Spielman M, Kerr A, Tissier AF, Dickinson HG, Jones JDG, Smeekens SC, Graham IA (2002) Trehalose-6-phosphate synthase 1, which catalyses the first step in trehalose synthesis, is essential for *Arabidopsis* embryo maturation. Plant J 29:225–235

Echevarría C, Garcia-Mauriño S, Alvarez R, Soler A, Vidal J (2001) Salt stress increases the Ca^{2+}-independent phospho*enol*pyruvate carboxylase kinase activity in *Sorghum* leaves. Planta 214:283–287

Eicks M, Maurino V, Knappe S, Flügge U-I, Fischer K (2002) The plastidic pentose phosphate translocator represents a link between the cytosolic and the plastidic pentose phosphate pathways in plants. Plant Physiol 128:512–522

Emmerlich V, Linka N, Reinhold T, Hurth MA, Traub M, Martinoia E, Neuhaus HE (2003) The plant homolog to the human sodium/dicarboxylic cotransporter is the vacuolar malate carrier. Proc Natl Acad Sci USA 100:11122–11126

Entus R, Poling M, Herrmann KM (2002) Redox regulation of *Arabidopsis* 3-Deoxy-D-*arabino*-heptulosonate 7-phosphate synthase. Plant Physiol 129:1866–1871

Fellermeier M, Sagner S, Spiteller P, Spiteller M, Zenk MH (2003) Early steps of deoxyxylulose phosphate pathway in chromoplasts of higher plants. Phytochemistry 64:199–207

Franco AC, Lüttge U (2002) Midday depression in savanna trees: coordinated adjustments in photochemical efficiency, photorespiration, CO_2 assimilation and water use efficiency. Oecologia 131:356–365

Frommer WB, Schulze WX, Lalonde S (2003) Hexokinase, jack-of-all-trades. Science 300:261–263

Garg AK, Kim J-K, Owens TG, Ranwala AP, Choi YD, Kochian LV, Wu RJ (2002) Trehalose accumulation in rice plants confers high tolerance levels to different abiotic stresses. Proc Natl Acad Sci USA 99:15898–15903

Gerhardt S, Echt S, Busch M, Freigang J, Auerbach G, Bader G, Martin WF, Bacher A, Huber R, Fischer M (2003) Structure and properties of an engineered transketolase from maize. Plant Physiol 132:1941–1949

Ghoshal D, Husic HD, Goyal A (2002) Dissolved inorganic carbon concentration mechanism in *Chlamydomonas moewusii*. Plant Physiol Biochem 40:299–305

Giegé P, Heazlewood JL, Roessner-Tunali U, Millar AH, Fernie AR, Leaver CJ, Sweetlove LJ (2003) Enzymes of glycolysis are functionally associated with the mitochondrion in *Arabidopsis* cells. Plant Cell 15:2140–2151

Gómez-Casati DF, Cortassa S, Aon MA, Iglesias AA (2003) Ultrasensitive behavior in the synthesis of storage polysaccharides in cyanobacteria. Planta 216:969–975

Graciet E, Lebreton S, Camadro J-M, Gontero B (2002) Thermodynamic analysis of the emergence of new regulatory properties in a phosphoribulokinase-glyceraldehyde 3-phosphate dehydrogenase complex. J Biol Chem 277:12697–12702

Green BJ, Li W-Y, Manhart JR, Fox TC, Summer EJ, Kennedy RA, Pierce SK, Rumpho ME (2000) Mollusc–algal chloroplast endosymbiosis. Photosynthesis, thylakoid protein maintenance, and chloroplast gene expression continue for many months in the absence of the algal nucleus. Plant Physiol 124:331–342

Griffiths H, Helliker B, Roberts A, Haslam RP, Girnus J, Robe WE, Borland AM, Maxwell K (2002) Regulation of Rubisco activity in Crassulacean acid metabolism plants: better late than never. Funct Plant Biol 29:689–696

Grzymski J, Schofield OM, Falkowski PG, Bernhard JM (2002) The function of plastids in the deep-sea benthic foraminifer, *Nonionella stella*. Limnol Oceanogr 47:1569–1580

Hafke JB, Hafke Y, Smith JAC, Lüttge U, Thiel G (2003) Vacuolar malate uptake is mediated by an anion-selective inward rectifier. Plant J 35:116–128

Harrington GN, Bush DR (2003) The bifunctional role of hexokinase in metabolism and glucose signaling. Plant Cell 15:2493–2496

Hatch MD, Slack CR (1966) Photosynthesis by sugar-cane leaves. A new carboxylation reaction and the pathway of sugar formation. Biochem J 101:103–111

Hauschild R, von Schaewen A (2003) Differential regulation of glucose-6-phosphate dehydrogenase isoenzyme activities in potato. Plant Physiol 133:47–62

Hellblom F, Axelsson L (2003) External HCO_3^- dehydration maintained by acid zones in the plasma membrane is an important component of the photosynthetic carbon uptake in *Ruppia cirrhosa*. Photosyn Res 77:173–181

Hendriks JHM, Kolbe A, Gibon Y, Stitt M, Geigenberger P (2003) ADP-glucose pyrophosphorylase is activated by posttranslational redox-modification in response to light and to sugars in leaves of *Arabidopsis* and other plant species. Plant Physiol 133:838–849

Herfort L, Thake B, Roberts J (2002) Acquisition and use of bicarbonate by *Emiliania huxleyi*. New Phytol 156:427–436

Herter S, Fuchs G, Bacher A, Eisenreich W (2002) A bicyclic autotrophic CO_2 fixation pathway in *Chloroflexus aurantiacus*. J Biol Chem 277:20277–20283

Hibberd JM, Quick WP (2002) Characteristics of C_4 photosynthesis in stems and petioles of C_3 flowering plants. Nature 415:451–454

Holbrook GP, Keys AJ (2003) Evidence for recycling of inorganic phosphate by wheat chloroplasts during photosynthesis at air levels of CO_2 and O_2. J Plant Physiol 160:1351–1360

Holtum JAM (2002) Crassulacean acid metabolism: plasticity in expression, complexity of control. Funct Plant Biol 29:657–661

Huertas IE, Colman B, Espie GS (2002) Mitochondrial-driven bicarbonate transport supports photosynthesis in a marine microalga. Plant Physiol 130:284–291

Hügler M, Krieger RS, Jahn M, Fuchs G (2003) Characterization of acetyl-CoA/propionyl-CoA carboxylase in *Metallosphaera sedula*. Carboxylating enzyme in the 3-hydroxypropionate cycle for autotrophic carbon fixation. Eur J Biochem 270:736–744

Igamberdiev AU, Lea PJ (2002) The role of peroxisomes in the integration of metabolism and evolutionary diversity of photosynthetic organisms. Phytochemistry 60:651–674

Ishijima S, Uchibori A, Takagi H, Maki R, Ohnishi M (2003) Light-induced increase in free Mg^{2+} concentration in spinach chloroplasts: measurement of free Mg^{2+} by using a fluorescent probe and necessity of stromal alkalinization. Arch Biochem Biophys 412:126–132

Ito H, Iwabuchi M, Ogawa K (2003) The sugar-metabolic enzymes aldolase and triose-phosphate isomerase are targets of glutathionylation in *Arabidopsis thaliana*: detection using biotinylated glutathione. Plant Cell Physiol 44:655–660

Jiang H, Xu D-Q (2001) The cause of the difference in leaf net photosynthetic rate between two soybean cultivars. Photosynthetica 39:453–459

Kanao T, Kawamura M, Fukui T, Atomi H, Imanaka T (2002) Characterization of isocitrate dehydrogenase from the green sulfur bacterium *Chlorobium limicola*. A carbon dioxide-fixing enzyme in the reductive tricarboxylic acid cycle. Eur J Biochem 269:1926–1931

Kiirats O, Lea PJ, Franceschi VR, Edwards GE (2002) Bundle sheath diffusive resistance to CO_2 and effectiveness of C_4 photosynthesis and refixation of photorespired CO_2 in a C_4 cycle mutant and wild-type *Amaranthus edulis*. Plant Physiol 130:964–976

Kubien DS, von Caemmerer S, Furbank RT, Sage RF (2003) C_4 photosynthesis at low temperature. A study using transgenic plants with reduced amounts of Rubisco. Plant Physiol 132:1577–1585

Kühn C (2003) A comparison of the sucrose transporter systems of different plant species. Plant Biol 5:215–232

Laby RJ, Kim D, Gibson SI (2001) The *ram1* mutant of *Arabidopsis* exhibits severely decreased β-amylase activity. Plant Physiol 127:1798–1807

Lara MV, Casati P, Andreo CS (2002) CO_2-concentrating mechanisms in *Egeria densa*, a submersed aquatic plant. Physiol Plant 115:487–495

Law RD, Crafts-Brandner SJ, Salvucci ME (2001) Heat stress induces the synthesis of a new form of ribulose-1,5-bisphosphate carboxylase/oxygenase activase in cotton leaves. Planta 214:117–125

Lebreton S, Graciet E, Gontero B (2003) Modulation, via protein–protein interactions, of glyceraldehyde-3-phosphate dehydrogenase activity through redox phosphoribulokinase regulation. J Biol Chem 278:12078–12084

Lemaire SD, Collin V, Keryer E, Quesada A, Miginiac-Maslow M (2003) Characterization of thioredoxin *y*, a new type of thioredoxin identified in the genome of *Chlamydomonas reinhardtii*. FEBS Lett 543:87–92

Lilley RMcC, Wang XQ, Krausz E, Andrews TJ (2003) Complete spectra of the far-red chemiluminescence of the oxygenase reaction of Mn^{2+}-activated ribulose-bisphosphate carboxylase/oxygenase establish excited Mn^{2+} as the source. J Biol Chem 278:16488–16493

Loreto F, Velikova V, di Marco G (2001) Respiration in the light measured by $^{12}CO_2$ emission in $^{13}CO_2$ atmosphere in maize leaves. Aust J Plant Physiol 28:1103–1108

Lu C, Qiu N, Lu Q, Wang B, Kuang T (2003) PSII photochemistry, thermal energy dissipation, and the xanthophyll cycle in *Kalanchoë daigremontiana* exposed to a combination of water stress and high light. Physiol Plant 118:173–182

Lunn JE, Gillespie VJ, Furbank RT (2003) Expression of a cyanobacterial sucrose-phosphate synthase from *Synechocystis* sp. PCC 6803 in transgenic plants. J Exp Bot 54:223–237

Lüttge U (2002) CO_2-concentrating: consequences in Crassulacean acid metabolism. J Exp Bot 53:2131–2142

Lytovchenko A, Fernie AR (2003) Photosynthetic metabolism is severely impaired on the parallel reduction of plastidial and cytosolic isoforms of phosphoglucomutase. Plant Physiol Biochem 41:193–200

Lytovchenko A, Sweetlove L, Pauly M, Fernie AR (2002) The influence of cytosolic phosphoglucomutase on photosynthetic carbohydrate metabolism. Planta 215:1013–1021

Mamedov TG, Suzuki K, Miura K, Kucho K-i, Fukuzawa H (2001) Characteristics and sequence of phosphoglycolate phosphatase from a eukaryotic green alga *Chlamydomonas reinhardtii*. J Biol Chem 276:45573–45579

Markham JE, Kruger NJ (2002) Kinetic properties of bifunctional 6-phosphofructo-2-kinase/fructose-2,6-bisphosphatase from spinach leaves. Eur J Biochem 269:1267–1277

Martinsuo P, Pursiheimo S, Aro E-M, Rintamäki E (2003) Dithiol oxidant and disulfide reductant dynamically regulate the phosphorylation of light-harvesting complex II proteins in thylakoid membranes. Plant Physiol 133:37–46

McRae SR, Christopher JT, Smith JAC, Holtum JAM (2002) Sucrose transport across the vacuolar membrane of *Ananas comosus*. Funct Plant Biol 29:717–724

Millenaar FF, Lambers H (2003) The alternative oxidase: in vivo regulation and function. Plant Biol 5:2–15
Milligan AJ, Morel FMM (2002) A proton buffering role for silica in diatoms. Science 297:1848–1850
Miyao M (2003) Molecular evolution and genetic engineering of C_4 photosynthetic enzymes. J Exp Bot 54:179–189
Møller IM (2002) A new dawn for plant mitochondrial NAD(P)H dehydrogenases. Trends Plant Sci 7:235–237
Moore B, Zhou L, Rolland F, Hall Q, Cheng W-H, Liu Y-X, Hwang I, Jones T, Sheen J (2003) Role of the *Arabidopsis* glucose sensor HXK1 in nutrient, light, and hormonal signaling. Science 300:332–336
Moroney JV, Bartlett SG, Samuelsson G (2001) Carbonic anhydrases in plants and algae. Plant Cell Environ 24:141–153
Murchie EH, Yang J, Hubbart S, Horton P, Peng S (2002) Are there associations between grain-filling rate and photosynthesis in the flag leaves of field-grown rice? J Exp Bot 53:2217–2224
Naidu SL, Moose SP, AL-Shoaibi AK, Raines CA, Long SP (2003) Cold tolerance of C_4 photosynthesis in *Miscanthus* × *giganteus*: adaptation in amounts and sequence of C_4 photosynthetic enzymes. Plant Physiol 132:1688–1697
Nielsen TH, Stitt M (2001) Tobacco transformants with strongly decreased expression of pyrophosphate:fructose-6-phosphate 1-phosphotransferase expression in the base of their young growing leaves contain much higher levels of fructose-2,6-bisphosphate but no major changes in fluxes. Planta 214:106–116
Niittylä T, Messerli G, Trevisan M, Chen J, Smith AM, Zeeman SC (2004) A previously unknown maltose transporter essential for starch degradation in leaves. Science 303:87–89
Ocón E, Hermoso R, Chueca A, Lázaro JJ (2001) Cross-linking between chloroplast fructose-1,6-bisphosphatase and thioredoxin *f*. Physiol Plant 113:452–460
Ogle K (2003) Implications of interveinal distance for quantum yield in C_4 grasses: a modeling and meta-analysis. Oecologia 136:532–542
Ogren WL (2003) Affixing the O to Rubisco: discovering the source of photorespiratory glycolate and its regulation. Photosyn Res 76:53–63
Okano Y, Mizohata E, Xie Y, Matsumura H, Sugawara H, Inoue T, Yokota A, Kai Y (2002) X-ray structure of *Galdieria* Rubisco complexed with one sulfate ion per active site. FEBS Lett 527:33–36
Olsson T, Thelander M, Ronne H (2003) A novel type of chloroplast stromal hexokinase is the major glucose-phosphorylating enzyme in the moss *Physcomitrella patens*. J Biol Chem 278:44439–44447
Pagnussat GC, Fiol DF, Salerno GL (2002) A CDPK type protein kinase is involved in rice SPS light modulation. Physiol Plant 115:183–189
Parry MAJ, Andralojc PJ, Mitchell RAC, Madgwick PJ, Keys AJ (2003) Manipulation of Rubisco: the amount, activity, function and regulation. J Exp Bot 54:1321–1333
Pastore D, Trono D, Laus MN, di Fonzo N, Passarella S (2001) Alternative oxidase in durum wheat mitochondria. Activation by pyruvate, hydroxypyruvate and glyoxylate and physiological role. Plant Cell Physiol 42:1373–1382
Peisker M, Apel H (2001) Inhibition by light of CO_2 evolution from dark respiration: comparison of two gas exchange methods. Photosyn Res 70:291–298
Penna S (2003) Building stress tolerance through over-producing trehalose in transgenic plants. Trends Plant Sci 8:355–357
Peñuelas J, Llusià J (2002) Linking photorespiration, monoterpenes and thermotolerance in *Quercus*. New Phytol 155:227–237
Pfanz H, Aschan G, Langenfeld-Heyser R, Wittmann C, Loose M (2002) Ecology and ecophysiology of tree stems: corticular and wood photosynthesis. Naturwissenschaften 89:147–162

Pierce S, Winter K, Griffiths H (2002) The role of CAM in high rainfall cloud forests: an in situ comparison of photosynthetic pathways in Bromeliaceae. Plant Cell Environ 25:1181–1189

Pinelli P, Loreto F (2003) $^{12}CO_2$ emission from different metabolic pathways measured in illuminated and darkened C_3 and C_4 leaves at low, atmospheric and elevated CO_2 concentration. J Exp Bot 54:1761–1769

Porchia AC, Fiol DF, Salerno GL (1999) Differential synthesis of sucrose and trehalose in *Euglena gracilis* cells during growth and salt stress. Plant Sci 149:43–49

Portis AR (2003) Rubisco activase – Rubisco's catalytic chaperone. Photosyn Res 75:11–27

Price GD, Badger MR (2002) Advances in understanding how aquatic photosynthetic organisms utilize sources of dissolved inorganic carbon for CO_2 fixation. Funct Plant Biol 29:117–121

Qi J, Isupov MN, Littlechild JA, Anderson LE (2001) Chloroplast glyceraldehyde-3-phosphate dehydrogenase contains a single disulfide bond located in the C-terminal extension to the B subunit. J Biol Chem 276:35247–35252

Raghavendra AS, Padmasree K (2003) Beneficial interactions of mitochondrial metabolism with photosynthetic carbon assimilation. Trends Plant Sci 8:546–553

Raines CA (2003) The Calvin cycle revisited. Photosyn Res 75:1–10

Raven JA (2002) Evolutionary options. Nature 415:375–377

Raven JA (2003) Carboxysomes and peptidoglycan walls of cyanelles: possible physiological functions. Eur J Phycol 38:47–53

Raven JA, Girard-Bascou J (2001) Algal model systems and the elucidation of photosynthetic metabolism. J Phycol 37:943–950

Raven JA, Walker DI, Jensen KR, Handley LL, Scrimgeour CM, McInroy SG (2001) What fraction of the organic carbon in sacoglossans is obtained from photosynthesis by kleptoplastids? An investigation using the natural abundance of stable carbon isotopes. Mar Biol 138:537–545

Regierer B, Fernie AR, Springer F, Perez-Melis A, Leisse A, Koehl K, Willmitzer L, Geigenberger P, Kossmann J (2002) Starch content and yield increase as a result of altering adenylate pools in transgenic plants. Nature Biotech 20:1256–1260

Ritte G, Lloyd JR, Eckermann N, Rottmann A, Kossmann J, Steup M (2002) The starch-related R1 protein is an α-glucan, water dikinase. Proc Natl Acad Sci USA 99:7166–7171

Robinson JJ, Scott KM, Swanson ST, O'Leary MH, Horken K, Tabita FR, Cavanaugh CM (2003) Kinetic isotope effect and characterization of form II RubisCO from the chemoautotrophic endosymbionts of the hydrothermal vent tubeworm *Riftia pachyptila*. Limnol Oceanogr 48:48–54

Rolletschek H, Weber H, Borisjuk L (2003) Energy status and its control on embryogenesis of legumes. Embryo photosynthesis contributes to oxygen supply and is coupled to biosynthetic fluxes. Plant Physiol 132:1196–1206

Rosenstiel TN, Fisher AJ, Fall R, Monson RK (2002) Differential accumulation of dimethylallyl diphosphate in leaves and needles of isoprene- and methylbutenol-emitting and nonemitting species. Plant Physiol 129:1276–1284

Rost B, Riebesell U, Burkhardt S, Sültemeyer D (2003) Carbon acquisition of bloom-forming marine phytoplankton. Limnol Oceanogr 48:55–67

Sage RF (2002) Variation in the k_{cat} of Rubisco in C_3 and C_4 plants and some implications for photosynthetic performance at high and low temperature. J Exp Bot 53:609–620

Sage RF, Sage TL (2002) Microsite characteristics of *Muhlenbergia richardsonis* (Trin.) Rydb., an alpine C_4 grass from the White Mountains, California. Oecologia 132:501–508

Sainis JK, Dani DN, Dey GK (2003) Involvement of thylakoid membranes in supramolecular organization of Calvin cycle enzymes in *Anacystis nidulans*. J Plant Physiol 160:23–32

Salamone PR, Kavakli IH, Slattery CJ, Okita TW (2002) Directed molecular evolution of ADP-glucose pyrophosphorylase. Proc Natl Acad Sci USA 99:1070–1075

Salerno GL, Curatti L (2003) Origin of sucrose metabolism in higher plants: when, how and why? Trends Plant Sci 8:63–69

Sánchez R, Cejudo FJ (2003) Identification and expression analysis of a gene encoding a bacterial-type phospho*enol*pyruvate carboxylase from *Arabidopsis* and rice. Plant Physiol 132:949–957

Sayed OH (2001) Crassulacean acid metabolism 1975–2000, a check list. Photosynthetica 39:339–352

Saze H, Ueno Y, Hisabori T, Hayashi H, Izui K (2001) Thioredoxin-mediated reductive activation of a protein kinase for the regulatory phosphorylation of C_4-form phosphoenolpyruvate carboxylase from maize. Plant Cell Physiol 42:1295–1302

Scheibe R, Wedel N, Vetter S, Emmerlich V, Sauermann S-M (2002) Co-existence of two regulatory NADP-glyceraldehyde 3-P dehydrogenase complexes in higher plant chloroplasts. Eur J Biochem 269:5617–5624

Scheidig A, Fröhlich A, Schulze S, Lloyd JR, Kossmann J (2002) Downregulation of a chloroplast-targeted β-amylase leads to a starch-excess phenotype in leaves. Plant J 30:581–591

Schloss JV (2002) Oxygen toxicity from plants to people. Planta 216:38–43

Schluepmann H, Pellny T, van Dijken A, Smeekens S, Paul M (2003) Trehalose 6-phosphate is indispensable for carbohydrate utilization and growth in *Arabidopsis thaliana*. Proc Natl Acad Sci USA 100:6849–6854

Schmidt J, Batic F, Pfanz H (2000) Photosynthetic performance of leaves and twigs of evergreen holly (*Ilex aquifolium* L.). Phyton 40:179–190

Schneider A, Häusler RE, Kolukisaoglu Ü, Kunze R, van der Graaff E, Schwacke R, Catoni E, Desimone M, Flügge U-I (2002) An *Arabidopsis thaliana* knock-out mutant of the chloroplast triose phosphate/phosphate translocator is severely compromised only when starch synthesis, but not starch mobilisation is abolished. Plant J 32:685–699

Servaites JC, Geiger DR (2002) Kinetic characteristics of chloroplast glucose transport. J Exp Bot 53:1581–1591

Sharkey TD (1996) Isoprene synthesis by plants and animals. Endeavour 20:74–78

Sinha AK, Hofmann MG, Römer U, Köckenberger W, Elling L, Roitsch T (2002) Metabolizable and non-metabolizable sugars activate different signal transduction pathways in tomato. Plant Physiol 128:1480–1489

Smidansky ED, Clancy M, Meyer FD, Lanning SP, Blake NK, Talbert LE, Giroux MJ (2002) Enhanced ADP-glucose pyrophosphorylase activity in wheat endosperm increases seed yield. Proc Natl Acad Sci USA 99:1724–1729

Smidansky ED, Martin JM, Hannah LC, Fischer AM, Giroux MJ (2003) Seed yield and plant biomass increases in rice are conferred by deregulation of endosperm ADP-glucose pyrophosphorylase. Planta 216:656–664

Snoeijs P, Klenell M, Choo KS, Comhaire I, Ray S, Pedersén M (2002) Strategies for carbon acquisition in the red marine macroalga *Coccotylus truncatus* from the Baltic Sea. Mar Biol 140:435–444

Sparla F, Pupillo P, Trost P (2002) The C-terminal extension of glyceraldehyde-3-phosphate dehydrogenase subunit B acts as an autoinhibitory domain regulated by thioredoxins and nicotinamide adenine dinucleotide. J Biol Chem 277:44946–44952

Spreitzer RJ, Salvucci ME (2002) Rubisco: structure, regulatory interactions, and possibilities for a better enzyme. Annu Rev Plant Biol 53:449–475

Strand Å, Foyer CH, Gustafsson P, Gardestrom P, Hurry V (2003) Altering flux through the sucrose biosynthesis pathway in transgenic *Arabidopsis thaliana* modifies photosynthetic acclimation at low temperatures and the development of freezing tolerance. Plant Cell Environ 26:523–535

Sudo E, Makino A, Mae T (2003) Differences between rice and wheat in ribulose-1,5-bisphosphate regeneration capacity per unit of leaf-N content. Plant Cell Environ 26:255–263

Tchernov D, Silverman J, Luz B, Reinhold L, Kaplan A (2003) Massive light-dependent cycling of inorganic carbon between oxygenic photosynthetic microorganisms and their surroundings. Photosyn Res 77:95–103

Terashima I, Ono K (2002) Effects of $HgCl_2$ on CO_2 dependence of leaf photosynthesis: evidence indicating involvement of aquaporins in CO_2 diffusion across the plasma membrane. Plant Cell Physiol 43:70–78

Tiessen A, Hendriks JHM, Stitt M, Branscheid A, Gibon Y, Farré EM, Geigenberger P (2002) Starch synthesis in potato tubers is regulated by post-translational redox modification of ADP-glucose pyrophosphorylase: a novel regulatory mechanism linking starch synthesis to the sucrose supply. Plant Cell 14:2191–2213

Tiessen A, Prescha K, Branscheid A, Palacios N, McKibbin R, Halford NG, Geigenberger P (2003) Evidence that SNF1-related kinase and hexokinase are involved in separate sugar-signalling pathways modulating post-translational redox activation of ADP-glucose pyrophosphorylase in potato tubers. Plant J 35:490–500

Tortell PD, Morel FMM (2002) Sources of inorganic carbon for phytoplankton in the eastern subtropical and equatorial Pacific Ocean. Limnol Oceanogr 47:1012–1022

Trevanion SJ (2002) Regulation of sucrose and starch synthesis in wheat (*Triticum aestivum* L.) leaves: role of fructose 2,6-bisphosphate. Planta 215:653–665

Uehlein N, Lovisolo C, Siefritz F, Kaldenhoff R (2003) The tobacco aquaporin NtAQP1 is a membrane CO_2 pore with physiological functions. Nature 425:734–737

Van Heerden PDR, Krüger GHJ, Loveland JE, Parry MAJ, Foyer CH (2003) Dark chilling imposes metabolic restrictions on photosynthesis in soybean. Plant Cell Environ 26:323–337

Vogel G, Fiehn O, Jean-Richard-dit-Bressel L, Boller T, Wiemken A, Aeschbacher RA, Wingler A (2001) Trehalose metabolism in *Arabidopsis*: occurrence of trehalose and molecular cloning and characterization of trehalose-6-phosphate synthase homologues. J Exp Bot 52:1817–1826

von Caemmerer S (2003) C_4 photosynthesis in a single C_3 cell is theoretically inefficient but may ameliorate internal CO_2 diffusion limitations of C_3 leaves. Plant Cell Environ 26:1191–1197

von Caemmerer S, Furbank RT (2003) The C_4 pathway: an efficient CO_2 pump. Photosyn Res 77:191–207

Voznesenskaya EV, Franceschi VR, Kiirats O, Artyusheva EG, Freitag H, Edwards GE (2002) Proof of C_4 photosynthesis without Kranz anatomy in *Bienertia cycloptera* (Chenopodiaceae). Plant J 31:649–662

Voznesenskaya EV, Franceschi VR, Artyusheva EG, Black CC, Pyankov VI, Edwards GE (2003) Development of the C_4 photosynthetic apparatus in cotyledons and leaves of *Salsola richteri* (Chenopodiaceae). Int J Plant Sci 164:471–487

Walker DA (1995) Manipulating photosynthetic metabolism to improve crops: an inversion of ends and means. J Exp Bot 46:1253–1259

Walker DA (2003) Chloroplasts in envelopes: CO_2 fixation by fully functional intact chloroplasts. Photosyn Res 76:319–327

Wang Y-L, Zhou J-H, Wang Y-F, Bao J-S, Chen H-B (2001) Properties of hybrid enzymes between *Synechococcus* large subunits and higher plant small subunits of ribulose-1,5-bisphosphate carboxylase/oxygenase in *Escherichia coli.* Arch Biochem Biophys 396:35–42

Whitney SM, Andrews TJ (2001) Plastome-encoded bacterial ribulose-1,5-bisphosphate carboxylase/oxygenase (RubisCO) supports photosynthesis and growth in tobacco. Proc Natl Acad Sci USA 98:14738–14743

Whitney SM, Andrews TJ (2003) Photosynthesis and growth of tobacco with a substituted bacterial Rubisco mirror the properties of the introduced enzyme. Plant Physiol 133:287–294

Wingler A (2002) The function of trehalose biosynthesis in plants. Phytochemistry 60:437–440

Winter K, Holtum JAM (2002) How closely do the $\delta^{13}C$ values of Crassulacean acid metabolism plants reflect the proportion of CO_2 fixed during day and night? Plant Physiol 129:1843–1851

Winter H, Robinson DG, Heldt HW (1993) Subcellular volumes and metabolite concentrations in barley leaves. Planta 191:180–190
Wyka TP, Lüttge UE (2003) Contribution of C_3 carboxylation to the circadian rhythm of carbon dioxide uptake in a Crassulacean acid metabolism plant *Kalanchoë daigremontiana.* J Exp Bot 54:1471–1479
Xiao W, Sheen J, Jang J-C (2000) The role of hexokinase in plant sugar signal transduction and growth and development. Plant Mol Biol 44:451–461
Yoshimura Y, Kubota F, Hirao K (2001) Estimation of photorespiration rate by simultaneous measurements of CO_2, gas exchange rate, and chlorophyll fluorescence quenching in the C_3 plant *Vigna radiata* (L.) Wilczek and the C_4 plant *Amaranthus mongostanus* L. Photosynthetica 39:377–382
Zeeman SC, Tiessen A, Pilling E, Kato KL, Donald AM, Smith AM (2002) Starch synthesis in *Arabidopsis.* Granule synthesis, composition, and structure. Plant Physiol 129:516–529
Zhang N, Kallis RP, Ewy RG, Portis AR (2002) Light modulation of Rubisco in *Arabidopsis* requires a capacity for redox regulation of the larger Rubisco activase isoform. Proc Natl Acad Sci USA 99:3330–3334
Zhou R, Cheng L, Wayne R (2003) Purification and characterization of sorbitol-6-phosphate phosphatase from apple leaves. Plant Sci 165:227–232

Grahame J. Kelly
School of Life Sciences
Queensland University of Technology
Brisbane, Queensland 4000, Australia

Note added in proof (2005):
The report of nitric oxide (NO) synthesis by a photorespiratory enzyme (Sect. 5.5) has been retracted [see Travis J (2004) Science 306:960].

Subject Index

Note: This index is designed to be useful but not excessively long. Therefore, many items that appear repeatedly throughout the reviews, and are given attention in Sections where they would be expected to appear, are **not** listed. These include the 17 central items presented below (each of these appears within the Section headings of at least five of the 16 reviews), and most metabolites, enzymes and processes that would be expected to appear in appropriate Sections. For example, "glycolate" is addressed in all reviews within the Section on photorespiration; "glyceraldehyde-3-P dehydrogenase" is addressed in most reviews within the Section devoted to Calvin cycle enzymes; and "translocation" is addressed in many reviews within the Section devoted to sucrose.

C_3–C_4 intermediate plants
C_4 metabolism
C_4 plants; C_4 photosynthesis: C_4 pathway
Calvin cycle
carbon, inorganic (Ci) uptake
chloroplast
crassulacean acid metabolism (CAM)
fatty acids
light-mediated activation of enzymes
nitrogen assimilation
photoinhibition
photorespiration
respiration (mitochondrial)
ribulose-1,5-P_2 (RuBP) carboxylase/oxygenase (Rubisco)
starch
sucrose
transporters, of chloroplast envelope

Y

Z